Application of Polymer Materials in Pavement Design

Application of Polymer Materials in Pavement Design

Editors

Wei Jiang
Quantao Liu
Jose Norambuena-Contreras
Yue Huang

MDPI

Basel • Beijing • Wuhan • Barcelona • Belgrade • Novi Sad • Cluj • Manchester

Editors
Wei Jiang
Chang'an University
Xi'an, China

Quantao Liu
Wuhan University of Technology
Wuhan, China

Jose Norambuena-Contreras
University of Bío-Bío
Concepcion, Chile

Yue Huang
University of Leeds
Leeds, UK

Editorial Office
MDPI
St. Alban-Anlage 66
4052 Basel, Switzerland

This is a reprint of articles from the Special Issue published online in the open access journal *Polymers* (ISSN 2073-4360) (available at: https://www.mdpi.com/journal/polymers/special_issues/Appl_Polym_Mater_Pavement_Des).

For citation purposes, cite each article independently as indicated on the article page online and as indicated below:

Lastname, A.A.; Lastname, B.B. Article Title. *Journal Name* **Year**, *Volume Number*, Page Range.

ISBN 978-3-0365-9235-0 (Hbk)
ISBN 978-3-0365-9234-3 (PDF)
doi.org/10.3390/books978-3-0365-9234-3

Contents

About the Editors

Wei Jiang

Wei Jiang is a professor and deputy director of Key Laboratory for Special Area Highway Engineering of Ministry of Education, Chang'an University. He focuses on eco-friendly and intelligent perceptive pavement materials, the self-consistent use of green energy, road intelligent detection, and digital operation and maintenance. Dr. Jiang has published over 120 high-quality academic papers, and received 17 national, provincial, and ministerial awards, including the Second Prize of the National Scientific and Technological Progress Award. He has received several prestigious awards, including the National Science Fund for Excellent Young Scholars and the Fok Ying-Tong Education Foundation for Young Teachers award, and the Youth Top-Notch Talent award as part of the Support Program of Shaanxi Province and the Rising Star of Science and Technology in Shaanxi Province. He has also been honored with the China Highway Young Scientist Medal and the IAAM Young Scientist Medal for his contributions to the field. He actively contributes to the academic community as a member of several editorial boards, including those of the *Journal of Chang'an University* (Natural Science Edition) and *Transport Research*, and as a youth editorial member and guest editor for various renowned journals in the field of materials science and engineering.

Quantao Liu

Quantao Liu is a professor at the State Key Laboratory of Silicate Materials for Architectures, Wuhan University of Technology. His research interests include functional road materials, self-healing construction materials, and anti-UV ageing technologies of asphalt materials. Specifically, he developed a kind of stress-responsive polymer capsule for asphalt pavements that can release internal healing agent upon vehicle loading. He also developed antioxidant intercalated UV-shielding material, which demonstrated desirable UV-shielding properties and free radical capture properties, and successfully enhanced the anti-thermal oxygen and anti-ultraviolet aging properties of asphalt binder. He has published more than 120 SCI indexed articles in international journals. He has won eight provincial awards for his research and held a dozen grants from NSFC, MoST, MoE, and Hubei province. He has delivered more than 10 invited presentations at international conferences in Italy, USA, Japan, and the Netherlands.

Jose Norambuena-Contreras

Dr. Jose Norambuena-Contreras is an Associate Professor in Civil Engineering at the University of Bio-Bio (UBB-Chile). At UBB-Chile, he leads the Research Group LabMAT focused on Self-Healing Construction Materials for Pavement Infrastructure, integrating Material Science, Civil Engineering, and Chemical Engineering disciplines. Previously, he has held Scientific Positions as a Researcher and Postdoctoral Fellow at GITECO-Construction Technology Applied Research Group at the University of Cantabria (Spain), EMPA-Swiss Federal Laboratories for Materials Science and Technology at the ETH-Zürich (Switzerland), and NTEC-Nottingham Transportation Engineering Centre at the University of Nottingham (UK). Dr. Norambuena-Contreras has performed high-level multidisciplinary research in the field of Advanced Construction Materials for Civil Engineering (h-index 31, 3000+ citations). He has published 65+ peer-reviewed journal publications (indexed in WoS), 8+ peer-reviewed book chapter publications, and 50+ peer-reviewed conference publications. In addition, he is co-editor of the book "Self-Healing Construction Materials". As a result of his research reputation, he regularly contributes as an external reviewer in 10+ top peer-reviewed journals on Material Science and Engineering. Finally, he has been a Principal Investigator (PI) or co-PI on 15+ Research Projects supported by several industries, as well as national and international funding agencies. He was recently awarded the IAAM Scientist Medal from the International Association of Advanced Materials (IAAM).

Yue Huang

Yue Huang is an Associate Professor at the Institute for Transport Studies (ITS), University of Leeds. His research areas include LCA, pavement recycling, and road safety. He is the Programme Leader of MSc Transport Planning and Engineering, as well as the Deputy Director of Postgraduate Research Studies at ITS. Dr. Huang obtained his Bachelor and Master Degrees from Chang'an University, China. After completing his PhD at Newcastle University in 2007, he started his career as a Research Engineer at Scott Wilson (now AECOM) until 2011 when he became a Research Fellow at the University of Nottingham. He became a Senior Lecturer at Liverpool John Moores University in 2012 and joined ITS in Leeds in 2018. He has been involved in a number of research projects in the UK, EU, and overseas. Dr. Huang has more than 10 years of experiences in teaching undergraduate and postgraduate courses, leading modules, developing programs (including MOOC), and supervising PhD students. Dr. Huang is a Chartered Engineer (CEng) and a Fellow of the Higher Education Academy (FHEA).

Preface

With the growing demand for passenger transport and freight, the design, construction, and maintenance of transportation infrastructure now face new development opportunities and challenges. Under the combined effects of traffic load and environmental factors, the performance and serviceability of road pavement deteriorate over time. Improving the durability and surface characteristics of road pavement remains an important research goal. In addition, emerging technologies and societal changes put forward a high demand for green pavement, environmentally friendly construction and maintenance, and smart and intelligent infrastructure. Traditional pavement materials cannot meet the needs of our evolving society. In recent years, the advantages of new materials in improving road performance have attracted extensive interest from industry professionals and academia. Polymer materials exhibit high thermal stability, superior mechanical properties, as well as corrosion and chemical resistance. They are widely used in important fields of the economy such as aerospace, biomedicine, transportation, electronics, etc. With the development of material science and technology, polymer materials demonstrate great potential for road engineering applications.

In order to succinctly summarize recent advancements in polymer road materials research and to promote innovation in this field, Professor Wei Jiang of Chang'an University collaborated with Professor Quantao Liu of Wuhan University of Technology, Associate Professor Jose Norambuena-Contreras of the University of Bío-Bío, and Associate Professor Yue Huang of the University of Leeds to co-organize the Special Issue titled "Application of Polymer Materials in Pavement Design".

This Special Issue comprises 17 research articles and 2 review articles, covering studies on both individual and composite polymer modifications. Topics include high-content polymer-modified bitumen, epoxy resin adhesives, SBS-modified bitumen, PE-modified bitumen, crumb-rubber-modified bitumen, and biopolymeric capsules and bitumen rejuvenators.

We wish to extend our appreciation to the *Polymers* editorial team for their valuable technical support and to the scholarly community for their great contributions to this Special Issue.

Wei Jiang, Quantao Liu, Jose Norambuena-Contreras, and Yue Huang
Editors

polymers

MDPI

Article

Viscoelastic Behavior and Phase Structure of High-Content SBS-Modified Asphalt

Dongdong Yuan [1,2], Chengwei Xing [1,2], Wei Jiang [1,2,*], Jingjing Xiao [3], Wangjie Wu [1,2], Pengfei Li [1,2] and Yupeng Li [1,2]

1 Key Laboratory for Special Area Highway Engineering of Ministry of Education, Chang'an University, Xi'an 710064, China; ddy@chd.edu.cn (D.Y.); xingcw@chd.edu.cn (C.X.); wwj@chd.edu.cn (W.W.); 2020021080@chd.edu.cn (P.L.); liyupeng@chd.edu.cn (Y.L.)

2 School of Highway, Chang'an University, South 2nd Ring Road Middle Section, Xi'an 710064, China

3 School of Civil Engineering, Chang'an University, Xi'an 710064, China; xiaojj029@sina.com

* Correspondence: jiangwei@chd.edu.cn

Abstract: To investigate the effect of styrene-butadiene-styrene (SBS) modifier content on the viscoelastic behavior of SBS-modified asphalt (SBSMA) at different temperatures and phase structures, the star SBS modifier was chosen to fabricate seven types of SBSMA with different contents. Multiple stress creep recovery (MSCR), linear amplitude sweep (LAS), and low-temperature frequency sweep tests were adopted to study the influence of SBS modifier content on the viscoelastic performance of SBSMA at high to low temperatures. The SBSMA's microstructure with different contents was investigated using a fluorescence microscope. The results indicated that the change in non-recoverable creep compliance and creep recovery rate was bounded by 4.5% content at high temperatures, with an apparent turning point. The changing slope of content at less than 4.5% was much higher than that of the content greater than 4.5%. At medium temperatures, the fatigue life of SBSMA increased exponentially with the rising modifier content. The rate of increase in fatigue life was the largest as the content increased from 4.5% to 6.0%. At low temperatures, the low-temperature viscoelastic property index G (60 s) of SBSMA decreased logarithmically as the modifier content increased. In terms of the microscopic phase structure, the SBS modifier gradually changed from the dispersed to the continuous phase state with the increasing SBS modifier content.

Keywords: high-content polymer modified asphalt; SBS; viscoelastic behavior; phase structure

Citation: Yuan, D.; Xing, C.; Jiang, W.; Xiao, J.; Wu, W.; Li, P.; Li, Y. Viscoelastic Behavior and Phase Structure of High-Content SBS-Modified Asphalt. *Polymers* **2022**, *14*, 2476. https://doi.org/10.3390/polym14122476

Academic Editor: Alexander Malkin

Received: 21 May 2022
Accepted: 16 June 2022
Published: 17 June 2022

1. Introduction

Asphalt pavement consists of aggregates, fillers, and asphalt binders; the design of asphalt binders and research on its related properties remain at the core of asphalt pavement [1,2]. Using an asphalt binder with excellent properties can remarkably improve the quality of asphalt pavement [3,4]. With the increasing traffic load, a neat asphalt binder will not be enough to fulfill the requirements of traffic development [5,6]. Styrene-butadiene-styrene (SBS)–modified asphalt (SBSMA) is universally applied because of its properties such as excellent durability, anti-ageing, fatigue resistance, and water damage resistance [7,8]. SBSMA has been utilized extensively on national highways in China [9,10]. SBS is a block copolymer created from the anionic polymerization of 1,3-butadiene, styrene (monomer), tetrahydrofuran (activator), and n-butyllithium (initiator) in the cyclohexane solvent [11]. According to the different contents of polystyrene and polybutadiene, as well as the difference in the molecular structure, SBS can be divided into the linear and star structures, as shown in Figure 1 [12]. In general, the molecular weight of the star structure is higher than that of the linear structure. Previous research and applications have suggested that the SBSMA content in many countries is 3.0–6.0%, considering the limitations of asphalt pavement construction costs and construction technologies [13,14]. Yet, no in-depth investigation has been conducted on the higher dosage of this material.

However, porous asphalt concrete has been more and more used in pavement materials with the advancements in pavement green technology [15,16]. Therefore, modified asphalt with high viscosity is commonly applied in porous asphalt concrete [17,18]. At present, modified asphalt with high viscosity is mainly fabricated using a high content of SBSMA (6.0–12.0%) [19].

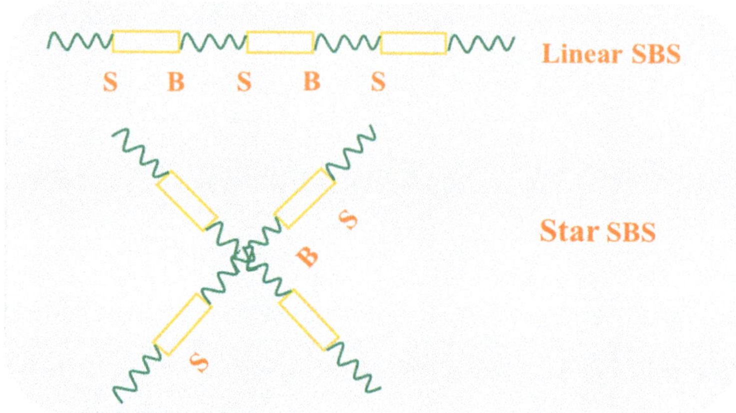

Figure 1. Linear SBS and Star SBS.

Zhang et al. examined the rheological behavior of high-content SBSMA after the addition of a plasticizer (fural exact oil) and a crosslinker (sulfur). Their results illustrated that the plasticizer reduced the anti-rutting performance of SBSMA. The inclusion of a crosslinker created a polymer network, a network structure with better ageing, which increased SBSMA's ageing resistance [20]. This is because the ageing of the polymer depends on the actual 3D structure (more chaotic or more structured) and the type of crosslinks [21]. However, the SBS modifier content in their study was just 6.0%, remarkably lower than what is used in practice. Yan et al. evaluated the ageing properties of high-content SBSMA, whose findings revealed that the SBS modifier breakdown occurs during the early stages of ageing. In addition, short-term ageing at elevated temperatures can severely degrade high-content SBSMA's anti-rutting capabilities [22]. Lin et al. investigated the rheological properties of high-content SBSMA. Their results suggested that SBSMA with higher content would possess superior rheological qualities; however, 9% is the ideal dosage for economic reasons [23]. Zhang et al. explored the composition of high viscosity–modified asphalt and discovered that increasing the SBS modifier concentration is one of the most effective approaches to maintaining the physical qualities of high viscosity–modified asphalt [24]. Giacomo et al. observed that a high-content SBS modifier was able to lessen the ageing-induced stiffening of SBSMA. Compared to typical SBSMA with a lower polymer concentration, the SBS modifier network in the high-content SBSMA could present a barrier to the oxidation of the binder, leading to better anti-ageing performance [3,25].

In a nutshell, limited studies on high-content SBSMA have been conducted so far. Furthermore, most research focuses on the influence of ageing on the properties of high-content SBSMA. At present, research on the viscoelastic properties and phase structure of high-content SBSMA is not detailed enough. Therefore, to investigate the viscoelastic behavior of high-content SBSMA at high to low temperatures, we employed three new dynamic shear rheometer (DSR) test methods. That is, the multiple stress creep recovery (MSCR) test, linear amplitude sweep (LAS) test, and 4 mm low-temperature frequency sweep test. Meanwhile, the microstructure of SBSMA with different contents was analyzed by a fluorescence microscope. The flowchart of this study is shown in Figure 2. This study clarified the viscoelastic properties and microscopic phase structure of high-content

SBS modified asphalt at different temperatures, which provides a reference for the wider application of high-content SBS modified asphalt.

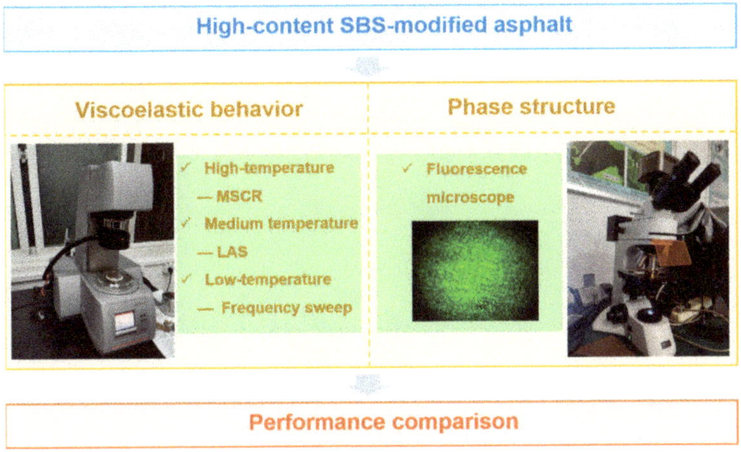

Figure 2. Flowchart of this study.

2. Materials and Methods

2.1. Raw Materials

Shell 70# neat asphalt was utilized to prepare high-content SBSMA. Its basic attributes are tabulated in Table 1. The adopted SBS modifiers were provided by Baling Petrochemical Company of Sinopec Group. Table 2 lists the basic parameters of the modifier. The modifier content was chosen as 3.0%, 4.5%, 6.0%, 7.5%, 9.0%, 10.5% and 12.0% of the mass of the neat asphalt, respectively. The sulfur powder with purity of more than 99% provided by Shanghai Qunkang Asphalt Technology Co., Ltd. (Shanghai, China) was selected as the stabilizer. The content of the stabilizer was chosen as 0.15% of the mass of the neat asphalt.

Table 1. Basic property of neat asphalt.

Item	Shell 70#
Penetration at 25 °C, 0.1 mm	70.8
Softening point, °C	49.2
Ductility at 5 °C, 5 cm min^{-1}, cm	74.3

Table 2. Technical performance of SBS modifier.

Type	Star Type
Specific gravity, g cm^{-3}	0.94
Elongation at break, %	680
Tensile strength, MPa	21.2
Melt index, g (10 min^{-1})	7.0

2.2. Experimental Methods

2.2.1. Preparing SBSMA

SBSMA can be prepared as Figure 3. First, the corresponding quality of neat asphalt, SBS modifier, and stabilizer was weighed. We heated the neat asphalt to the molten state in an oven. The weighed SBS modifier was added into it and stirred with a vane stirrer at 170 °C and 1000 r/min for half an hour. The blended neat asphalt and SBS modifier were then sheared by a shear emulsifying machine at 170 °C and 3000 rotations per min for 10 min. Next, the stabilizer was added, and the mixture was sheared for an hour at 170 °C and 5000 rotations per min. Finally, the sheared SBS modifier was heated at 170 °C for 90 min to allow the prepared SBSMA to fully develop and escape the air bubbles created during the preparation process.

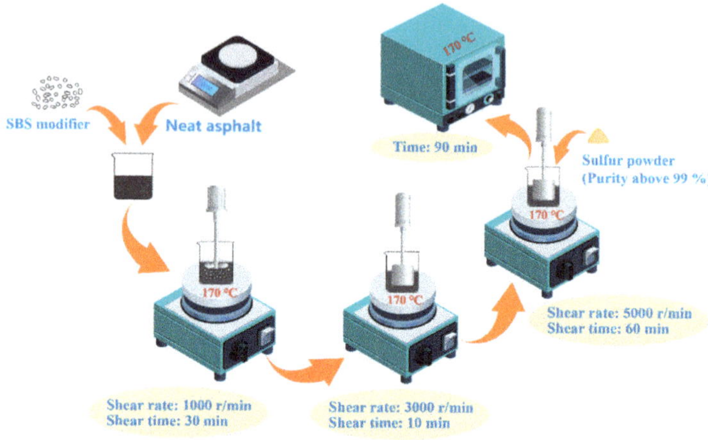

Figure 3. Preparation process of SBSMA.

2.2.2. MSCR Test

The high-temperature viscoelastic behavior of SBSMA was investigated using MSCR tests. In this test, Smartpave 102 DSR (Anton Paar company, Graz, Austria) was utilized. Two round parallel plates with a diameter of 25 mm were utilized, and the gap between them was 1 mm. To simulate the high temperature experienced by the asphalt pavement in summer, referring to the AASHTO M332-19 [26] classification standard, 64 °C was selected for the MSCR test. The creep and recovery tests were performed with the constant stress (0.1 kPa and 3.2 kPa), respectively. During the test, the stress was first loaded for 1 s. Afterward, zero stress was recovered for 9 s. First, the test was run for 20 cycles at 0.1 kPa. Then, it ran for 10 cycles at 3.2 kPa. Finally, it ran 30 creep and recovery cycles for 300 s [27,28].

2.2.3. LAS Test

The LAS test can assess SBSMA's fatigue properties at medium temperatures. Smartpave 102 DSR (Anton Paar company, Graz, Austria) was used in this test, with an 8 mm-parallel plate die and a 2 mm-gap. The LAS test was performed in a loading mode (controlled strain) with a design test time of 300 s. During the test, the sine wave dynamic load amplitude rose linearly from 0.1% to 30% [29,30]. Herein, 25 °C was chosen as the LAS test temperature.

2.2.4. Low-Temperature Frequency Sweep Test

The low-temperature frequency sweep test was performed by DSR, which could overcome the disadvantages of too many materials and the long test time of the low-temperature bending beam rheometer (BBR) test. The 4 mm parallel plate low-temperature frequency sweep test can substitute the BBR test to assess the asphalt property at low temperatures [31]. Therefore, a 4 mm parallel plate for the low-temperature frequency sweep test was selected herein to explore SBSMA's low-temperature viscoelastic behavior. The tests used a SmartPave Model 102 DSR with a 4 mm-parallel plate die placed at the 3 mm-gap. The loading model was controlled strain, and the strain was 1%; the sweep frequency was 0.1–100 rad/s, and the test temperature was −5 °C and −15 °C [32].

2.2.5. Fluorescence Microscope Test

An LW300LFT fluorescence microscope (Nikon Corporation, Tokyo, Japan) was employed to determine SBSMA's distribution and morphological characteristics in the asphalt phase. The magnification of the eyepiece of the fluorescence microscope and that of the objective lens was 10.

3. Results and Discussion

3.1. Viscoelastic Behavior at High Temperatures

3.1.1. Non-Recoverable Creep Compliance and Creep Recovery Rate

Generally, a high-temperature rutting phenomenon will occur on the asphalt pavement due to the accumulation of the asphalt binder's non-recoverable strain. Therefore, the non-recoverable creep compliance (J_{nr}) is a critical evaluation indicator of the MSCR test (Equation (1)). Under a certain recovery time, the greater the recovery deformation and the smaller the amount of non-deformation of asphalt, the less likely high-temperature rutting will occur. Therefore, the creep recovery rate (R) can also represent the high-temperature viscoelastic behavior of asphalt (Equation (2)). J_{nr} and R are calculated by average values in 10 creep recovery cycles, respectively [31]. The J_{nr} at 0.1 kPa and 3.2 kPa are denoted as $J_{nr}0.1$ and $J_{nr}3.2$. Furthermore, the R at corresponding stress levels are represented as $R0.1$ and $R3.2$.

$$J_{nr} = \frac{\varepsilon_u}{\sigma} \tag{1}$$

where, ε_u is the adjusted strain value after the recovery period, σ is the value of the applied stress level.

$$R = \frac{\varepsilon_p - \varepsilon_u}{\varepsilon_p} \times 100\% \tag{2}$$

where, ε_p is the adjusted strain value when the creep loading ends, ε_u is the adjusted strain value after recovery period.

Figure 4 presents the J_{nr} and R of the eight kinds of asphalt at 64 °C. Under different stress levels, $J_{nr}3.2 > J_{nr}0.1$, $R3.2 < R0.1$. This order indicates that the increase in stress will worsen the high-temperature viscoelastic properties of SBSMA. With the rising modifier content, the J_{nr} of SBSMA dropped and the R increased, suggesting that increasing the modifier content can improve the rheological properties of SBSMA. When the modifier content increased by 1.5%, an obvious turning point was noticed in the change in J_{nr} and R, which was bounded by 4.5% content. The changing slope of the content less than 4.5% is much higher than that of the content greater than 4.5%, indicating that when the modifier content exceeds 4.5%, it can enhance the SBSMA's high-temperature viscoelastic behavior with a limited improvement effect. This is because when the content of SBS modifier is 4.5%, the viscosity of the asphalt phase and the elasticity of the SBS phase in SBSMA reach an equilibrium state. Although increasing the content of the modifier can increase the mechanical strength of the SBS phase, the effect is not obvious. Figure 5 shows that for Shell 70# neat asphalt, the addition of the star SBS modifier to 3.0% content caused a decrease in $J_{nr}0.1$ and $J_{nr}3.2$ by 82.44% and 78.91%. $R0.1$ and $R3.2$ rose by 955.00% and 8690.90%. It is shown that adding the SBS modifier can considerably boost the neat asphalt's high-temperature viscoelastic performance. For SBSMA, when it rises from 3.0% to 4.5%, the change rate of J_{nr} and R is the largest, followed by the change rate when the content increases from 4.5% to 6.0%. Thereafter, with the increasing content, the change rate of J_{nr} and R becomes smaller, which may be related to the microstructure of SBSMA.

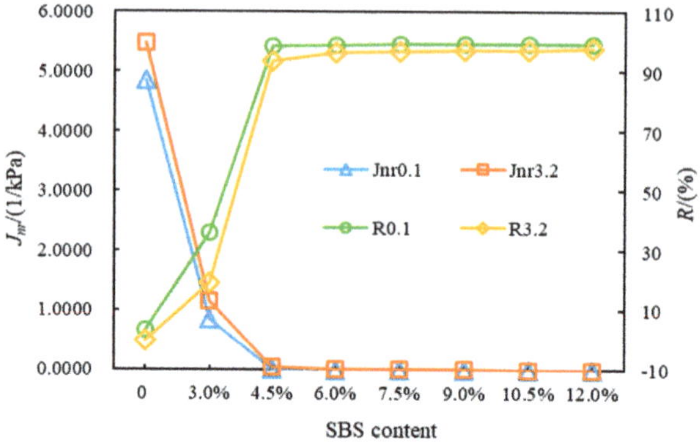

Figure 4. Non-recoverable creep compliance and creep recovery rate of asphalt.

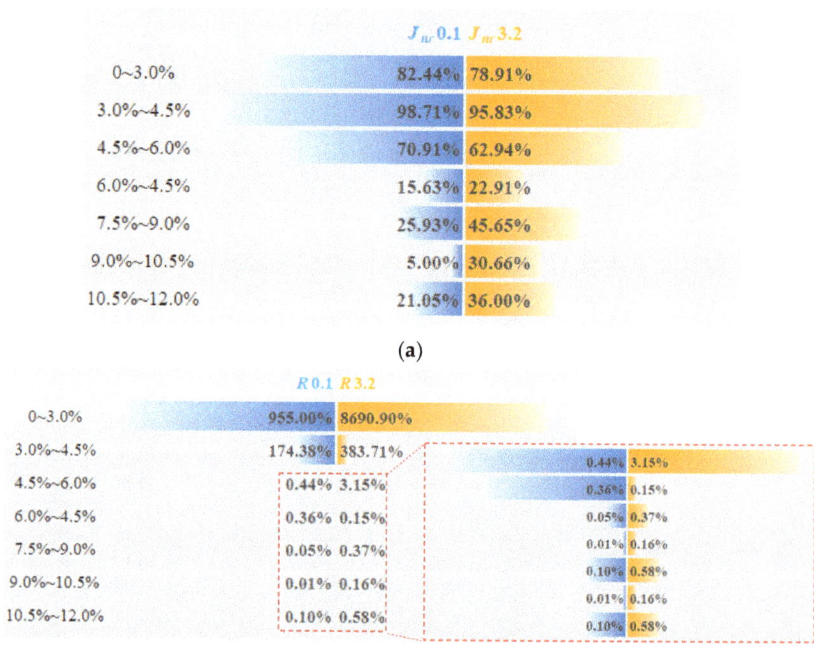

Figure 5. The change rate of J_{nr} and R. (**a**) non-recoverable creep compliance; (**b**) Creep recovery rate.

3.1.2. Stress Sensitivity

The SBSMA's sensitivity to stress can be expressed by the difference in the J_{nr} under 3.2 kPa and 0.1 kPa, which is calculated according to Equation (3).

$$J_{nr\text{-}diff} = [(J_{nr}3.2 - J_{nr}0.1)/J_{nr}0.1] \times 100\% \quad (3)$$

where, $J_{nr\text{-}diff}$ is the stress sensitivity, $J_{nr}0.1$ and $J_{nr}3.2$ are the values of J_{nr} of asphalt at 0.1 kPa and 3.2 kPa. The $J_{nr\text{-}diff}$ of eight types of asphalt at 64 °C are plotted in Figure 6.

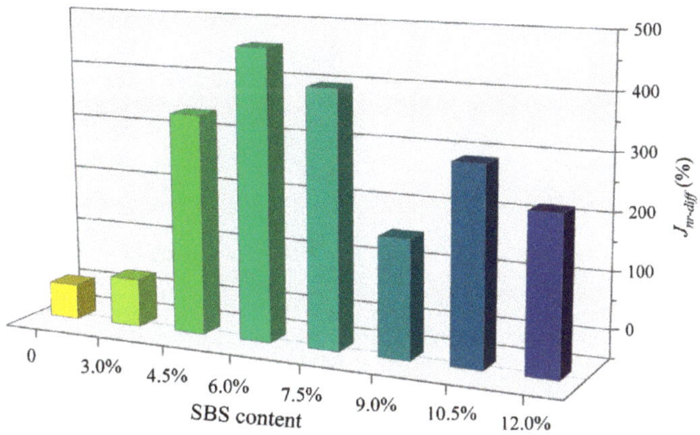

Figure 6. $J_{nr\text{-}diff}$ of eight types of asphalt.

Figure 6 shows that the $J_{nr\text{-}diff}$ of the neat asphalt is the smallest among the eight types of asphalt, and the $J_{nr\text{-}diff}$ of SBSMA is much larger than that of the neat asphalt. For SBSMA, $J_{nr\text{-}diff}$ has no evident change rule with the rising star SBS modifier content. The $J_{nr\text{-}diff}$ of 3.0% content is the smallest, and the $J_{nr\text{-}diff}$ of 6.0% content is the largest. This may be attributed to a large number of polymer chain segments inside the SBSMA with complex mechanical behavior and different phase structures. The literature has shown that the viscoelastic performance of modified asphalt could be determined according to the notion of whether the $J_{nr\text{-}diff}$ of the modified asphalt is greater than 5% [30]. When $J_{nr\text{-}diff}$ is greater than 5%, it is a nonlinear viscoelastic state. Similarly, when $J_{nr\text{-}diff}$ is less than 5%, it is a linear viscoelastic state. At 3.2 kPa and 64 °C, the viscoelastic performance of star SBSMA with a content of more than 3.0% is all nonlinear.

3.2. *Viscoelastic Properties at Medium Temperatures*

3.2.1. Stress–Strain Response

Asphalt undergoes elastic and plastic deformations in the LAS test under repeated loading. The shear stress gradually decreases when the applied load reaches a particular point; however, the shear strain increases in that case. A peak value of the shear stress is observed in the LAS test's stress–strain curve. The AASHTO TP 101-12 specification defines this peak value as the asphalt yield stress, and its shear strain the yield strain. Figure 7 presents the stress–strain curves of LAS tests of eight kinds of asphalt. The neat asphalt has the highest yield stress and the smallest yield strain. Only the SBSMA with 3.0%, 4.5% and 6.0% contents demonstrated a peak in the stress–strain curve for the SBSMA. The order of yield stress was 3.0% > 6.0% > 4.5%, and the order of yield strain was 6.0% > 4.5% > 3.0%. The ranking of yield stress and yield strain is inconsistent. When the content was greater than 6.0%, the stress increases relatively slowly in the loading process. With the increase in strain, the stress gradually becomes flat and the stress–strain curve does not show a peak. This finding indicated that the yield stress or yield strain can only characterize the stress–strain response of asphalt under medium temperature conditions and repeated loadings but cannot characterize the fatigue properties of asphalt.

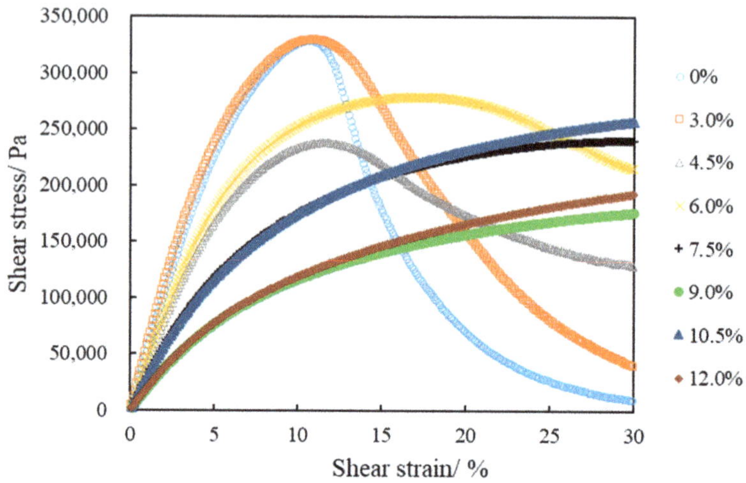

Figure 7. Stress–strain response of asphalt.

3.2.2. Fatigue Life

The results of the LAS tests were further analyzed using the Viscoelastic Continuous Damage Mechanics theoretical model [31]. Notably, the LAS test is composed of frequency sweep and amplitude sweep. First, the parameter α is acquired via the frequency sweep test and is calculated according to Equations (4)–(6) [32,33].

$$G' = |G^*| \times \cos\delta \quad (4)$$

$$\lg G' = m(\lg\omega) + b \quad (5)$$

$$a = \frac{1}{m} \quad (6)$$

where G' is the storage modulus (MPa), G^* is the complex shear modulus (MPa), δ is the phase angle (°), m and b are the fitting parameters, a is a parameter for asphalt's viscoelastic behavior.

Secondly, the damage variable (D) is calculated, as shown in Equation (7) [34].

$$D\,(t) \cong \sum_{i=1}^{N} [\pi\gamma_0^2\,(C_{i-1} - C_i)]^{a/(1+a)}\,(t_i - t_{i-1})^{1/(1+a)} \quad (7)$$

where t is the test time (s), t_i is the current test time, t_{i-1} is the previous test time, N is the total number of tests, $C(t)$ is the integrity parameters, which is calculated by Equation (8).

$$C\,(t) = |G^*|(t)/|G^*|_{\text{initial}} \quad (8)$$

where, $|G^*|(t)$ is the complex shear modulus with test time in the amplitude sweep test (MPa), $|G^*|_{\text{initial}}$ is the complex shear modulus when the test starts (MPa), γ_0 is the test strain (%).

For the loss variable (D (t)) and integrity parameter (C (t)), there is the following relationship, as shown in Equation (9).

$$C\,(t) = C_0 - C_1[D\,(t)]^{C_2} \quad (9)$$

where, $C_0 = 1$, C_1 and C_2 are the fitting parameters.

Finally, the fatigue life is calculated, as shown in Equation (10).

$$N_f = A\left(\gamma_{\max}\right)^{-B} \tag{10}$$

where, γ_{max} is the expected maximum strain (%) and A and B are the fatigue correlation coefficients, which are calculated according to Equation (11) to Equation (12).

$$A = \frac{f\left(D_f\right)^{[1+(1-C_2)a]}}{[1+(1-C_2)a](\pi C_1 C_2)^a} \tag{11}$$

where, f = 10 Hz, $D_f = (\frac{C_0 - C_{peak}}{C_1})^{1/C_2}$.

$$B = 2a \tag{12}$$

The above calculation results present the damage characteristic curves and fatigue life curves of eight kinds of asphalt (Figures 8 and 9).

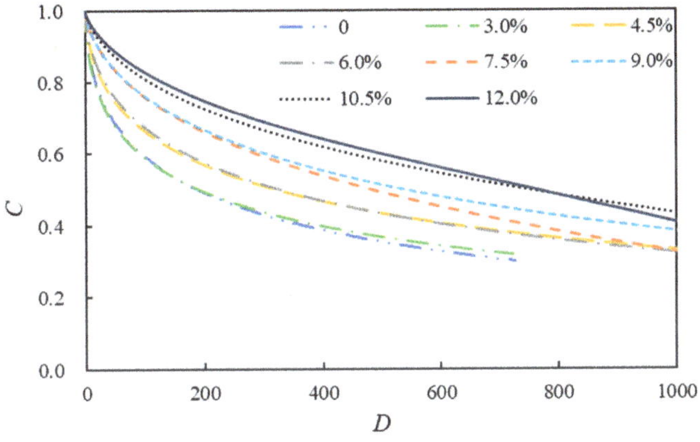

Figure 8. Damage characteristic curves of asphalt.

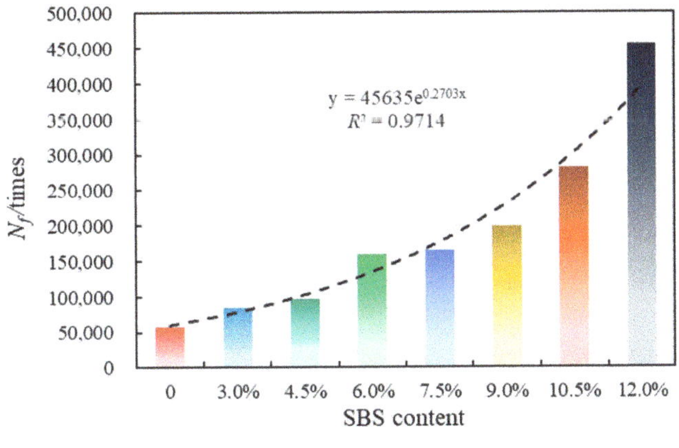

Figure 9. Fatigue life of asphalt.

Figure 8 shows that C represents the integrity parameter of asphalt and D represents the cumulative damage parameter. The expression $C = 1$ denotes that the asphalt is intact, and the expression $C = 0$ denotes that it has been completely damaged. Figure 8 shows that when the cumulative damage parameter has a certain value, the asphalt's integrity improves with the rising modifier content. Among them, the neat asphalt's integrity is the worst, and the integrity of the 12.0% SBSMA is the best. However, the damage characteristic curves of 12.0% and 10.5% SBSMA appear staggered, suggesting that with the increase in the damage intensity, the damage resistance of 10.5% SBSMA is higher than that of 12.0%.

Figure 9 shows different asphalts' fatigue life at 2.5% strain. The fatigue life of SBSMA increases exponentially with the increasing modifier content. This is because with the increase in SBS modifier content, the distribution of the SBS phase in asphalt gradually forms a cross-linking state, which increases the fatigue life. Specifically, for Shell 70# neat asphalt, N_f increased by 47.3% after adding 3.0% of the star SBS modifier. For SBSMA, when the content rises from 4.5% to 6.0%, the change rate of N_f is the largest (63.9%), followed by the change rate from 10.5% to 12.0%, and the change rate from 6.0% to 7.5% is the smalles (2.8%).

3.3. Viscoelastic Properties at Low Temperatures

3.3.1. Modulus and Phase Angle

Figures 10 and 11 show the variations in G^* and δ, with eight kinds of asphalt frequencies at −5 °C and −15 °C. The G^* of eight types of asphalt increased when the frequency was gradually augmented from 0.1 rad/s to 100 rad/s, which conformed to Generalized Maxwell model [35]. However, the phase angle change is more complicated. At −5 °C, the phase angle of SBSMA decreases with the increasing frequency when the content of SBSMA is less than 10.5%. Although the SBSMA with contents of 10.5% and 12.0%, the phase angle decreased and the phenomenon of "first increase and then decrease" will appear in the change process. At −15 °C, the phase angle of SBSMA has no obvious pattern when the frequency enlarges. With the rising modifier content, the G^* and δ of SBSMA had little change. The order of the G^* and δ of the eight types of asphalt at −5 °C and −15 °C was inconsistent. This may be because in the preparation of high-content SBSMA, the same shear time as the low-content SBSMA was used, resulting in a portion of the SBS of the high-content SBSMA being too late to swell and cross-link. Additionally, each high-content SBSMA is too late to swell and the cross-link of the SBS is different, so the SBSMA with the increase in the amount of low-temperature viscoelastic properties is also different.

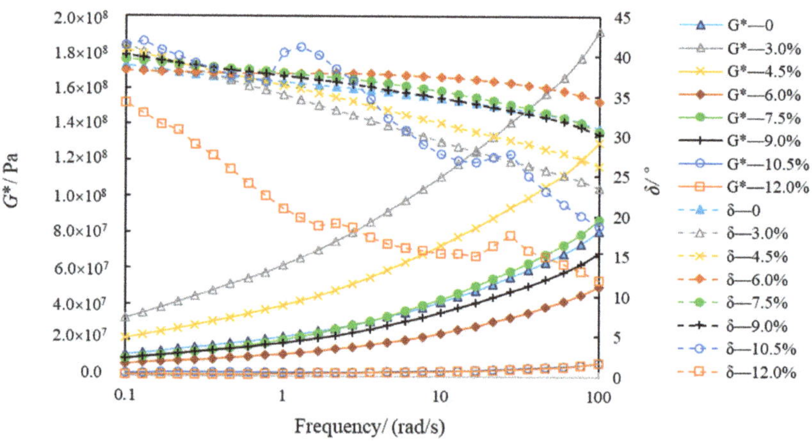

Figure 10. Modulus and phase angle of asphalt at −5 °C.

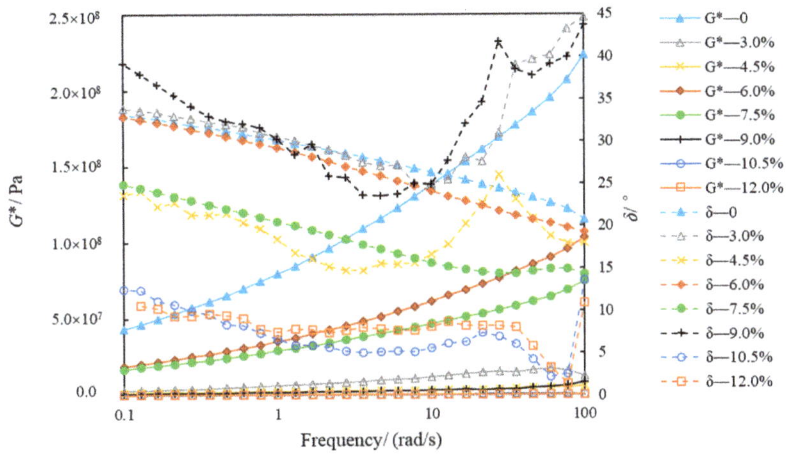

Figure 11. Modulus and phase angle of asphalt at −15 °C.

3.3.2. Evaluating Index

Sui et al. [36] compared the relaxation modulus $G\ (t)$ master curve according to the 4 mm DSR test with the creep stiffness modulus $S\ (t)$ master curve based on BBR tests. They showed that G (60 s) and m_r (60 s) had an excellent linear correlation with S (60 s) and m_c (60 s). Therefore, G (60 s) and m_r (60 s) are recommended for 4 mm DSR as the evaluation indicator of low-temperature viscoelastic properties of asphalt. While calculating G (60 s) and m_r (60 s), the primary curve of storage modulus (G') of asphalt was first derived by fitting the time-temperature equivalence principle. Equation (13) describes the transformation between the primary curves of the $G'\ (\omega)$ and the $G\ (t)$. Then, Equation (14) was adopted to fit the master curve of the $G\ (t)$. Finally, G (60 s) and m_r (60 s) were counted according to Equations (15) and (16), respectively.

$$G\ (t)\ \approx\ G'\ (\omega)\ |_{\omega=2/\pi t} \tag{13}$$

where, $G\ (t)$ is the relaxation modulus (Pa), $G'\ (\omega)$ is the storage modulus (Pa), t is the test time (s), ω is the angular frequency (rad/s).

$$y = ax^2 + bx + c \tag{14}$$

where, a, b, and c are parameters for fitting.

$$G\ (60\ \mathrm{s}) = ax^2 + bx + c\ |_{x=1.78} \tag{15}$$

$$m_r\ (60\ \mathrm{s}) = 2ax + b\ |_{x=1.78} \tag{16}$$

Using −15 °C as the reference temperature, the master curve of the relaxation modulus of eight kinds of asphalt is drawn in this study (Figure 12). Next, G (60 s) and m_r (60 s) were counted (Figures 13 and 14).

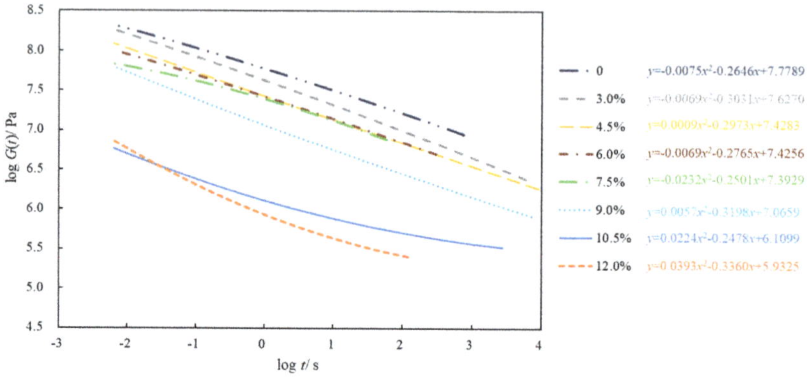

Figure 12. Main curve of relaxation modulus of asphalt at low temperatures (−15 °C).

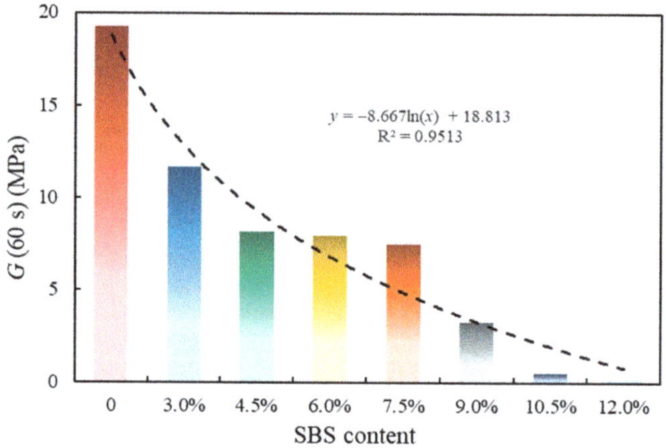

Figure 13. Evaluation index G (60 s) of asphalt at low temperatures.

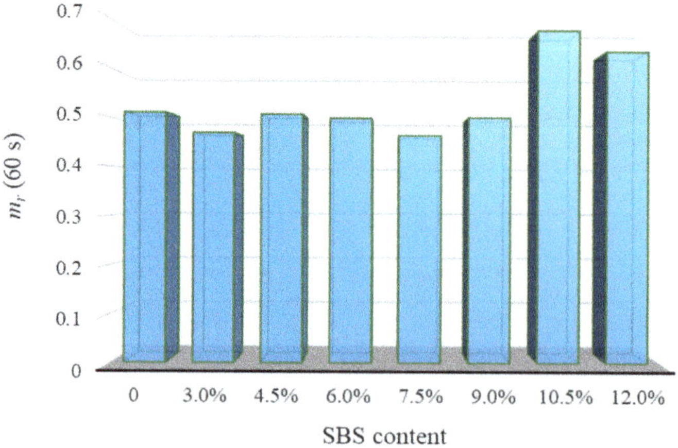

Figure 14. Evaluation index m_r (60 s) of asphalt at low temperatures.

Figure 12 plots that the relaxation modulus level of neat asphalt is obviously higher than that of SBSMA, whereas the relaxation modulus of 3.0% is the largest and that of 12.0% is the smallest. Furthermore, the SBS modifier reduced the asphalt's relaxation modulus level, lowering the temperature stress accumulation inside the asphalt and improving its cracking resistance at low temperatures.

Figure 13 shows the low-temperature property evaluation index G (60 s) of asphalt. It is shown that the G (60 s) of SBSMA decreased logarithmically with the increasing modifier content. For Shell 70# neat asphalt, G (60 s) decreased by 39.5% after adding 3.0% content. For SBSMA, when it rises from 9.0% to 10.5%, the change rate of G (60 s) is the largest. In contrast, the change rate is the smallest when the content increases from 4.5% to 6.0%. Figure 14 shows the asphalt's low-temperature property evaluation index m_r (60 s). With the increasing SBS modifier content, m_r (60 s) shows no apparent change law, which is attributed to the different slopes of relaxation modulus curves of different types of asphalt in 60 s. Importantly, using m_r (60 s) to determine the low-temperature property of SBSMA is still worthy of in-depth discussion and research.

3.4. Phase Structure of SBSMA

Figure 15 illustrates the microscopic phase structures of the eight kinds of asphalt under a 100-fold fluorescence microscope.

The microscopic phase structure of SBSMA is a two-phase (SBS phase and asphalt phase) coexistence of the co-blended structural system compared to the neat asphalt. The phase structure varies with the SBS modifier content. When the SBS modifier content rises from 3.0% to 12.0%, the distribution state of SBS phase in asphalt phase gradually forms a cross-linking state; that is, the SBS modifier gradually changes from the dispersed to the continuous phase state. This is because as the modifier increases, the number of particles of the modifier increases, and its specific surface area increases significantly. Under the action of surface tension, the modifier particles are more likely to agglomerate. Specifically, when this content is 3.0% and 4.5%, the SBS modifier is in the dispersed phase state, whereas the neat asphalt is in the continuous phase state. The 4.5% content of the SBS modifier dispersion is more uniform than the 3.0% content of the SBS modifier. When it reaches 6.0%, the SBS modifier is also present in the dispersed phase. Moreover, a particular network structure appears and the SBS modifier gradually changes from the dispersed to the continuous phase state. When the SBS modifier content is 7.5% and 9.0%, the two phases of the blend are continuous, and the modifier-formed network structure is intertwined with each other. Compared with the 7.5% content of the SBS modifier, the mesh structure formed by the 9.0% content is more closely intertwined. When this content reaches 10.5% and 12.0%, the SBS modifier becomes a continuous phase state, whereas the neat asphalt becomes a dispersed phase state, and the formed network structure exhibits a rough surface and large interwoven "leaf". In contrast, the 12.0% SBSMA had a rougher surface and larger interwoven "blade" than the 10.5% SBSMA.

(a) (b)

Figure 15. *Cont.*

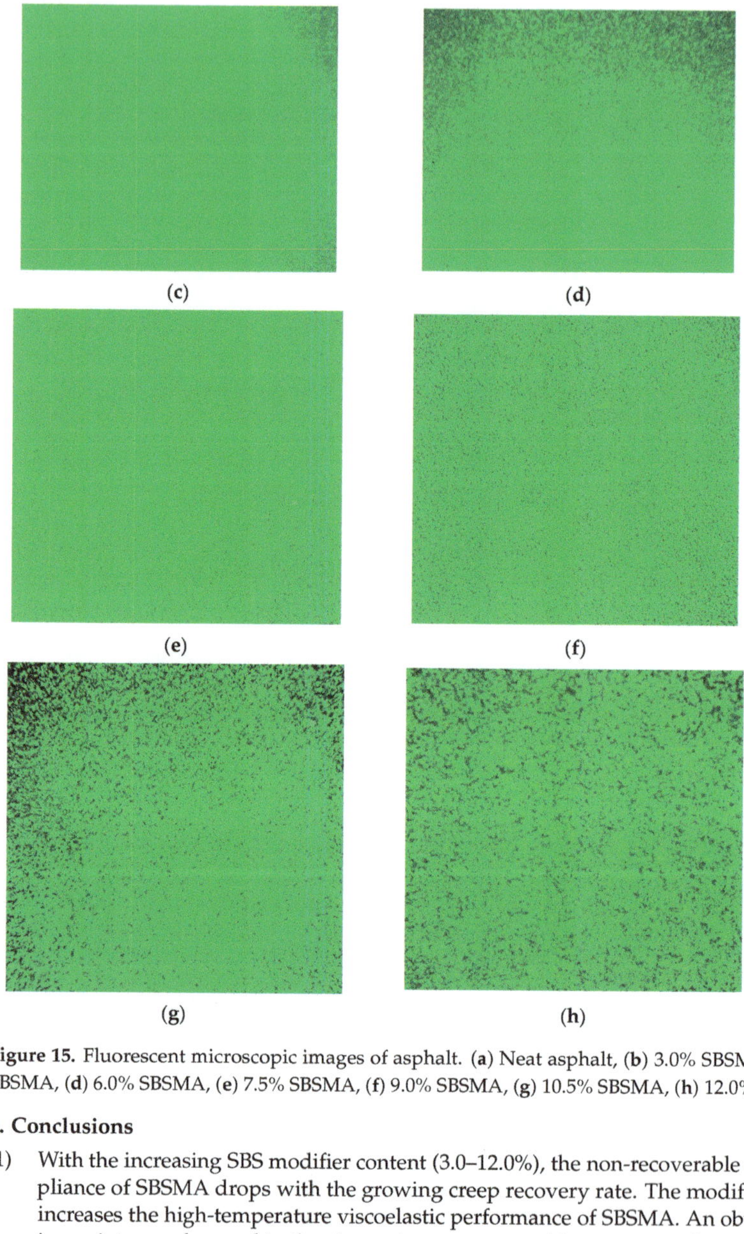

(c) (d) (e) (f) (g) (h)

Figure 15. Fluorescent microscopic images of asphalt. (**a**) Neat asphalt, (**b**) 3.0% SBSMA, (**c**) 4.5% SBSMA, (**d**) 6.0% SBSMA, (**e**) 7.5% SBSMA, (**f**) 9.0% SBSMA, (**g**) 10.5% SBSMA, (**h**) 12.0% SBSMA.

4. Conclusions

(1) With the increasing SBS modifier content (3.0–12.0%), the non-recoverable creep compliance of SBSMA drops with the growing creep recovery rate. The modifier content increases the high-temperature viscoelastic performance of SBSMA. An obvious turning point was observed in the change in non-recoverable creep compliance and creep recovery rate, which is bounded by the 4.5% content. The changing slope of the content less than 4.5% is much higher than that of the content greater than 4.5%.

(2) The fatigue life of SBSMA increases exponentially with the increasing modifier content. Moreover, the growth rate of fatigue life is the largest (63.9%) when the content increases from 4.5% to 6.0%.

(3) The 4 mm DSR test can evaluate the viscoelastic performance of SBSMA at low temperatures. G (60 s) and m_r (60 s) were selected as evaluation indicators. The G (60 s) of SBSMA decreases logarithmically with the increasing modifier content

(3.0–12.0%). However, m_r (60 s) has no noticeable change with the rising content. The SBS modifier cut down the low-temperature relaxation modulus of asphalt.

(4) In terms of the microscopic phase structure, the microscopic phase structure of SBSMA is a two-phase (SBS modifier and neat asphalt) coexistence of the co-blended structural system. With the increasing SBS modifier content, the SBS modifier gradually changes from a dispersed to a continuous phase state. When the modifier amount is less than 6.0%, the SBS modifier is present in a dispersed phase. Similarly, when the modifier is more than 6.0%, the SBS modifier is present in the continuous phase.

(5) The viscoelastic properties of high-content SBS modified asphalt at different temperature were investigated, and the phase structure of different content SBS modified asphalt were clarified. The chemical composition of high-content SBS modified asphalt and its aging characteristics will be studied in the next step.

Author Contributions: Conceptualization, W.J. and J.X.; methodology, D.Y.; software, C.X.; validation, D.Y., C.X. and J.X.; formal analysis, D.Y.; investigation, D.Y.; resources, W.J.; data curation, P.L.; writing—original draft preparation, D.Y.; writing—review and editing, Y.L. and W.W.; visualization, W.W.; supervision, C.X. and W.J.; project administration, W.J.; funding acquisition, W.J. All authors have read and agreed to the published version of the manuscript.

Funding: This research was funded by the National Natural Science Foundation of China, grant number 52122809 and 52038001, the Fok Ying-Tong Education Foundation, grant number 161072, the Youth Top-notch Talent Support Program of Shaanxi Province, the Fundamental Research Funds for the Central Universities-Excellent doctoral cultivation project of Chang'an University, grant number 300203211215, the China Scholarship Council (CSC), grant number 202006560054.

Institutional Review Board Statement: Not applicable.

Informed Consent Statement: Not applicable.

Data Availability Statement: Access to any other materials can be requested by writing to the corresponding authors.

Conflicts of Interest: The authors declare no conflict of interest.

References

1. Sha, A.M.; Liu, Z.Z.; Jiang, W.; Qi, L.; Hu, L.Q.; Jiao, W.X.; Barbieri, D.M. Advances and development trends in eco-friendly pavements. *J. Road Eng.* **2021**, *1*, 1–42. [CrossRef]
2. Jiang, W.; Yuan, D.D.; Shan, J.H.; Ye, W.L.; Lu, H.H.; Sha, A.M. Experimental study of the performance of porous ultra-thin asphalt overlay. *Int. J. Pavement Eng.* **2022**, *23*, 2049–2061. [CrossRef]
3. Liu, J.N.; Zhang, T.H.; Guo, H.Y.; Wang, Z.J.; Wang, X.F. Evaluation of self-healing properties of asphalt mixture containing steel slag under microwave heating: Mechanical, thermal transfer and voids microstructural characteristics. *J. Clean. Prod.* **2022**, *342*, 130932. [CrossRef]
4. Wang, C.H.; Wang, M.H.; Chen, Q.; Zhang, L. Basic performance and asphalt smoke absorption effect of environment-friendly asphalt to improve pavement construction environment. *J. Clean. Prod.* **2022**, *333*, 130142. [CrossRef]
5. Gao, J.; Yao, Y.Q.; Song, L.; Xu, J.; Yang, J.G. Determining the maximum permissible content of recycled asphalt pavement stockpile in plant hot-mix recycled asphalt mixtures considering homogeneity: A case study in China. *Case Stud. Constr. Mater.* **2022**, *16*, e00961. [CrossRef]
6. JTTE Editorial Office; Chen, J.; Dan, H.; Ding, Y.; Gao, Y.; Guo, M.; Guo, S.; Han, B.; Hong, B.; Hou, Y.; et al. New innovations in pavement materials and engineering: A review on pavement engineering research 2021. *J. Traffic Transp. Eng.* **2021**, *8*, 815–999. [CrossRef]
7. Jiang, W.; Xiao, J.J.; Yuan, D.D.; Lu, H.H.; Xu, S.D.; Huang, Y. Design and experiment of thermoelectric asphalt pavements with power-generation and temperature-reduction functions. *Energy Build.* **2018**, *169*, 39–47. [CrossRef]
8. Chen, M.Y.; Geng, J.G.; Xia, C.Y.; He, L.L.; Liu, Z. A review of phase structure of SBS modified asphalt: Affecting factors, analytical methods, phase models and improvements. *Constr. Build. Mater.* **2021**, *294*, 123610. [CrossRef]
9. Yu, H.; Bai, X.; Qian, G.; Wei, H.; Gong, X.; Jin, J.; Li, Z. Impact of ultraviolet radiation on the aging properties of SBS-modified asphalt binders. *Polymers* **2019**, *11*, 1111. [CrossRef]
10. Tan, G.J.; Wang, W.S.; Cheng, Y.C.; Wang, Y.; Zhu, Z.Q. Master curve establishment and complex modulus evaluation of SBS-modified asphalt mixture reinforced with basalt fiber based on generalized sigmoidal model. *Polymers* **2020**, *12*, 1586. [CrossRef]

11. Kozak, R.; Matlengiewicz, M. Effect of enthalpy of polar modifiers interaction with n-butyllithium on the reaction enthalpy, kinetics and chain microstructure during anionic polymerization of 1, 3-butadiene. *Polym. Test.* **2017**, *64*, 20–37. [CrossRef]
12. Li, H.B.; Li, W.B.; Sheng, Y.P.; Lv, H.L. Influence of compound action of rubber powder and SBS on high-temperature performance of asphalt pavement surface. *J. Mater. Civil. Eng.* **2021**, *33*, 04021126. [CrossRef]
13. Zhang, Z.W.; Sha, A.M.; Liu, X.; Luan, B.; Gao, J.; Jiang, W.; Ma, F. State-of-the-art of porous asphalt pavement: Experience and considerations of mixture design. *Constr. Build. Mater.* **2020**, *262*, 119998. [CrossRef]
14. Li, J.; Xiao, F.P.; Zhang, L.F.; Amirkhanian, S.N. Life cycle assessment and life cycle cost analysis of recycled solid waste materials in highway pavement: A review. *J. Clean. Prod.* **2019**, *233*, 1182–1206. [CrossRef]
15. Lu, R.; Jiang, W.; Xiao, J.J.; Xing, C.W.; Ruan, C.; Li, Y.P.; Wu, W.J. Temperature characteristics of permeable asphalt pavement: Field research. *Constr. Build. Mater.* **2022**, *332*, 127379. [CrossRef]
16. Cai, J.; Song, C.; Gong, X.B.; Zhang, J.P.; Pei, J.Z.; Chen, Z.W. Gradation of limestone-aggregate-based porous asphalt concrete under dynamic crushing test: Composition, fragmentation and stability. *Constr. Build. Mater.* **2022**, *323*, 126532. [CrossRef]
17. Hu, J.Y.; Ma, T.; Zhu, Y.H.; Huang, X.M.; Xu, J.; Chen, L.B. High-viscosity modified asphalt mixtures for double-layer porous asphalt pavement: Design optimization and evaluation metrics. *Constr. Build. Mater.* **2021**, *271*, 121893. [CrossRef]
18. Cai, J.; Song, C.; Zhou, B.C.; Tian, Y.F.; Li, R.; Zhang, J.P.; Pei, J.Z. Investigation on high-viscosity asphalt binder for permeable asphalt concrete with waste materials. *J. Clean. Prod.* **2019**, *228*, 40–51. [CrossRef]
19. Luo, Y.F.; Zhang, K.; Li, P.L.; Yang, J.H.; Xie, X.B. Performance evaluation of stone mastic asphalt mixture with different high viscosity modified asphalt based on laboratory tests. *Constr. Build. Mater.* **2019**, *225*, 214–222. [CrossRef]
20. Zhang, F.; Hu, C.B. Preparation and properties of high viscosity modified asphalt. *Polym. Compos.* **2017**, *38*, 936–946. [CrossRef]
21. De Keer, L.; Kilic, K.I.; Van Steenberge, P.H.M.; Daelemans, L.; Kodura, D.; Frisch, H.; De Clerck, K.; Reyniers, M.-F.; Barner-Kowollik, C.; Dauskardt, R.H.; et al. Computational prediction of the molecular configuration of three-dimensional network polymers. *Nat. Mater.* **2021**, *20*, 1422–1430. [CrossRef] [PubMed]
22. Yan, C.Q.; Huang, W.D.; Lin, P.; Zhang, Y.; Lv, Q. Chemical and rheological evaluation of aging properties of high content SBS polymer modified asphalt. *Fuel* **2019**, *252*, 417–426. [CrossRef]
23. Lin, P.; Yan, C.; Huang, W.; Li, Y.; Zhou, L.; Tang, N.; Xiao, F.; Zhang, Y.; Lv, Q. Rheological, chemical and aging characteristics of high content polymer modified asphalt. *Constr. Build. Mater.* **2019**, *207*, 616–629. [CrossRef]
24. Zhang, F.; Hu, C.B. The composition and ageing of high-viscosity and elasticity asphalts. *Polym. Compos.* **2017**, *38*, 2509–2517. [CrossRef]
25. Cuciniello, G.; Leandri, P.; Filippi, S.; Lo, P.D.; Losa, M.; Airey, G. Effect of ageing on the morphology and creep and recovery of polymer-modified bitumens. *Mater. Struct.* **2018**, *51*, 136. [CrossRef]
26. *AASHTO M332-19*; Performance-Graded Asphalt Binder Using Multiple Stress Creep Recovery (MSCR) Test. AASHTO: Washington, DC, USA, 2019.
27. Huang, W.D.; Tang, N.P. Characterizing SBS modified asphalt with sulfur using multiple stress creep recovery test. *Constr. Build. Mater.* **2015**, *93*, 514–521. [CrossRef]
28. Gaspar, M.S.; Nogueira, B.; Vasconcelos, K.L.; Leite, L.; Bernucci, L. Effect of different creep and recovery times on the MSCR test for highly modified asphalt binder. *J. Test. Eval.* **2021**, *49*, 1. [CrossRef]
29. Gao, J.F.; Wang, H.N.; You, Z.P.; Yang, X. Gray relational entropy analysis of high temperature performance of bio-asphalt binder and its mixture. *Int. J. Pavement Res. Technol.* **2018**, *11*, 698–708. [CrossRef]
30. Ameri, M.; Mirzaiyan, D.; Amini, A. Rutting resistance and fatigue behavior of gilsonite-modified asphalt binders. *J. Mater. Civil. Eng.* **2018**, *30*, 04018292. [CrossRef]
31. Cao, W.; Wang, C. A new comprehensive analysis framework for fatigue characterization of asphalt binder using the Linear Amplitude Sweep test. *Constr. Build. Mater.* **2018**, *171*, 1–12. [CrossRef]
32. Mannan, U.A.; Islam, M.R.; Tarefder, R.A. Effects of recycled asphalt pavements on the fatigue life of asphalt under different strain levels and loading frequencies. *Int. J. Fatigue* **2015**, *78*, 72–80. [CrossRef]
33. Ashish, P.K.; Singh, D.; Bohm, S. Investigation on influence of nanoclay addition on rheological performance of asphalt binder. *Road Mater. Pavement Des.* **2017**, *18*, 1007–1026. [CrossRef]
34. Yuan, D.D.; Jiang, W.; Xiao, J.J.; Tong, Z.; Jia, M.; Shan, J.H.; Ogbon, A.W. Assessment of the Aging Process of Finished Product–Modified Asphalt Binder and Its Aging Mechanism. *J. Mater. Civ. Eng.* **2022**, *34*, 04022174. [CrossRef]
35. Hajj, R.; Filonzi, A.; Rahman, S.; Bhasin, A. Considerations for using the 4 mm plate geometry in the dynamic shear rheometer for low temperature evaluation of asphalt binders. *Trans. Res. Rec.* **2019**, *2673*, 649–659. [CrossRef]
36. Sui, C.; Farrar, M.J.; Tuminello, W.H.; Turner, T.F. New technique for measuring low-temperature properties of asphalt binders with small amounts of material. *Trans. Res. Rec.* **2010**, *2179*, 23–28. [CrossRef]

polymers

MDPI

Article

Rheological Behaviors and Damage Mechanism of Asphalt Binder under the Erosion of Dynamic Pore Water Pressure Environment

Wentao Wang [1], Kang Zhao [1], Tingting Xie [1], Huifang Liu [1], Guanyi Zhao [1] and Linbing Wang [2,*]

1 National Center for Materials Service Safety, University of Science and Technology Beijing, Beijing 100083, China
2 The Sensing and Perception Lab, School of Environmental, Civil, Agricultural and Mechanical Engineering, University of Georgia, Athens, GA 30602, USA
* Correspondence: linbing.wang@uga.edu

Abstract: Asphalt binder plays an important role in the overall resistance of asphalt mixture to the moisture damage induced by a dynamic pore water pressure environment. This study evaluates the moisture sensitivity of asphalt binder from the perspective of rheological behaviors using the dynamic shear rheometer (DSR) and the bending beam rheometer (BBR) methods at high, medium, and low temperatures. The damage mechanism is further discussed quantitatively based on the Fourier transform infrared spectroscopy (FTIR) method. The results indicate that a longer conditioning duration is beneficial for asphalt binder to recover its adhesion at 60 °C in multiple stress creep recover (MSCR) tests, but the increasing pore water pressure magnitude of 60 psi held an opposite effect in this study. The asphalt binder's fatigue life at 20 °C in linear amplitude sweep (LAS) tests decreased obviously with conditioning duration and environmental severity, but the reducing rate gradually slowed down, while the groups of 50 psi—4000 cycles and 60 psi—4000 cycles held a comparable erosion effect. Both the stiffness and relaxation moduli at −12 °C in the BBR tests exhibited an obvious decreasing trend with conditioning duration and environmental severity. The erosion effect on the asphalt binder was gradually enhanced, but it also exhibited a slightly more viscous performance. Water conditioning induced several obvious characteristic peaks in the FTIR absorbance spectra of the asphalt binder. The functional group indexes presented a trend of non-monotonic change with conditioning duration and environmental severity, which made the asphalt binder show complicated rheological behaviors, such as non-monotonic variations in performance and the abnormal improving effect induced by dynamic pore water pressure conditioning.

Keywords: asphalt binder; moisture damage; dynamic pore water pressure; rheological behaviors; damage mechanism

Citation: Wang, W.; Zhao, K.; Xie, T.; Liu, H.; Zhao, G.; Wang, L. Rheological Behaviors and Damage Mechanism of Asphalt Binder under the Erosion of Dynamic Pore Water Pressure Environment. *Polymers* **2022**, *14*, 4731. https://doi.org/10.3390/polym14214731

Academic Editors: Wei Jiang, Quantao Liu, Jose Norambuena-Contreras and Yue Huang

Received: 17 October 2022
Accepted: 1 November 2022
Published: 4 November 2022

1. Introduction

Asphalt binder plays a vital role in bonding aggregate particles with different sizes, mineral powder, and additives tightly together to form an asphalt mixture, such as hot mix asphalt (HMA). For asphalt pavement located in rainy regions, its service performance is severely challenged by the environment of dynamic pore water pressure formed by the interaction between moving vehicle tires, surface runoff, and asphalt pavement [1], which causes moisture damage and affects its service life. Specifically, the comprehensive service performance of the eroded asphalt pavement is undoubtably affected at different temperatures [2]. Different from traditional water environments, such as freeze–thaw and static water immersion [3], the erosion processes in a dynamic pore water pressure environment exhibit quite a unique damage mechanism on asphalt binder. In this case, it is meaningful to explore the performance of asphalt binder, such as rheological behaviors, in the erosion of a dynamic pore water pressure environment and make efforts to investigate this relevant damage mechanism.

Multiple techniques have been adopted to explore the moisture damage of asphalt binder. The dynamic shear rheometer (DSR) technique is a quite useful method to inspect the properties of asphalt binder, due not only to its rich mechanical loading modes but also to the advantage of less consumption of asphalt binder samples. Jing et al. [4] investigated the effect of aging on the viscoelastic characteristics of asphalt binder which mainly include rheological, fatigue, and relaxation properties. Zhang and Gao [5] predicted fatigue crack growth in viscoelastic asphalt binder and presented a new model named DSR-C for prediction. Das and Singh [6] studied the influence of nano-size hydrated lime filler on the rutting performance of asphalt mastic. Ziade et al. [7] combined DSR and numerical techniques to investigate rheological behavior in asphalt binder composites. It is convenient to adopt the DSR method to evaluate the rheological behaviors of asphalt binder before and after being conditioned in a dynamic water environment.

The bending beam rheometer (BBR) test can measure the flexural creep performance of asphalt binder at a relative low temperature below 0 °C. Basically, creep stiffness S and the flexural creep rate *m-value* are often used to assess the performance of asphalt binder [8–10]. However, more viscoelastic properties can be further explored, such as master curve of modulus, creep compliance, and relaxation modulus, based on the measured raw data in a BBR test [11]. In particular, the indicator of the ratio of creep rate *m-value* to creep stiffness S was found to represent the change rate of creep compliance [12], which could be applied to evaluate the erosion degree of the moisture damage of asphalt binder.

The Fourier transform infrared spectroscopy (FTIR) technique can detect the change in the functional groups and chemical bonds of asphalt binder before and after a certain environment conditioning. Ahmad et al. [13] used the FTIR method to assess the chemical and mechanical changes in asphalt binder due to moisture conditioning and found an increase in the hydroxyl group. Mannan et al. [14] combined the FTIR and DSR methods to investigate the healing properties of asphalt binder after being eroded by a dynamic water environment.

Based on the analysis discussed above, it could be found that systematic research on moisture damage related to a dynamic pore water pressure environment on asphalt binder's rheological behaviors was not sufficient, especially at the various ranges of low, medium, and high temperatures. For the FTIR technique, it is essential to quantitatively analyze characteristic peaks with their nearby regions together, but not only to compare the relative position of the infrared spectra qualitatively, which might be helpful to explore the mechanism of moisture damage.

The main objective of this study is to evaluate the moisture sensitivity of asphalt binder from the perspective of rheological behaviors and its damage mechanism. The parameters of conditioning duration and pore water pressure magnitude were taken into account to provide different types of dynamic water conditioning environments. The DSR method was first applied to evaluate the high-temperature property and the medium-temperature fatigue performance for the conditioned asphalt binder. Then, the BBR method was adopted to explore the retained low-temperature creep properties and the change in the viscous performance of the asphalt binder. Finally, the FTIR method was selected to quantitatively investigate the fluctuations of the characteristic peaks in the infrared spectra.

2. Materials and Sample Preparation

In this study, a typical base asphalt binder with the performance grade (PG) of 64−16 was selected to be conditioned in a dynamic pore water pressure environment, and then samples were prepared for the DSR, BBR, and FTIR experiments. The basic physical properties of the asphalt binder were inspected, and the relevant results are summarized in Table 1, which meet the requirements of the specification [15].

Table 1. Physical properties of the base asphalt binder sample.

Items		Units	Requirements	Results
Penetration (25 °C, 100 g, 5 s)		0.1 mm	80–100	85
Softening point		°C	≥45	48.5
Ductility (5 cm/min)		cm	≥45 (10 °C)	53
Residue after rolling thin film oven test (163 °C, 85 min)	Mass loss	%	≤±0.8	−0.450
	Residual penetration ratio (25 °C, 100 g, 5 s)	%	≥57	65
	Residual ductility (5 cm/min)	cm	≥8 (10 °C)	20
Performance Grade (PG)		-	64–16	

To thoroughly erode the samples, asphalt binder was first poured into a shallow disc container with the thickness controlled at about 4 mm and then placed in a dynamic pore water pressure environment for full conditioning. The conditioned asphalt binder was then heated in a 100 °C oven environment to evaporate possible retained water, followed by sample fabrications in a flow state for further residual performance inspection. For the DSR tests, the asphalt binder was poured into the molds with diameters of 8 mm and 25 mm, and two parallel samples were applied for each DSR test. For the BBR tests, thin beam samples were made, and their dimensions were controlled at 127 ± 2 mm in length, 12.7 ± 0.05 mm in width, and 6.35 ± 0.05 mm in thickness, while three parallel samples were applied. Further, asphalt-binder-coated film slides were made for the FTIR inspection.

3. Test and Analysis Methods

3.1. Dynamic Pore Water Pressure Conditioning Method

The device named Moisture Induced Sensitivity Tester (MIST) was selected to provide the water conditioning environment of dynamic pore water pressure in this study [16]. A typical bladder located in the sealed cylinder container of the MIST device is driven to be repeatedly inflated and restituted, as shown in Figure 1, which can provide the cyclic water conditioning environment of dynamic pore water pressure. Three variable parameters can be adjusted in the MIST method, which include pore water pressure magnitude, water temperature, and conditioning duration. In this study, the water temperature was controlled at 60 °C to ensure a stable and typical high-temperature water environment for all sample conditioning tests. The variable parameter of conditioning duration was adjusted to 2000, 3000, and 4000 cycles for the water environment, with the pore water pressure magnitude controlled at 50 psi. In the MIST test, it takes about 4 s for each conditioning cycle of the dynamic pore water pressure environment. Further, an experimental group was set to the 60 psi pore water pressure magnitude with the conditioning duration controlled at 4000 cycles.

3.2. The DSR Method

The DSR method was applied to evaluate the rheological behaviors of asphalt binder before and after being conditioned in a dynamic pore water pressure environment. Two types of typical DSR tests were conducted in high and medium temperatures, which included the multiple stress creep recover (MSCR) test and the linear amplitude sweep (LAS) test. The DSR device used in this study and the experimental details are shown in Figure 2.

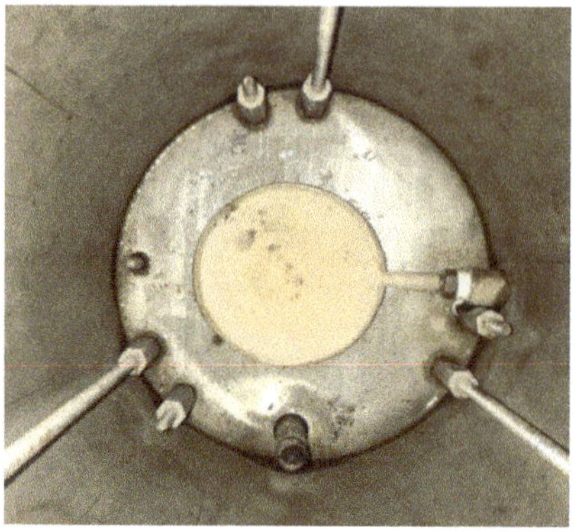

Figure 1. A bladder inside the sealed container of MIST.

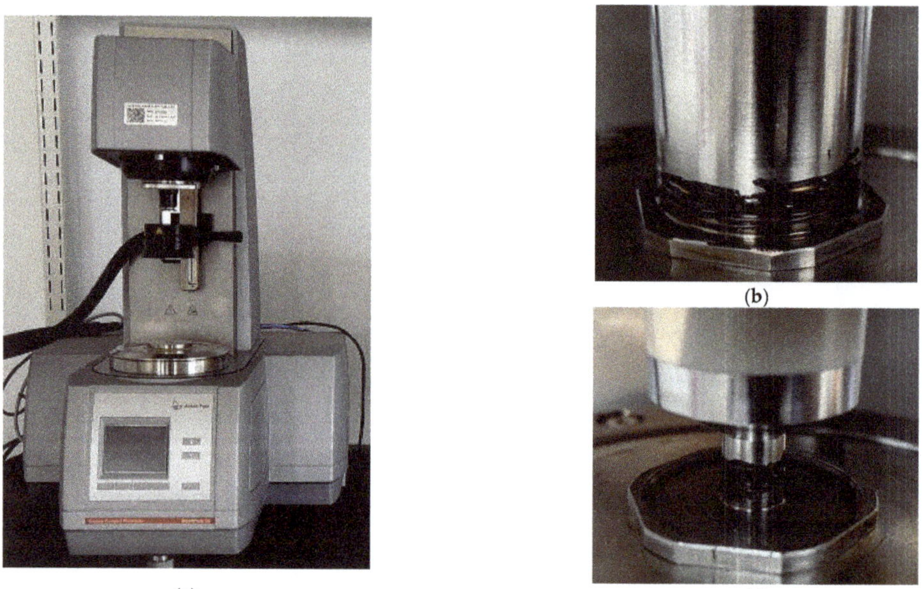

(**a**) (**b**) (**c**)

Figure 2. The DSR test: (**a**) The DSR device; (**b**) A sample with 25 mm dia.; (**c**) A sample with 8 mm dia.

3.2.1. MSCR Test

The MSCR tests were conducted to evaluate the change in the asphalt binder's performance at the high temperature of 60 °C after being conditioned in a dynamic water environment. A MSCR test contains 20 cyclic cycles, in which the first 10 cycles apply a creep stress of 0.1 kPa and the remaining 10 cycles load at 3.2 kPa. Each cycle includes a loading period of 1 s followed by a recovering period of 9 s, while the entire MSCR test lasts for 200 s [17]. A typical shear strain curve in a MSCR test is shown in Figure 3. The indicator of unrecoverable creep compliance J_{nr} for each creep recovery cycle can be calculated based on Equation (1), where γ_{nr} and γ_0 are the initial strain and residual strain for each creep-recovery cycle and τ is shear stress. The average value of J_{nr} is obtained based on 10 cycles for each shear stress level.

$$J_{nr} = \frac{\gamma_{nr} - \gamma_0}{\tau} \quad (1)$$

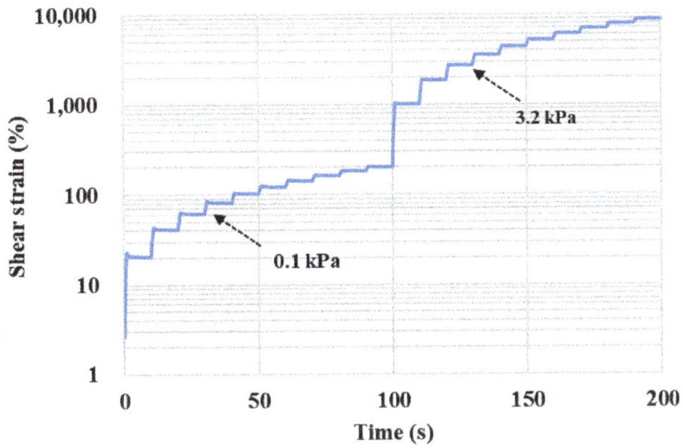

Figure 3. A typical shear strain curve in a MSCR test.

3.2.2. LAS Test

The LAS tests were adopted to investigate the variation in fatigue behavior at the medium temperature of 20 °C for the asphalt binder before and after being conditioned in a dynamic water environment. The applied shear strain with a sinusoidal wave is controlled in the LAS test, which increases linearly from 0.1% to 30% [18]. The load frequency was set to 10 Hz. A typical curve between shear strain and stress is shown in Figure 4. A simplified viscoelastic continuum damage (S−VECD) model was selected to obtain the damage characteristic curve to evaluate the evolution of the fatigue damage of the asphalt binder. Equation (2) can be applied to fit the fatigue damage characteristic curve between the damage variable S and pseudo-modulus C, in which C_0 is often set at 1 and both C_1 and C_2 are curve-fit coefficients.

$$C = C_0 - C_1(S)^{C_2} \quad (2)$$

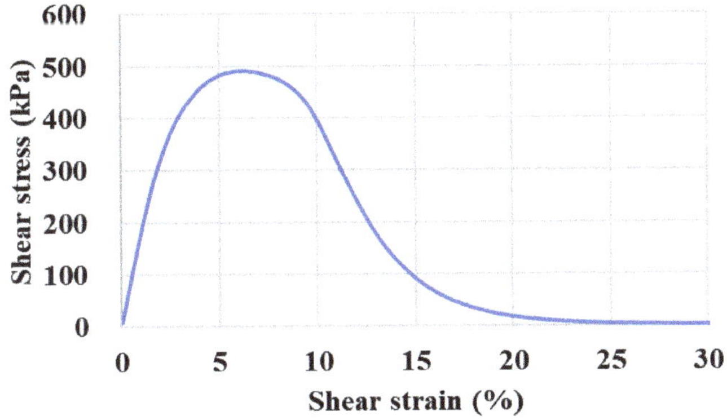

Figure 4. A typical shear strain-stress curve in a LAS test.

3.3. The BBR Method

The BBR method was applied to inspect the low-temperature rheological behaviors of asphalt binder before and after being eroded by a dynamic pore water pressure environment. According to the ASTM D6648 standard [19], as shown in Figure 5, a thin beam sample of asphalt binder is placed on two supports with a constant load of 980 ± 50 mN located at its mid-span, while the flexural deformation can be further measured. In this study, the low temperature of the BBR tests was controlled at −12 °C according to the PG 64−16 of the asphalt binder.

Figure 5. A sample of asphalt binder in the BBR test.

In the BBR method, the basic indicators of the flexural creep stiffness S and the flexural creep rate *m-value* can be calculated based on the measured creep mechanical stress and strain data. The flexural creep stiffness $S(t)$ is determined using Equation (3), where $\delta(t)$ is the deformation at the mid-span of the thin beam sample, P is the constant load, and h, L, and b are the thickness, length, and width of the thin beam sample. The flexural creep rate $m(t)$ is the absolute value of the slope of the flexural creep stiffness $S(t)$ versus time curve in the double logarithm coordinate system. In particular, creep compliance D shares a countdown relationship with flexural creep stiffness modulus S, which can be further applied to explore more rheological behaviors of asphalt binder via the BBR test.

$$S(t) = \frac{PL^3}{4bh^3\delta(t)} \tag{3}$$

As shown in Equations (4) and (5), both the Burgers model and the second-order Prony series expression can represent the creep compliance $D(t)$ of asphalt binder according to the BBR test. The Burgers model takes characteristics of the Maxwell and Kelvin models into account, while the second-order Prony series expression was constructed based on the generalized Kelvin model. As shown in Figure 6, both the Burgers and Prony models can well fit the creep compliance curve in the BBR test.

$$D(t) = \frac{\varepsilon(t)}{\sigma_o} = \frac{1}{E_1} + \frac{t}{\eta_1} + \frac{1}{E_2}\left(1 - e^{-t\cdot\frac{E_2}{\eta_2}}\right) \tag{4}$$

$$D(t) = D_0 + D_1\left[1 - \exp\left(-\frac{t}{\rho_1}\right)\right] + D_2\left[1 - \exp\left(-\frac{t}{\rho_2}\right)\right] \tag{5}$$

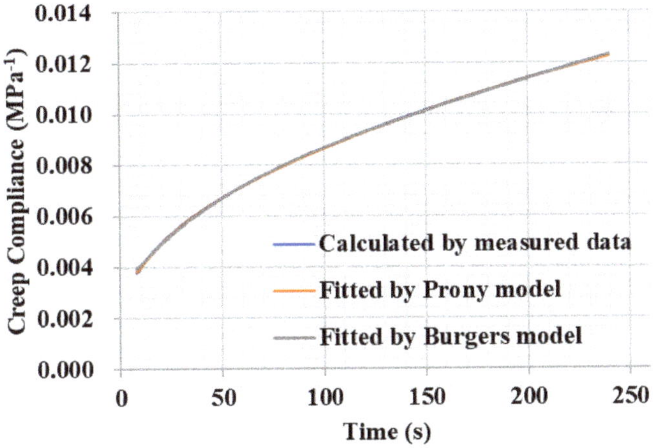

Figure 6. Creep compliance curves determined using different methods in a BBR test.

In the Burgers model, E_1 and E_2 are the instantaneous elastic modulus of the Maxwell model and the delayed elastic modulus of the Kelvin model, respectively, while both η_1 and η_2 are viscous coefficients. These viscoelastic parameters can further calculate the indexes of relaxation time λ and the delay time τ, which are determined using Equations (6) and (7). Relaxation time λ shows the ability of asphalt binder to release internal stress, and delay time τ often indicates creep and delayed elastic performance. A shorter relaxation time λ often means a better viscous performance, while a larger τ usually indicates a much more viscous deformation, which means a good anti-cracking ability at a low temperature.

$$\lambda = \frac{\eta_1}{E_1} \tag{6}$$

$$\tau = \frac{\eta_2}{E_2} \tag{7}$$

In the second-order Prony series expression, D_0 is the compliance of the individual spring element in the Kelvin model, D_1 and D_2 are the compliances of the spring element in a Kelvin element, and ρ_1 and ρ_2 are ratios of the viscosity coefficient of a dashpot element to the modulus of the spring element in a Kelvin element. As creep compliance D has a reciprocal relationship with relaxation modulus E in the Laplace domain, as shown in Equation (8), the relaxation modulus E can be thus derived, and the detailed derivation processes can be found in reference [20].

$$\int_0^t E(t-\tau)\frac{\partial D(t)}{\partial \tau}d\tau = 1 \tag{8}$$

In particular, an indicator m/S can be derived to reflect the growth rate of creep compliance $D'(t)$, as shown in Equation (9), which combines characteristics of both the creep stiffness S and creep rate *m-value*. The detailed derivation processes of the m/S indicator can be found in the previous research [12]. Time t is often set at a representative 60th s, and a larger $m(t)/S(t)$ often indicates a larger $D'(t)$, which means a slower increase in creep stiffness.

$$\frac{m(t)}{S(t)} \approx D'(t) \times t \tag{9}$$

3.4. The FTIR Method

The FTIR method can effectively inspect the change in the functional groups and chemical bonds of asphalt binder after being eroded by external service environments such as a dynamic pore water pressure environment. The moisture damage mechanism can be illustrated to a certain extent based on the several appeared obvious characteristic peaks in several ranges of wavenumbers of the obtained infrared spectrum.

Based on the measured transmissivity curves, the absorbance curves with a wavenumber range from 4000 to 400 cm^{-1} for the different types of conditioned asphalt binder samples can be determined. After being calibrated based on the baseline and subsequently smoothed, the area values of these obvious characteristic peaks for the functional groups and chemical bonds of the asphalt binder in an absorbance curve were calculated. In particular, the area of a characteristic peak was determined according to the tangent line connected by its lowest points on the left and right sides. In order to eliminate the influence of possible experimental errors as much as possible, the relative area rather than the raw area was selected for further analysis. As the characteristic peak around the 2923 cm^{-1} wavenumber appeared in the infrared spectra for all types of asphalt binder samples, the area of this characteristic peak was selected as the base functional group area to determine the relative area values of the other functional groups and chemical bonds for all types of asphalt binder. Therefore, a characteristic functional group index I_i for asphalt binder can be thus determined using Equation (10), in which A_i is the relative area for a characteristic peak and ΣA_i is the sum of these relative areas of all occurred characteristic peaks.

$$I_i = {A_i}\Big/{\sum A_i} \tag{10}$$

4. Results and Discussion

4.1. High-Temperature Rheological Behavior

The shear strain curves for the asphalt binder conditioned in different dynamic water environments are summarized in Figure 7. It could be found that the curves for the conditioned asphalt binder are located in relatively higher positions compared with the control group, which indicates the decline in high-temperature performance after being eroded by dynamic water. In the dynamic water environment of 60 °C−50 psi, the shear strain curves gradually moved down with the increase in conditioning duration, which showed a decreasing accumulated permanent deformation. Compared between the groups of 50 psi and 60 psi, it could be found that the shear strain curve of the 50 psi group was located in a relatively lower position, which means an increasing deformation and a worse high-temperature creep recovery ability of the asphalt binder with the increase in environment severity.

As a standard indicator, J_{nr} can well evaluate the ability to resist the high-temperature rutting problem of asphalt binder. The variations of unrecoverable creep compliance J_{nr} are summarized in Figure 8, which include $J_{nr0.1}$ and $J_{nr3.2}$ for the two shear stress levels. It could be found that both $J_{nr0.1}$ and $J_{nr3.2}$ decreased with the conditioning duration, which exhibited an improved effect on the high-temperature performance of the asphalt binder. A longer conditioning duration would be beneficial for asphalt binder to recover its adhesion. However, the increasing pore water pressure magnitude of 60 psi held an opposite effect. Therefore, a suitable combination of pore water pressure magnitude and conditioning duration would be helpful for asphalt binder to increase its cohesion to aggregate.

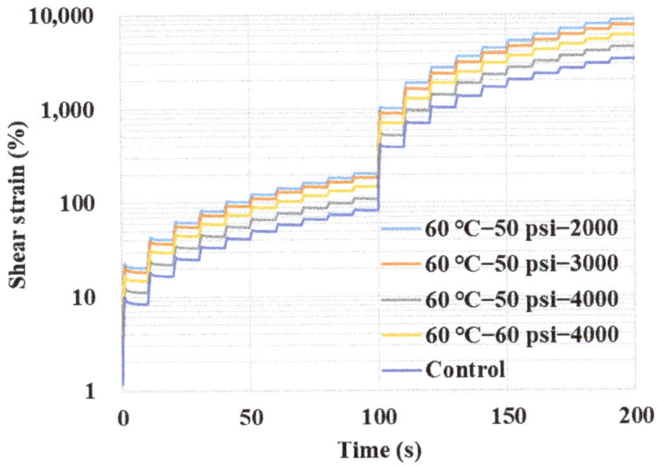

Figure 7. Shear strain curves in MSCR tests.

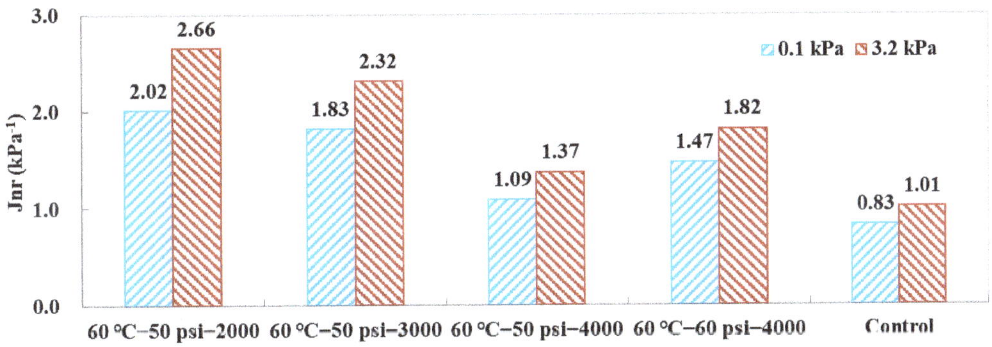

Figure 8. Variation of unrecoverable creep compliance *Jnr* in MSCR tests.

4.2. Medium-Temperature Fatigue Behavior

After being conditioned in dynamic pore water pressure, the retained fatigue behaviors of the asphalt binder were evaluated at the medium temperature of 20 °C using the LAS method. The damage characteristic curves of these samples of asphalt binder are presented in Figure 9. Compared with the control group, the indicator of pseudo-modulus C decreased with the conditioning duration and environmental severity at a certain same damage variable S, which indicated an increasing evolution of fatigue damage. To reach a certain pseudo-modulus C, the group of 60 °C−60 psi−4000 cycles held the relative smallest damage variable S, which showed the severe erosion of the dynamic water environment on the asphalt binder.

The C-S curve can illustrate the basic characteristics of the fatigue damage evolution of asphalt binder during a LAS load, while fatigue life N_f can be thus determined to evaluate the fatigue performance in different levels of shear strains. Fatigue life N_f of these water-conditioned samples are predicted and summarized in Figure 10. At the shear strains of both 2.5% and 5.0%, the indicators of fatigue life N_f for the asphalt binder samples conditioned in different dynamic water environments exhibited an obvious decreasing trend with conditioning duration and environmental severity. The percent reduction of fatigue life N_f was also determined. It was found that the rate of reduction slowed down, especially for the groups of 50 psi−4000 cycles and 60 psi−4000 cycles, which means a comparable erosion effect by the dynamic water environments.

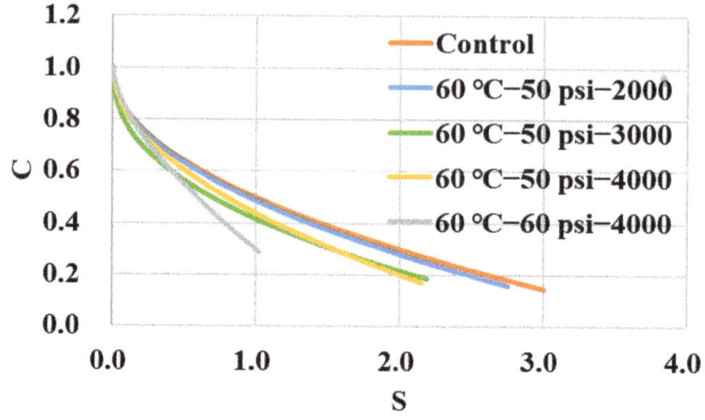

Figure 9. Damage characteristic curves of asphalt binder in LAS tests.

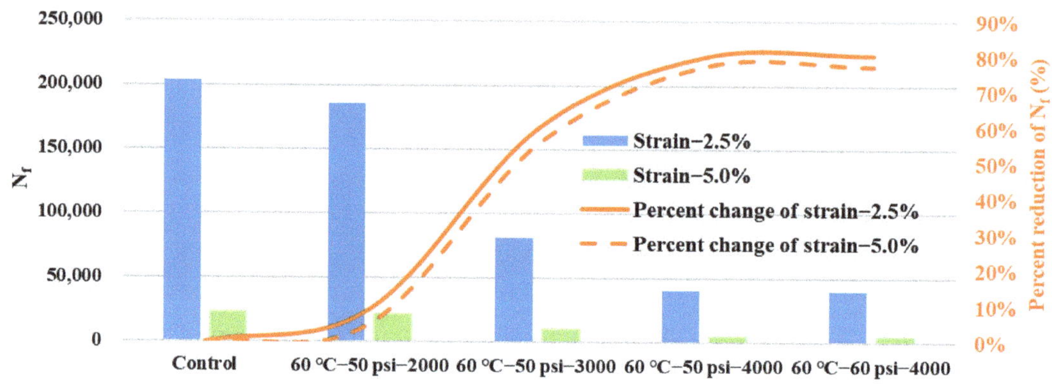

Figure 10. Fatigue life of asphalt binder in LAS tests.

4.3. Low-Temperature Flexural Creep Behavior

The flexural creep behavior of asphalt binder before and after being conditioned in a dynamic pore water pressure environment was inspected using the BBR method at a low temperature of −12 °C. The flexural creep stiffness modulus *S* and relaxation modulus *E* are not the same, and the value of stiffness modulus *S* is usually obviously larger than relaxation modulus *E*, while their variation curves with time thus do not coincide completely in a BBR test [20]. The variations in retained stiffness modulus *S* and relaxation modulus *E* are shown in Figure 11. It can be clearly found that the retained modulus ratios for these two indicators declined with the parameters of conditioning duration and pore water pressure magnitude. Compared between the conditioning duration groups of 2000 and 3000 cycles in the dynamic water environment of 60 °C−50 psi, the percent change in the stiffness modulus and relaxation modulus were 2.09% and 2.91% in the group of 2000 cycles, respectively, but this value relationship was reversed in the group of 3000 cycles, which were 6.13% and 4.26%, respectively. In this case, more rheological behaviors can be explored for asphalt binder based on different indicators, such as relaxation modulus in this study.

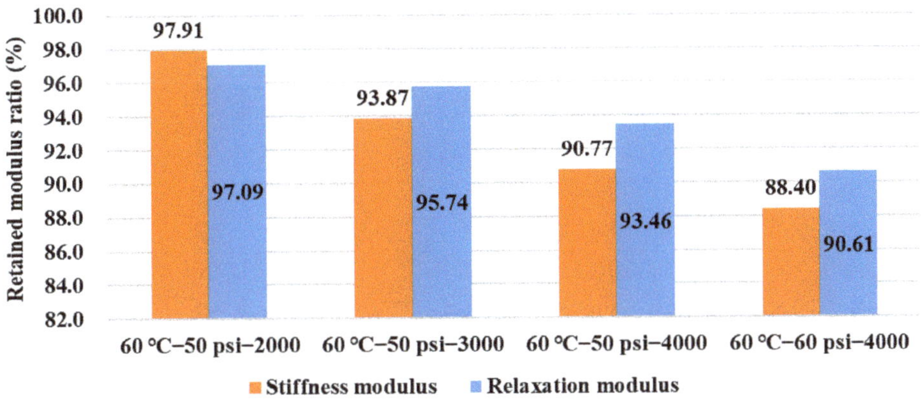

Figure 11. Variation of stiffness modulus and relaxation modulus in BBR tests.

Compared with the virgin asphalt binder, the indicator *m/S @ 60th* increased with conditioning duration and pore water pressure magnitude, and their percent change ratios also exhibited an increasing trend, as shown in Figure 12, which indicates a more severe erosion effect in a harsh dynamic pore water pressure environment. It can be concluded that the indicator *m/S @ 60th* can well evaluate the erosion degree of asphalt binder before and after being conditioned in a dynamic water environment.

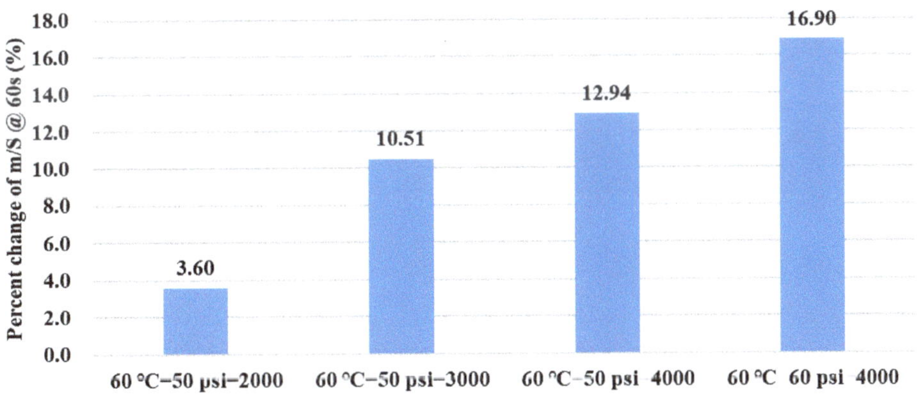

Figure 12. Percent change of m/S at 60th s in BBR tests.

The percent change in the viscoelastic indicators is shown in Figure 13. It can be found that the relaxation time λ of the asphalt binder continually shortened after being conditioned in the dynamic pore water pressure environment, and its increasing variation trend is obvious with the increase in conditioning duration and environment severity. Delay time τ slightly increased after being conditioned, but its variation trend in percent change ratio showed a reversal phenomenon in the 60 °C–60 psi group. It indicated a slightly more viscous performance for the asphalt binder after being conditioned in the dynamic pore water pressure environment. The resistance ability to moisture damage of an asphalt mixture, such as HMA, highly counts on the combined performance of its components, which include asphalt binder and fine aggregate mixture (FAM). After being eroded by a dynamic water environment, although the asphalt binder component exhibited a slightly improved viscous performance, the FAM component was eroded greatly, which together induced the deterioration in comprehensive performance for HMA [12]. To further explore the obvious change in viscoelastic behavior for asphalt binder, it is necessary to set a larger interval for the parameter of conditioning duration in a future study.

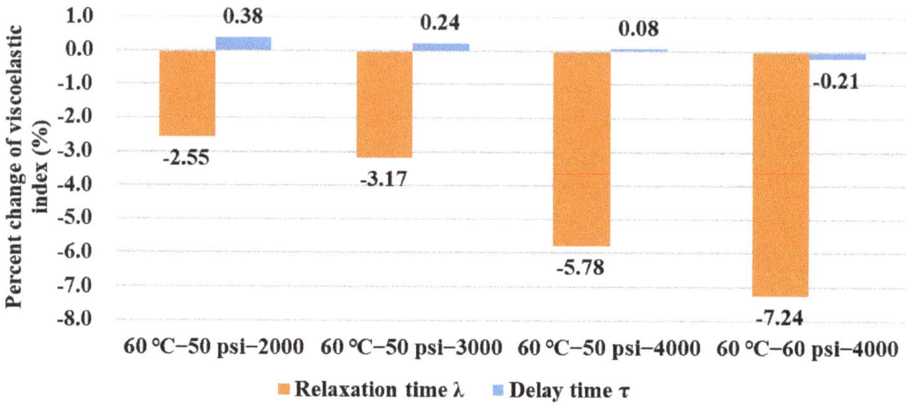

Figure 13. Percent change of viscoelastic indexes in BBR tests.

4.4. Damage Mechanism Analysis via FTIR Spectrums

The FTIR absorbance spectra of the five types of asphalt binder samples which were conditioned in the different water environments of dynamic pore water pressure are summarized in Figure 14. Obvious characteristic peaks can be found in several of the same ranges of wavenumbers, such as 3600–3200, 3100–2800, 1800–1200, and 1000–600 cm^{-1}, and the absorbance curves of the asphalt binder exhibited basically regular fluctuations with the increase in conditioning duration and environmental severity [21]. For example, the control group showed a relative higher absorbance value in the wavenumber range of 3600–3200 cm^{-1} but held lower absorbance values in other wavenumber ranges. The group of 60 °C−60 psi−4000 cycles basically showed the highest absorbance values within the wavenumber range of 3000–1200 cm^{-1}.

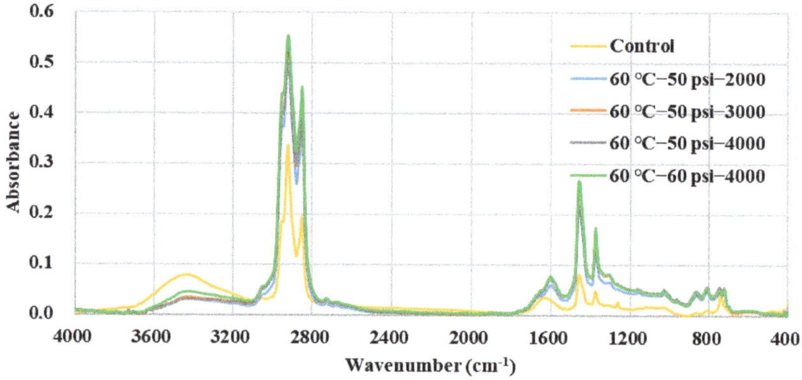

Figure 14. FTIR spectra of different types of asphalt binder.

The functional group index I_i was applied to quantitatively evaluate the comprehensive features of the different obvious characteristic peaks, which could help to explore the change in the detailed functional groups related to the moisture damage of asphalt binder. Obvious characteristic peaks with their functional group indexes are summarized in Figure 15a. It could be clearly found that the methylene group (=CH_2) held the largest absolute values of the functional group index which were mainly due to the consideration of the two characteristic peaks of 2923 and 2852 cm^{-1}. For the hydroxyl group (−OH) at 3425 cm^{-1}, great declines in the absolute values of the functional group index occurred inside the asphalt binder after water conditioning, while unremarkable absolute values of I_i for the other functional groups appeared as shown in Figure 15a.

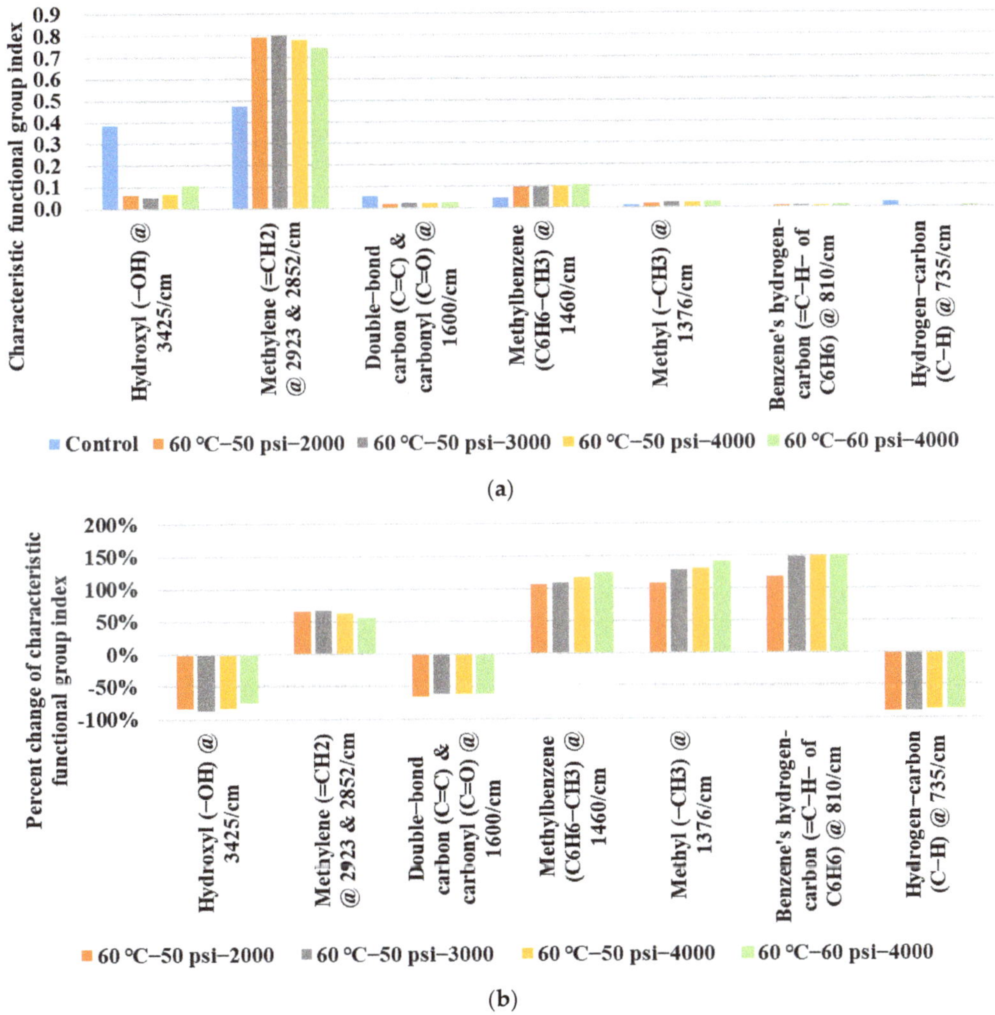

Figure 15. Characteristic functional group index of different types of asphalt binder: (**a**) Characteristic functional group index; (**b**) Percent change in characteristic functional group index.

Compared with the control group, the percent changes in the functional group indexes of the water-conditioned asphalt binder were further calculated to quantify the detailed changes. As shown in Figure 15b, the increase (percent change in $I_i > 0$) and the decrease (percent change in $I_i < 0$) in the functional group index values for the different functional groups can be clearly seen. Great changes occurred in methylbenzene ($C_6H_6-CH_3$) @ 1460 cm^{-1}, methyl ($-CH_3$) @ 1376 cm^{-1}, and benzene's hydrogen-carbon (=C−H− of C_6H_6) @ 810 cm^{-1}, because their percent changes in I_i were larger than 100%, and their growth degree basically increased with conditioning duration and environmental severity. Similarly, the decrease in I_i indexes for the hydrogen-carbon bond (C−H) @ 735 cm^{-1}, double-bond carbon bond (C=C) and carbonyl groups (C=O) @ 1600 cm^{-1} also exhibited this kind of trend. For the percent changes in the I_i indexes in the hydroxyl group (−OH) @ 3425 cm^{-1} and methylene group (=CH_2) @ 2923 & 2852 cm^{-1}, their fluctuations showed a different trend which increased first but decreased later with conditioning duration and environmental severity. It might be one of the possible reasons why dynamic water environment conditioning made a trend of non-monotonic change on the asphalt binder's rheological behaviors.

5. Conclusions

This study assessed the rheological behaviors and the damage mechanism of asphalt binder under the erosion of a dynamic pore water pressure environment using the DSR, BBR, and FTIR methods. Based on the analysis discussed above, the following conclusions can be obtained:

(a) A longer conditioning duration is beneficial for the asphalt binder to recover its adhesion at the high temperature of 60 °C in MSCR tests, but the increasing pore water pressure magnitude of 60 psi held an opposite effect in this study.
(b) The fatigue life of the asphalt binder at the medium temperature of 20 °C in the LAS tests decreased obviously with the conditioning duration and environmental severity of the dynamic pore water pressure, but the reducing rate gradually slowed down, while the groups of 50 psi−4000 cycles and 60 psi−4000 cycles held a comparable erosion effect.
(c) The flexural stiffness modulus and relaxation modulus at the low temperature of −12 °C in the BBR tests exhibited an obvious decreasing trend with conditioning duration and environmental severity. The erosion effect on the asphalt binder was gradually enhanced, but it also exhibited a slightly more viscous performance.
(d) Dynamic water environment conditioning induced obvious appearances of several main characteristic peaks in the FTIR absorbance spectra of the asphalt binder. The functional group indexes presented a trend of non-monotonic change with conditioning duration and environmental severity, which made the asphalt binder show complicated rheological behaviors such as non-monotonic variations in performance and the abnormal improving effect induced by dynamic pore water pressure conditioning.

Author Contributions: Conceptualization, W.W.; methodology, W.W., K.Z., T.X. and H.L.; resources, G.Z.; data curation, W.W., K.Z. and T.X.; writing—original draft preparation, W.W.; writing—review and editing, L.W.; supervision, L.W. All authors have read and agreed to the published version of the manuscript.

Funding: The authors acknowledge the financial support of the China Postdoctoral Science Foundation (Project No. 2021M690356).

Institutional Review Board Statement: Not applicable.

Informed Consent Statement: Not applicable.

Data Availability Statement: Not applicable.

Conflicts of Interest: The authors declare no conflict of interest.

References

1. Wang, W.; Zhao, K.; Li, J.; Luo, R.; Wang, L. Characterization of dynamic response of asphalt pavement in dry and saturated conditions using the full-scale accelerated loading test. *Constr. Build. Mater.* **2021**, *312*, 125355. [CrossRef]
2. Wang, W.; Wang, L.; Miao, Y.; Cheng, C.; Chen, S. A survey on the influence of intense rainfall induced by climate warming on operation safety and service life of urban asphalt pavement. *J. Infrastruct. Preserv. Resil.* **2020**, *1*, 4. [CrossRef]
3. Wang, W.; Wang, L.; Xiong, H.; Luo, R. A review and perspective for research on moisture damage in asphalt pavement induced by dynamic pore water pressure. *Constr. Build. Mater.* **2019**, *204*, 631–642. [CrossRef]

4. Jing, R.; Varveri, A.; Liu, X.; Scarpas, A.; Erkens, S. Rheological, fatigue and relaxation properties of aged bitumen. *Int. J. Pavement Eng.* **2020**, *21*, 1024–1033. [CrossRef]
5. Zhang, Y.; Gao, Y. Predicting crack growth in viscoelastic asphalt binder under a rotational shear fatigue load. *Road Mater. Pavement Des.* **2021**, *22*, 603–622. [CrossRef]
6. Das, A.; Singh, D. Influence of nano size hydrated lime filler on rutting performance of asphalt mastic. *Road Mater. Pavement Des.* **2021**, *22*, 1023–1043. [CrossRef]
7. Ziade, E.; Tehrani, F.; Beghin, A.; Petit, C.; Absi, J.; Millien, A.; Reynaud, P. Experimental and numerical investigation on the rheological behaviour of bituminous composites via DSR testing. *Road Mater. Pavement Des.* **2021**, *22*, 328–344. [CrossRef]
8. Laukkanen, O.; Soenen Winter, H.H.; Seppala, J. Low-temperature rheological and morphological characterization of SBS modified bitumen. *Constr. Build. Mater.* **2018**, *179*, 348–359. [CrossRef]
9. Chiara, R.; Augusto CF Di, W.; Michael, P.W. Effect of cooling medium on low-temperature properties of asphalt binder. *Road Mater. Pavement Des.* **2017**, *18*, 234–255.
10. Lucia, T.; Orazio, B.; Ezio, S. Low-temperature properties of bituminous nanocomposites for road applications. *Constr. Build. Mater.* **2018**, *171*, 397–403.
11. Findley, W.; Lai, J.; Onaran, K. *Creep and Relaxation of Nonlinear Viscoelastic Materials with an Introduction to Linear Viscoelasticity*; Dover Publications: New York, NY, USA, 1989; pp. 1–7. ISBN 0-486-66016-8.
12. Wang, W.; Luo, R.; Yan, G.; Wang, L. Evaluation on moisture sensitivity induced by dynamic pore water pressure for asphalt mixture and its components using the bending beam rheometer method. *Constr. Build. Mater.* **2020**, *251*, 118942. [CrossRef]
13. Ahmad, M.; Mannan, U.A.; Islam, M.R.; Tarefder, R.A. Chemical and mechanical changes in asphalt binder due to moisture conditioning. *Road Mater. Pavement Des.* **2017**, *19*, 1216–1229. [CrossRef]
14. Mannan, U.A.; Ahmad, M.; Tarefder, R.A. Influence of moisture conditioning on healing of asphalt binders. *Constr. Build. Mater.* **2017**, *146*, 360–369. [CrossRef]
15. Ministry of Transport of the People's Republic of China. *JTG F40-2004 Technical Specifications for Construction of Highway Asphalt Pavements*; China Communication Press: Beijing, China, 2004; pp. 9–15. ISBN 978-7-114-05328-2.
16. *ASTM D7870-20*; Standard Practice for Moisture Conditioning Compacted Asphalt Mixture Specimens by Using Hydrostatic Pore Pressure. ASTM: West Conshohocken, PA, USA, 2020.
17. *ASTM D7405-20*; Standard Test Method for Multiple Stress Creep and Recovery (MSCR) of Asphalt Binder Using a Rheometer. ASTM: West Conshohocken, PA, USA, 2020.
18. *AASHTO TP 101-14*; Standard Method of Test for Estimating Damage Tolerance of Asphalt Binders Using the Linear Amplitude Sweep. AASHTO: Washington, DC, USA, 2013.
19. *ASTM D6648-08*; Standard Test Method for Determining the Flexural Creep Stiffness of Asphalt Binder Using the Bending Beam Rheometer (BBR). ASTM: West Conshohocken, PA, USA, 2016.
20. Wang, W.; Luo, R.; Li, J.; Wang, L. Evaluation on the influence of dynamic water pressure environment on viscoelastic mechanical performance of asphalt mixture using the bending beam rheometer method. *Constr. Build. Mater.* **2022**, *321*, 126428. [CrossRef]
21. Wang, W.; Luo, R.; Yang, H.; Wang, L. Analysis of water erosion on asphalt binder using multi-scale experimental methods. *Int. J. Pavement Res. Technol.* **2022**, *15*, 485–495. [CrossRef]

polymers MDPI

Article

Effects of Bitumen Thickness on the Aging Behavior of High-Content Polymer-Modified Asphalt Mixture

Peng Lin [1,2], Xueyan Liu [2], Shisong Ren [2,*], Jian Xu [1], Yi Li [2] and Mingliang Li [1]

1 Key Laboratory of Transport Industry of Road Structure and Material, Research Institute of Highway, Ministry of Transport, Xitucheng Road No. 8, Beijing 100088, China
2 Section of Pavement Engineering, Faculty of Civil Engineering & Geosciences, Delft University of Technology, Stevinweg 1, 2628 CN Delft, The Netherlands
* Correspondence: shisong.ren@tudelft.nl

Abstract: The film thickness of asphalt mixtures is critical for determining their performance and aging durability. However, understanding of the appropriate film thickness and its influence on performance and aging behavior for high-content polymer-modified asphalt (HCPMA) mixtures is still limited. This research aims to examine the relationship between film thickness, performance, and aging behavior of HCPMA mixtures in order to establish an optimal film thickness that ensures satisfactory performance and aging durability. HCPMA specimens with film thicknesses ranging from 6.9 μm to 17 μm were prepared using a 7.5% SBS-content-modified bitumen. Various tests, including Cantabro, SCB, SCB fatigue, and Hamburg wheel-tracking tests, were conducted to evaluate raveling, cracking, fatigue, and rutting resistance before and after aging. The key findings indicate that insufficient film thickness negatively affects aggregate bonding and performance, while excessive thickness reduces mixture stiffness and resistance to cracking and fatigue. A parabolic relationship between the aging index and film thickness was observed, suggesting that increasing film thickness improves aging durability up to a point, beyond which excessive thickness adversely impacts aging durability. The optimal film thickness for HCPMA mixtures, considering performance before and after aging and aging durability, falls within the 12.9 to 14.9 μm range. This range ensures the best balance between performance and aging durability, offering valuable insights for the pavement industry in designing and utilizing HCPMA mixtures.

Keywords: film thickness; HCPMA; optimal film thickness; aging durability

Citation: Lin, P.; Liu, X.; Ren, S.; Xu, J.; Li, Y.; Li, M. Effects of Bitumen Thickness on the Aging Behavior of High-Content Polymer-Modified Asphalt Mixture. *Polymers* **2023**, *15*, 2325. https://doi.org/10.3390/polym15102325

Academic Editors: Wei Jiang, Quantao Liu, Jose Norambuena-Contreras and Yue Huang

Received: 17 March 2023
Revised: 8 May 2023
Accepted: 10 May 2023
Published: 16 May 2023

1. Introduction

Open-graded friction courses (OGFCs) have been utilized since the 1950s in Europe and the United States to enhance the frictional resistance of asphalt pavements [1–3]. Early research on porous asphalt pavement in China began in the 1980s [4], while Japan imported its initial porous asphalt experience from Europe in 1987. In Malaysia, early applications of porous asphalt occurred in the 1990s on JKR roads [5]. The Chinese government has recently promoted the 'Sponge City' strategy to encourage widespread adoption of OGFCs [6]. However, porous asphalt mixtures, such as OGFCs, pose challenges to the performance of asphalt binders due to their open-graded structure, which makes them more susceptible to environmental and traffic-related stresses. Conventional SBS-modified asphalt, containing 3% to 4.5% SBS content, performs adequately under certain conditions but may face limitations when it comes to resisting shear stress and heavy loads in specific applications, such as OGFCs. To address these performance demands, a novel asphalt binder known as high-content polymer-modified asphalt (HCPMA) was introduced in the early 2000s. HCPMA features an SBS concentration of 5% to 15%, offering improved properties over conventional SBS-modified asphalt, particularly when applied in OGFCs. The aging issues are especially pronounced in porous asphalt mixtures due to their open-graded structure, which results in the excessive oxidation of bitumen. HCPMA, with its high SBS content, exhibits certain challenges related to aging. The susceptibility of SBS polymers to aging can be attributed to the presence of unsaturated double-carbon bonds in its molecular structure in SBS polymers [7]. Over time, as the SBS polymer undergoes degradation, the polymer network within HCPMA may be compromised, resulting in a significant decline in the material's rheological properties.

Numerous studies have been carried out on HCPMA; however, a comprehensive understanding of its modification mechanism and aging properties remains elusive. Several researchers have examined the effects of incorporating a plasticizer (furfural extract oil) and a cross-linker (sulfur) on the rheological properties of HCPMA [8–10]. Nevertheless, the SBS content in their studies was only 6%, significantly lower than the concentration employed in real-world applications. In another investigation, Zhang [11] delved into the composition and aging of HCPMA asphalt, proposing that increasing the SBS polymer content serves as one of the most effective methods for maintaining its physical properties. Furthermore, Jing and Vaveri explored the aging properties of bitumen's microscale aging behavior, examining samples from both laboratory and field settings using chemical and rheological tests [12]. Wu and Hou, on the other hand, employed Fourier transform infrared (FTIR) spectroscopy and atomic force microscopy (AFM) techniques to analyze changes in the morphology and chemical structure of SBS-modified binders [13,14].

Despite the durability concerns surrounding HCPMA, limited investigations have been conducted on its aging behavior at the mixture level. Ruan et al. concluded that oxidative aging damages the polymer network in SBS-modified bitumen, resulting in decreased temperature susceptibility, a broader relaxation spectrum, and reduced polymer modification effectiveness for enhancing asphalt mixture ductility [15]. Cuciniello et al. suggested that SBS polymer degradation can counteract the stiffening of the bitumen phase in HCPMA during oxidative aging [16]. SBS polymer degradation plays a dominant role, particularly when the SBS content by weight of the total binder exceeds 6%, leading to increased nonrecoverable compliance. Mackiewicz et al. discovered that asphalt mixtures with a 7% polymer-modified binder were most resistant to permanent deformations [17]. Lin et al. noted that due to HCPMA's complex aging behavior, an overall evaluation of resistance to permanent deformation should be determined for asphalt mixtures after both short- and long-term aging [7]. Yan et al. examined the impact of HCPMA aging using the Cantabro loss test, semicircular bend (SCB) test, and Hamburg wheel-tracking (HWT) test [18]. They observed an inconsistency between the low nonrecoverable compliance value (Jnr3.2) of an extracted binder and the deep, rutting depth of the asphalt mixture after 2 h of aging at 210 °C, indicating that the Jnr3.2 from the MSCR test might overestimate HCPMA's rutting resistance [19].

The bitumen film thickness in asphalt mixtures significantly influences aging, permeability, and stability, prompting researchers to propose it as a criterion for ensuring the durability of asphalt mixtures. Campen et al. [20], Goode et al. [21], and Kumar and Goetz [22] studied the tensile strength and resilient modulus of asphalt mixtures with varying effective film thicknesses and aging degrees. They also analyzed the penetration, viscosity, complex modulus, and phase angle of recovered binders from these mixtures, concluding that film thickness significantly affects the aging hardening of bitumen and asphalt mixtures. Kandhal and Chakraborty [23] investigated the impact of bitumen film thickness on the short-term and long-term aging of asphalt mixtures to establish a relationship between film thickness and aging characteristics. They determined that aging accelerated when the bitumen thickness was less than 9 μm and proposed that a minimum average film thickness of 8 μm is more rational than specifying a minimum void in the mineral aggregate (VMA), as recommended by McLeod and adopted by Superpave [24,25].

Sengoz and Topal [26] all found that an optimum bitumen film thickness of about 9–10 μm is necessary to prevent accelerated aging and ensure adequate resilient modulus and indirect tensile-strength values in hot-mix asphalt, with a minimum VMA of 15.2% required to achieve this thickness. Dong et al. [27] observed that the asphalt film failure mode changes from adhesion to cohesion failure when the bitumen film thickness increases from less than 23.61 μm to over 219 μm. Kandhal and Chakraborty [28], Sengoza and Agarb [29], Yang and Jiang [30]. Zhang [31–33] reported that the bitumen film thickness also has significant influence on the aging effect and fatigue performance. In summary, the bitumen film thickness in asphalt mixtures plays a crucial role in determining pavement durability and performance, with the optimum thickness depending on various factors, such as binder type, asphalt mixture, aggregate gradation, and environmental conditions.

As it is challenging to measure the actual bitumen film in asphalt mixtures, the bitumen film thickness is a conceptual average value, with different methods developed for its calculation. Initially, calculations were based on an aggregate surface-area determination, which assumed that the effective bitumen binder coated aggregate particles with an equal film thickness [20]. However, this method did not consider air voids or voids of mineral aggregate (VMA). Boris Radovskiy proposed a new definition of film thickness, as well as analytical formulas for calculating the film thickness considering VMA [34]. Heitzman [35] developed a method to include the particle-shape effect of the aggregate in bitumen film-thickness calculations and conducted a sensitivity analysis of various factors for different methods. Despite the limitations of the aggregate shape investigation, the most widely used bitumen film-thickness calculation remains based on aggregate surface area factors, as described in the Asphalt Institute Manual, Series No. 2 [36].

2. Objective

Previous research has demonstrated the critical role of asphalt mixture film thickness in determining performance and aging behavior, and optimum film thicknesses have been established for various mixtures. However, the application of HCPMA in porous asphalt is relatively new, and the appropriate film thickness and its influence on performance and aging behavior remain unclear. Durability is a vital factor in determining the performance of HCPMA porous asphalt, and ensuring an adequate bitumen film thickness is essential for acceptable durability. Current pavement specifications (EN 13108-1:2016) have minimum bitumen content requirements to ensure durability, which may not be suitable for HCPMA porous asphalt. This study aims to identify the optimum bitumen film thickness for HCPMA porous asphalt, ensuring satisfactory durability based on aging characteristics. The primary objectives of this investigation are:

(1) To evaluate the impact of bitumen film thickness on HCPMA porous asphalt mixture performance by assessing its indirect tensile strength, rutting resistance, fatigue resistance, and raveling resistance.
(2) To examine the aging behavior of an HCPMA porous asphalt mixture and establish the correlation between bitumen film thickness and its aging durability.
(3) To recommend an optimum bitumen film thickness for an HCPMA porous asphalt mixture that ensures satisfactory performance and aging durability, based on the characterization of its performance and aging behavior.

3. Materials and Methods

3.1. Bitumen and Aggregate

In this study, a high-content polymer-modified asphalt (HCPMA) containing 7.5% linear SBS polymer (30% styrene) was used as a representative sample. The HCPMA was prepared in the laboratory based on the literature and previous research [3,7], using a shearing machine and a blending machine through the following process: First, the SBS polymer was added to the base bitumen at 185 °C and sheared at 4000 rpm for 30 min. Next, the sheared binder was mixed using a mechanical blender at 500 rpm for an hour. Finally, 0.15 wt% sulfur was added to enhance the storage stability, and the resulting mixture was blended for an additional 30 min. The conventional and rheological properties of the HCPMA are presented in Table 1.

Table 1. Conventional and rheological properties of HCPMA.

Parameters	HCPMA
SBS content	7.50%
Penetration at 25 °C, 0.1 mm	41
Softening point, °C	>100
Ductility at 5 °C, cm	44.1
135 °C, viscosity, Pa.s	5.72
70 °C complex modulus, Pa	5281
70 °C phase angle, °	48
70 °C Jnr3.2, kPa^{-1}	0.011
70 °C R3.2, %	98.8
Elastic recovery at 25 °C, %	90.3
48 h softening-point difference, °C	1.8

In order to evaluate the influence of bitumen film thickness on the aging characteristics of asphalt mixture, the basalt aggregates and limestone filler were used to prepare the mixture specimens. The aggregate properties of the basalt aggregate and limestone filler can be seen in Table 2.

Table 2. Basic properties of aggregate and filler.

Basic Properties	Basalt Aggregates/Limestone Filler
Flat and elongated particles of aggregate (%)	10.20
Fine aggregate angularity (%)	55.80
LA abrasion (%)	10.70
Fracture 1 face (%)	99.20
Fracture 2 face (%)	98.70
Water absorption of limestone filler (%)	1.21
Sand equivalent of fine aggregate (0.063–2.00 mm) (%)	72.00

3.2. Mixture Design and Preparation

The typical open-graded friction course with 13 mm nominal maximum aggregate sizes (OGFC-13) was investigated, with three distinct gradations—fine, mid, and coarse—being selected to represent the different particle size distributions within the HCPMA mixtures. Additionally, three bitumen content values (4%, 5%, and 6%) were chosen for preparing the HCPMA-mixture specimens, maintaining the air void at 20%. Detailed gradations can be found in Table 3. In drainage-wearing courses, where HCPMA is employed as the binder, fibers are not commonly used in porous asphalt mixtures due to the high strength and viscosity of HCPMA. Furthermore, the primary objective of this research is to examine the influence of film thickness on the performance of porous asphalt. Consequently, no fibers were incorporated into the mixtures in this study.

The mixture specimens were prepared using the superpave gyratory compactor (SGC) at 185 °C, following field-experience guidelines. Film thickness, a crucial factor affecting the performance and durability of HCPMA mixtures, was calculated using the Asphalt Institute Manual, Series No. 2 (MS-2) [36]. This method relies on the surface-area (SA) factors for each sieve size specified in the aggregate gradation. The average film thickness can be determined using the following equation:

$$T_F = 1000 \times \frac{P_{be}}{SA \times P_s \times G_b} \quad (1)$$

where:

T_F = average film thickness (μm);
P_{be} = percentage (by weight) of effective asphalt binder in the mix;
SA = surface area of aggregate gradation (m^2/kg);

P_s = percentage (by weight) of aggregate;
G_b = specific gravity of asphalt binder.

Table 3. Asphalt mixture gradation and bitumen film thickness calculation.

Sieve Size	SA Factors	HCPMA-6.9 μm		HCPMA-10.3 μm		HCPMA-12.9 μm		HCPMA-14.9 μm		HCPMA-17.0 μm	
		Fine Gradation		Mid Gradation		Mid Gradation		Mid Gradation		Coarse Gradation	
(mm)	(m^2/kg)	% Passing	SA (m^2/kg)	% Passing	SA (m^2/kg)	% Passing	SA (m^2/kg)	% Passing	SA (m^2/kg)	% Passing	SA (m^2/kg)
19	0.41	99.9	0.41	99.9	0.41	99.9	0.41	99.9	0.41	99.9	0.41
12.5	-	92.3	-	92.3	-	92.3	-	92.3	-	92.3	-
9.5	-	65.1	-	65.0	-	65.0	-	65.0	-	65.0	-
4.75	0.41	19.0	0.08	18.7	0.08	18.7	0.08	18.7	0.08	18.4	0.08
2.36	0.82	13.6	0.11	12.1	0.10	12.1	0.10	12.1	0.10	10.2	0.08
1.18	1.64	11.5	0.19	9.3	0.15	9.3	0.15	9.3	0.15	7.3	0.12
0.6	2.87	10.0	0.29	7.2	0.21	7.2	0.21	7.2	0.21	5.2	0.15
0.3	6.14	8.8	0.54	5.6	0.35	5.6	0.35	5.6	0.35	3.7	0.22
0.15	12.29	8.4	1.03	5.0	0.62	5.0	0.62	5.0	0.62	3.0	0.37
0.075	32.77	7.9	2.62	4.9	1.64	4.9	1.64	4.9	1.64	2.5	0.83
∑SA (m^2/kg)		5.27		3.54		3.54		3.54		2.27	
Pbe (%)		3.63		3.64		4.56		5.28		3.84	
FT (μm)		6.9		10.24		12.87		14.90		16.94	

According to the Asphalt Institute MS-2, the surface area of the aggregates can be calculated from gradation based on the percentage, passing a set of sieves as follows:

$$SA = 0.01\sum_{i}^{N} PP_i \times CP_i \tag{2}$$

where:

N = number of sieves considered in the surface area calculation;
PP_i = percentage of aggregates passing sieve i (defined for sieves of 9.5, 4.75, 2.36, 1.18, 0.60, 0.30, 0.15, and 0.075 mm);
CP_i = surface area factor outlined in Asphalt Institute MS-2.

Utilizing Equations (1) and (2), the bitumen film thickness can be determined, and the corresponding results are presented in Table 3. To categorize the HCPMA mixtures, they were labeled based on their approximate bitumen film thickness. For instance, HCPMA-12.9 μm refers to an asphalt mixture with an approximate film thickness of 12.9 μm.

3.3. *Aging Process of Asphalt Mixture*

In this study, both short-term and long-term aging procedures were conducted in accordance with AASHTO R30. The blended loose mixture was placed on a steel pan lined with aluminum foil to prevent the asphalt mixture from adhering to the pan. The thickness of the asphalt mixture was maintained below 1 inch (25.4 mm) to ensure consistent and uniform aging (Figure 1). The standard short-term aging temperature of 135 °C was deemed unsuitable for this study, as it is significantly lower than the mixing and transport temperature of HCPMA and does not accurately represent field conditions. Consequently, short-term aging was carried out at the compaction temperature of HCPMA (163 °C) for 2 h to simulate the aging process during the mixing, transportation, and paving stages [19]. The mixture was stirred every 60 min to ensure uniform aging throughout the process.

Long-term aging was also performed in compliance with AASHTO R30 to emulate the in-field aging of HCPMA. Following the short-term aging, the loose HCPMA was compacted using a superpave gyratory compactor (SGC). Once cooled to an ambient temperature, the compacted HCPMA specimens were placed in an oven set at 85 °C for 120 h to undergo long-term aging. Subsequently, the HCPMA specimens were allowed to cool to environmental temperature for 16 h, after which the long-term aged specimens were prepared for subsequent testing.

Figure 1. Schematic diagram of aging pan and loose HCPMA mixture.

3.4. Test Methods

3.4.1. Cantabro Loss Test

In order to evaluate the raveling resistance of the porous asphalt mixture, the Cantabro test was applied in this research. Following the standard of AASHTO TP 108-14, a cylindrical asphalt mixture specimen with a 100 mm diameter and a 63.5 mm height was cured in a 25 °C bath for 20 h. Then the mixture specimen was placed into the Los Angeles abrasion machine, and the machine was rotated for 300 revolutions at a speed of 30 rpm at 25 °C. The percentage of the mass loss was calculated to represent the raveling of the asphalt mixture. For each type of mixture, four replicates were conducted.

3.4.2. Semicircular Bending (SCB) Strength Test

The cracking resistance of HCPMA was evaluated using the semicircular bend (SCB) strength test. In accordance with previous research and preparation methodologies [37], cylindrical specimens measuring 135 mm in height and 150 mm in diameter were prepared using a superpave gyratory compactor. Subsequently, each cylindrical specimen was cut into four semicircular specimens with dimensions of 50 mm in height and 150 mm in diameter. A notch, 15 mm in length and 1.5 mm in width, was cut at the center bottom of each specimen to ensure the appropriate cracking mode. The mixture specimens were preconditioned in a test chamber at 25 °C for a minimum of 4 h before conducting the SCB strength test at a displacement rate of 50 mm/min and a temperature of 25 °C. Based on the force-displacement curve, tensile strength (σ_{max}) and fracture energy (G_f) were calculated using the following equations.

Tensile strength (σ_{max}) represents the asphalt mixture's strength and is calculated using the equation:

$$\sigma_{max} = \frac{4.263 \times F_{max}}{D \times t} \quad (3)$$

where F_{max} is the maximum force in N, D is the specimen's diameter in mm, and t is the specimen's thickness in mm. The constant 4.263 is derived from research by Van de Ven et al. by using 3D finite element analysis, assuming a support span of 80% of the specimen's diameter [37].

Fracture energy (G_f) represents the amount of energy consumed to create cracks per unit area and is calculated using the equation:

$$G_f = \frac{W_f}{A_{lig}} \tag{4}$$

where (W_f) (fracture work) is the work done during the fracture process, calculated as the area under the force-displacement curve. A_{lig} is the ligament area, calculated as:

$$A_{lig} = (W - a) \times T \tag{5}$$

where W is the specimen's height in mm, a is the depth of the specimen's notch in mm, and T is the specimen's thickness in mm.

3.4.3. Semicircular Bending (SCB) Fatigue Test

The fatigue resistance of HCPMA was assessed using the SCB fatigue test. As documented in the literature, the stress-controlled SCB fatigue test is a widely utilized method, and in this study, four stress amplitudes (stress ratio (σ_{ratio}) of 0.3, 0.4, 0.5, and 0.6) were chosen and applied in the SCB fatigue tests [38–40]. In this research, the stress level is defined as the ratio of stress amplitude to the tensile strength of the specimen [39].

Prior to the SCB fatigue test, the SCB strength test was performed on four replicate specimens to determine their tensile strengths at 15 °C. Once the tensile strength values were obtained, stress amplitudes were calculated using the selected stress ratios (0.3, 0.4, 0.5, and 0.6) and then applied to the SCB fatigue test. The frequency of the repeated compressive load was set at 10 Hz, comprising a 0.1 s half-sine load and no rest period under different stress levels. Four replicate SCB fatigue tests were conducted at 15 °C for each HCPMA type at each stress level.

To further analyze the SCB fatigue results, a function can be utilized to describe the relationship between the fatigue life and the stress ratio. The literature shows a linear relationship between stress ratio and fatigue life in double logarithmic coordinates [39]. Thus, the fatigue life can be described with Equations (6) and (7). The expression between $lg\left(N_f\right)$ and $\lg(\sigma_{ratio})$ is a simple linear relationship. Based on the SCB fatigue results, the least-squares equation can be obtained, and then the parameters a and b are determined correspondingly.

$$N_f = a(\sigma_{ratio})^b \tag{6}$$

$$lg\left(N_f\right) = lga + b \times \lg(\sigma_{ratio}) \tag{7}$$

where

N_f is the fatigue life in the SCB fatigue test;
σ_{ratio} is the stress ratio, which is the ratio between the loading stress in the SCB fatigue test and the peak stress in the SCB strength test;
a is a regression parameter that describes the fatigue life of the specimen;
b is a regression parameter that describes the stress sensitivity of the specimen.

3.4.4. Hamburg Wheel-Tracking (HWT) Test

The rutting resistance of the HCPMA mixture was evaluated using the Hamburg wheel-tracking (HWT) test. The HWT test was performed with a double wheel track (DWT) device from the Controls Group, Milan, Italy, at 60 °C under moist conditions, in accordance with AASHTO T324-11. Two cylindrical specimens, each measuring 150 mm in height and 62 mm in diameter, were loaded into the HWT device as a single test sample. The HCPMA mixture specimens were subjected to steel wheels rolling at a speed of 52 passes/min, while a linear variable differential transformer (LVDT) was employed to record the specimens' relative vertical deformation. The test concluded either when the vertical deformation reached 20 mm or the specimen had undergone 20,000 load passes.

Following the HWT test, a rutting depth-load number curve was generated for each test sample, with the results representing the average of two test repetitions. Rutting resistance can be characterized by the slope of the creep stage, while stripping resistance can be characterized by the slope of the stripping stage. It should be noted that the stripping stage is considered only when the stripping stage slope is at least twice the creep slope. The consistency and reliability of the two replicates were observed, and any variations between the repetitions were reported in the manuscript.

3.4.5. Aging Index of Performance Parameters

To assess the impact of bitumen film thickness on the aging durability of HCPMA porous asphalt mixtures, it is crucial to establish a reliable and robust metric capable of effectively quantifying the aging process. For this purpose, a series of aging indices are proposed, which are based on the ratio of performance evaluation parameters before and after aging. These parameters include the Cantabro loss ratio, fracture strength, fracture energy, fatigue parameters, and creep slope. These aging indices offer a standardized measure of the degradation extent resulting from the aging process, enabling a comprehensive comparison of the performance of different mixtures with varying bitumen film thicknesses. The aging index can be defined as follows:

$$Aging\ Index\ (AI) = \frac{Performance\ Parameter_{afteraging}}{Performance\ Parameter_{beforeaging}} \times 100\% \tag{8}$$

The significance of these aging indices lies in their ability to quantify the impact of aging on the HCPMA porous asphalt mixtures and correlate this degradation with the bitumen film thickness. The aging indices will provide a valuable tool for pavement engineers and researchers to assess and monitor the performance of HCPMA porous asphalt mixtures under different aging conditions, ultimately contributing to the design and construction of more durable and sustainable asphalt pavements.

4. Results and Discussion

4.1. Cantabro Loss Test Results

The Cantabro test is designed to evaluate the adhesive and cohesive properties of high-content polymer-modified asphalt (HCPMA) and assess its resistance to raveling. The results of HCPMA before and after aging can be seen in Figure 2.

In the unaged HCPMA analysis, a noticeable relationship between film thickness and performance was observed in the Cantabro loss results. As film thickness increased from 6.89 μm to 12.9 μm, Cantabro loss declined from 18.2% to 11.0%. However, upon further increasing the film thickness to 17.0 μm, the Cantabro loss rose to 23.3%. This suggests that 12.9 μm is the optimal film thickness for achieving the best adhesive, cohesive, and raveling-resistance properties. When compared to the CROW guidelines (Netherlands Centre for Research on Road Engineering) standard, which stipulates a maximum Cantabro loss of 20%, HCPMA mixture samples with film thicknesses of 14.9 μm or less met the requirement. This acceptable performance can likely be attributed to the high styrene–butadiene–styrene (SBS) content in HCPMA. The analysis of unaged samples indicates that insufficient film thickness leads to ineffective aggregate bonding, while excessive film thickness results in reduced stiffness and compromised adhesive and cohesive performance. Consequently, a film thickness of 12.9 μm is deemed optimal.

For aged HCPMA samples, the film thickness's impact on Cantabro loss is akin to that of the unaged samples. The Cantabro loss decreased from 31.0% to 14.6% when the film thickness increased from 6.89 μm to 12.9 μm. However, further increasing the film thickness to 17.0 μm resulted in a Cantabro loss of 29.4%. Thus, the optimal film thickness of 12.9 μm for adhesive, cohesive, and raveling-resistance properties remains consistent after aging. In comparison to the CROW-2015 standard, HCPMA samples with film thicknesses of 6.89 μm and 17 μm did not satisfy the 20% Cantabro loss requirement after aging, possibly due to HCPMA aging and the reduced effectiveness of its high-SBS content. After aging, the influence of film thickness on adhesive performance remained in line with the observations from unaged samples, with 12.9 μm remaining as the optimal thickness.

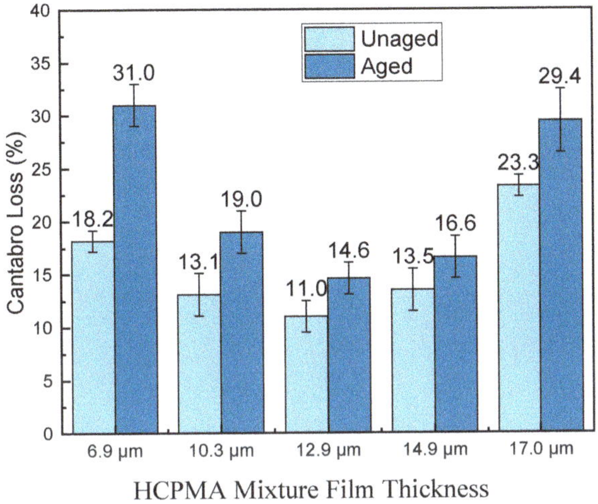

Figure 2. Influence of bitumen film thickness on the raveling resistance of HCPMA.

The aging index of Cantabro loss is defined as the ratio between the Cantabro loss after aging and the Cantabro loss before aging. The results can be seen in Figure 3. Lower Cantabro loss values indicate a better raveling resistance, while higher aging index values suggest reduced aging durability. As observed in Figure 3, the aging index decreases considerably as the film thickness increases, following a parabolic relationship with film thickness.

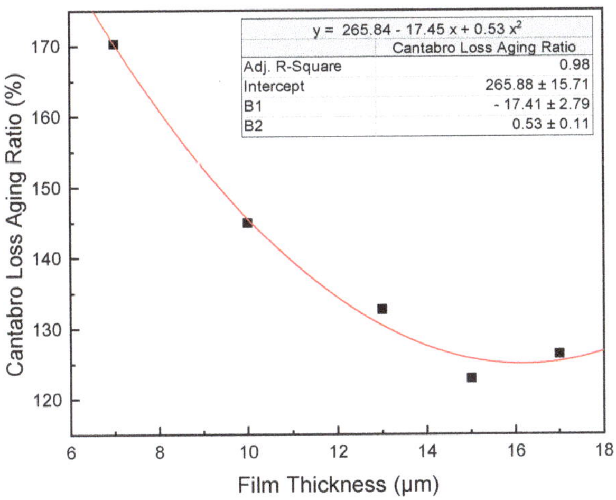

Figure 3. Influence of bitumen film thickness on the aging ratio of Cantabro loss.

In summary, increasing film thickness enhances aging durability up to a point, but beyond that point, it offers diminishing returns. The optimal film thickness before and after aging for adhesive and cohesive performance is 12.9 μm. Film thickness beyond 12.9 μm provides limited improvements in aging durability.

4.2. SCB Cracking Test

The semicircular bend (SCB) test aims to characterize the cracking resistance properties of HCPMA and the influence of aging on its cracking resistance. The SCB cracking results of HCPMA before and after aging can be seen in Figure 4a,b.

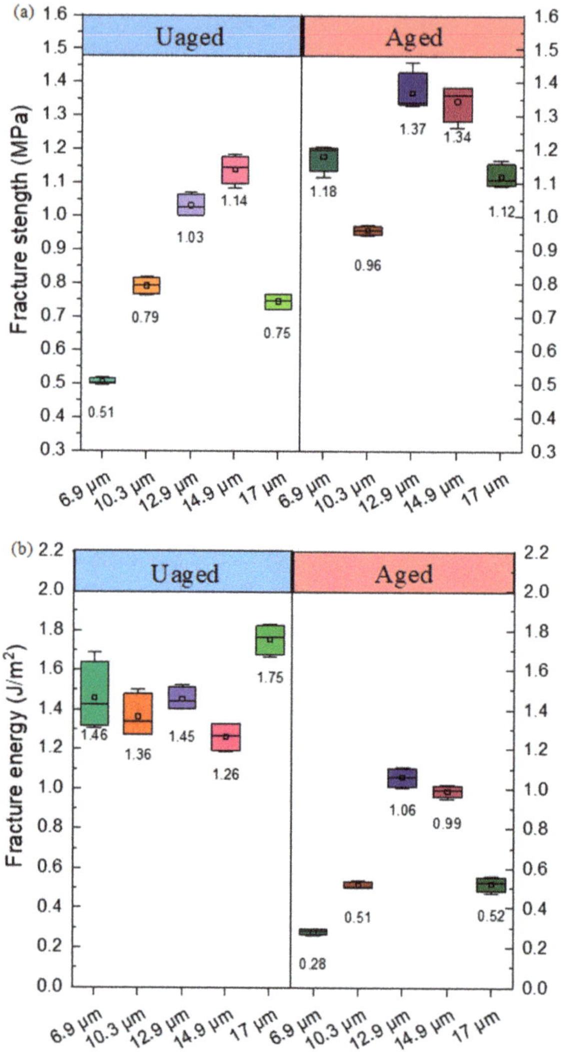

Figure 4. Influence of bitumen film thickness on SCB results of HCPMA: (**a**) fracture strength and (**b**) fracture energy.

Regarding the fracture strength, for unaged samples, the strength increased from 0.51 MPa to 1.14 MPa when the film thickness increased from 6.9 μm to 14.9 μm, and the fracture strength decreased to 0.75 MPa when the film thickness reached 17 μm. The optimal film thickness considering the fracture strength of HCPMA before aging was

14.9 μm. For aged samples, the fracture strength increased significantly during aging, likely due to bitumen oxidation, which increased the stiffness of HCPMA. The fracture strength increased from 1.18 MPa to 1.37 MPa when the film thickness increased from 6.9 μm to 12.9 μm and decreased to 1.12 MPa when the film thickness reached 17 μm. The optimal film thickness, considering the fracture strength of HCPMA after aging, was 14.9 μm.

As for the fracture energy, before aging, the influence of film thickness on fracture energy was not significant, with values ranging from 1.26 J/m^2 to 1.75 J/m^2 when the film thickness increased from 6.9 to 17 μm. There was no clear trend between film thickness and fracture energy. After aging, the fracture energy decreased significantly due to bitumen oxidation and SBS polymer degradation in HCPMA, leading to a decreased ability to absorb energy during cracking. The fracture strength after aging increased from 0.28 MPa to 1.06 MPa when the film thickness increased from 6.9 μm to 12.9 μm and decreased to 0.52 MPa when the film thickness reached 17 μm. The optimal film thickness, considering the fracture strength of HCPMA after aging, was 12.9 μm.

The aging indices of fracture strength and fracture energy provide a quantitative assessment of the change in cracking resistance properties of HCPMA before and after aging, and the results can be seen in Figure 5. By comparing these indices, we can evaluate the impact of aging on the performance of HCPMA mixtures and the influence of film thickness on the aging durability.

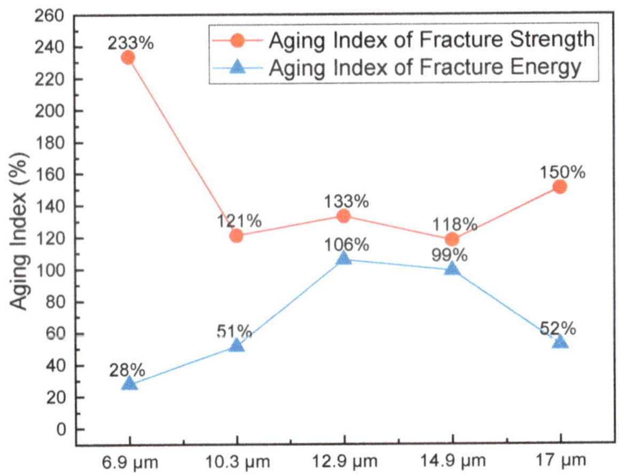

Figure 5. Influence of bitumen film thickness on the fracture strength of HCPMA.

For the fracture strength, the aging index decreased from 233% to 118% when the film thickness increased from 6.9 μm to 14.9 μm. However, when the film thickness reached 17 μm, the aging index of fracture strength increased to 150%. This indicates that an increase in film thickness up to 14.9 μm improves the aging durability in terms of fracture strength, but beyond that point, the benefits are limited. On the other hand, the aging index of fracture energy showed an inverse relationship with film thickness. The index increased from 28% to 106% when the film thickness increased from 6.9 μm to 12.9 μm and decreased to 52% when the film thickness was 17 μm. This suggests that the aging durability, in terms of fracture energy, is more sensitive to changes in the film thickness and is best at 12.9 μm.

While the aging indices of fracture strength and energy show contrary responses to the increase in film thickness, it is important to note that fracture energy is a more critical parameter for cracking resistance. This is because cracking resistance is not determined solely by the strength and stiffness of the mixture but also by the ductility and ability to absorb energy during cracking. Considering both the fracture strength and fracture energy aging indices, a film thickness of 12.9 μm provides the best balance for aging durability and cracking resistance in HCPMA mixtures.

In light of the SCB test results and discussions, the optimum film thickness for the HCPMA mixture's cracking resistance and aging durability has been identified. Considering both unaged and aged HCPMA mixtures, the performance perspective suggests that the optimum film thickness for fracture strength and fracture energy lies within the range of 12.9 to 14.9 μm. Meanwhile, from an aging durability perspective, a 12.9 μm film thickness is found to be ideal for maintaining the cracking resistance of the HCPMA mixture as it demonstrates a better balance between the aging indices of fracture strength and fracture energy, which are essential indicators of the mixture's long-term performance.

4.3. SCB Fatigue Test Results

The SCB fatigue test is designed to characterize the fatigue performance of HCPMA, as well as to analyze the influence of aging on its fatigue performance. This can be observed through the SCB performance results of HCPMA before and after aging, as depicted in Figure 6a,b.

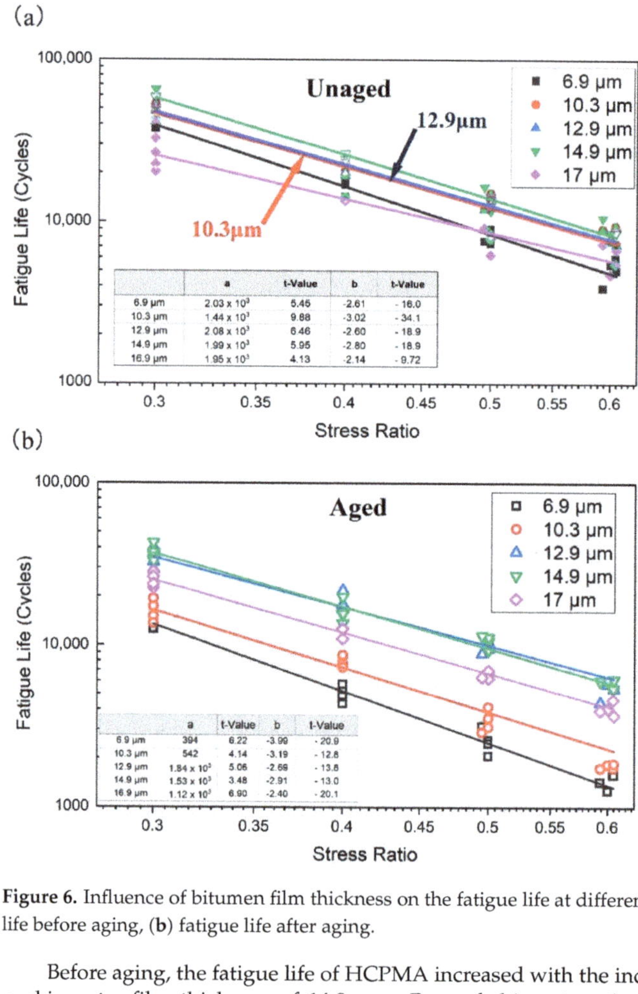

	a	t-Value	b	t-Value
6.9 μm	2.03 x 10³	5.45	-2.61	- 16.0
10.3 μm	1.44 x 10³	9.88	-3.02	- 34.1
12.9 μm	2.08 x 10³	6.46	-2.60	- 18.9
14.9 μm	1.99 x 10³	5.95	-2.80	- 18.9
16.9 μm	1.95 x 10³	4.13	-2.14	- 9.72

	a	t-Value	b	t-Value
6.9 μm	394	6.22	-3.99	- 20.9
10.3 μm	542	4.14	-3.19	- 12.8
12.9 μm	1.84 x 10³	5.06	-2.69	- 13.8
14.9 μm	1.53 x 10³	3.48	-2.91	- 13.0
16.9 μm	1.12 x 10³	6.90	-2.40	- 20.1

Figure 6. Influence of bitumen film thickness on the fatigue life at different stress ratios, (**a**) fatigue life before aging, (**b**) fatigue life after aging.

Before aging, the fatigue life of HCPMA increased with the increase in film thickness, peaking at a film thickness of 14.9 μm. Beyond this point, the fatigue life decreased significantly. Furthermore, it was observed that the slope of the fatigue life decreased as the film thickness increased, indicating that the fatigue performance of asphalt mixture is less sensitive to the increase in the stress ratio. In contrast, after aging, the fatigue life of HCPMA decreased significantly, likely due to the oxidation of the bitumen and degradation of the SBS polymer, which reduces the flexibility of the HCPMA. When comparing HCPMA with different thicknesses after aging, it was observed that HCPMA, with a thickness of 6.9 μm, had the lowest fatigue life, and as the film thickness increased, the fatigue life also increased significantly. The fatigue life is no longer as sensitive to the increase in the stress ratio, reaching a maximum when the film thickness is 12.9 μm or 14.9 μm, and then decreasing when the film thickness reaches 17.0 μm.

The SCB fatigue results highlight that, when the film thickness is too thin, the HCPMA cannot provide sufficient coating, leading to poor bonding between the aggregates and the binder. This results in a weaker mixture that is more susceptible to cracking and fatigue failure under repeated loading. On the other hand, when the film thickness is too high, the excess binder can negatively impact the mechanical properties of the mixture, decreasing its stiffness and making it more susceptible to deformation under repeated loading, ultimately reducing its fatigue life.

Fatigue parameters are essential for quantitatively characterizing the fatigue life of specimens, especially in the context of the SCB test results, which can be seen in Figure 7. In this analysis, the fatigue life is calculated using Equations (6) and (7). Two regression parameters are considered: fatigue life parameter a, which describes the fatigue life of the specimen, and stress sensitivity parameter b, which represents the stress sensitivity of the specimen.

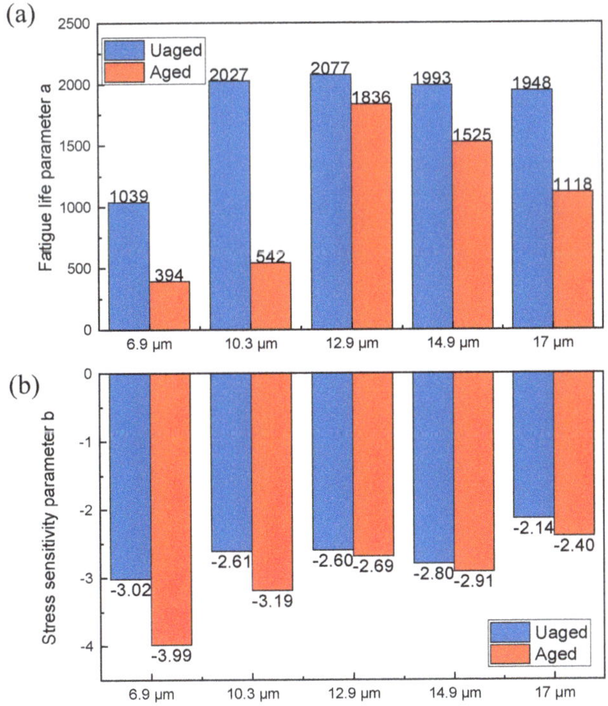

Figure 7. Influence of bitumen film thickness on the fatigue parameters of HCPMA, (**a**) influence on fatigue life parameter a, (**b**) influence on stress sensitivity b.

Before aging, as the film thickness increased, fatigue life parameter a rose from 1039 to reach its highest value of 2077 at a film thickness of 12.9 μm. However, when the film thickness was further increased, this parameter decreased slightly to 1948 at 17 μm. After aging, fatigue life parameter a declined significantly, particularly for HCPMA with film thicknesses of 6.9 μm and 10.3 μm. This behavior suggests that, after aging, the fatigue parameter a becomes more sensitive to changes in film thickness. The high content of SBS in HCPMA contributes to the material's excellent performance before aging, increasing its tolerance to variations in film thickness. However, after aging, the oxidation of the bitumen phase and degradation of the SBS polymer rendered the HCPMA more sensitive to film thickness.

The stress sensitivity parameter b exhibited a different trend. Before aging, this parameter increased from −3.02 to −2.14 as the film thickness increased from 6.9 μm to 17.0 μm. This indicates that a more considerable film thickness significantly reduces the stress sensitivity of HCPMA during fatigue loading. After aging, the stress sensitivity parameter b decreased significantly, particularly for HCPMA with film thicknesses of 6.9 μm and 10.3 μm. Similar to the parameter a, b also declined with increased film thickness after aging, suggesting that a greater film thickness decreases the stress sensitivity of HCPMA even after aging.

The aging index plays a crucial role in understanding the fatigue behavior of HCPMA specimens after undergoing the aging process. Defined as the ratio between fatigue parameters before and after aging, the aging index results can be observed in Figure 8.

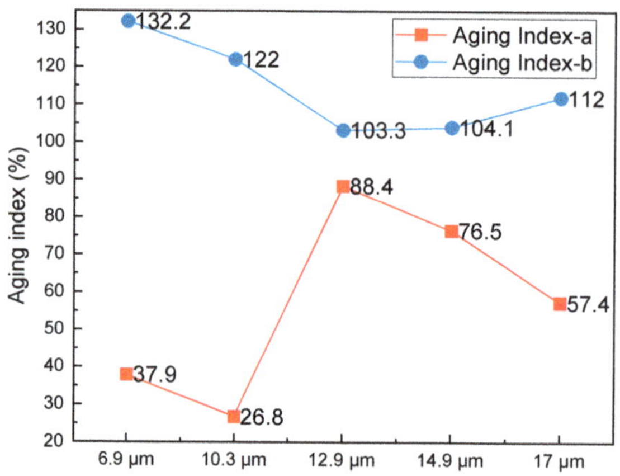

Figure 8. Influence of bitumen film thickness on the aging indices of fatigue parameters.

Upon examining Figure 8, it is evident that the aging index for parameter a (Aging Index-a) increased from 37.9% to 88.4% as the film thickness rose from 6.9 μm to 14.9 μm. However, when the film thickness reaches 17 μm, aging index a increased to 57.4%. In contrast, the aging index for parameter 'b' (aging index b) decreased from 132.2% to 103.3% as the film thickness increased from 6.9 μm to 12.9 μm, before increasing again to 112% when the film thickness reached 17 μm. This contrasting influence of film thickness on the aging indices of parameters 'a' and 'b' is noteworthy. A lower aging index a and a higher aging index b indicates that the HCPMA has improved fatigue resistance after aging. Taking both aging indices into consideration, a film thickness of 12.9 μm appears to be optimal for achieving the best aging durability.

In summary, the fatigue performance of HCPMA can be effectively quantified using fatigue life parameter a and stress sensitivity parameter b. The behavior of both parameters is influenced by the film thickness, with the fatigue life parameter a increasing and then decreasing with film thickness, reaching its peak value at 12.9 μm for both before and after aging. The stress sensitivity parameter b decreased with increasing film thickness before and after aging, suggesting that a more significant film thickness contributes to reduced stress sensitivity. By considering both aging index a and aging index b, it is determined that a film thickness of 12.9 μm is optimal for ensuring the highest aging durability.

4.4. HWT Test Results

The Hamburg wheel-tracking test (HWTT) was employed to assess the rutting and moisture susceptibility of HCPMA both before and after aging, by replicating real-world pavement conditions such as heavy traffic loads, repeated wheel passes, and water exposure. The results of the HWTT for HCPMA can be found in Figures 9 and 10.

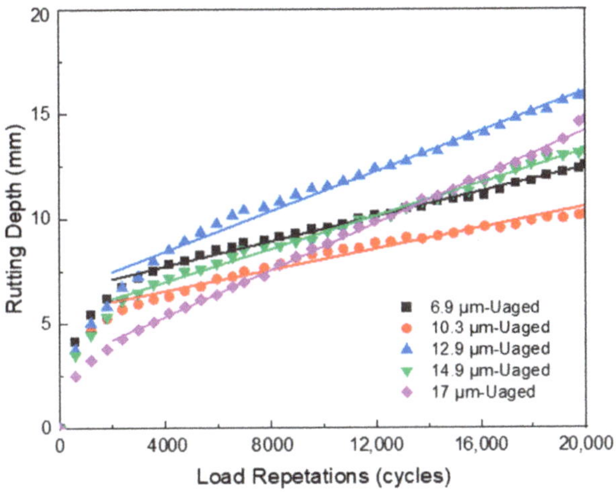

Figure 9. Influence of bitumen film thickness on the rutting resistance of HCPMA before aging.

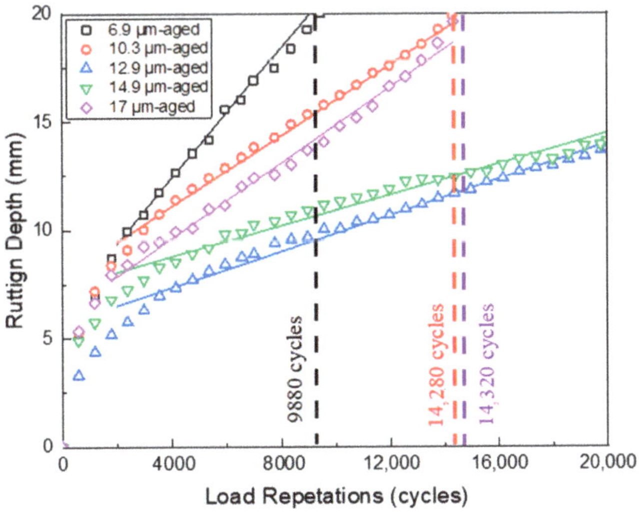

Figure 10. Influence of bitumen film thickness on the rutting resistance of HCPMA after aging.

A typical rutting curve derived from the HWTT was divided into three distinct stages: postcompaction, creep, and stripping phases [41–43]. The post compaction phase involves the specimen's consolidation as the wheel load compacts the mixture. During the creep phase, the deformation occurs primarily due to the viscous flow of the asphalt mixtures, which is characterized by a constant rate of increase in rut depth per load cycle (creep slope). The stripping phase commences when the bond between the asphalt binder and the aggregate weakens, leading to visible damage, such as stripping or raveling, as more load cycles are applied. The stripping inflection point (SIP) represents the number of load cycles at which a sudden increase in rut depth is observed, primarily as a result of asphalt binder stripping from the aggregate. The stripping stage is typically considered when the slope in the potential stripping region is twice the creep slope.

The HWTT results of the unaged HCPMA mixture, as depicted in Figure 9, demonstrate that the postcompaction phase occurs within the initial 2000 cycles. Subsequently, a clear and extended creep phase was observed, lasting for 20,000 cycles, until the conclusion of the test. A significant stripping phase is not present, as the slope does not increase to twice the creep slope, indicating that the HCPMA binder maintains a good bonding performance and resists the stripping of the binder from the aggregate. After 20,000 loading cycles, the rutting depths of all HCPMA samples remain under 20 mm, without a distinct relationship between rutting depth and film thickness. The highest rutting depth, measuring 16.5 mm, was attained when the HCPMA samples exhibited a film thickness of 12.9 μm.

The creep slope, which signifies the rate at which the asphalt mixture deforms under continuous loading, is of significant importance. The results reveal that, as film thickness increases, the creep slope also rises correspondingly. This relationship may be due to the vital role the aggregate skeleton plays in porous asphalt mixtures, offering effective load distribution, particle interlock, reduced reliance on binder, and increased stiffness, all of which contribute to enhanced rutting resistance. When film thickness increases, the mixture's stiffness diminishes, which subsequently leads to an increased creep slope.

The aged HCPMA Hamburg wheel-tracking test (HWTT) results, as illustrated in Figure 10, provide insights into the performance and durability of the HCPMA porous asphalt mixtures after aging. The postcompaction phase of HCPMA samples after aging showed consistent loading cycles, ranging between 1 and 2000 cycles. This observation indicates that the aging process does not significantly alter the HCPMA mixtures' response to the applied load cycles.

The HWTT results reveal a significantly steeper creep slope for the HCPMA samples after aging. With varying film thicknesses (6.9 μm, 10.3 μm, and 17 μm), the HCPMA mixtures reached a rutting depth of 20 mm after completing 9880, 14,280, and 14,320 load cycles, respectively. The steeper creep slope suggests a more rapid deformation of the asphalt mixture under sustained loading after aging, highlighting the impact of the aging process on the material's performance. Furthermore, no significant stripping phase was observed after aging. This observation implies that the HCPMA binder maintains its excellent bonding properties, resisting the stripping of the binder from the aggregate even after the aging process. The absence of a stripping phase demonstrates the durability of the HCPMA binder in preserving adhesion to the aggregate throughout the aging process.

The analysis of the creep slope reveals a complex relationship between film thickness, performance, and aging durability. As the film thickness increased, the creep slope decreased considerably until a film thickness of 14.9 μm was reached. Beyond this point, the film thickness increased significantly when the thickness changed from 14.9 μm to 17 μm. When the film thickness is too low, the aging durability of the HCPMA cannot be guaranteed, resulting in a large creep slope after aging. Conversely, when the film thickness is too high (greater than 14.9 μm), the excess bitumen reduces the mixture's stiffness, leading to a decrease in rutting resistance. These findings suggest that an optimal film thickness range between 12.9 and 14.9 μm allows the HCPMA to strike a balance between performance and aging durability, ensuring a reasonable creep slope, both before and after aging.

To quantitatively analyze the rutting curve, the data after 2000 cycles was fitted into a linear function, and the creep slope was obtained. This analysis provides a deeper understanding of the HCPMA's performance in terms of rutting resistance, and the results can be seen in Figure 11.

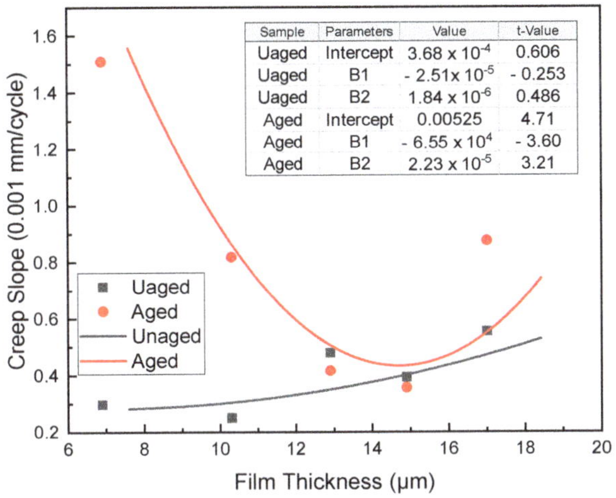

Figure 11. Influence of bitumen film thickness on the creep slopes before and after aging.

Before aging, the creep slope remained at a very low level, indicating that the HCPMA demonstrates excellent rutting resistance. Additionally, the creep slope showed a slight increase with the rise in film thickness. After aging, the creep slope exhibited a significant increase, highlighting the effect of aging on the rutting resistance. The creep slope decreased from 1.52 to 0.42 (0.001 mm/cycle) when the film thickness grew from 7.9 to 14.9 μm. Conversely, the creep slope ascended from 0.42 to 0.92 (0.001 mm/cycle) when the film thickness expanded from 14.9 μm to 17 μm. A parabolic relationship was observed between the creep slope and film thickness, with the parameters of the fitting curve shown in Figure 11.

It is essential to note that the creep slope is not sensitive to changes in the film thickness before aging. However, after aging, the creep slope exhibited a parabolic relationship with film thickness, reaching its lowest point when the film thickness was approximately 14.9 μm. Taking into account the rutting resistance before and after aging, the optimal film thickness was determined to be 14.9 μm. This analysis provides valuable insights into the material's performance and can inform future research and development efforts in the field of asphalt mixtures.

Figure 12 displays the relationship between film thickness and creep slopes before and after aging, as well as the aging index for each respective film thickness. By examining Figure 12, it becomes apparent that the aging index of the creep slope, which is calculated as the ratio of creep slopes after aging to before aging, displays a nonlinear relationship with the film thickness.

To further explore this relationship, we first calculated the fitting function to model the parabolic relationship between the aging index and film thickness. The fitted function can be seen in Figure 12. The parabolic relationship between the aging index and the film thickness implies that there is a range of film thickness values that deliver good performances in terms of rutting resistance and aging durability. However, the performance deteriorates significantly for film thickness values outside of this range. As demonstrated in Figure 12, the optimal film thickness for HCPMA mixtures is approximately 14.9 μm.

In conclusion, our detailed analysis of the table and the derived fitting function indicates that the optimal film thickness for HCPMA mixtures is approximately 14.9 μm. This value provides a balance between rutting resistance and aging durability, ensuring satisfactory performance throughout the pavement's service life. It is essential to maintain the film thickness within the optimal range to minimize the impact of the aging process on the performance of the asphalt mixture.

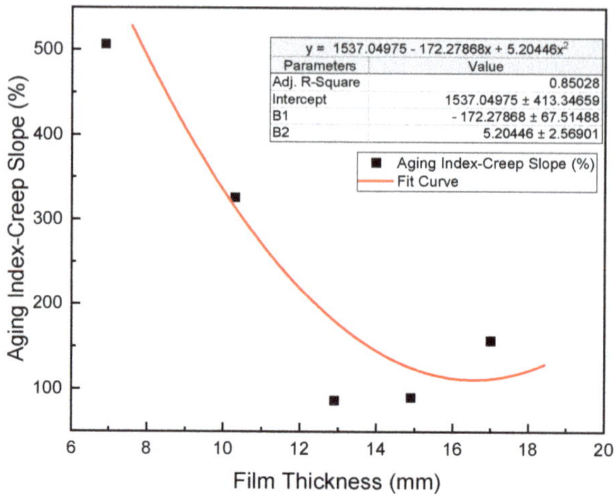

Figure 12. Influence of bitumen film thickness on Agin Index of creep slope.

5. Conclusions and Recommendation

This study aimed to explore the relationship between film thickness and performance variations before and after aging and determine the optimal bitumen film thickness for high-content polymer-modified asphalt (HCPMA) mixtures to ensure satisfactory performance and aging durability by evaluating various properties. A range of experiments, including the Cantabro test, SCB test, SCB fatigue test, and Hamburg wheel-tracking test, were conducted, and the conclusions are as follows:

(1) Considering all of the tests, an insufficient film thickness was found to impair the bonding between aggregates and negatively affect performance, while an excessive film thickness reduced mixture stiffness, cracking resistance and fatigue performance. The optimal film thickness ranges between 12.9 μm and 14.9 μm for various properties, such as adhesive and cohesive performance, raveling resistance, fracture strength, and fatigue performance before and after aging.

(2) The experimental results obtained from the Cantabro test, SCB test, SCB fatigue test, and HWTT revealed a parabolic relationship between the aging index and film thickness, indicating that the increase in film thickness improved the aging durability, but a too-thick film thickness still harmed to the aging durability.

(3) The optimal film thickness, considering performance before and after aging and aging durability, is summarized in Table 4. Based on the results of all of the tests, the optimum film thickness for high-content polymer-modified asphalt (HCPMA) mixtures is within the range of 12.9 to 14.9 μm. This range ensures the best balance between performance before and after aging and aging durability.

Table 4. Optimal Film Thickness for Performance and Aging Durability in Various Test Methods.

Test Method	Optimal Film Thickness		
	Performance before Aging	Performance after Aging	Aging Durability
Cantabro test	12.9 μm	12.9 μm	14.9 μm
SCB cracking test	14.9 μm	12.9 μm	12.9 μm
SCB fatigue test	14.9 μm	12.9 μm	12.9 μm
HWTT test	-	14.9 μm	14.9 μm

Future research endeavors should address the limitations of this study so as to obtain a more comprehensive understanding of HCPMA porous asphalt mixtures. This would involve extending the investigation to encompass a wider variety of HCPMA materials, which would enhance the generalizability of the findings. Additionally, it would be beneficial to consider the impact of fiber reinforcement on the performance and aging behavior of the mixtures. Furthermore, a more in-depth analysis of failure modes during SCB tests could be conducted by using advanced imaging techniques, enabling the differentiation between adhesive and cohesive failures. By addressing these limitations and incorporating these future research directions, it will be possible to optimize the mixture designs, ultimately improving the performance and durability of HCPMA porous asphalt mixtures.

Author Contributions: X.L. and P.L. conceived and designed the experiments; P.L. participated in the experiments and measurements; J.X. and M.L. participated in the discussion of the results; P.L. drafted the manuscript. Monitoring and review were carried out by P.L., S.R., Y.L. and X.L. All authors have read and agreed to the published version of the manuscript.

Funding: This research is supported by the opening funding in 2020 was supported by the Key Laboratory of Transport Industry of Road Structure and Material (Research Institute of Highway, Ministry of Transport).

Institutional Review Board Statement: Not applicable.

Informed Consent Statement: Not applicable.

Data Availability Statement: The data presented in this study are available on request from the corresponding author.

Conflicts of Interest: The authors declare no conflict of interest.

References

1. Kandhal, P.S.; Mallick, R.B. *Open Graded Friction Course: State of the Practice*; Transportation Research Board, National Research Council: Washington, DC, USA, 1998.
2. Lu, X. Chemical and Rheological Evaluation of Ageing Properties of SBS Polymer Modified Bitumens. *Fuel* **1998**, *77*, 961–972. [CrossRef]
3. Yan, C.; Huang, W.; Lin, P.; Zhang, Y.; Lv, Q. Chemical and Rheological Evaluation of Aging Properties of High Content SBS Polymer Modified Asphalt. *Fuel* **2019**, *252*, 417–426. [CrossRef]
4. Liu, Q.; Cao, D. Research on Material Composition and Performance of Porous Asphalt Pavement. *J. Mater. Civ. Eng.* **2009**, *21*, 135–140. [CrossRef]
5. Nakanishi, H.; Hamzah, M.O.; Hasan, M.M.; Karthigeyan, P.; Shaur, O. Mix design and application of porous asphalt pavement using Japanese technology. In *IOP Conference Series: Materials Science and Engineering*; IOP Publishing: Selangor, Malaysia, April 2019; Volume 512, p. 012026.
6. Xia, J.; Zhang, Y.; Xiong, L.; He, S.; Wang, L.; Yu, Z. Opportunities and Challenges of the Sponge City Construction Related to Urban Water Issues in China. *Sci. China Earth Sci.* **2017**, *60*, 652–658. [CrossRef]
7. Lin, P.; Yan, C.; Huang, W.; Li, Y.; Zhou, L.; Tang, N.; Xiao, F.; Zhang, Y.; Lv, Q. Rheological, Chemical and Aging Characteristics of High Content Polymer Modified Asphalt. *Constr. Build. Mater.* **2019**, *207*, 616–629. [CrossRef]
8. Zhang, W.; Jia, Z.; Wang, F. Effect and prediction of aromatic oil on swelling degree of direct-to-plant SBS modifier in bitumen. *Pet. Sci. Technol.* **2019**, *37*, 1033–1040. [CrossRef]
9. Chen, J.-S.; Huang, C.C. Fundamental Characterization of SBS-Modified Asphalt Mixed with Sulfur. *J. Appl. Polym. Sci.* **2007**, *103*, 2817–2825. [CrossRef]
10. Rathi, A.; Elsayed, M.; Krause-Rehberg, R.; Dierkes, W.K.; Noordermeer, J.W.; Bergmann, C.; Trimbach, J.; Blume, A. Effect of aromatic oil on the S-SBR/BR blend components revealed using BDS and PALS. In *Deutsche Kautschuk Tagung*; DKT: Shenzhen, China, 2018.
11. Zhang, F.; Hu, C. Preparation and Properties of High Viscosity Modified Asphalt. *Polym. Compos.* **2017**, *38*, 936–946. [CrossRef]
12. Jing, R.; Varveri, A.; Liu, X.; Scarpas, A.; Erkens, S. Laboratory and Field Aging Effect on Bitumen Chemistry and Rheology in Porous Asphalt Mixture. *Transp. Res. Rec.* **2019**, *2673*, 365–374. [CrossRef]
13. Wu, S.; Pang, L.; Mo, L.; Chen, Y.; Zhu, G. Influence of Aging on the Evolution of Structure, Morphology and Rheology of Base and SBS Modified Bitumen. *Constr. Build. Mater.* **2009**, *23*, 1005–1010. [CrossRef]
14. Hou, Y.; Wang, L.; Wang, D.; Guo, M.; Liu, P.; Yu, J. Characterization of Bitumen Micro-Mechanical Behaviors Using AFM, Phase Dynamics Theory and MD Simulation. *Materials* **2017**, *10*, 208. [CrossRef]
15. Ruan, Y.; Davison, R.R.; Glover, C.J. The Effect of Long-Term Oxidation on the Rheological Properties of Polymer Modified Asphalts. *Fuel* **2003**, *82*, 1763–1773. [CrossRef]
16. Cuciniello, G.; Leandri, P.; Filippi, S.; Lo Presti, D.; Losa, M.; Airey, G. Effect of Ageing on the Morphology and Creep and Recovery of Polymer-Modified Bitumens. *Mater. Struct.* **2018**, *51*, 136. [CrossRef]

17. Mackiewicz, P.; Szydło, A. Viscoelastic Parameters of Asphalt Mixtures Identified in Static and Dynamic Tests. *Materials* **2019**, *12*, 2084. [CrossRef]
18. Yan, C.; Huang, W.; Lv, Q.; Lin, P. Investigating the Field Short-Term Aging of High Content Polymer-Modified Asphalt. *Int. J. Pavement Eng.* **2019**, *22*, 1263–1272. [CrossRef]
19. Yan, C.; Huang, W.; Zheng, M.; Zhang, Y.; Lin, P. Influence of Ageing on High Content Polymer Modified Asphalt Mixture Stripping, Cracking and Rutting Performances. *Road Mater. Pavement Des.* **2020**, *22*, 1824–1841. [CrossRef]
20. Campen, W.H.; Smith, J.R.; Erickson, L.G.; Mertz, L.R. The Relationships between Voids, Surface Area, Film Thickness and Stability in Bituminous Paving Mixtures. *Proc. Assoc. Asph. Paving Technol.* **1959**, *28*, 149–178.
21. Goode, J.F.; Lufsey, L.A. Voids, Permeability, Film Thickness vs. Asphalt Hardening. *Proc. Assoc. Asph. Paving Technol.* **1965**, *34*, 214–222.
22. Kumar, A.; Goetz, W.H. Asphalt Hardening as Affected by Film Thickness, Voids and Permeability in Asphaltic Mixtures. In Proceedings of the Association of Asphalt Paving Technologists Proc, San Antonio, TX, USA, 21 February 1977; Volume 46.
23. Kandhal, P.S.; Chakraborty, S. *Evaluation of Voids in the Mineral Aggregate for HMA Paving Mixtures*; Report No. 9; National Center for Asphalt Technology: Auburn, AL, USA, 1996.
24. Kandhal, P.S.; Foo, P.S.; Mallick, R.B. *A Critical Review of VMA Requirements in Superpave: NCAT Report No. 98-1*; National Center for Asphalt Technology: Tuscaloosa, AL, USA, 1998.
25. McLeod, N. Void requirements for dense-graded bituminous paving mixtures. In *Bituminous Paving Materials*; ASTM International: West Conshohocken, PA, USA, 1959.
26. Sengoz, B.; Topal, A. Minimum voids in mineral aggregate in hot-mix asphalt based on asphalt film thickness. *Build. Environ.* **2007**, *42*, 3629–3635. [CrossRef]
27. Dong, M.; Sun, W.; Li, L.; Gao, Y. Effect of Asphalt Film Thickness on Shear Mechanical Properties of Asphalt-Aggregate Interface. *Constr. Build. Mater.* **2020**, *263*, 120208. [CrossRef]
28. Kandhal, P.S.; Chakraborty, S. Effect of Asphalt Film Thickness on Short- and Long-Term Aging of Asphalt Paving Mixtures. *Transp. Res. Rec.* **1996**, *1535*, 83–90. [CrossRef]
29. Sengoz, B.; Agar, E. Effect of Asphalt Film Thickness on the Moisture Sensitivity Characteristics of Hot-Mix Asphalt. *Build. Environ.* **2007**, *42*, 3621–3628. [CrossRef]
30. Yang, S.; Jiang, J.; Leng, Z.; Ni, F. Feasibility and Performance of the Semi-Circular Bending Test in Evaluating the Low-Temperature Performance of Asphalt Mortar. *Constr. Build. Mater.* **2021**, *269*, 121305. [CrossRef]
31. Zhang, R.; Sias, J.E.; Dave, E.V.; Rahbar-Rastegar, R. Impact of aging on the viscoelastic properties and cracking behavior of asphalt mixtures. *Transp. Res. Rec.* **2019**, *2673*, 406–415. [CrossRef]
32. Zhang, R.; Sias, J.E.; Dave, E.V. Evaluation of the cracking and aging susceptibility of asphalt mixtures using viscoelastic properties and master curve parameters. *J. Traffic Transp. Eng. (Engl. Ed.)* **2022**, *9*, 106–119. [CrossRef]
33. Zhang, R.; Sias, J.E.; Dave, E.V. Comparison and correlation of asphalt binder and mixture cracking parameters incorporating the aging effect. *Constr. Build. Mater.* **2021**, *301*, 124075. [CrossRef]
34. Radovskiy, B. Analytical Formulas for Film Thickness in Compacted Asphalt Mixture. *Transp. Res. Rec.* **2003**, *1829*, 26–32. [CrossRef]
35. Heitzman, M. New Approaches for Computing Film Thickness. *J. Assoc. Asph. Paving Technol.* **2006**, *75*, 1120–1168.
36. Institute, A. *Asphalt Mix Design Methods (2015) MS-2*, 7th ed.; AI: Lexington, KY, USA, 2015; ISBN 978-1-934154-70-0.
37. Van de Ven, M.; Smit, A.d.F.; Krans, R.L. Possibilities of a Semi-Circular Bending Test. In Proceedings of the Eighth International Conference on Asphalt Pavements Federal Highway Administration, Seattle, WA, USA, 10–14 August 1997.
38. Gao, L.; Ni, F.; Ling, C.; Yan, J. Evaluation of Fatigue Behavior in Cold Recycled Mixture Using Digital Image Correlation Method. *Constr. Build. Mater.* **2016**, *102*, 393–402. [CrossRef]
39. Jiang, J.; Ni, F.; Dong, Q.; Wu, F.; Dai, Y. Research on the Fatigue Equation of Asphalt Mixtures Based on Actual Stress Ratio Using Semi-Circular Bending Test. *Constr. Build. Mater.* **2018**, *158*, 996–1002. [CrossRef]
40. Jiang, J.; Ni, F.; Dong, Q.; Zhao, Y.; Xu, K. Fatigue Damage Model of Stone Matrix Asphalt with Polymer Modified Binder Based on Tensile Strain Evolution and Residual Strength Degradation Using Digital Image Correlation Methods. *Measurement* **2018**, *123*, 30–38. [CrossRef]
41. Al-Khateeb, G.; Basheer, I. A Three-Stage Rutting Model Utilising Rutting Performance Data from the Hamburg Wheel-Tracking Device (WTD). *Road Transp. Res. J. Aust. New Zealand Res. Pract.* **2009**, *18*, 12.
42. Lv, Q.; Bahia, H.U.; Huang, W.; Tang, N.; Zheng, M. Effects of Modifiers and Binder Properties on the Performance of Asphalt Mixtures in the Hamburg Wheel-Tracking Device Test. *Asph. Paving Technol.* **2018**. [CrossRef]
43. Zhang, J.; Huang, W.; Zhang, Y.; Lv, Q.; Yan, C. Evaluating Four Typical Fibers Used for OGFC Mixture Modification Regarding Drainage, Raveling, Rutting and Fatigue Resistance. *Constr. Build. Mater.* **2020**, *253*, 119131. [CrossRef]

polymers

MDPI

Article

Multi-Objective Optimization of Epoxy Resin Adhesive for Pavement Toughened by Self-Made Toughening Agent

Huaiqiang Ba [1], Luxin Guo [2], Haiyang Huan [2], Shibo Zhang [2] and Zhiwei Lin [2,*]

1 CRCC Yunnan Investment Co., Ltd., Kunming 650299, China
2 Faculty of Civil Engineering and Mechanics, Kunming University of Science and Technology, Kunming 650500, China
* Correspondence: lzw@kmust.edu.cn

Abstract: Epoxy resin adhesive for pavement is often insufficient in flexibility and toughness. Therefore, a new type of toughening agent was prepared to overcome this shortcoming. To achieve the best toughening effect of a self-made toughening agent on an epoxy resin adhesive, its ratio to the epoxy resin needs to be optimally selected. A curing agent, a toughening agent, and an accelerator dosage were chosen as independent variables. The epoxy resin's adhesive tensile strength, elongation at break, flexural strength, and flexural deflection were used as response values to establish a single-objective prediction model of epoxy resin mechanical property indexes. Response surface methodology (RSM) was used to determine the single-objective optimal ratio and analyze the effect of factor interaction on epoxy resin adhesive's performance indexes. Based on principal component analysis (PCA), multi-objective optimization was performed using gray relational analysis (GRA) to construct a second-order regression prediction model between the ratio and gray relational grade (GRG) to determine the optimal ratio and to validate it. The results showed that the multi-objective optimization using response surface methodology and gray relational analysis (RSM-GRA) was more effective than the single-objective optimization model. The optimal ratio of epoxy resin adhesive was 100 parts of epoxy resin, 160.7 parts curing agent, 16.1 parts toughening agent, and 3.0 parts accelerator. The measured tensile strength was 10.75 MPa, elongation at break was 23.54%, the bending strength was 6.16 MPa, and the bending deflection was 7.15 mm. RSM-GRA has excellent accuracy for epoxy resin adhesive ratio optimization and can provide a reference for the epoxy resin system ratio optimization design of complex components.

Keywords: asphalt pavement; pavement materials; epoxy resin; toughening agent; mechanical properties; response surface method; gray relational analysis

Citation: Ba, H.; Guo, L.; Huan, H.; Zhang, S.; Lin, Z. Multi-Objective Optimization of Epoxy Resin Adhesive for Pavement Toughened by Self-Made Toughening Agent. *Polymers* **2023**, *15*, 1946. https://doi.org/10.3390/polym15081946

Academic Editors: Wei Jiang, Quantao Liu, Jose Norambuena-Contreras and Yue Huang

Received: 13 March 2023
Revised: 15 April 2023
Accepted: 18 April 2023
Published: 19 April 2023

1. Introduction

The steel bridge deck pavement system is an important part of the steel box girder bridge, whose waterproof adhesive layer is critical. Its failure is one of the main factors causing the problem of steel deck pavement [1]. The failure of the waterproof adhesive layer leads to the separation or slippage between the pavement layer and the bridge deck. This allows the steel structure of the steel box girder bridge to be eroded by water, thus causing serious damage to the main structure of the bridge. Using materials with excellent mechanical properties, stability, and deformability in relation to the waterproof bonding layer is a critical measure to avoid adhesive failure and to prolong the life of the steel bridge deck pavement system [2]. At present, two kinds of waterproof adhesives for steel bridge deck pavement are commonly used, namely, epoxy resin adhesive and methacrylate adhesive [3,4]. Epoxy resin adhesive has excellent mechanical properties, high bonding strength, good impermeability, strong corrosion resistance, and other advantages [5,6], and it has been extensively studied.

Zeng et al. [7] studied the components of epoxy resin binder by single factor control variable method. With the increase in the amount of toughening agent, the elongation at break increased, but the tensile strength and elastic modulus decreased. Chen et al. [8] studied the effects of different raw-material compositions on the performance of epoxy resins for road surfaces. They determined reasonable material compositions, as well as ratios, and found that the epoxy resin raw-material composition had a significant impact on the performance of epoxy resin, and its bonding performance index was proportional to the mechanical performance index. Xi et al. [9] studied three types of epoxy resins for waterproof binding and found that the tensile strength and elongation at break data for all three were quite different. Liu et al. [10] studied the basic properties of epoxy resin adhesive for steel bridge deck pavement under different environmental conditions. The tensile strength of the self-made epoxy resin adhesive was 3.7 MPa, the elongation at break could reach 180%, and it had good high- and low-temperature performance, bonding performance, and water permeability resistance. Zhou et al. [11] summarized the results of relevant research on epoxy resin adhesives. They found that the elongation at break of epoxy resin adhesives with high tensile strength is low, and their strength and deformability are often negatively correlated. How to balance the interaction of each component and how to achieve the balance of strength and deformability, two important mechanical indexes, as well as how to prepare epoxy resin with high strength, strong deformability, and good toughness as a waterproof binder, have become the foci of this research.

Response surface methodology (RSM) is an optimization method, integrating experimental design and mathematical modeling proposed by mathematicians Box and Wilson [12], which can be used to solve multifactor and multilevel continuous response problems. Compared with orthogonal experimental and uniform designs, it has the advantage of high accuracy and can also analyze the interaction between influencing factors [13–15]. Epoxy resin binder has several performance indicators with different dimensions to be evaluated that are difficult to balance, which leads to the challenge of determining their optimal ratio. Gray relational analysis (GRA) is introduced to solve multi-objective response problems; it is suitable for solving those involving complex relationships among multiple objectives and factors and can optimize multi-objective responses. Compared with most scholars' single-factor control variable method [16,17], the RSM–GRA method has the advantages of analyzing the interactions of multiple factors and obtaining the best material parameters quickly and accurately [18].

In this study, the dosage of the curing agent, self-made toughening agent, and accelerator were taken as independent variables, and the tensile strength, elongation at break, bending strength, and bending deflection of epoxy resin adhesive were taken as response values. The response surface optimization test was designed by the response surface methodology Box–Behnken design (RSM-BBD) method. RSM was used to establish the single-objective prediction model of each response. The influence of factor interactions on the response values was analyzed, and the single objective optimal ratio was determined. On this basis, the gray relational analysis (GRA) method was introduced to solve the problem of multi-objective response optimization, and the gray relational degree (gray relational grade (GRG)) prediction model was established to optimize the above indicators to obtain the optimal proportion scheme of epoxy resin binder with comprehensive properties. Our results provide a reference for the design of an epoxy resin system with complex components and may guide the further development and engineering application of self-made acrylate copolymer toughening agents.

2. Materials and Methods

2.1. Raw Materials

The epoxy resin used in the study was E-51 bisphenol A epoxy resin, whose epoxy equivalent is 184~195 g/eq, and its viscosity is 10,000~16,000 mPa·s (25 °C), and this was produced by Baling Petrochemical Company, Sinopec., Yueyang, China. The curing agent was tung oil anhydride (TOA) with a viscosity of 5000~15,000 mPa·s (25 °C), and the anhydride equivalent is 120, and it was produced by Shandong Jiaying Chemical Technology Co., Ltd., Qingdao, China. The accelerator was 2,4,6-tris (dimethylaminomethyl) phenol

(DMP-30) with an amine value of 600~630 mg/g and viscosity of 100~300 mPa·s (25 °C), which was produced by Changzhou Shanfeng Chemical Co., Ltd., Changzhou, China.

2.2. Synthetic Toughening Agent

The synthesis reaction was carried out in two steps. Step 1: polyether glycol (PTMG) was dehydrated and vacuumed at 110 °C for 2 h. Then, when the temperature was reduced to 60 °C, a certain amount of toluene diisocyanate (TDI) was added, and the molar ratio of TDI to PTMG was 2:1. The reaction was carried out at 80 °C for 2 h to obtain the NCO-terminated polyurethane prepolymer. Step 2: after the prepolymer was prepared, 1,4-butanediol diglycidyl ether (BDDGE) and a small amount of 2-ethyl-4-methylimidazole (EMI) were added successively, and the reaction temperature was kept at about 160 °C. Samples were taken every 1 h for infrared testing, and the reaction was continued until the NCO infrared absorption peak disappeared. The epoxy value of the final product was 0.23, and it was marked as the self-made epoxy resin toughening agent for pavements (SM-EPT).

2.3. Sample Preparation

The epoxy resin and the toughening agent were preheated in a vacuum drying oven (DZF-1, Beijing Yongguangming Medical Instrument Co., Ltd. Beijing, China) at 80 °C and dried in a vacuum for 30 min, then mixed, and stirred at 80 °C for 5 min at 1000 R/min to obtain component A. The curing agent and the accelerator were mixed and stirred for 5 min at 80 °C and 1000 R/min to obtain component B. The components A and B were mixed and stirred for 10 min at 60 °C and 500 R/min to obtain the epoxy resin binder. Then, the binder was placed in a vacuum drying oven, defoamed at 60 °C for 20 min, and poured into a PTFE mold. The geometric size of the specimen is shown in Figure 1. It was cured at 100 °C/12 h + 120 °C/12 h in a constant-temperature drying oven (101-0ES, Beijing Yongguangming Medical Instrument Co., Ltd. Beijing, China) and then naturally cooled to room temperature for demolding and performance testing. The prepared specimen is shown in Figure 2.

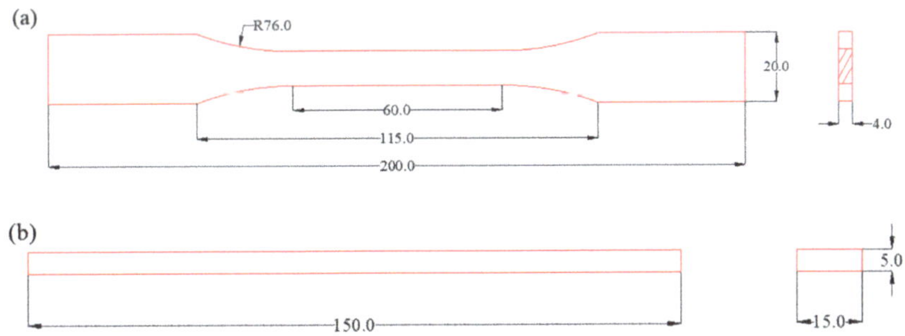

Figure 1. The geometric size of the specimen: (**a**) tensile test specimen, (**b**) bending test specimen.

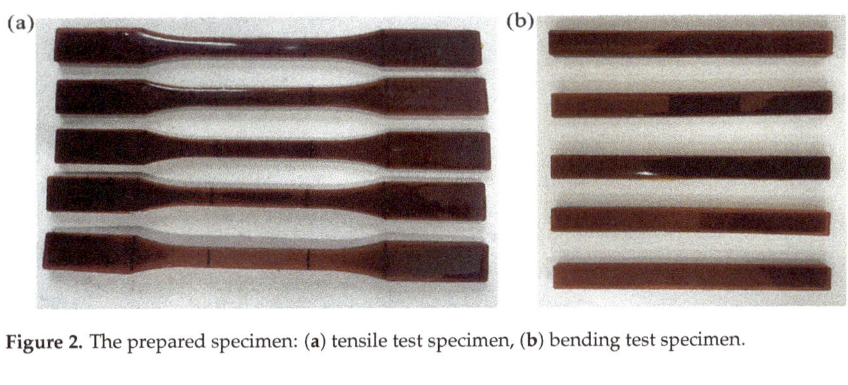

Figure 2. The prepared specimen: (**a**) tensile test specimen, (**b**) bending test specimen.

2.4. Test Method

The mechanical properties of the specimens were tested according to the test method for properties of resin castings (GB/T 2567-2021). An electronic universal testing machine (DDL100, Changchun Research Institute of Mechanical Sciences Co., Ltd., Changchun, China) was used to continuously record the tensile and bending loads. Extensometer (YYU-25/50, Steel Yanke Testing Technology Co., Ltd., Beijing, China) was used to measure the elongation within the gauge length of the tensile test piece. The deflection at the mid-span of the bending specimen was measured by a displacement meter (JC-LVDT type, Liyang Jincheng Testing Instrument Factory, Liyang, China). The loading speed and the bending tests were 10 mm/min. Five samples were tested in each group to obtain the average value.

2.5. Response Surface Experimental Design

Consulting with the relevant research results [19–21], the dosage range of the curing agent was 150~170%, the dosage range of toughening agent was 10~20%, and the dosage of the accelerator was 2~4%. Following the RSM-BBD method, the amount of curing agent, toughening agent, and accelerator (calculated by the mass percentage of epoxy resin, represented by A, B, and C, respectively) were taken as the independent variables, and the tensile strength (σ_t), elongation at break (ε_t), bending strength (σ_f), and bending deflection (S) of epoxy resin adhesive were taken as the response values. A three-factor, three-level response surface test was designed. The factor code and level design are shown in Table 1. The data were analyzed by Design-Expert8.0 statistical software, and the response surface design and test results are shown in Table 2.

Table 1. Test factors and levels.

Level	*A*	*B*	*C*
	Curing Agent/%	Toughening Agent/%	Accelerator/%
−1	150	10	2
0	160	15	3
1	170	20	4

Table 2. Response surface test design and test results.

NO.	Level			Test Results			
	A	*B*	*C*	Tensile Strength/MPa	Elongation at Break/%	Flexural Strength/MPa	Flexural Deflection/mm
1	1	1	0	7.70	22.58	5.03	6.30
2	−1	1	0	6.16	21.35	3.57	6.60
3	0	−1	1	7.77	20.34	4.93	5.28
4	0	0	0	10.70	21.68	6.26	5.86
5	−1	0	−1	6.27	14.70	2.70	4.50
6	−1	0	1	7.96	16.99	4.65	4.70
7	0	−1	−1	8.47	13.79	3.85	3.62
8	0	0	0	10.22	21.66	6.20	5.93
9	1	−1	0	8.40	14.16	4.67	3.94
10	0	0	0	10.40	23.17	5.96	6.32
11	0	0	0	10.39	22.42	6.09	5.80
12	1	0	−1	9.42	12.82	5.14	3.50
13	0	0	0	10.50	22.13	5.95	5.55
14	−1	−1	0	6.47	19.26	3.70	5.20
15	0	1	−1	7.18	22.97	3.75	7.00
16	1	0	1	7.34	13.64	4.08	4.20
17	0	1	1	6.82	23.23	4.21	6.22

3. Results and Discussion

3.1. Construction of the Single-Objective Optimization Model

3.1.1. Model Construction and Model Verification

Multivariate quadratic regression fitting was carried out on the test data to obtain the regression models of σ_t, ε_t, σ_f, and S, as shown in Equations (1)–(4).

$$\sigma_t = -468.105 + 5.302 \times A + 2.248 \times B + 21.595 \times C - 0.002 \times A \times B - 0.094 \times A \times C + 0.170 \times B \times C - 0.015 \times A^2 - 0.069 \times B^2 - 1.159 \times C^2 \quad (1)$$

$$\varepsilon_t = -1024.470 + 12.993 \times A - 5.158 \times B + 32.626 \times C + 0.032 \times A \times B - 0.037 \times A \times C - 0.315 \times B \times C - 0.042 \times A^2 - 0.053 \times B^2 - 3.465 \times C^2 \quad (2)$$

$$\sigma_f = -294.172 + 3.270 \times A + 0.770 \times B + 18.830 \times C - 0.003 \times A \times B - 0.075 \times A \times C - 0.031 \times B \times C - 0.0095 \times A^2 - 0.036 \times B^2 - 1.004 \times C^2 \quad (3)$$

$$S = -198.850 + 2.251 \times A - 0.075 \times B + 4.994 \times C + 0.005 \times A \times B + 0.013 \times A \times C - 0.122 \times B \times C - 0.084 \times A^2 + 0.0018 \times B^2 - 0.824 \times C^2 \quad (4)$$

where σ_t is tensile strength in MPa; ε_t is elongation at break in %; σ_f is bending strength in MPa; S is bending deflection in mm; A is the amount of curing agent,%; B is the amount of toughening agent in %; and C is the dosage of accelerator in %;

To explore the significance of the influence between the factors (independent variables) of the response surface regression model and the response values, the regression model was subjected to analysis of variance, as shown in Table 3. The model was evaluated using the F test. The larger the F value is, the smaller the p value is, which means that the probability of the invalid hypothesis of the model is also smaller, and the model is more significant [22]. The F values of the regression models σ_t, ε_t, σ_f, and S were 66.50, 49.57, 46.32, and 40.49, respectively, and the corresponding p values were less than 0.0001, indicating that the statistical significance of the four models was very high. The p values, corresponding to the F values of the four regression models, were greater than 0.05, indicating that the lack of fit caused by error was not significant. The four models could well describe the relationship between response and factors.

Table 3. Results of ANOVA of response surface model.

	Model σ_t		Model ε_t		Model σ_f		Model S	
	F	p	F	p	F	p	F	p
Model	66.50	<0.0001	49.57	<0.0001	46.32	<0.0001	40.49	<0.0001
A	65.71	<0.0001	19.06	0.0033	52.65	0.0002	23.36	0.0019
B	19.28	0.0032	117.34	<0.0001	0.99	0.3526	162.85	<0.0001
C	3.84	0.0909	22.65	0.0021	16.81	0.0046	7.90	0.0261
AB	0.56	0.4804	18.44	0.0036	1.37	0.2805	4.60	0.0692
AC	51.89	0.0002	0.99	0.3518	51.59	0.0002	1.25	0.3009
BC	0.42	0.5367	18.21	0.0037	2.19	0.1825	29.70	0.0010
A^2	145.06	<0.0001	137.38	<0.0001	85.83	<0.0001	59.78	0.0001
B^2	182.64	<0.0001	13.82	0.0075	78.29	<0.0001	17.90	0.0039
C^2	82.52	<0.0001	93.06	<0.0001	96.58	<0.0001	56.98	0.0001
Lack of Fit	3.83	0.1135	1.93	0.2670	3.95	0.1086	0.47	0.9113

Table 3 shows the relationship between three independent variables (factors) and materials' response values. It can be seen that the order of influence of the three factors A (curing agent dosage), B (toughening agent dosage), and C (accelerator dosage) on the tensile strength model was $A > B > C$, the order of influence of the three factors on the bending strength model was $A > C > B$, and factor A was the most significant in both models. The results show that the amount of curing agent was the main factor affecting the strength of epoxy resin adhesive. *A*, *B*, and *C* influenced the elongation at break model in the order of *B* > C > *A* and influenced the bending deflection model in the order of *B* > *A* > C. Factor *B* was the most significant factor in both models, which indicates that the content of toughening agent was the main factor affecting the deformation ability of epoxy resin adhesive. In the two models of tensile strength and flexural strength, *AC* was significant ($p < 0.05$), while *AB* and *BC* were not ($p > 0.05$), indicating that the interaction of curing agent dosage *A* and accelerator dosage *C* had significant effects on tensile strength and flexural strength. The order of factor interaction significance in the elongation at break model was *AB* > *BC* > *AC*, and the order of factor interaction significance in the bending deflection model was *BC* > *AB* > *AC*.

Table 4 shows the statistical analysis of the fitting accuracy of the four regression models σ_t, ε_t, σ_f, and *S*. The R^2 of the four models is close to 1, indicating that the correlation between the predicted value and the actual value of the four regression models was good. The difference between the calibration coefficient of determination (R^2_{Adj}) and the prediction coefficient of determination (R^2_{Pred}) of the four models was less than 0.2, and the coefficient of variation (CV) for all of them was less than 10%. The signal-to-noise ratio was far greater than 4, which further shows that the fitting error of the four regression models σ_t, ε_t, σ_f, and *S* was small, and the model fitting effect was good [23,24]. Figure 3 shows the comparison results of actual values and predicted values of models σ_t, ε_t, σ_f, and *S*. The predicted values of the four regression models are close to the actual values, and the average deviations between the actual values and the predicted values are 1.68%, 2.23%, 2.57%, and 1.60%, respectively, indicating that the reliability of the model fitting is high. The above analysis shows that the four regression models can accurately describe the functional relationship between the response values and the factors and analyze and predict the test results.

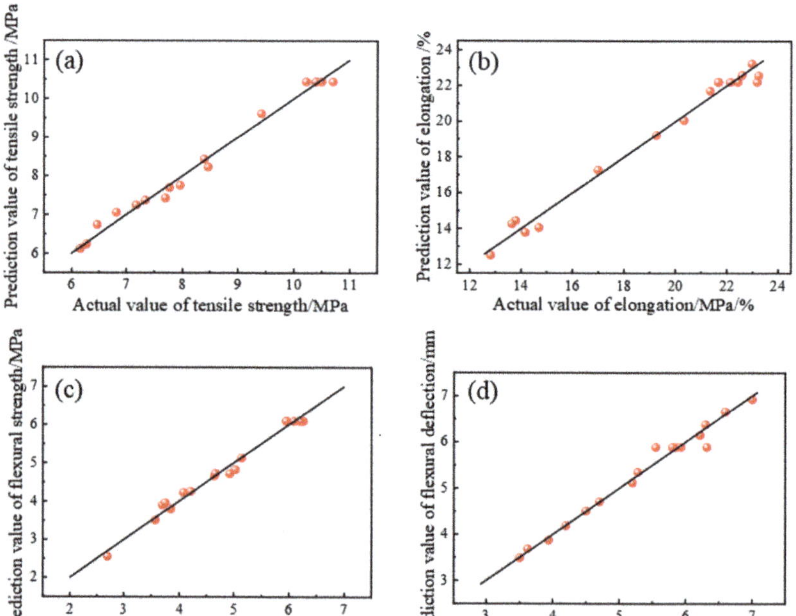

Figure 3. Comparison of actual and prediction values: (**a**) model σ_t, (**b**) model ε_t, (**c**) model σ_f, and (**d**) model *S*.

Table 4. Fit statistics for response surfaces of the model.

Model	R^2	R^2_{Adj}	R^2_{Pred}	CV/%	Adeq. Precision
σ_t	0.9884	0.9736	0.8581	3.13	21.516
ε_t	0.9846	0.9647	0.8441	3.83	18.947
σ_f	0.9835	0.9623	0.7959	4.41	22.049
S	0.9812	0.9569	0.9398	4.20	19.992

3.1.2. Analysis of Three-Dimensional Response Surface Interaction Effect

According to the test results and the variance analysis results in Table 3, the tensile strength σ_t, elongation at break ε_t, bending strength σ_f, and bending deflection S of the epoxy resin adhesive were all affected by the interaction between the factors.

In the regression model of tensile strength σ_t, the F value of the interaction term AC was the largest, and the p value was the smallest, which indicates that the interaction of curing agent dosage A and accelerator dosage C had the most significant effect on the tensile strength of epoxy resin adhesive. Figure 4 is a three-dimensional response surface showing the effect of the interaction between the curing agent dosage A and the accelerator dosage C on the tensile strength (σ_t), with the toughening agent dosage B of 15%. When the amount of accelerator was constant, the tensile strength increased first and then decreased with the increase in the amount of curing agent. The main reason is that when the amount of curing agent is too small, the proportion of epoxy is too high, and a large number of ether bonds are formed in the reaction process, resulting in many irreversible cross-linking points. Thus, the transesterification reaction does not occur easily in the curing system. When the cross-linking density and uniformity of the curing system were reduced, strength decreased [25,26]. When the amount of curing agent was constant, the tensile strength of epoxy resin binder increased first and then decreased with the increase in the amount of accelerator, which is consistent with the research results of Liang Ming et al. [19]. The reason is that a proper amount of DMP-30 accelerator can catalyze the anhydride group, promote the curing and cross-linking of epoxy resin, and relatively improve its mechanical properties. However, the excessive amount of accelerator will make the curing system release too much heat in unit time, resulting in phase separation of the curing system, reducing the cross-linking density, and, thus, reducing the mechanical properties of the resin [27–29]. In summary, simultaneously increasing the amount of curing agent and accelerator within a certain range can effectively improve the tensile strength of epoxy resin binder.

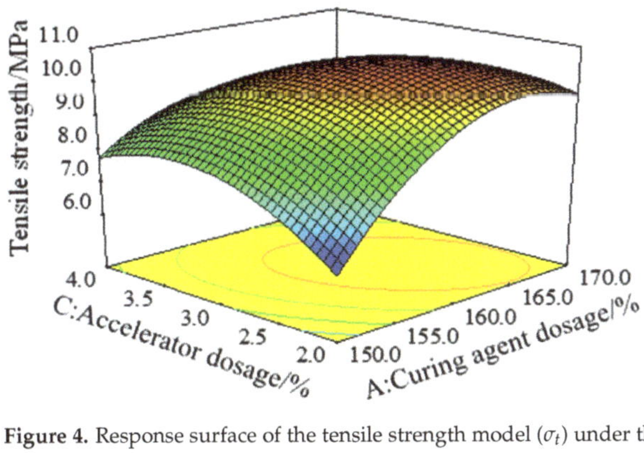

Figure 4. Response surface of the tensile strength model (σ_t) under the interaction of factors A and C.

Figure 5a is a three-dimensional response surface, showing the effect of the interaction between the curing agent dosage A and the toughening agent dosage B on the elongation at break (ε_t), with the accelerator dosage C of 3%. Figure 5b is a three-dimensional response surface of the effect of the interaction between the toughening agent content B and the accelerator content C on the elongation at break (ε_t), with a curing agent content of 160%. When the dosage of toughening agent *B* was constant, the elongation at break increased first and then decreased with the increase in the dosage of curing agent *A* or the dosage of accelerator *C*. When the dosage of curing agent *A* or accelerator *C* was constant, the elongation at break increased with the increase in the dosage of the toughening agent. The variance analysis of the model ε_t results from Table 3 obtained the *p* values of 0.0036 and 0.0037 for *AB* and *BC*, respectively, indicating that the interaction of toughening agent B and curing agent *A*, as well as the interaction between toughening agent *B* and accelerator *C*, had significant effects on the elongation at break. The interaction between toughening agent dosage *B* and curing agent dosage *A* had the most significant effect on the elongation at break. *A* and *B*, as well as *B* and *C* interactions, had significant effects on elongation at break, which may be due to the decrease in cross-link density of the curing system when the toughening agent is used, but the increase in cross-link density can be caused by the addition of proper amounts of curing agent and accelerator. The interactions between the amount of toughening agent and the amount of curing agent, and between the amount of toughening agent and the accelerator, significantly affected the elongation at break [30,31]. Thus, in summary, to improve the elongation at break of epoxy resin adhesive, it is necessary to comprehensively consider the effects of the interaction of the amounts of toughening agent and curing agent, as well as the quantities of toughening agent and accelerator on the elongation at break of epoxy resin adhesive in the process of ratio optimization.

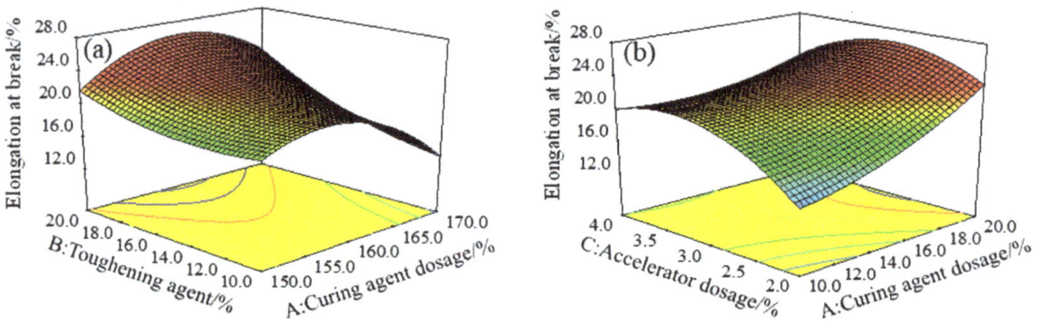

Figure 5. Response surface of elongation at break (ε_t) under the (**a**) interaction of factors *A* and *B*, as well as the (**b**) interaction of factors *A* and *C*.

Figure 6 is a three-dimensional response surface with the effect of the interaction of the curing agent dosage *A* and the accelerator dosage C on the flexural strength (σ_f), with the toughening agent dosage *B* of 15%. It can be seen from Figure 6 that, with the increase in the amount of curing agent or accelerator, the bending strength of the epoxy resin binder increased first and then decreased. According to the variance analysis results of the σ_f model in Table 3, the *AC* term was significant ($p = 0.0002 < 0.05$), and the response surface was steep when the curing agent dosage *A* and the accelerator dosage C changed at the same time, indicating that the interaction of the curing agent dosage *A* and the accelerator dosage C had a significant effect on the bending strength. The reason is that the tertiary amine produced by the decomposition of the DMP-30 accelerator can react with the anhydride group in the anhydride curing agent to produce carboxylate anion, which can catalyze the anhydride curing agent and make the system more easily cross-linked and cured. However, the excessive accelerator will quickly cross-link and cure the system, resulting in the unreacted chain segment that cannot continue participating in the reaction. Macroscopically, the interaction between the amount of DMP-30 accelerator and the amount of anhydride curing agent affects the mechanical properties of epoxy resin adhesive [19,27,28].

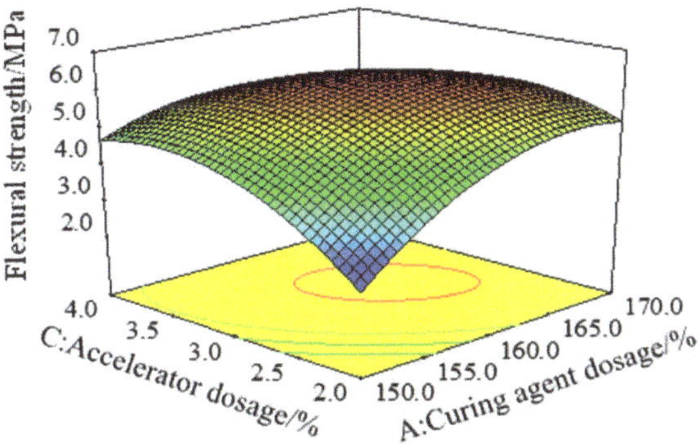

Figure 6. Response surface of bending strength (σ_f) under the interaction of factors A and factor C.

Figure 7 is a three-dimensional response surface with the effect of the interaction of the toughening agent dosage B and the accelerator dosage C on the bending deflection (S) of the epoxy resin adhesive with the curing agent dosage A of 160%. With the increase in the content of toughening agent, the range of change in bending deflection was greater when the content of accelerator was 2% than when it was 4%, indicating that the sensitivity of bending deflection of epoxy resin adhesive to the amount of toughening agent decreased with the increase in the amount of accelerator. According to the variance analysis results of the bending deflection S model in Table 3, the BC term was significant ($p = 0.001 < 0.05$), indicating that the interaction of the toughening agent B and the accelerator C had a significant effect on the bending deflection.

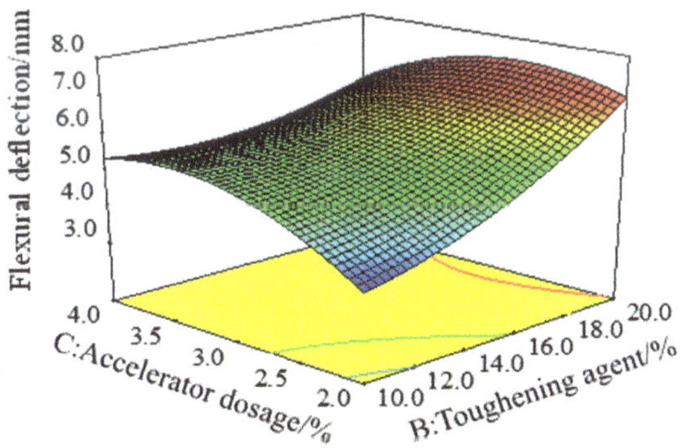

Figure 7. Response surface of bending deflection (S) under the interaction of factor B and factor C.

3.1.3. Single-Objective Optimization Results

The single-objective response was optimized by Design-Expert 8.0 software. The maximum tensile strength of epoxy resin adhesive was 10.61 MPa, the corresponding amount of curing agent was 163.1%, the amount of toughening agent was 14.3%, and the amount of accelerator was 2.8%. The maximum elongation at break of the epoxy resin adhesive was 24.60%, and the corresponding amounts of curing agent, toughening agent, and accelerator were 156.9%, 18.8%, and 3.1%, respectively. The maximum bending strength of the epoxy resin adhesive was 6.17 MPa, and the amounts of curing agent, toughening agent, and accelerator were 162.5%, 14.9%, and 3.1%, respectively. The maximum bending deflection of epoxy resin adhesive was 7.42 mm, the corresponding amount of curing agent was 159.0%, the amount of toughening agent was 20.0%, and the amount of accelerator was 2.8%.

3.2. Construction of Multi-Objective Optimization Model

3.2.1. Calculation Process of Gray Correlation Degree

Gray relational analysis (GRA) can convert multiple optimization objectives into gray relational values by reducing dimensions and then optimizing the gray relational values [32,33]. The larger the gray correlation value is, the better the corresponding response is [34]. The calculation process of the gray correlation degree is as follows:

(1) Normalization. The tensile strength (σ_t), elongation at break (ε_t), bending strength (σ_f), and bending deflection (S) of the epoxy resin adhesive were normalized to eliminate the effect of dimensions on the analysis. The bigger all four are, the better, and the normalization formula is shown in Equation (5).

$$N = \frac{y - \min(y)}{\max(y) - \min(y)} \tag{5}$$

where N is the normalized value of each response, $\max(y)$ is the maximum value of the actual response, $\min(y)$ is the minimum value of the actual response, and y is the actual value of each group of tests.

(2) Calculation of gray correlation coefficient. The gray relational coefficient (GRC) represents the relationship between the test result and the optimal solution [35], and the calculation formulas are shown in Equations (6) and (7)

$$\text{GRC} = \frac{\Delta_{\min} + \xi\Delta_{\max}}{\Delta + \xi\Delta_{\max}} \tag{6}$$

$$\Delta = 1 - N \tag{7}$$

where Δ represents the deviation sequence, ξ is the judgment coefficient, $\xi \in [0, 1]$, and, in this study, ξ is 0.5.

(3) Response weight calculation. To obtain the gray correlation coefficient of each response, it is necessary to calculate the influence weight. Principal component analysis (PCA) quantitatively analyzes the weight of the contribution rate of each target to the response by reducing the dimensionality [36]. With the help of the PCA analysis module of Minitab software, the influence weights of tensile strength σ_t, elongation at break ε_t, bending strength σ_f, and bending deflection S on gray correlation degree are calculated.

(4) Calculation of gray correlation degree. Gray relational degree (GRG) is the weighted sum of gray relational coefficients. The higher the gray relational degree is, the better the corresponding response is. The calculation formula is presented in Equation (8)

$$\text{GRG} = \sum_{i=1}^{n} \beta_i \text{GRC} \tag{8}$$

where $\sum_{i=1}^{n} \beta_i = 1$, and β_i is the weight of the ith response, calculated by PCA.

3.2.2. Calculation Result and Analysis of Gray Correlation Degree

The test results are normalized by Equation (5), and GRC is calculated by Equations (6) and (7). The influence weights of tensile strength σ_t, elongation at break ε_t, bending strength σ_f, and bending deflection S on the gray correlation degree are further calculated by PCA. The results are shown in Table 5. Table 6 shows the results of the gray correlation degree calculated according to Equation (8).

Table 5. Weightiness of response.

Principal Component	Eigenvalue	Weightiness/%
σ_t	2.393	59.813
ε_t	1.482	37.038
σ_f	0.089	2.221
S	0.037	0.928
Sum	4	100

Table 6. Gray correlation degree calculation results.

NO.	GRC				GRG
	σ_t	ε_t	σ_f	S	
1	0.431	0.889	0.591	0.714	0.607
2	0.333	0.735	0.398	0.814	0.488
3	0.437	0.643	0.572	0.504	0.517
4	1.000	0.771	1.000	0.606	0.911
5	0.339	0.379	0.333	0.412	0.354
6	0.453	0.455	0.525	0.432	0.455
7	0.504	0.355	0.425	0.341	0.446
8	0.825	0.768	0.967	0.621	0.806
9	0.497	0.365	0.528	0.364	0.447
10	0.883	0.989	0.856	0.720	0.920
11	0.880	0.865	0.913	0.593	0.873
12	0.639	0.333	0.614	0.333	0.523
13	0.919	0.826	0.852	0.547	0.879
14	0.349	0.567	0.410	0.493	0.433
15	0.392	0.952	0.415	1.000	0.606
16	0.403	0.352	0.449	0.385	0.385
17	0.369	1.000	0.465	0.692	0.608

3.2.3. Construction and Optimization of GRG Response Model

In order to optimize the proportion of epoxy resin binder, the mapping relationship between curing agent dosage A, toughening agent dosage B, accelerator dosage C, and GRG must be established. In this paper, Design-Expert8.0 software was used to establish the second-order mathematical prediction model of GRG, as shown in Equation (9). The comparison between the actual and predicted values of GRG is shown in Figure 8. It can be seen that all data are evenly distributed on a straight line and on both sides, and the average deviation between the fitting value and the actual value was 3.42%, indicating that the fitting effect of the GRG prediction model is good. According to the variance analysis results of the GRG prediction model in Table 7, the model had $p < 0.0001$, indicating that the model was highly significant. The R^2 of the GRG prediction model was 98.39%, the difference between R^2_{Adj} and R^2_{Pred} was less than 0.2, and the coefficient of variation CV was less than 5%, indicating that the model had high reliability and good fitting degree and could be used for subsequent prediction and optimization.

$$\begin{aligned} GRG = & -68.414 + 0.811 \times A + 0.100 \times B + 2.204 \times C + 5.211 \times 10^{-4} \times A \times B - 5.963 \times 10^{-3} \\ & \times A \times C - 3.427 \times 10^{-3} \times B \times C - 2.495 \times 10^{-3} \times A^2 - 5.388 \times 10^{-3} \times B^2 - 0.199 \times C^2 \end{aligned} \quad (9)$$

Table 7. ANOVA results for response surface model of GRG.

Source	Sum of Squares	Mean Square	F	p
Model	0.610	0.068	47.64	<0.0001
A	6.709×10^{-3}	6.709×10^{-3}	4.70	0.0668
B	0.027	0.027	18.98	0.0033
C	1.628×10^{-4}	1.628×10^{-4}	0.11	0.7455
AB	2.715×10^{-3}	2.715×10^{-3}	1.90	0.2103
AC	0.014	0.014	9.96	0.0160
BC	1.175×10^{-3}	1.175×10^{-3}	0.82	0.3945
A^2	0.26	0.26	183.57	<0.0001
B^2	0.076	0.076	53.50	0.0002
C^2	0.17	0.17	116.85	<0.0001
Error	9.991×10^{-3}	1.428×10^{-3}	-	-
Lack of Fit	1.820×10^{-3}	6.068×10^{-4}	0.30	0.8269
	$R^2 = 0.9839$	$R^2_{Adj} = 0.9633$	$R^2_{Pred} = 0.9327$	CV/% = 6.26

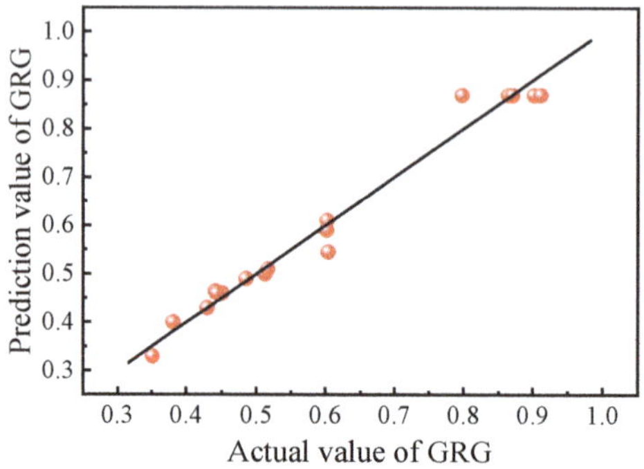

Figure 8. Comparison of actual and prediction values of GRG.

To reflect the influence of the interaction of each factor on the GRG response value, the three-dimensional response surface of GRG under the interaction of different factors was established using the mathematical prediction model, as shown in Figure 9. The shape of the response surface of the GRG model under the interaction of different factors is a quadratic paraboloid with a downward opening, indicating a maximum value of GRG in the test range. According to the variance analysis results of the GRG response model in Table 7, the influence of the three factors on the GRG model was in the order of $B > A > C$. The interaction of curing agent dosage A and accelerator dosage C had the most significant influence on GRG. The optimized GRG response surface model was analyzed using Design-Expert 8.0 software, and the optimal gray correlation degree was 0.885, corresponding to a curing agent dosage of 160.7%, toughening agent dosage of 16.1%, and accelerator dosage of 3.0%.

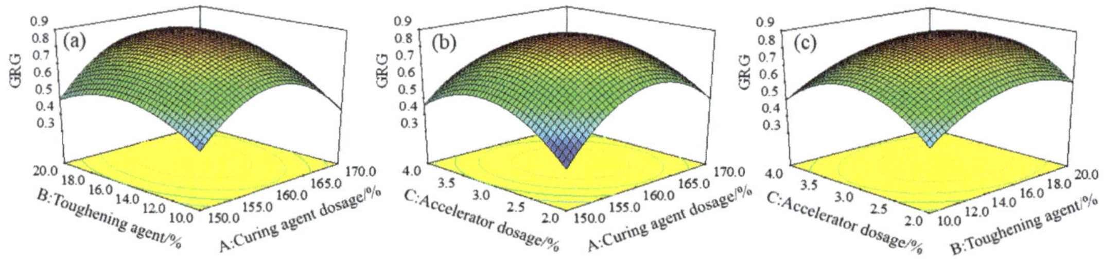

Figure 9. GRG three-dimensional response surface under the interaction of factors (**a**) A and B, (**b**) A and C, and (**c**) C and B.

3.3. Experimental Verification

Through the analysis of the results, the proportion of the optimal gray correlation degree (GRG_{max}) was determined, and the proportion corresponding to the optimal gray correlation degree (GRG_{max}) was selected for comparison with the proportion corresponding to the maximum tensile strength σ_{t-max}, the maximum elongation at break ε_{t-max}, the maximum bending strength σ_{f-max}, and the maximum bending S_{max} deflection obtained by the single-objective prediction model.

The ratio obtained from the single-objective prediction model in Section 3.1.3 and the ratio obtained from the multi-objective optimization model in Section 3.2.3 were experimentally verified, and the experimental verification results are shown in Table 8. Figure 10 shows the stress–strain curve for the tensile test and the load-deflection curve for the bending test. The analysis revealed that the tensile strength obtained by the ratio σ_{t-max} was the largest, the elongation at break obtained by the ratio ε_{t-max} was the largest, the bending strength obtained by the ratio σ_{f-max} was the maximum, and the bending deflection obtained by the ratio S_{max} was the maximum. It can be found that the actual values of tensile strength, elongation at break, bending strength, and bending deflection deviate from the predicted values by 2.7%, 2.0%, 2.8%, and 1.1%, respectively, with actual values higher than the predicted values. This proved the feasibility of the single objective prediction model.

Table 8. Response optimal ratio experimental comparison table.

NO.	A/%	B/%	C/%	Tensile Strength/MPa	Elongation at Break/%	Flexural Strength/MPa	Flexural Deflection/mm	GRG
GRG_{max}	160.7	16.1	3.0	10.75	23.54	6.16	7.15	0.830
σ_{t-max}	163.1	14.3	2.8	10.90	19.73	6.05	5.44	0.794
ε_{t-max}	156.9	18.8	3.1	8.62	25.10	5.30	6.85	0.503
σ_{f-max}	162.5	14.9	3.1	10.21	21.32	6.34	5.63	0.626
S_{max}	159.0	20.0	2.8	8.24	22.20	5.01	7.50	0.408

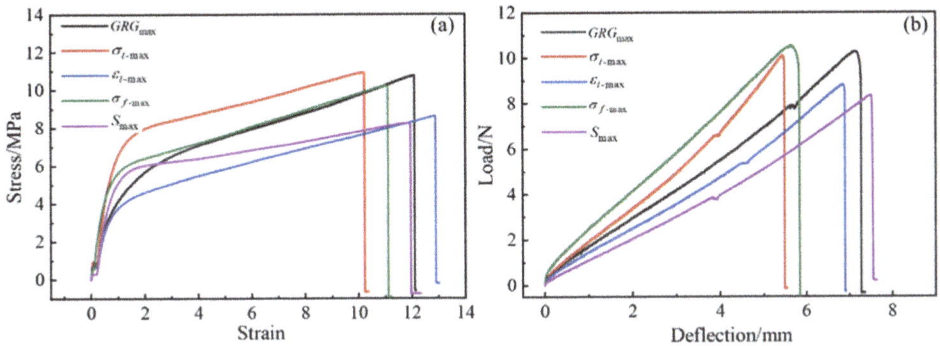

Figure 10. (**a**) Stress–strain curve of tensile test, (**b**) load–deflection curve of bending test.

GRG_{max} was compared to σ_{t-max}, elongation at break increased by 19.31%, bending strength increased by 1.81%, and bending deflection increased by 31.43%. GRG_{max} was compared to ε_{t-max}, tensile strength increased by 24.70%, bending strength increased by 16.22%, and bending deflection increased by 4.38%. GRG_{max} was compared to σ_{f-max}, the tensile strength was increased by 5.29%, the elongation at break was increased by 10.41%, and the bending deflection was increased by 27.00%. GRG_{max} was compared to S_{max}, the tensile strength was improved by 30.46%, the elongation at break was improved by 6.04%, and the bending strength was improved by 22.95%.

The gray relational degrees of GRG_{max}, σ_{t-max}, ε_{t-max}, σ_{f-max}, and S_{max} were 0.830, 0.794, 0.503, 0.626, and 0.408, respectively, and the gray relational degree of GRG_{max} was the highest. The results showed that the optimum ratio of epoxy resin binder was obtained by multi-objective optimization of response surface methodology and gray relational grade. The optimum ratio was 160.7% for the curing agent, 16.1% for the toughening agent, and 3.0% for the accelerator.

4. Conclusions

(1) The amount of curing agent is the main factor affecting the strength of epoxy resin adhesive, and the amount of toughening agent is the main factor affecting its deformability. The interaction of curing agent dosage and accelerator dosage has the most significant effect on the tensile strength and bending strength of epoxy resin binder. The interaction of toughening agent dosage and curing agent dosage has the most significant effect on the elongation at break of the epoxy resin binder, and the interaction of toughening agent dosage and accelerator dosage has a significant effect on bending deflection.

(2) The optimal ratio of epoxy resin binder obtained by RSM-GRA multi-objective optimization was a curing agent dosage of 160.7%, toughening agent dosage of 16.1%,

and an accelerator dosage of 3.0%. An epoxy resin binder with a tensile strength of 10.75 MPa, elongation at break of 23.54%, bending strength of 6.16 MPa, and a bending deflection of 7.15 mm can be prepared with this ratio. The test results show that the RSM-GRA multi-objective optimization model is accurate, effective, and has future application significance for optimizing the epoxy resin binder ratios.

(3) Compared with the single-objective optimization model, in this study, the RSM-GRA multi-objective optimization model was used to obtain the largest gray relational grade (GRG) of the performance indicators of the epoxy resin binder, and the corresponding epoxy resin binder had the best comprehensive performance. The RSM-GRA multi-objective optimization method used in this paper can not only optimize the proportion of epoxy resin binder, but it also provide a reference for the proportion optimization of other complex epoxy resin components.

Author Contributions: H.B.: conceptualization, methodology, writing—original draft, and investigation. L.G.: data curation and writing—review and editing. H.H.: conceptualization, methodology, data curation. S.Z.: conceptualization, methodology. Z.L.: funding acquisition and project administration, supervision. All authors have read and agreed to the published version of the manuscript.

Funding: Science and Technology Project of Transportation Department of Yunnan Province: Science and Education Division [2021] No.39.

Institutional Review Board Statement: Not applicable.

Informed Consent Statement: Not applicable.

Data Availability Statement: Not applicable.

Acknowledgments: The authors thank all project research team members for their help with technical support.

Conflicts of Interest: The authors declare no conflict of interest.

References

1. Zhang, Z.Q.; Tao, J.; Zhang, S.T. Experiment and evaluation on performance of epoxy asphalt waterproof cohesive layer on bridge deck pavement. *J. Chang. Univ. (Nat. Sci. Ed.)* **2011**, *31*, 1–6. (In Chinese)
2. Xu, Y.; Lv, X.; Ma, C.; Liang, F.; Qi, J.; Chou, Z.; Xu, S. Shear Fatigue Performance of Epoxy Resin Waterproof Adhesive Layer on Steel Bridge Deck Pavement. *Front. Mater.* **2021**, *7*, 469. [CrossRef]
3. Kumar, P.; Patnaik, A.; Chaudhary, S. A review on application of structural adhesives in concrete and steel-concrete composite and factors influencing the performance of composite connections. *Int. J. Adhes. Adhes.* **2017**, *77*, 1–14. [CrossRef]
4. Zhang, M.; Hao, P.; Men, G.; Liu, N.; Yuan, G. Research on the compatibility of waterproof layer materials and asphalt mixture for steel bridge deck. *Constr. Build. Mater.* **2021**, *269*, 121346. [CrossRef]
5. Xiang, Q.; Xiao, F. Applications of epoxy materials in pavement engineering. *Constr. Build. Mater.* **2020**, *235*, 117529. [CrossRef]
6. Paluvai, N.R.; Mohanty, S.; Nayak, S.K. Synthesis and Modifications of Epoxy Resins and Their Composites: A Review. *Polym. Plast. Technol. Eng.* **2014**, *53*, 1723–1758. [CrossRef]
7. Zeng, G.D.; Wen, G.X.; Li, J.C.; Ouyang, T.Z.; Yuan, M.; Zhou, Y.; Li, Q.; Wu, C.F. Development of waterproof binder for road and bridge curing flexible epoxy resin at room temperature. *Highway* **2021**, *66*, 310–315. (In Chinese)
8. Chen, Q.; Lu, Y.; Wang, C.; Han, B.; Fu, H. Effect of raw material composition on the working performance of waterborne epoxy resin for road. *Int. J. Pavement Eng.* **2022**, *23*, 2380–2391. [CrossRef]
9. Xi, L.; Huang, W.R. Study on the application of epoxy resin in waterproofing layer of bridge deck pavement. *J. China Foreign Highw.* **2016**, *36*, 292–295. (In Chinese)
10. Liu, P.; Liu, Y.G.; Hao, Z.H.; Sheng, X.Y.; Li, L. Effect of environmental conditions on bonding performance of self-made epoxy resin adhesive. *Appl. Chem. Ind.* **2019**, *48*, 2032–2035. (In Chinese)
11. Zhou, L.; Zhang, D.; Li, X.; Gao, Z.; Chen, Q.; Wang, C. Overview: Application of Resin Waterproof Adhesive Materials in Bridge Deck Pavement in China. *Adv. Civ. Eng.* **2022**, *2022*, 2320374. [CrossRef]
12. Box, G.E.; Wilson, K.B. On the experimental attainment of optimum conditions. In *Breakthroughs in Statistics: Methodology and Distribution*; Springer: Berlin/Heidelberg, Germany, 1992; pp. 270–310.
13. Li, L.; Zhang, Q.; He, Q.; Hu, X.B. Application of response surface methodology in experiment design and optimization. *Res. Explor. Lab.* **2015**, *34*, 41–45. (In Chinese)
14. Cheng, J.L.; Zheng, M.; Lou, J.Q. Comparison of several common optimal experimental design methods. *Res. Explor. Lab.* **2012**, *31*, 7–11. (In Chinese)

15. Li, Z.; Lu, D.; Gao, X. Optimization of mixture proportions by statistical experimental design using response surface method—A review. *J. Build. Eng.* **2021**, *36*, 102101. [CrossRef]
16. Chen, X.Y.; Huang, W.D.; Zhang, W.J.; Lai, Z.P.; Lian, G.F. Multiple targets technology optimization based grey relative analysis of 18Ni300 die steel formed by selective laser melting. *Chin. J. Lasers* **2020**, *47*, 341–351. (In Chinese)
17. Singh, O.P.; Kumar, G.; Kumar, M. Role of Taguchi and grey relational method in optimization of machining parameters of different materials: A review. *Acta Electron. Malays. (AEM)* **2019**, *3*, 19–22. [CrossRef]
18. Deshmukh, S.S.; Jadhav, V.S.; Shrivastava, R. Review on Single and Multi-objective Optimization Process Parameters of EDM Using Taguchi Method and Grey Relational Analysis. In Proceedings of the 9th International Conference of Materials Processing and Characterization (ICMPC), Hyderbad, India, 8–10 March 2019; pp. 3856–3866.
19. Liang, M.; Su, L.P.; Qiu, Z.M.; Xin, X.; Yao, Z.Y.; Ma, C.Y.; Ding, X.M. Effects of DMP-30 on curing kinetics and mechanical properties of epoxy resin/anhydride system. *J. China Univ. Pet. (Ed. Nat. Sci.)* **2021**, *45*, 175–181. (In Chinese)
20. Fu, P.; Tan, X.; Xiao, L.H.; Nie, X.A.; Huang, J.R.; Zhang, L. Study on high-performance tung oil-based anhydride epoxy curing agent. *Resin* **2021**, *36*, 22–30. (In Chinese)
21. Zou, S.L.; Jiang, S.F.; Shen, R.D. Properties of phenolic/epoxy resin curing system modified with acrylic block copolymer. *Adhesion* **2016**, *37*, 51–56. (In Chinese)
22. Wang, J.W.; Wang, W. Response surface based multi-objective optimization of basalt fiber reinforced foamed concrete. *Mater. Rep.* **2019**, *33*, 4092–4097. (In Chinese)
23. He, D.P.; Pan, Z.Q.; Wang, H.G.; Du, W.X. Optimizing the content of activated crumb rubber and SBR composite modified asphalt by response surface methodology-grey relation analysis. *N. Chem. Mater.* **2022**, *50*, 187–191. (In Chinese)
24. Van Den Bergh, D.; Van Doorn, J.; Marsman, M.; Draws, T.; van Kesteren, E.-J.; Derks, K.; Dablander, F.; Gronau, Q.F.; Kucharský, Š.; Gupta, A.R.K.N.; et al. A Tutorial on Conducting and Interpreting a Bayesian ANOVA in JASP. *Annee Psychol.* **2020**, *120*, 73–96. [CrossRef]
25. Zhou, D.W.; Zhai, J.X.; Li, S.; Quan, X.J.; Huo, F.; Li, G. Study on curing behavior and solubility of epoxy/anhydride/DMPA system. *Chem. Res. Appl.* **2017**, *29*, 1760–1764. (In Chinese)
26. Smallenburg, F.; Leibler, L.; Sciortino, F. Patchy particle model for vitrimers. *Phys. Rev. Lett.* **2013**, *111*, 188002. [CrossRef] [PubMed]
27. Han, Y.; Wang, Z.; Zhao, S.; Wang, J. AC impedance function of electrochemical working station as novel curing degree monitor method: A model curing system of epoxy/anhydride/DMP-30. *Measurement* **2019**, *145*, 600–610. [CrossRef]
28. Huang, C.; Ge, Z.; Zhao, B.B.; Wang, Z.; Luo, Y.J. Effects of DMP-30 on curing behavior of epoxy resin/maleicanhydride systems. *J. Chem. Eng. Chin. Univ.* **2017**, *31*, 197–204. (In Chinese)
29. Fan, M.; Liu, J.; Li, X.; Jue, C.; Zhang, J. Curing behaviors and properties of an extrinsic toughened epoxy/anhydride system and an intrinsic toughened epoxy/anhydride system. *Thermochim. Acta* **2013**, *554*, 39–47. [CrossRef]
30. Liu, Z.L.; Li, H.F.; Gu, J.Y.; Wang, D.Z.; Qu, C.Y.; Yang, H.D. Properties of epoxy resin modified with acrylic block copolymer. *Polym. Mater. Sci. Eng.* **2018**, *34*, 24–31+36. (In Chinese)
31. Heng, Z.; Chen, Y.; Zou, H.; Liang, M. Simultaneously enhanced tensile strength and fracture toughness of epoxy resins by a poly(ethylene oxide)-block-carboxyl terminated butadiene-acrylonitrile rubber dilock copolymer. *RSC Adv.* **2015**, *5*, 42362–42368. [CrossRef]
32. Huang, W.; Mao, X.; Wu, Q.; Zhang, J. Experimental Investigation on the Shear Characteristics of Frozen Silty Clay and Grey Relational Analysis. *Sustainability* **2023**, *15*, 180. [CrossRef]
33. Wen, P.; Wang, C.; Chen, M.; Chai, Z. Engineering Property Evaluation and Multiobjective Parameter Optimization of Argillaceous Gangue–Filled Subgrade Based on Grey Relational Analysis. *J. Mater. Civ. Eng.* **2023**, *35*, 04023007. [CrossRef]
34. Hong, Q.; Shi, Y.Y.; Lu, D.N.; Guo, Y.M. Multi-response parameter optimization for the composite tape winding process based on grey relational analysis and response surface methodology. *Acta Mater. Compos. Sin.* **2019**, *36*, 2822–2832. (In Chinese)
35. Xie, Z.; Liu, X.; Liu, Z.; Chen, P.; Lai, B.; Zhan, B.; Lao, J. Human Reliability Analysis of Virus Detection Equipment Based on Entropy Weighting Method and GRA. In Proceedings of the 7th International Conference on Condition Monitoring of Machinery in Non-Stationary Operations (CMMNO), Guangzhou, China, 11–13 June 2021; pp. 263–266.
36. Jiang, R.C.; Ci, S.K.; Liu, D.W.; Sun, H.X.; Wang, D.F. Ply optimization of carbon fiber reinforced plastic control arm based on grey relational analysis. *Acta Mater. Compos. Sin.* **2022**, *39*, 390–398. (In Chinese)

polymers MDPI

Article

Developing Performance-Based Mix Design Framework Using Asphalt Mixture Performance Tester and Mechanistic Models

Jong-Sub Lee [1], Sang-Yum Lee [2,*] and Tri Ho Minh Le [3]

[1] Pavement R&D Office, Korea Expressway Corporation Research Institute, Dongbu-daro 922, Dongtan-myeon, Hwaseong-si 18489, Republic of Korea
[2] Department of Civil Engineering, Induk University, 12 Choansan-ro, Nowon-gu, Seoul 01878, Republic of Korea
[3] Faculty of Civil Engineering, Nguyen Tat Thanh University, 300A Nguyen Tat Thanh Street, District 4, Ho Chi Minh City 70000, Vietnam; lhmtri@ntt.edu.vn
* Correspondence: yummy0220@induk.ac.kr

Abstract: This paper proposes a performance-based mix design (PBMD) framework to support performance-related specifications (PRS) needed to establish relationships between acceptable quality characteristics (AQCs) and predicted performance, as well as to develop fatigue-preferred, rutting-preferred, and performance-balanced mix designs. The framework includes defining performance tests and threshold values, developing asphalt mix designs, identifying available performance levels, conducting sensitivity analysis, establishing the relationships between AQCs and predicted performance, and determining performance targets and AQC values for the three PBMDs using predicted performance criteria. Additionally, the framework recommends selecting the PBMD category for each asphalt layer to minimize pavement distresses. In this study, the proposed PBMD protocol was applied to FHWA accelerated loading facility (ALF) materials using asphalt mixture performance tester (AMPT) equipment coupled with mechanistic models. The study developed nine mix designs with varying design VMAs and air voids using the Bailey method. The cracking and rutting performance of the mix designs were determined by direct tension cyclic (DTC) fatigue testing, triaxial stress sweep (TSS) testing, and viscoelastic continuum damage (S-VECD) and viscoplastic shift models for temperature and stress effects. The study found that adjusting the design VMA was the primary way to achieve required performance targets. For fatigue-preferred mix design, the recommended targets were a cracking area of 0 to 1.9%, a rut depth of 10 mm, and a design VMA of 14.6 to 17.6%. For rutting-preferred mix design, the recommended targets were a cracking area of 18%, a rut depth of 0 to 3.8 mm, and a design VMA of 10.1 to 13.1%. For performance-balanced mix design, the recommended targets were a cracking area of 8.1 to 10.7%, a rut depth of 4.6 to 6.4 mm, and a design VMA of 12.6 to 14.3%. Finally, pavement simulation results verified that the proposed PBMD pavement design with fatigue-preferred mix in the bottom layer, performance-balanced mix in the intermediate layer, and rutting-preferred mix in the surface mix could minimize bottom-up cracking propagation without exceeding the proposed rutting performance criterion for long-life.

Keywords: performance-based; asphalt mix design; performance-related specification; viscoelastic continuum damage model; viscoplastic shift model

Citation: Lee, J.-S.; Lee, S.-Y.; Le, T.H.M. Developing Performance-Based Mix Design Framework Using Asphalt Mixture Performance Tester and Mechanistic Models. *Polymers* **2023**, *15*, 1692. https://doi.org/10.3390/polym15071692

Academic Editors: Wei Jiang, Quantao Liu, Jose Norambuena-Contreras and Yue Huang

Received: 22 January 2023
Revised: 10 March 2023
Accepted: 27 March 2023
Published: 29 March 2023

1. Introduction

The Moving Ahead for Progress in the 21st Century Act (MAP-21) transportation bill emphasizes performance and new innovations and technologies for the transportation system's growth and development [1–3]. The Superpave® mix design system was developed under the strategic highway research program (SHRP). However, there has been a need for performance tests to ensure the satisfactory performance of the asphalt mixtures under in-service conditions [4–8]. A performance-based mix design (PBMD) determines the optimal proportions based on predicted performance, which balances competing demands for cracking and permanent deformation [9,10] because these two distresses are somewhat oppositely driven by the relative quantities of aggregate and asphalt binder [11–13]. This study suggests a PBMD framework that can support performance-related specification (PRS) required for the mathematic relationships between volumetric asphalt mix designs and predicted performance and develop a fatigue-preferred, rutting-preferred, performance-balanced mix designs that can reduce the critical structural distresses for long-life [7,11,13,14].

The PRS is a specification that describes the desired levels of key materials and construction quality characteristics that have been found to correlate with fundamental engineering properties that predict performance [15]. In order to support a PRS system, a PBMD framework can provide the mathematical models that explain the relationship between the acceptance quality characteristics (AQCs) and predicted performance for selected mixtures [2,11,12,16]. The volumetric and/or mechanical AQCs that can be measured during construction are also identified through the PBMD framework [1,7,16–19]. Here, the volumetric AQCs are asphalt mix design requirements (i.e., design air void, binder content, design voids in mineral aggregate (VMA)) measured by volumetric tests, whereas mechanical AQCs are simplified performance indices measured by a mechanical performance tester at the time of construction [3,20]. Through the mathematical models, the role of PBMD in PRS is to provide asphalt mix designers guidance on how to adjust asphalt material and volumetric designs to obtain the required performance targets [7,14,16].

Several state departments of transportation (DOTs) have modified current volumetric asphalt mix designs to better control performance-based on mechanical tests (TRB Circular, 2014). In California, polymer-modified binder in the surface layer and rich asphalt bottom layers were constructed for long-life rehabilitation of the I-710 highway [4,18,21]. Polymer-modified binder mixtures are chosen to improve both rutting and fatigue cracking (one might consider this a performance-balanced mix design), whereas the rich binder mixture is focused on better compaction and more resistance to bottom-up cracking and moisture susceptibility (one might consider this a fatigue-preferred mix design) [22]. Furthermore, New Jersey DOT developed structurally oriented asphalt mix designs to prevent critical pavement distresses [15,17,23–25]: (1) high performance thin overlay mix design using polymer-modified PG 76-22 binder for rutting performance (one might consider this a rutting-preferred mix design), (2) binder-rich intermediate course mix design using at least PG-70-28 with more binder for reflective cracking performance (fatigue-preferred mix design), (3) bridge deck waterproofing surface course mix design using polymer-modified asphalt binder or concentrated thermoplastic-polymeric asphalt modifier for rut and fatigue performances (performance-balanced mix design), and (4) bottom-rich base course mix design using PG 76-28 binder with a minimum of 5% asphalt binder by weight for bottom-up cracking performance (fatigue-preferred mix design). Most DOTs have already recognized the importance of developing performance-based asphalt mix design systems that can address structural distresses for long-life pavements [15,17,23–25].

In pavement engineering, accurate and reliable pavement performance models are crucial for the design and maintenance of roads. However, the calibration and improvement of these models require comprehensive field performance data that are often neglected. Despite the presence of some studies, the importance of in-situ testing and analysis of pavement cracking and rutting performance has not been adequately recognized. Hence, there is a need to emphasize the significance of field performance data in developing and refining pavement performance models. In this regard, numerous studies have highlighted the importance of real-scale measures and performance analysis, as evidenced by recent research [26–28].

The concept of a perpetual pavement aims to practice performance-based mix design (and structural design) that can avert bottom-up cracking propagation so that the periodic surface milling and resurfacing can maintain the overall thick pavement over 50 years [29]. A typical perpetual pavement design might consist the following asphalt layers: (1) a surface layer of 40 to 80 mm thickness for rutting resistance, friction, and permeability (possibly a rutting-preferred mix design), (2) an intermediate layer of 100 to 180 mm thickness for rutting and fatigue cracking resistances (performance-balanced mix design), and (3) a bottom layer of 75 to 100 mm thickness for bottom-up cracking resistance (fatigue-preferred mix design).

In this study, the PBMD framework is developed using asphalt mixture performance tester (AMPT) coupled with mechanistic simplified viscoelastic continuum damage (S-VECD) and viscoplastic shift models. Furthermore, asphalt pavement structural analysis program, layered viscoelastic pavement analysis for critical distress (LVECD) developed at North Carolina State University, was used to model fatigue cracking propagation and permanent deformation.

The objective of this paper is to develop the performance-based mix design framework that can provide guidance on how to adjust asphalt materials and designs to achieve required performance targets based on mechanistically predicted performance results that support performance-related specifications for longer life.

2. Methods

2.1. Current Asphalt Mix Design Methods and Their Requirements

The Hveem mix design method was developed to select the highest asphalt content without exceeding the minimum stability. Hveem evaluates the performance of asphalt mixture by Hveem stabilometer and cohesiometer to measure a resistance to deformation and ravelling. The compacted specimen is fabricated using kneading compactor [30]. Furthermore, the Marshall mix design method was introduced to find optimum asphalt content and density ultimately realized under design traffic. The test specimens are compacted using drop hammer (FHWA, 1988) and the density and optimum binder content were determined by satisfying the criteria of stability and flow tests. Recently, the Superpave® mix design method was developed to increase requirements in aggregate and asphalt binder selection process in addition to the volumetric criteria. Furthermore, it tries to provide rational mix designs applicable for various traffic volumes, axle loads, and climates. Performance tests and models were developed to predict rutting, fatigue cracking, and thermal cracking performance, but were not ready to be implemented at the time. A newer gyratory compaction method was provided because it was found that it orients aggregate particles in the same manner which occurs in the field (Brown, 1989).

The volumetric requirements of the design VMA, and voids filled with asphalt (VFA), and air void have been utilized to control the quality of asphalt mixtures for asphalt mix design for over 25 years. Here, the design VMA is affected by aggregate gradation, aggregate surface texture, aggregate shape, use of manufactured sand, fines and dust content, and so forth. A comparison and contrast of the aggregate and binder selection method, compaction method, and volumetric requirements of the three asphalt mix designs is shown in Table 1.

Table 1. Comparison of asphalt mix designs currently used in the U.S.

Component	Hveem Mix Design	Marshall Mix Design	Superpave Mix Design
Aggregate selection method [31,32]	LA abrasion, sulfate soundness, polishing, crushed face count, flat and elongated particle count	LA abrasion, sulfate soundness, polishing, crushed face count, flat and elongated particle count	Angularity for internal friction, flat and elongated particles for aggregate breakage, clay content for adhesive bond, toughness by LA abrasion test, soundness by sodium or magnesium sulfate test, gradation control points

Table 1. *Cont.*

Component	Hveem Mix Design	Marshall Mix Design	Superpave Mix Design
Asphalt binder selection method [31,33]	Asphalt cement grade for type and geographical location	Asphalt cement grade for type and geographical location	Performance grade by LTPP Bind software and AASHTO Superpave program, for original binder, flash point, rotational viscosity, and dynamic shear rheometer for rolling thin-film oven-aged binder, mass loss and dynamic shear rheometer, for pressure-aging vessel-aged binder, dynamic shear rheometer and bending beam rheometer
Compaction method [7,31]	Kneading	Drop Hammer	Gyratory
Volumetric mix design requirement [7,31]	Hveem stability and air void	Marshall stability, flow, air void, VMA	Air void, VMA, VFA, dust-to-binder ratio

2.2. Development of Performance-Based Mix Design Framework

2.2.1. Performance-Based Mix Design to Support Performance-Related Specification

The first role of the PBMD framework is to support a PRS system by finding the mathematical models of relationships between the volumetric AQCs and their performance before construction. During off-paving seasons, it would be helpful to conduct sensitivity analyses of the volumetric and/or mechanical AQCs on predicted performance so that the most sensitive AQCs and mathematical models can be determined. Full factorials of volumetric mix design scenarios, such as those explored in this study, need not be investigated. However, several alternative mix designs based on the experience and judgment of the contractor can be evaluated. It provides agencies and contractors guidance on how to adjust the AQCs in order to reach the performance targets, as well as a rational resource to calculate incentive or pay factor in the PRS system. Without the PBMD support in PRS, the uncertainty of performance predictions may result in a reluctant agreement between agency and contractor. If possible, they may select limited amounts of mechanical AQCs to improve the accuracy of performance predictions, but there may be significant reluctance to incorporate mechanical performance tests during production and construction.

Figure 1 demonstrates the role of the PBMD as a part of the PRS system in chronological order. At first, PRS performance models and the corresponding performance criteria and pay factor calculation method need to be defined by agency and contractor. Second, the mathematical models of relationships between the volumetric AQCs (i.e., design VMA and air void) in job mix formula or mechanical AQCs (i.e., performance index) measured by equipment such as an asphalt mixture performance tester (AMPT) and the predicted performance should be developed before construction. Not only does this provide guidance on how to adjust the asphalt mix design criteria to reach performance targets in the laboratory, but also on the acceptable variations of the volumetric or mechanical AQCs to control asphalt mixture's quality during production. Lastly, the incentive or pay factor for the PRS system is calculated by comparing as-constructed predicted life with as-targeted predicted life based on the mathematical models in the PBMD phase.

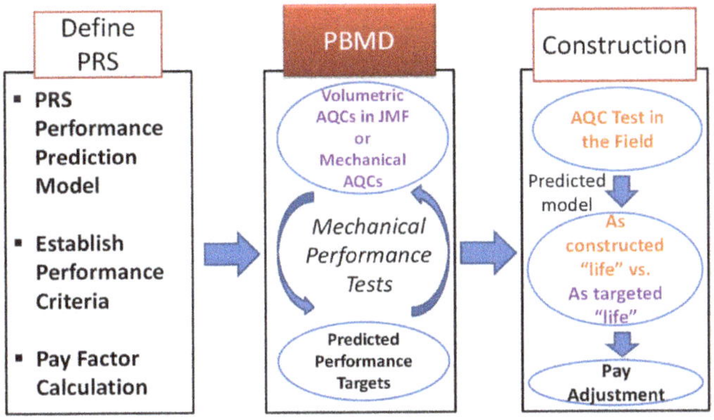

Figure 1. Role of performance-based mix design under the performance-related specification system.

2.2.2. Performance-Based Mix Design to Support Long-Life Pavement

After understanding the means to adjust the volumetric mix design requirements for required performance targets, the second role of the PBMD framework is to develop three PBMD categories of the fatigue-preferred, rutting-preferred, and performance-balanced mix designs. Asphalt pavement structures consist multiple asphalt layers above the subgrade and base unbound layers. Structural analysis programs can consider the effect of the boundary conditions (i.e., pavement stress and strain responses due to traffic loading, temperature gradients along the pavement depth) of each asphalt layer on fatigue cracking and rutting performance. These programs allow asphalt mix designers to select the PBMD category of each asphalt layer that can address critical pavement distresses. To be specific, a binder-rich mix of the fatigue-preferred mix design could be selected to prevent the bottom-up cracking propagation with a lesser consideration of rutting issue in the bottom layer. In contrast, a less binder-rich and a strong aggregate skeleton for rutting-preferred mix design could be chosen to have rutting resistance, more friction, and less permeability in the surface layer. Lastly, based on the related research suggestions and the author's team experience [17], polymer-modified binder mix with a strong aggregate skeleton of the performance-balanced mix design could be placed to prevent reflective bottom-up cracking and permanent deformation in the intermediate layer.

In this paper, the fatigue-preferred mix design is defined as the mix design that sacrifices but does not fail the rutting performance at the threshold value of rut depth, but largely improves the fatigue performance. In the same manner, the rutting-preferred mix design is defined as the mix design that shows the fatigue performance at the threshold value of cracking, but leads to a much better rutting performance. Lastly, the fatigue-rutting performance balance mix design is defined as the mix design that has the best trade-off between fatigue and rutting performance within possible performance targets that asphalt mix designers can develop.

Figure 2 describes the analytical procedure to determine the performance targets and their AQC values for the three mix design types using the fatigue and rutting criteria. The indication ①–④ represents the relation between intersection points of the criterion lines to the Design VMA. First, the relationship between the volumetric AQC and predicted fatigue and rutting performance are cross-plotted. The results of fatigue and rutting performance are separately expressed at the primary and secondary y-axis along with the same AQC values at the x-axis. By adjusting the scales of fatigue and rutting performance results at the primary and secondary y-axis, the fatigue cracking and rutting criteria can be placed at the same horizontal location so that a failure criterion line parallel to the x-axis can be drawn. This is because the normalized scale of fatigue and rutting performance by the failure criteria provides a balanced performance at the cross-point of the fatigue–AQC and rutting–AQC relationship lines. Second, the threshold value of the AQC for the rutting-preferred mix design is determined at the point where the failure criterion line and the fatigue cracking–AQC relationship line intercept. Third, in the same way, those of the fatigue-preferred mix design are determined at the cross-point between the failure criterion line and the rutting–AQC mathematical model. Fourth, the balanced performance targets of fatigue and rutting performance and their corresponding AQC values are found at the cross-point between the fatigue–AQC and rutting–AQC relationship lines.

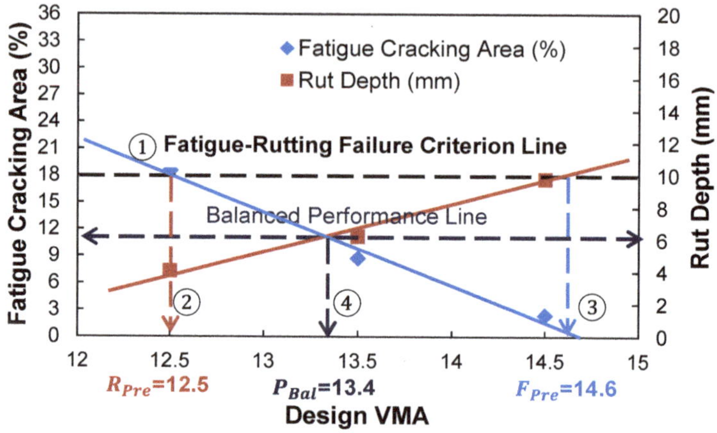

Figure 2. Analytical procedure to determine the performance targets and their volumetric AQC targets of three PBMD categories using the failure criterion and balanced performance lines.

2.2.3. Suggestion of Performance-Based Mix Design Protocol

In order to accomplish the aforementioned roles of the PBMD framework, this study suggests a protocol graphically summarized in Figure 3. The agency and contractor should first define which performance tests and analysis methods will be conducted to predict the fatigue and rutting performance of asphalt pavement. Then, the corresponding performance criteria need to be defined. After selecting asphalt materials, volumetric mix designs need to be developed to vary the design VMA and design air void that asphalt mix designers have under their control. After that, performance tests and analyses are conducted to predict the fatigue and rutting performance of the developed mix designs. In order to identify available performance targets that asphalt mix designer can develop through volumetric changes, the predicted performance of fatigue cracking and rutting are cross-plotted. Then, in order to determine which mix design parameter needs to be adjusted for performance targets by mix designers, a sensitivity analysis of individual volumetric design parameters on predicted performance is conducted. With the sensitivity results, the relationships between the sensitive volumetric AQC and its corresponding predicted performance need to be developed, and then the performance indices are identified by finding their correlations with the mechanistically predicted performance. After investigating the most sensitive fatigue and rutting performance index, the mathematical models of relationship between the determined mechanical AQCs and their corresponding predicted performance need to be optionally developed if the performance testing is available at the time of construction. The volumetric or mechanical AQC-predicted performance models can provide direction to mix designers toward the kinds of adjustment that should be made to achieve the required performance targets.

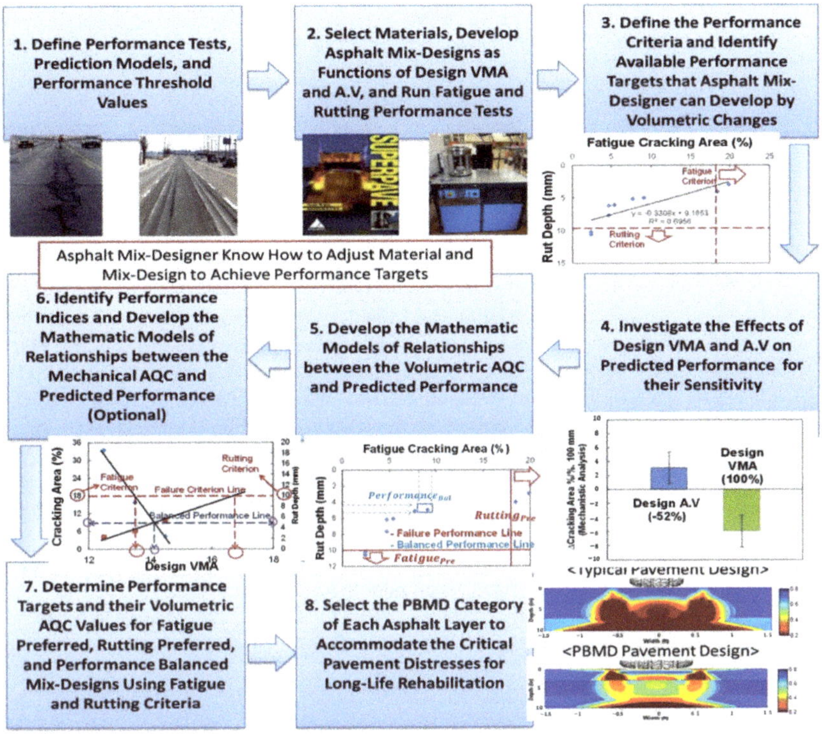

Figure 3. Suggested Performance-based mix design protocol.

The performance targets and their AQC values of the fatigue-preferred, rutting-preferred, and performance-balanced mix designs need to be determined by following the analytical protocol explained in Figure 2. Finally, the PBMD category of each asphalt layer to accommodate the critical pavement distresses for long life can then be selected and the corresponding location within the pavement structure can then be determined. Complete bottom-up cracking propagation that results in a reconstruction of the entire asphalt pavement, along with severe rut depth that causes safety issues, such as hydroplaning and dangerous driving conditions, are considered the critical pavement distresses to be avoided.

3. Application of Suggested Performance-Based Mix Design Protocol

For the application of the suggested PBMD protocol, this study utilizes component materials from the recently reconstructed FHWA accelerated loading facility (ALF) and asphalt mixture performance tester (AMPT) equipment developed under the NCHRP Project 9–29, "Simple Performance Tester for Superpave Mix design." These tools are combined with performance test methods, mechanistic models, and the three dimensional (3-D) pavement structural analysis program developed under the FHWA PRS project "Hot Mix Asphalt Performance-Related Specifications Based on Viscoelastoplastic Continuum Damage (VEPCD) Models" [34]. Using the ALF component materials, systematic experimental volumetric designs with three different VMA and air-void contents were developed using Superpave mix design principles and the Bailey method for aggregate gradation adjustments for VMA targets. After defining the performance threshold values from LVECD software output and developing the relationships between the AQCs and predicted performance, the performance targets and volumetric AQC values of three PBMD categories were determined. Finally, the appropriate PBMD categories of three asphalt layers were determined to avoid significant predicted distresses by comparing the performance results of PBMD pavement design with those of conventionally designed Superpave mixtures.

3.1. Define Performance Tests, Analysis Methods, and Performance Thresholds

Research under the FHWA PRS project has resulted in fatigue and rutting performance test methods and the corresponding mechanistic tests to feed models that can be performed on the AMPT. The fatigue cracking performance is determined by direct tension cyclic (DTC) testing and the simplified viscoelastic continuum damage (S-VECD) model, whereas rutting performance is characterized by triaxial stress sweep (TSS) testing and a viscoplastic shift model. The efficiency of these test methods and models allows the fatigue and rutting performance evaluation of asphalt mixture under a wide range of service conditions in four to five days. The LVECD program is the pavement structural performance prediction model that is compatible with the output from the AMPT. In order to efficiently reduce the computational time of the bottom-up and top-down cracking and rut depth predictions of asphalt pavement with seasonal temperature variation and traffic, a Fourier transform-based layered viscoelastic structural model has been adopted for the development of the LVECD program [35]. It was demonstrated that the LVECD program is capable of differentiating the top-down and bottom-up cracking pattern due to the viscoelastic material properties and boundary conditions of unbound layers under asphalt layers in a literature [15].

The performance threshold values used with LVECD results are based on the rulemaking program of national performance management measures as part of FHWA's the Moving Ahead for Progress in the 21st Century Act [17]. In the rulemaking program, the pavement condition rating threshold values to classify three levels of performance (good, fair, and poor) are suggested as 5 and 10% for surface cracking area and 5 and 10 mm for rut depth.

3.1.1. Testing Protocol

Uniaxial compressive dynamic modulus and the DTC fatigue tests were conducted to describe linear viscoelastic properties and viscoelastic damage characteristics of asphalt mixtures for S-VECD modeling in accordance with AASHTO TP79 and AASHTO TP107, respectively [25,36]. The dynamic modulus tests were completed at 5 °C, 22 °C, and 54.4 °C and at frequencies of 25, 10, 5, 1, 0.5, and 0.1 Hz. A test condition at 22 °C and at 0.01 Hz was added to better facilitate the construction of dynamic modulus master curves. Furthermore, the DTC fatigue tests were completed to fail at approximately 1000, 4000, 16,000, and 64,000 cycles at 18 °C and 10 Hz. Vertical deformations are measured using spring-loaded linear variable differential transformers (LVDTs) during dynamic modulus and DTC fatigue testing.

Triaxial repeated load permanent deformation (TRLPD) and triaxial stress sweep (TSS) tests were carried out to calibrate a reference permanent strain evolution curve (hereinafter called reference curve), reduced load time (time-temperature) shift factors, and vertical stress (time-stress) shift factors for viscoplastic shift modeling [24]. In addition to the rigorous mechanistic rutting analysis, incremental repeated load permanent deformation (iRLPD) tests were performed to construct minimum strain rate (MSR) master-curves used for rutting performance indices [8]. Three different deviatoric stresses are applied within one specimen to reduce the number of tests for the TSS and iRLPD tests. All permanent deformation tests were completed to apply cyclic haversine load followed by a rest period under confining pressure where the test specimens were enclosed with a latex membrane. The TRLPD data present a single loading block, including six hundred cycles with a single deviatoric stress and load time, while the TSS and iRLPD data demonstrate three and four loading blocks containing two hundred cycles of each loading block with different deviatoric stresses and a single load time in each. Detailed permanent deformation test conditions used in this study are provided in Table 2 [25,31,36]. Vertical deformations were measured using an AMPT actuator LVDT during iRLPD, TRLPD, and TSS testing and a steel ball was placed between the top platen and reaction frame which is different from the conventional flow number test protocol. The air-void target of the cylindrical specimens used for all fatigue and rutting tests was 7% +/− 0.5%.

Table 2. TRLPD, TSS, and iRLPD test conditions.

Test Type	Viscoplastic Shift Modeling		MSR Master-Curve
	TRLPD	TSS	iRLPD
Testing temperature (°C)	54	54, 40, 20	54
Confine pressure (kPa)		68.95 (10 psi)	
Pulse time (s)	0.4	0.4	0.1
Rest period	10	10 at 54 °C, 1.6 at 40 °C, 20 °C	0.9
Deviatoric Stress (kPa)	689.5 (100 psi)	482.6 (70 psi), 689.5 (100 psi), 896.3 (130 psi)	200 (29 psi), 400 (58 psi), 600 (87 psi), 800 (116 psi)
Number of cycles for each loading block	600	200	500
Testing time (min)	104	104 at 54 °C, 20 at 40 °C and 20 °C	35

3.1.2. Fatigue Cracking Analysis Methods

Traditionally, bending beam fatigue tests have been conducted to determine classical fatigue relationships between several tensile strain levels at the bottom of the beam specimen and numbers of cycles to failure at three different temperatures. The main advantage of rigorous mechanistic S-VECD model coupled with the DTC fatigue tests allows the reliable prediction of fatigue relationships within one to two days. The S-VECD model is based on Schapery's work of potential theory that uses internal state variable of damage S as following damage evaluation law [30].

$$\dot{S} = \left(-\frac{\partial W^R}{\partial S}\right)^{\alpha} \tag{1}$$

where,

S = internal state variable (damage),
W^R = total dissipated pseudo strain energy, and
α = damage evolution rate.

The main output from S-VECD is a damage characteristic curve that explains a relationship between pseudo-stiffness and a quantified damage state regardless of a strain level and temperature. The damage curve expresses the integrity of asphalt mixture as damage grows. The damage state at failure is determined at a peak phase angle during the DTC testing. Figure 4 demonstrates the mechanistic fatigue cracking analysis method used in this study to calculate the fatigue cracking damage percentage of the pavement cross-section cut through a vertical plane normal to the direction of traffic loading. The LVECD program can simulate the pseudo-stiffness (material's integrity) accumulating damage reduction at all nodal points of the asphalt pavement depth and width. The pseudo-stiffness values at the failed damage conditions determined by observing the peak phase angle are considered as fatigue cracking initiation based on the S-VECD theory. Thus, the fatigue cracking area can be quantified by calculating a ratio of a number of nodal points that contain pseudo-stiffness values below the failure point relative to all nodal points in the designated area of the pavement structure. In this study, it is the pavement cross-section of ±0.5 m in width from the center of the loading tire through the whole asphalt thickness.

Figure 5a,b describe fatigue performance indices that can be directly measured during the DTC tests without mechanistic analysis or post-processing of the data. In this study, two fatigue performance indices were evaluated: (1) the number of cycles to failure at a certain AMPT strain input value into the control software (this input may be different depending on a type of asphalt mixture) and (2) the average peak phase angle value at failure. These were investigated to identify the most sensitive fatigue index that has a strong correlation with mechanistic performance but requires significantly less time.

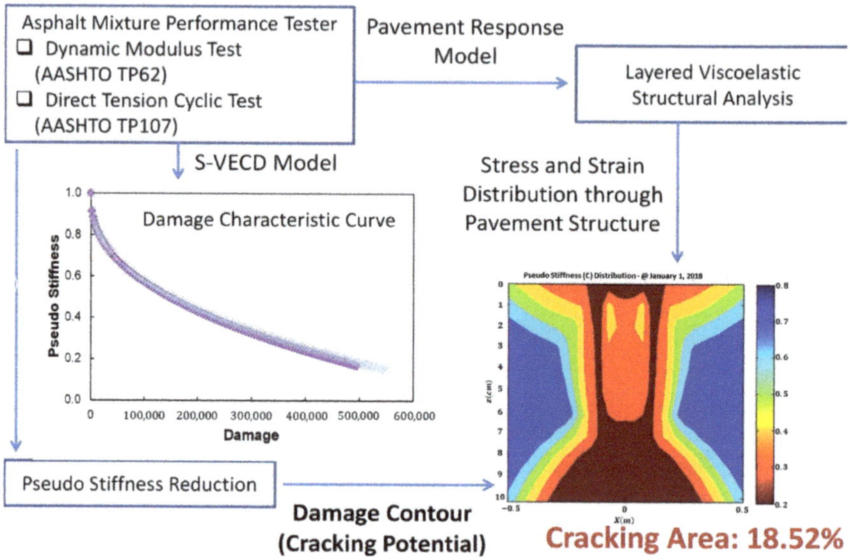

Figure 4. Mechanistic fatigue cracking area prediction method.

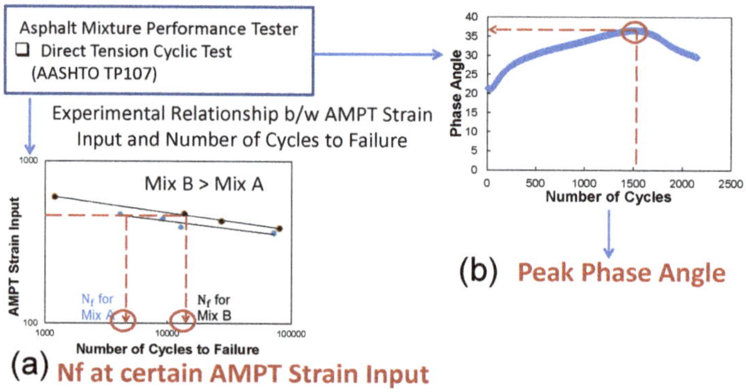

Figure 5. Fatigue cracking index method without mechanistic analysis.

3.1.3. Rutting Analysis Methods

In mechanistic-empirical pavement design and analysis, permanent to resilient strain ratio models have been used to predict pavement rut depth as functions of a temperature, number of loading cycle, and resilient strain calculated from dynamic modulus tests. Since the strain ratio model does not take into account the effect of different deviatoric stress caused by various traffic loads, there is a need to develop a new viscoplastic strain model that can consider the effects of temperature, load time, and deviatoric stress. Recently, the viscoplastic shift model has been developed to predict permanent strain at any temperature, load time, and deviatoric stress using time-temperature and time-stress superposition principles. The shift model consists a reference permanent strain growth master curve, a so-called reference curve, a reduced load time (time-temperature superposition) shift factors, and vertical stress (time-stress superposition) shift factors. The reference curve is fitted using an incremental model as expressed in Equation (2). Choi and Kim found that the slopes of incremental permanent strains in log–log scale measured from each loading block of TRLPD and TSS tests are identical regardless of reduced load time and deviatoric stress [24]. This observation allows nine loading blocks of TSS tests measured at three different temperatures and at three different deviatoric stresses horizontally translating to the reference curve so that the reduced time shift factors and stress shift factors can be calculated as expressed in Equation (3). Each shift factor is calculated as a ratio of a number of cycles at the reference curve (ΔN1) to a number of cycles at the TSS loading block that produces the identical amount of permanent strain (ΔN2), as shown in Figure 6. The detailed incremental model and shift factor functional forms and analysis procedure can be found in an article by Choi and Kim [24].

$$\varepsilon_{vp} = \frac{\varepsilon_0 N_{ref}}{(N_I + N_{ref})^{\beta}} \quad (2)$$

$$\begin{aligned} N_{ref} &= N \times 10^{a_{total}} \\ a_{total} &= a_{\xi_p} + a_{\sigma_v} \\ a_{\xi_p} = p_1 \log(\xi_p) + p_2,\ a_{\sigma_v} &= d_1 \left(\frac{\sigma_v}{P_a}\right)^{d_2} + d_3 \end{aligned} \quad (3)$$

where,

ε_{vp} = viscoplastic (permanent) strain,
$\varepsilon_0, N_I, \beta$ = incremental model coefficients
N_{ref} = number of cycles at the reference loading condition,
N = number of cycles at certain loading condition,
a_{total} = total shift factor,
a_{ξ_p} = reduced load time shift factor,
a_{σ_v} = vertical stress shift factor,
ξ_p = reduced load time,
p_1, p_2 = regression parameters for reduced load time shift factor,
σ_v = vertical stress,
p_a = atmospheric pressure (i.e., 14.7 psi or 101.3 kPa), and
d_1, d_2, d_3 = regression parameters for reduced stress shift factor.

The output of the viscoplastic shift model are direct inputs to the LVECD program to predict pavement rut depth along with the EICM climate data, traffic, pavement thickness, and stiffness of unbound layers.

As was described above for fatigue, simple rutting performance indices that can be directly measured by AMPT without a mechanical structural analysis and data processing were also investigated. First, Azari and Mohseni [8] suggested constructing a minimum strain rate (MSR) master-curve that can calculate the "b" coefficient as expressed in Equation (4). Here, the minimum strain rates are measured at the last 50 cycles of each loading group in the iRLPD tests. The "b" coefficient is a slope of relationship between the MSR and TP temperature-and-pressure.

$$MSR = a \times (T \times P)^{b} \quad (4)$$

where,

a, b = power function coefficients,
T = temperature (°C), and P = deviatoric stress (MPa).

The "b" coefficient is suggested as a performance indicator to characterize rutting performance. It is calculated at the MSR value at 600 kPa by assuming the "a" coefficient is 0.001, as shown in Figure 7a. In the same way, the "b" coefficient can be determined from the TSS test by assuming that the "a" coefficient is 0.001 at the MSR value of 689.5 kPa, as shown in Figure 7b. The other simple rutting index is the total accumulated permanent strains measured from the iRLPD, TRLPD, and TSS tests as described in Figure 7c–e.

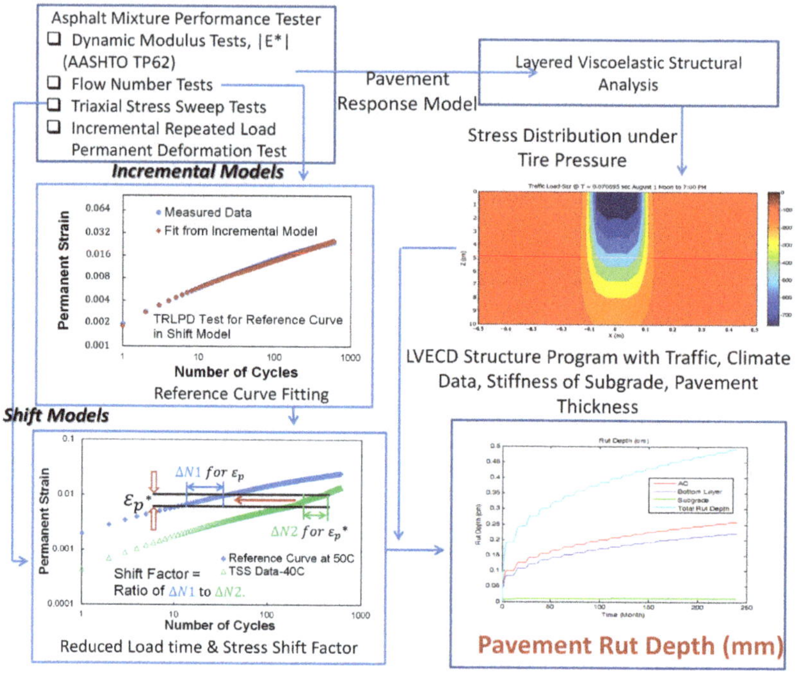

Figure 6. Mechanistic rut depth prediction method.

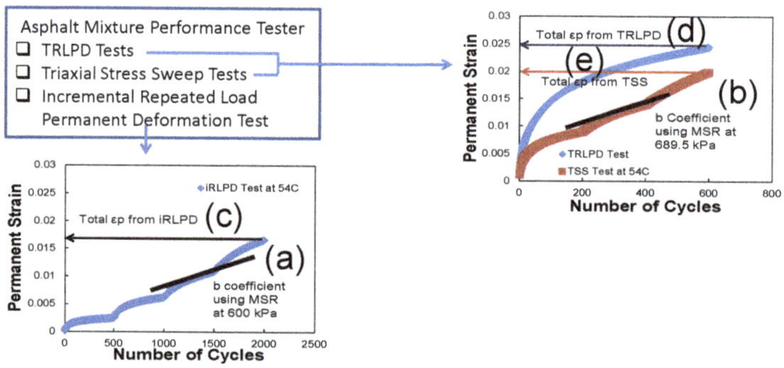

Figure 7. Rutting index method without mechanistic analysis: (a): b coefficient-MSR 600 kPa, (b): b coefficient-MSR at 689.5 kPa, total εp from (c): iRLPD, (d): TLRPD; (e): TSS.

3.1.4. Define Fatigue Cracking and Rut Depth Performance Thresholds

In this study, the rulemaking program of the national performance management measures as part of FHWA's response to the Moving Ahead for Progress in the 21st Century Act (MAP-21) is used to define the fatigue and rutting performance threshold values. The proposed pavement condition rating thresholds are shown in Table 3. Based on the full-scale empirical mechanistic test [17] and the road and pavement expert's suggestions, the extracted value were generated based on both asphalt pavement and asphalt pavement and jointed concrete pavement conditions under the AADT of 8000. Here, the failure criteria of rut depth (mm) and surface cracking area (%) of asphalt pavement are 10 mm and 10%, respectively.

Table 3. Proposed pavement condition rating thresholds from rulemaking.

Surface Pavement Type	Metric	Metric Range	Rating
Asphalt pavement	Rutting	<5 mm (0.2 in)	Good
		5 mm (0.2 in) to 10 mm (0.4 in)	Fair
		>10 mm (0.4 in)	Poor
Asphalt pavement and jointed concrete pavement	Surface cracking percentage	<5%	Good
		5 to 10%	Fair
		>10%	Poor

The LVECD output for rutting performance is the calculated rut depth (mm) which has identical units in the rulemaking program. Therefore, the 10 mm of rut depth is determined as the rutting failure criterion. However, the current LVECD output for fatigue performance is the cracking propagation percentage of pavement cross-section. Therefore, the fatigue cracking damage percentage computed from the damage contour within the asphalt layer simulated by the LVECD program is determined as a failure criterion. A transfer function study is under development to convert the cracking damage area (%) of pavement cross-section to surface cracking area (%).

3.2. Asphalt Mix Designs with a Wide Variation in Design VMA and Air Void

One of the FHWA ALF test sections reconstructed in 2013 is a Superpave 12.5 mm nominal maximum aggregate size (NMAS) asphalt mixture with PG 64-22 binder and 22% of reclaimed asphalt pavement (RAP) by weight (20% recycled binder ratio). This mix was selected to base the suggested performance-based mix design framework. Using the same component materials (large stocks are available), new mix designs were developed to vary the asphalt mix design requirements for design VMA and design air void that are ordinarily under an asphalt mix designer's control. Since the minimum volumetric criterion for a 12.5 mm NMAS mixture is 14% VMA, a slightly higher and lower design VMAs of 15% (as-constructed the ALF mix at design air void of 4%) and 14% were produced using the Bailey method. The Bailey method provides guidance to asphalt mix designers to determine different aggregate gradations that can predict design VMAs with aggregate packing concepts [29]. Loose unit weight and rodded unit weight tests are required to quantify the aggregate packing in accordance with AASHTO T 19 [37].

After determining three different trial aggregate gradations, they were produced with two binder content alternatives at design air-void contents of 3 and 5% that try to meet the VFA requirement of 65% to 78% for design traffic ESALs of 0.3 to 3 million. The three different aggregate gradations and nine mix designs developed are summarized in Figure 8 and Table 4. Details can also be found in another article by Lee et al. [17]. The alphabetic identifiers from Mix-A to Mix-I shown in Table 4 are used to distinguish the performance results of individual mix designs.

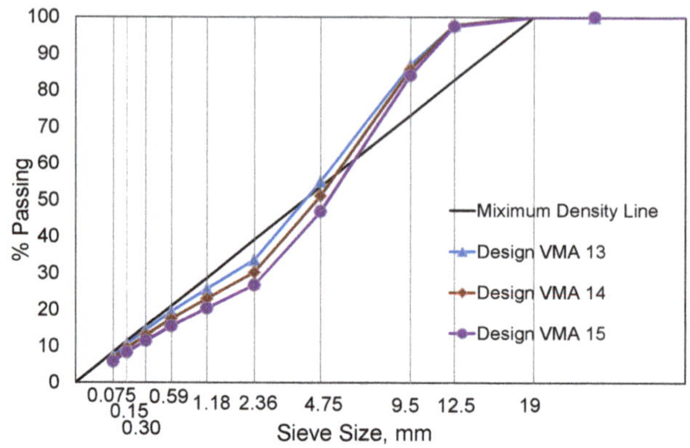

Figure 8. Three different aggregate gradations for the design VMAs.

Table 4. Volumetric mix designs for three different design VMAs and design air-void contents.

(%)	Aggregate Gradation 1 (VMA 15 Target)			Aggregate Gradation 2 (VMA 14 Target)			Aggregate Gradation 3 (VMA 13 Target)		
Design VMA by volume	15	14.5	14.7	14.1	13.5	13.7	12.9	12.5	12.8
Design AV by volume	5.3	3.8	3	4.9	3.7	2.9	5.1	3.9	3.1
Binder content by weight *	4.2	4.5	4.9	3.8	4.1	4.4	3.2	3.6	3.9
Gmm	2.769	2.754	2.735	2.775	2.760	2.746	2.803	2.783	2.769
VFA by volume	64.7	73.8	79.6	65.2	72.6	78.7	60.5	68.8	75.8
Performance specimen AV	7 Mix-A	7 Mix-B	7 Mix-C	7 Mix-D	7 Mix-E	7 Mix-F	7 Mix-G	7 Mix-H	7 Mix-I

* the specific gravity of the aggregate is very high at nearly 3.00, and thus, if it were a typical stone near 2.70, then the binder contents by weight would be approximately 0.5% higher.

3.3. *Define the Performance Criteria and Identify Achievable Performance Targets*

After completing the asphalt mix designs as functions of volumetric AQCs (design VMA and air void), the performance results of nine mix designs were investigated to identify achievable performance targets that asphalt mix designers can develop with reasonable volumetric changes. In this phase, the contrary relationship between fatigue and rutting performance may provide information on the best and worst performance targets the asphalt mix designers can achieve with a given set of materials. Furthermore, it may provide an indication to change aggregate gradation or modify virgin asphalt binder type to reach performance targets required by the agency and contractor. The LVECD program was utilized to predict the bottom-up and/or top-down fatigue cracking propagation as well as the rut depth of the nine mix designs. Seasonal temperature variations of the Washington D.C area were modeled with the Enhanced Integrated Climate Model (EICM) database and traffic was simulated with 4000 average annual daily truck traffic (AADTT) that provided about 4 million equivalent single axle loads (ESALs) in 5 years. The asphalt layer was 101.6 mm of (4 in) thick pavement. A single moving wheel load was simulated with a design velocity of 60 km/h, axle load of 45 kN (10 kips). The elastic moduli of for the subgrade were 70 MPa (10 ksi) and 150 MPa (22 ksi).

Figures 9 and 10 show the asphalt cross-section damaged area (%) and rut depth (mm) results modeled by the LVECD program. It is found that the results of the Mix-B, C, E, and F indicates that only bottom-up cracking propagations of 10 mm are observed when the fatigue damaged area was below 5%. Complete cracking damage was combined with bottom-up and top-down propagations for Mix-G and H when the fatigue damaged area was above 18%. Therefore, the fatigue cracking damaged area of 18% is selected as the fatigue cracking failure criterion from the LVECD program. In terms of rutting performance criterion, the accumulated asphalt layer rut depth of 10 mm is selected following the national rulemaking program.

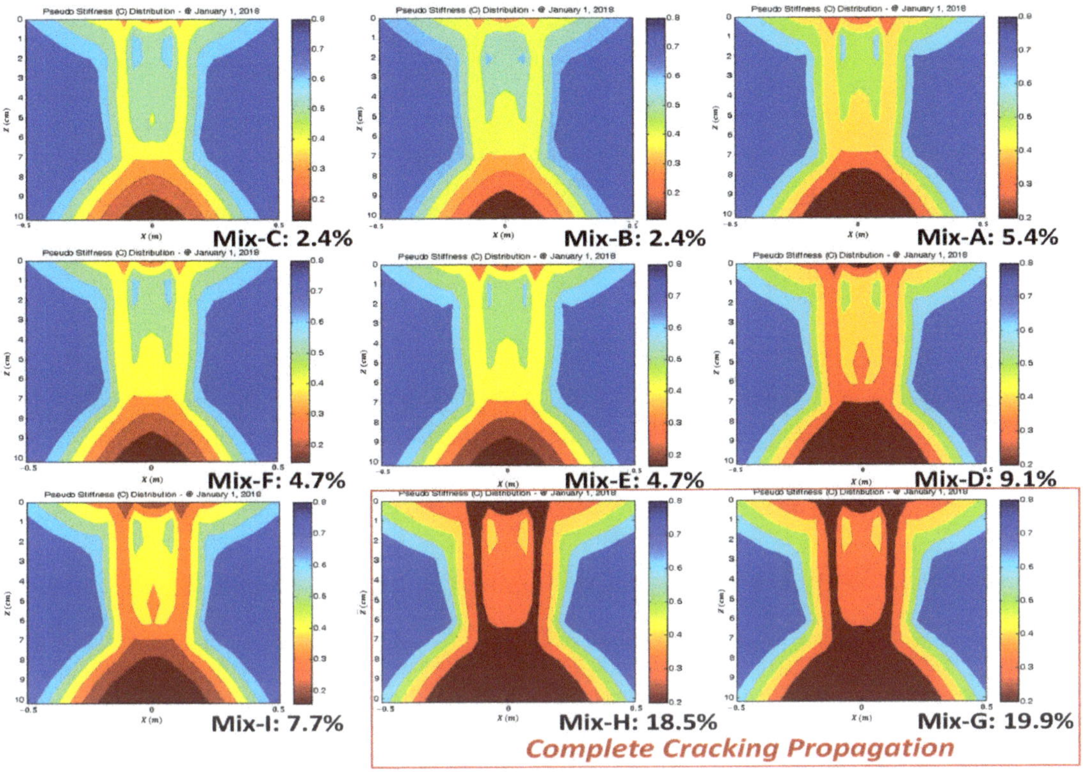

Figure 9. LVECD fatigue performance results: fatigue damaged area (%) results of nine mix designs at 101.6 mm (4 in) of thick pavement.

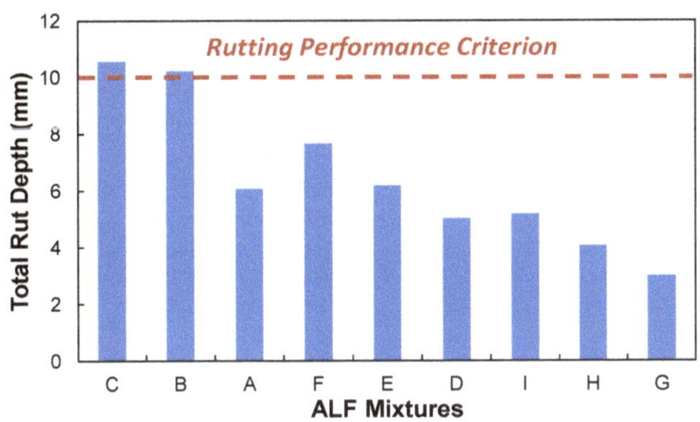

Figure 10. LVECD rutting performance results: rut depth (mm) of nine mix designs at 101.6 mm (4 in) of thick pavement.

Figure 11 presents the fatigue and rutting predictions cross-plotted that shows inverse relationships. The relationship can give information on the possible fatigue-preferred, rutting-preferred, and performance-balanced mix designs that are achievable by a mix designer with a given set of materials. Depending on circumstances, it may be concluded that the current asphalt mix design may need to modify the binder to obtain better engineering properties.

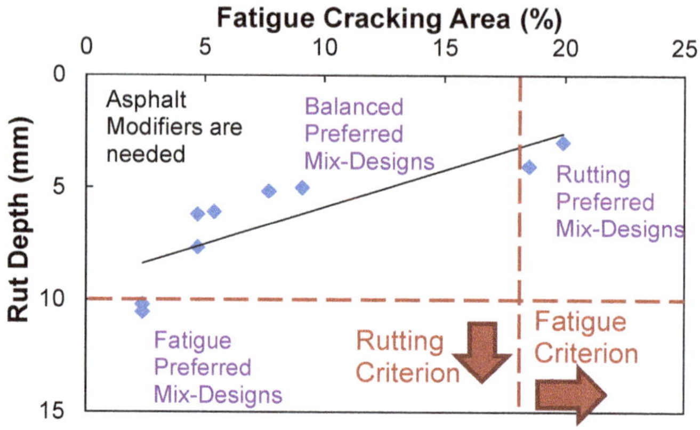

Figure 11. Fatigue and rutting performance cross-plot of the ALF mix designs.

In the next section, the sensitivity analysis results for each volumetric AQC will be discussed; it provides guidance on the extent in which design VMA and air void need to be changed to reach certain performance targets.

3.4. Investigate the Effects of Design VMA and Design Air Void on Predicted Performance for Their Sensitivity

A sensitivity analysis of design VMA and design air void was conducted to determine which volumetric parameter can be modified to meet required performance targets. The sensitivity was quantified by calculating the average slopes of linear relationships between the volumetric AQCs and predicted performance. The slope indicates how much the predicted fatigue and rutting performance can be changed with an increase of one percentage of the design VMA and air-void content. The sensitivity results of fatigue damage area (%) and asphalt rut depth (mm) are summarized in Figure 12. The most sensitive volumetric parameter is emphasized by the relative difference between the design VMA and air void, which is normalized to the greatest slope corresponding to 100%. Figure 12 illustrates that the design VMA is the more sensitive volumetric parameter that affects fatigue and rutting performance than design air void. It may be concluded that those consistent trends may imply that the deign VMA and air void can be used as volumetric AQCs that are related to fundamental engineering properties and their predicted performances in the PRS system.

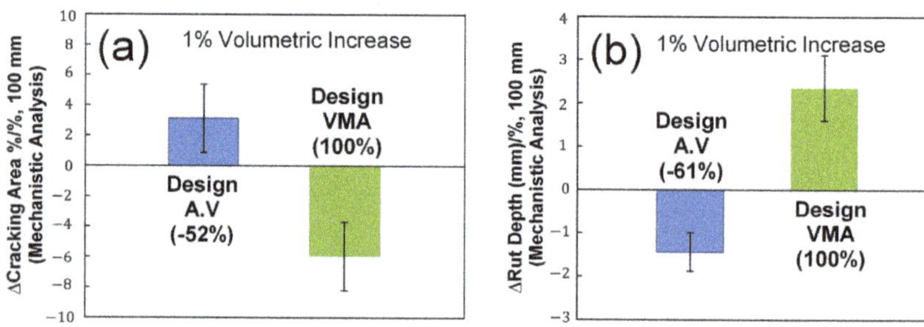

Figure 12. Sensitivity analysis results of individual volumetric AQCs on predicted mechanistic fatigue and rutting performance: (**a**) fatigue damage area (%) and (**b**) asphalt rut depth (mm).

3.5. Develop the Mathematical Models of Relationships between the Volumetric AQCs and Predicted Performance

Based on the sensitivity analysis results, the mathematical models of relationships between the design VMA (the most sensitive volumetric AQC) and the predicted fatigue and rutting performance were developed at design air voids of 3, 4, and 5% as shown in Figure 13. Since those volumetric AQCs are interrelated, the design VMA-predicted performance relationship should be considered with the design air void. Figure 13 indicates that the relationships are very linear. Furthermore, it is observed that the design VMA is even more sensitive on predicted fatigue performance at design air voids of 4 and 5%, whereas it is more sensitive on predicted rutting performance at design air voids of 3 and 4%. This implies the aggregate gradation plays a more important role on the fatigue performance with low binder content and rutting performance with high binder content.

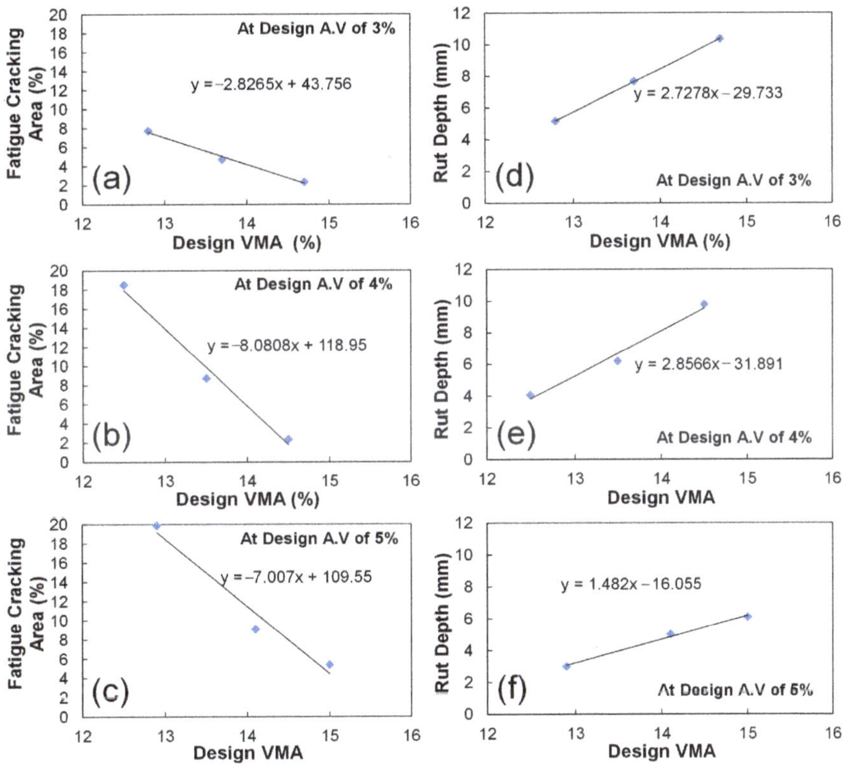

Figure 13. Development of mathematic relationships between the design VMA (%) and predicted performance (fatigue damage area (**a–c**), asphalt rut depth (**d–f**)) at design air voids of 3% (**a**,**d**), 4% (**b**,**e**), and 5% (**c**,**f**).

3.6. Identify Performance Indices (Mechanical AQCs) and Their Performance Threshold Values for Quality Assurance in PRS System

If asphalt mixture performance testers (AMPTs) are available to conduct the quality assurance during construction, the performance indices (mechanical AQCs in PRS system) may be used as a more accurate way to predict pavement performance. Otherwise, predictive relationships that are under development would have to be used and would have some unavoidable error. In this section, the relationship between proposed performance indices shown in previous sections that can be obtained by a faster or simpler interpretation of the fatigue and rutting mechanical tests and the already-completed mechanistic analysis results were quantified to find the indices that could more efficiently predict fatigue and rutting performance.

The number of cycles to failure at a certain AMPT software strain input and averaged phase angle value at failure were investigated to determine the best fatigue index as shown in Figure 14a–d. Here, the AMPT software strain input is the value the user provides in the software control prompt for the averaged on-specimen strain target value during the DTC fatigue testing. The advantage of this index is not having to run any S-VECD analytical structural simulations, only a quick interpretation of the physical test. It is found that the number of cycles to failure (fatigue life) at the AMPT software strain input of 250 microstrain has a poor correlation with the mechanistic fatigue cracking performance with an R-square of 0.32, while the fatigue life at the AMPT inputs of 350 and 450 microstrains have stronger correlations with the mechanistic cracking area with an R-square of 0.83 and 0.82, respectively, as shown in Figure 14a–c. Since a failure criterion of 1.2 loading cycles is impractical at the AMPT strain input of 450 microstrain, the fatigue life at the AMPT strain input of 350 microstrain is selected as a fatigue performance index candidate. For example, with this particular set of materials, using a 350 microstrain input to the control software, the cycles to failure was about 438 cycles, and the test would only need to be run for 44 s, less than a minute. In addition, Figure 14d illustrates a good relationship between the averaged phase angle at failure and predicted fatigue performance by an R-square of 0.64.

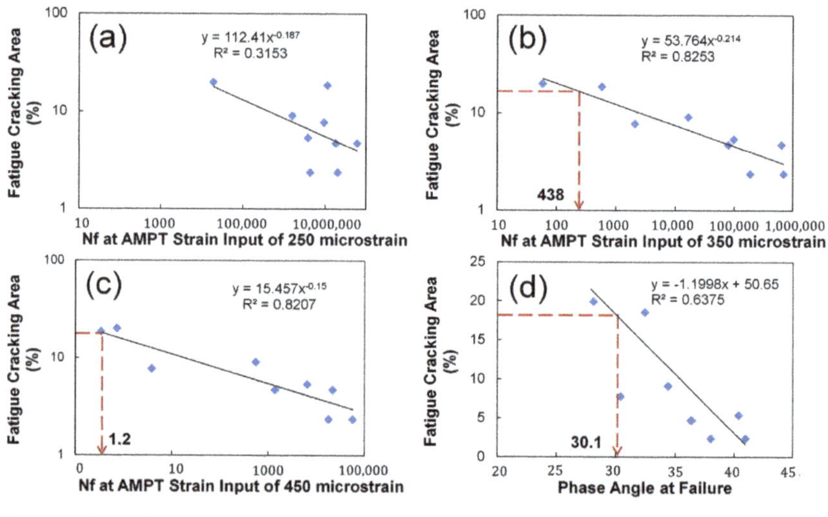

Figure 14. Identification of the optimal fatigue cracking index and its performance threshold value: (**a**) 250 micro-strain; (**b**) 350 micro-strain; (**c**) 450 micro-strain; (**d**) Phase angle at failure.

In the same way to identify the optimal fatigue index, an efficient rutting index is investigated. Figure 15a,b show the correlation of "b coefficient", the slope of minimum strain rate master-curve, from iRLPD and TSS tests at 54 °C with the mechanistic rut depth having an R-square of 0.73 and 0.67, respectively. Figure 15c–e present the linear relationship between the total accumulated permanent deformation (strain) from iRLPD, TSS, and TRLPD tests at 54 °C and the mechanistic rut depth. Figure 15f illustrates the negative correlation of the dynamic modulus at 54 °C and 10 Hz with the mechanistic rut depth by an R-square of 0.86. It is concluded that the total accumulated permanent strain is an optimal rutting index that can efficiently predict mechanistic rutting performance for these materials studied with a threshold value of 0.027.

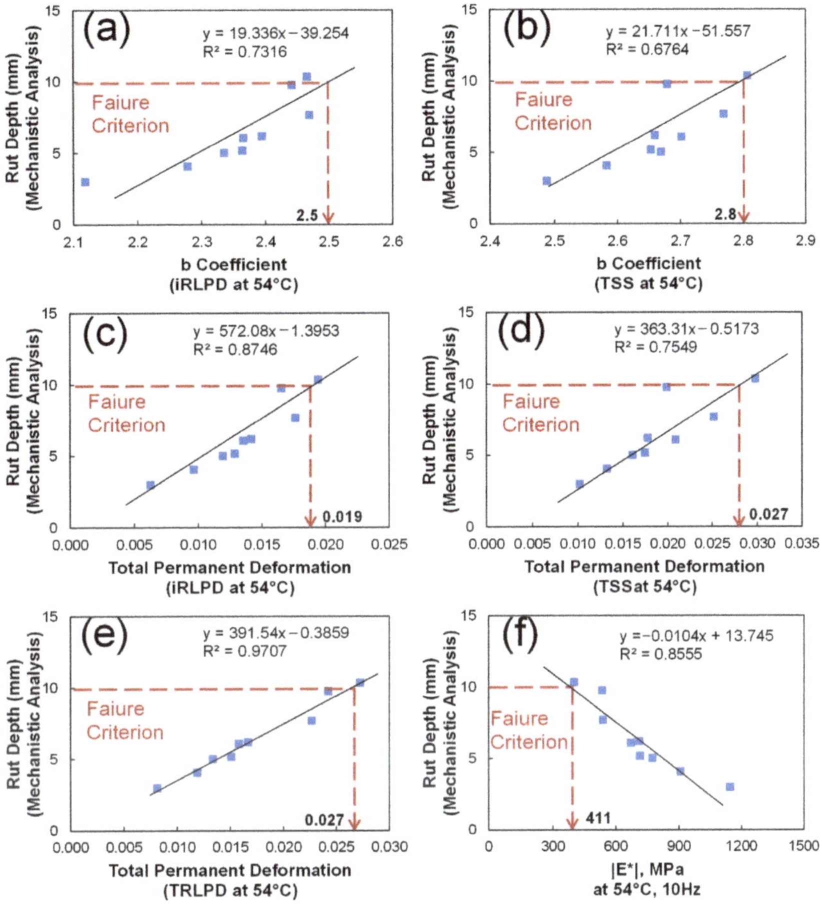

Figure 15. Identification of the optimal rutting index and its performance threshold values: b Coefficient (**a**) iRLPD, (**b**) TSS; Total permanent deformation at: (**c**) iRLPD, (**d**) TSS, (**e**) TRLPD; $|E^*|$: (**f**).

Relationships between the efficient performance indices and predicted mechanistic performance were compared in the same way as the volumetric AQC-predicted performance linear models as expressed in Figure 16. The results show that the model form of fatigue index is a power function, whereas that of rutting index is a linear function. The rutting index has a stronger relationship with the mechanistic performance rather than the fatigue index. Similar to the sensitivity of the design VMA on the mechanistically predicted performance, the design VMA is also more sensitive on the fatigue index at the design air voids of 4 and 5%, even though it is more sensitive on the rutting index at 3 and 4%. These volumetric or mechanical AQCs-predicted mechanistic performance mathematical models provide guidance on the extent in which the design VMA or performance indices need to be adjusted in order to reach the specific performance targets at the time of construction during production.

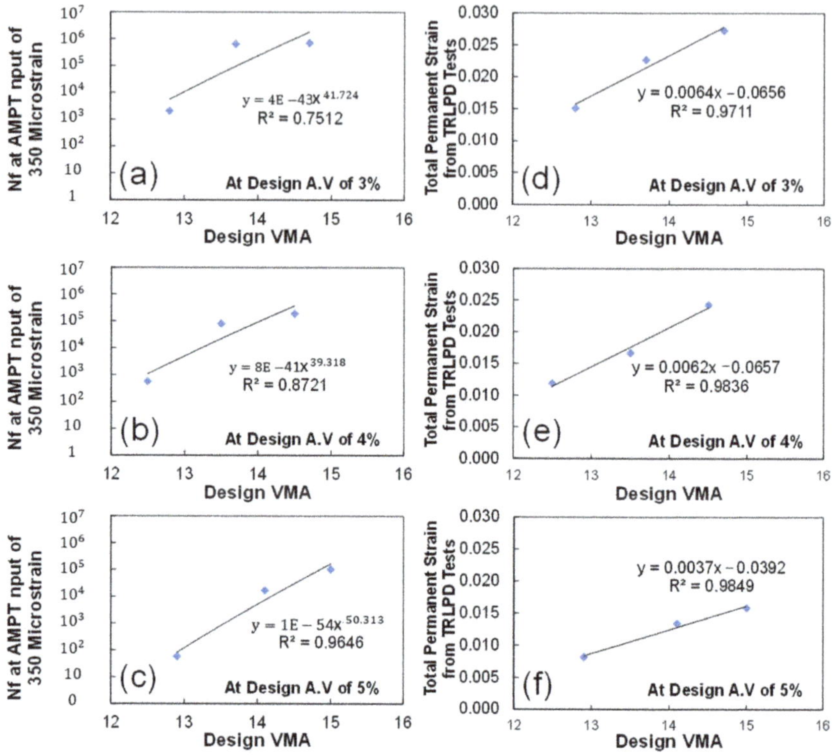

Figure 16. Development of the mathematical models on the relationship between the performance index and predicted performance (AMPT (**a**–**c**); TRLPD (**d**–**f**)) at design air voids of 3% (**a**,**d**), 4% (**b**,**e**), and 5% (**c**,**f**).

3.7. Determine Performance Targets and Their Volumetric AQC Values for Fatigue-Preferred, Rutting-Preferred, and Performance-Balanced Mix Designs

In order to develop the fatigue-preferred, rutting-preferred, performance-balanced mix designs, their performance targets need to be defined. In the phase for defining performance thresholds in the PBMD framework, the performance targets of fatigue and rutting-preferred mix designs for the LVECD output are determined, but the balanced performance target is not yet determined. Since the volumetric change of asphalt mixtures results in a reverse trend between fatigue and rutting performance, as shown in Figure 11, asphalt mix design that has the rutting performance at the threshold value of 10 mm may provide the fatigue-preferred mix design. In the same way, the fatigue performance at the threshold value of 18% may produce a rutting-preferred mix design. Finally, the performance-balanced mix design may need to produce the best balanced performance of fatigue and rutting within the available performance targets that mix designers can develop.

Figure 17 shows the way to determine the performance criteria and their volumetric AQC targets of three PBMD categories using the failure criterion and balanced performance lines. In order to determine these lines, the two relationships developed in Figure 13 need to be cross-plotted together. After normalizing the fatigue and rutting performance results in primary and secondary *y*-axis by the their threshold values of 18% and 10 mm, the failure criterion line connected between the fatigue cracking of 18% and rut depth criteria of 10 mm, and the balanced performance line parallel to the *x*-axis at the cross-point where the two relations intercept, could be drawn as shown in Figure 17.

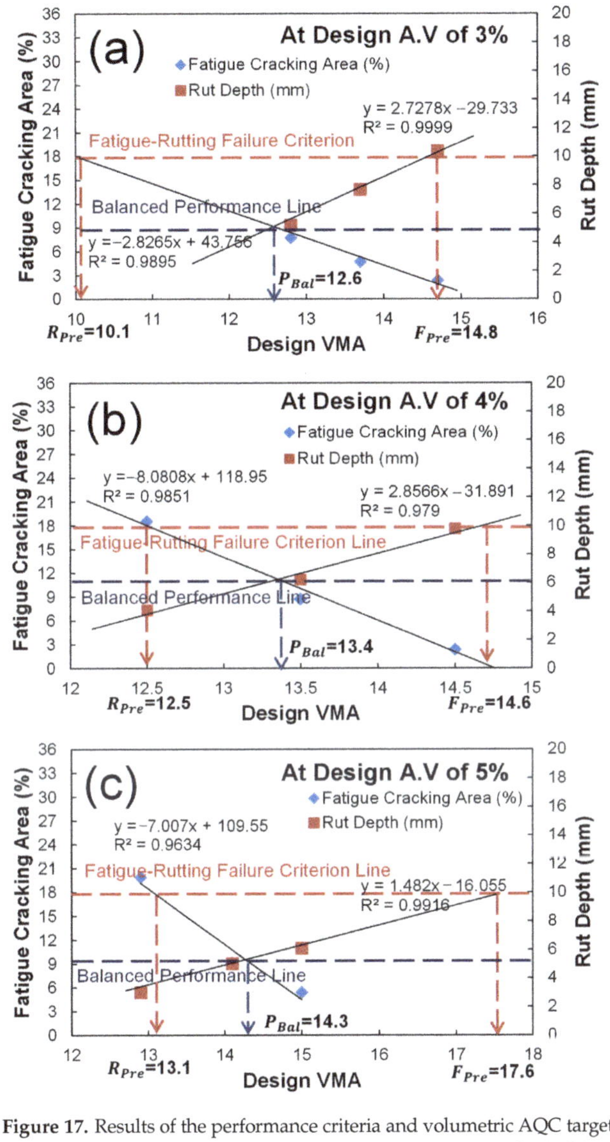

Figure 17. Results of the performance criteria and volumetric AQC targets of fatigue-preferred, rutting-preferred, and performance-balanced mix designs: (**a**) air void 3%; (**b**) air void 4%; (**c**) air void 5%.

Table 5 summarizes the results of fatigue and rutting performance targets and their corresponding design VMA values for the three PBMDs. The range of performance targets and AQC values are measured at design air voids of 3, 4, and 5%.The determined performance targets and corresponding volumetric AQC of fatigue-preferred mix design are fatigue cracking of 0 to 1.9%, rut depth of 10 mm, and design VMA of 14.8 to 17.6%, whereas those of rutting-preferred mix design are fatigue cracking of 18%, rut depth of 0 to 3.8 mm, and design VMA of 10.1 to 13.1%. In addition, those of performance-balanced mix design are fatigue cracking of 8.1 to 10.7%, rut depth of 4.6 to 6.4 mm, and design VMA of 12.6 to 14.3%. It is interesting to observe that the performance-based mix design has the best balanced performance at a design air void of 3%.

Table 5. Results of the performance targets and their volumetric AQC values of the three PBMDs at design air voids of 3, 4, and 5%.

PBMD		Fatigue-Preferred Mix Design			Rutting-Preferred Mix Design			Performance-Balanced Mix Design		
Design air void (%)		3	4	5	3	4	5	3	4	5
Performance targets	Cracking (%)	1.9	0	0	18	18	18	8.1	10.7	9.3
	rut depth (mm)	10	10	10	0	3.8	3.4	4.6	6.4	5.1
Design VMA (%)		14.8	14.6	17.6	10.1	12.5	13.1	12.6	13.4	14.3

3.8. Select the PBMD Category of Each Asphalt Layer to Accommodate Critical Pavement Distresses for Long-Life Rehabilitation

The final role in the PBMD framework is to select the category of each asphalt layer that can accommodate the critical distresses for longer life through the LVECD program. The inputs of each asphalt layer used in the LVECD program are the fundamental material properties that mechanistically predict fatigue and rutting performance under the in-service conditions. In order to determine those fundamental material properties of three PBMD categories, the determined performance targets of three PBMDs are expressed in the performance cross-plot as shown in Figure 18. Mix-C that slightly exceeds the rut depth of 10 mm and provides a fatigue cracking damage area of 2.4% is selected as the fatigue-preferred mix design. In the same way, the Mix-G is selected as the rutting-preferred mix design. Finally, Mix-D contained in the balanced performance target area is selected as the performance-balanced mix design for the pavement structural analysis example.

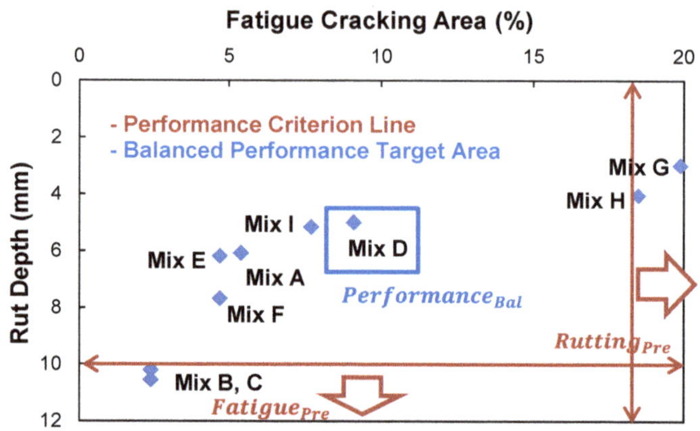

Figure 18. Selected PBMD mix designs based on performance levels for structural simulations.

A performance-optimized PBMD structural configuration is compared with full depth Superpave for a deep perpetual pavement scenario. The total asphalt thickness is 254 mm (10 in), consisting three layers. The LVECD program was run to evaluate fatigue and rutting performance in the Washington D.C area with a traffic loading of 8000 AADTT for 44 million ESALS over 30 years.

A Superpave structural design was analyzed to compare to a performance optimized structure. The Mix-E, having the minimum design VMA of 14, is selected for the 76.2 mm (3 in) thick surface layer. Mix-D, which is stiffer than Mix-E, is selected for the 101.6 mm (4 in) intermediate layer. Lastly, the 76.2 mm (3 in) thick bottom layer is chosen as Mix-G, which is stiffer than the Mix-D.

PBMD structural configuration should reduce complete bottom-up cracking propagation and resist rutting so that only surface milling and resurfacing can maintain the overall pavement structure in a sound condition over 50 years. Following the perpetual pavement concept, the fatigue-preferred mix design, Mix C, is selected for the 76.2 mm (3 in) thick bottom layer of the pavement. The performance-balanced mix design is selected to delay the bottom-up cracking propagation and have the rutting resistance for the 101.6 mm (4 in) intermediate layer. Finally, the rutting-preferred mix design is selected to have a rutting resistance and friction for the 76.2 mm (3 in) thick surface layer in accordance with the perpetual pavement concept.

In Figure 19, the simulation results from the LVECE program show that the PBMD structural configuration can reduce the bottom-up cracking propagation without exceeding the rutting performance criterion for a long-life pavement management system. The simulation results also imply that the asphalt mixture in the surface layer may need to have a higher performance asphalt binder with modifiers to better resist top-down fatigue performance.

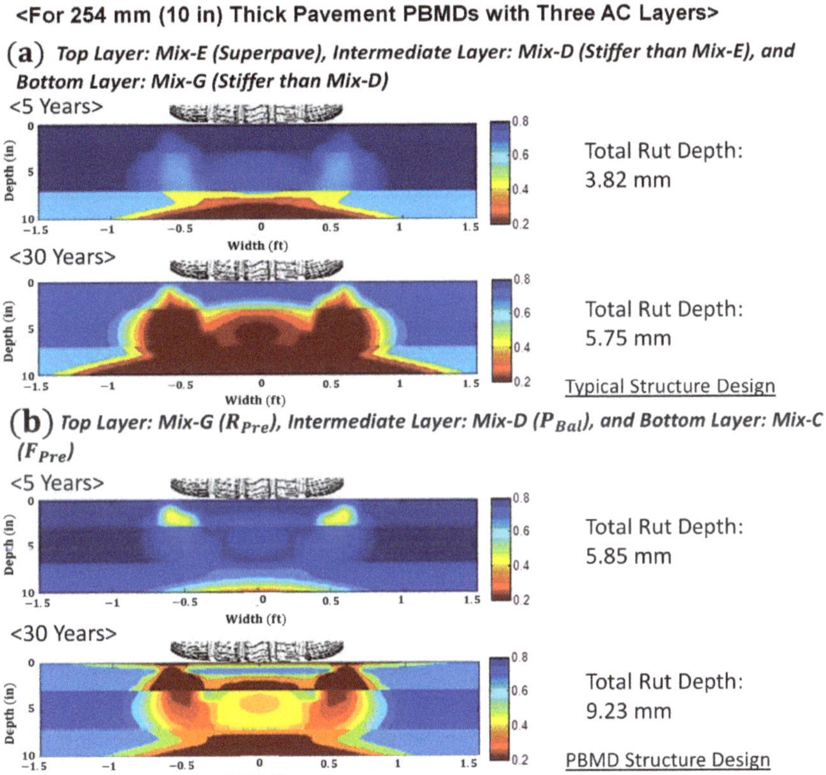

Figure 19. Comparison of full-depth Superpave with the PBMD structural configuration.

4. Conclusions

The proposed PBMD framework presents a comprehensive approach for supporting a PRS system and long-life pavements. The framework involves the development of various components, such as performance tests, prediction models, and threshold values. By doing so, it provides a systematic and standardized way of assessing the performance of asphalt pavements:

- One of the key features of the proposed framework is the identification of achievable performance levels through volumetric changes. Mix designers can develop asphalt mix designs that vary design voids VMA and design air void, which are factors that they have control over. By varying these design parameters, designers can achieve different performance levels, and the framework provides a way to identify these levels.

- Another critical aspect of the framework is the investigation of the effect of volumetric requirements on predicted performance for their sensitivity. The sensitivity analysis found that the design VMA is the most sensitive volumetric AQC that mix designers need to control for required performance targets. Moreover, mathematical models were developed to establish very linear relationships between the volumetric AQC of design VMA and the predicted performance to support the PRS.
- Efficient performance indices were also developed to facilitate the PRS system. The indices for fatigue and rutting were the number of cycles to failure at the AMPT software strain input of 350 microstrain and total accumulated permanent strain measured from the TRLPD tests. By using these indices, it is possible to evaluate the performance of pavements and determine if they meet the required criteria.
- Furthermore, the proposed framework determined the performance targets and their AQC values of three PBMD types using predicted performance criteria. For the fatigue-preferred mix design, the performance targets were a fatigue cracking area of 0 to 1.9% and a rut depth of 10 mm from a design VMA of 14.8 to 17.6%. The rutting-preferred mix design had a fatigue cracking area of 18% and a rut depth of 0 to 3.8 mm from design VMA as low as 10.1 to 13.1%. Additionally, the performance-balanced mix design criteria were a fatigue cracking area of 8.1 to 10.7% and a rut depth of 4.6 to 6.4 mm from design VMA of 12.6 to 14.3%. The performance-based mix design had the best-balanced performance at a design air void of 3%.
- Finally, the proposed PBMD pavement design with the fatigue-preferred mix design placed in the bottom layer, performance-balanced mix design in the intermediate layer, and rutting-preferred mix design in the surface can reduce the complete bottom-up cracking propagation without exceeding the rutting performance criteria. Simulation results from the LVECD structural analysis software verified the effectiveness of this design in achieving long-life pavements.
- Considering the limitation of this research, the need for further validation of the proposed framework through field testing and a verification of its effectiveness in different climatic and traffic conditions should be greatly considered. Additionally, future work could focus on incorporating environmental and economic factors into the framework to provide a more comprehensive approach to pavement design and maintenance.
- Overall, the proposed PBMD framework provides a robust and structured approach to support a PRS system and long-life pavements, enabling efficient and effective design and maintenance of asphalt pavements.

Author Contributions: J.-S.L., S.-Y.L. and T.H.M.L.: Conceptualization, methodology, writing—original draft, visualization, investigation, writing—review and editing, data curation, software. All authors have read and agreed to the published version of the manuscript.

Funding: This research received no external funding.

Institutional Review Board Statement: Not applicable.

Informed Consent Statement: Not applicable.

Data Availability Statement: Data will be provided upon request.

Acknowledgments: Not applicable.

Conflicts of Interest: The authors declare no conflict of interest.

References

1. Zhang, G.; Wu, H.; Li, P.; Qiu, J.; Nian, T. Pavement Properties and Predictive Durability Analysis of Asphalt Mixtures. *Polymers* **2022**, *14*, 803. [CrossRef] [PubMed]
2. Zhang, R.; Huang, J.; Zheng, W. Evaluation of the Significance of Different Mix Design Variables on Asphalt Mixtures' Cracking Performance Measured by Laboratory Performance Tests. *Constr. Build. Mater.* **2022**, *350*, 128693. [CrossRef]
3. Mivehchi, M.; Wen, H.; Cantrell, L. Systematic Evaluation of Effects of Recycled Materials and Mix Design Parameters on the Rutting and Cracking Performance of Asphalt Mixes. *J. Clean. Prod.* **2022**, *330*, 129693. [CrossRef]

4. Ren, S.; Hu, X. Fatigue Properties and Its Prediction of Polymer Concrete for the Repair of Asphalt Pavements. *Polymers* **2022**, *14*, 2941. [CrossRef]
5. Safaeldeen, G.I.; Al-Mansob, R.A.; Al-Sabaeei, A.M.; Yusoff, N.I.M.; Ismail, A.; Tey, W.Y.; Azahar, W.N.A.W.; Ibrahim, A.N.H.; Jassam, T.M. Investigating the Mechanical Properties and Durability of Asphalt Mixture Modified with Epoxidized Natural Rubber (ENR) under Short and Long-Term Aging Conditions. *Polymers* **2022**, *14*, 4726. [CrossRef]
6. Valdés-Vidal, G.; Calabi-Floody, A.; Duarte-Nass, C.; Mignolet, C.; Díaz, C. Development of a New Additive Based on Textile Fibers of End-of-Life Tires (ELT) for Sustainable Asphalt Mixtures with Improved Mechanical Properties. *Polymers* **2022**, *14*, 3250. [CrossRef]
7. Zhang, Y.; Cheng, H.; Sun, L. Performance-Based Design of Recycled Hot-Mix Asphalt (HMA) Incorporating Compaction Effort Variable. *Constr. Build. Mater.* **2021**, *303*, 124277. [CrossRef]
8. Azari, H.; Mohseni, A. Incremental Repeated Load Permanent Deformation Testing of Asphalt Mixtures. *Transp. Res. Board* **2012**. 12-4381. Available online: https://trid.trb.org/view/1130843 (accessed on 21 January 2023).
9. Javilla, B.; Fang, H.; Mo, L.; Shu, B.; Wu, S. Test Evaluation of Rutting Performance Indicators of Asphalt Mixtures. *Constr. Build. Mater.* **2017**, *155*, 1215–1223. [CrossRef]
10. Polaczyk, P.; Ma, Y.; Xiao, R.; Hu, W.; Jiang, X.; Huang, B. Characterization of Aggregate Interlocking in Hot Mix Asphalt by Mechanistic Performance Tests. *Road Mater. Pavement Des.* **2021**, *22*, S498–S513. [CrossRef]
11. Li, H.; Zhou, L.; Sun, J.; Wang, S.; Zhang, M.; Hu, Y.; Temitope, A.A. Analysis of the Influence of Production Method, Plastic Content on the Basic Performance of Waste Plastic Modified Asphalt. *Polymers* **2022**, *14*, 4350. [CrossRef]
12. Gao, G.; Sun, M.; Xu, C.; Qu, G.; Yang, Y. Interlaminar Shear Characteristics of Typical Polyurethane Mixture Pavement. *Polymers* **2022**, *14*, 3827. [CrossRef]
13. Sabouri, M. Evaluation of Performance-Based Mix Design for Asphalt Mixtures Containing Reclaimed Asphalt Pavement (RAP). *Constr. Build. Mater.* **2020**, *235*, 117545. [CrossRef]
14. Ghani, U.; Zamin, B.; Tariq Bashir, M.; Ahmad, M.; Sabri, M.M.S.; Keawsawasvong, S. Comprehensive Study on the Performance of Waste HDPE and LDPE Modified Asphalt Binders for Construction of Asphalt Pavements Application. *Polymers* **2022**, *14*, 3673. [CrossRef]
15. Park, H.J.; Kim, Y.R. Investigation into Top-down Cracking of Asphalt Pavements in North Carolina. *Transp. Res. Rec.* **2013**, *2368*, 45–55. [CrossRef]
16. Yousefi, A.A.; Haghshenas, H.F.; Shane Underwood, B.; Harvey, J.; Blankenship, P. Performance of Warm Asphalt Mixtures Containing Reclaimed Asphalt Pavement, an Anti-Stripping Agent, and Recycling Agents: A Study Using a Balanced Mix Design Approach. *Constr. Build. Mater.* **2022**, *363*, 129633. [CrossRef]
17. Lee, J.S.; Gibson, N.; Kim, Y.R. Use of Mechanistic Models to Investigate Fatigue Performance of Asphalt Mixtures: Effects of Asphalt Mix Design Targets and Compaction. *Transp. Res. Rec.* **2015**, *2507*, 108–119. [CrossRef]
18. Yu, J.; Chen, Y.; Wei, X.; Dong, N.; Yu, H. Performance Evaluation of Ultra-Thin Wearing Course with Different Polymer Modified Asphalt Binders. *Polymers* **2022**, *14*, 3235. [CrossRef]
19. Zaumanis, M.; Arraigada, M.; Wyss, S.A.; Zeyer, K.; Cavalli, M.C.; Poulikakos, L.D. Performance-Based Design of 100% Recycled Hot-Mix Asphalt and Validation Using Traffic Load Simulator. *J. Clean. Prod.* **2019**, *237*, 117679. [CrossRef]
20. Al-Khateeb, G.; Sukkari, A.; Zeiada, W.; Ezzat, H. Microscopy-Based Approach for Measuring Asphalt Film Thickness and Its Impact on Hot-Mix Asphalt Performance. *SSRN Electron. J.* **2022**, *18*, e01711. [CrossRef]
21. Ansar, M.; Sikandar, M.A.; Althoey, F.; Tariq, M.A.U.R.; Alyami, S.H.; Elsayed Elkhatib, S. Rheological, Aging, and Microstructural Properties of Polycarbonate and Polytetrafluoroethylene Modified Bitumen. *Polymers* **2022**, *14*, 3283. [CrossRef] [PubMed]
22. Hamid, A.; Baaj, H.; El-Hakim, M. Rutting Behaviour of Geopolymer and Styrene Butadiene Styrene-Modified Asphalt Binder. *Polymers* **2022**, *14*, 2780. [CrossRef] [PubMed]
23. Transportation Research Board. *Glossary of Transportation Construction Quality Assurance Terms*; TRB: Washington, DC, USA, 2018.
24. Choi, Y.T.; Richard Kim, Y. Development of Calibration Testing Protocol for Permanent Deformation Model of Asphalt Concrete. *Transp. Res. Rec.* **2013**, *2373*, 34–43. [CrossRef]
25. *AASHTO TP 107-14*; Determining the Damage Characteristic Curve of Asphalt Mixtures from Direct Tension Fatigue Tests. American Association of State Highway and Transportation Officials: Washington, DC, USA, 2016.
26. Gkyrtis, K.; Plati, C.; Loizos, A. Mechanistic Analysis of Asphalt Pavements in Support of Pavement Preservation Decision-Making. *Infrastructures* **2022**, *7*, 61. [CrossRef]
27. Dong, Q.; Huang, B.; Richards, S.H. Calibration and Application of Treatment Performance Models in a Pavement Management System in Tennessee. *J. Transp. Eng.* **2015**, *141*, 04014076. [CrossRef]
28. Haider, S.W.; Brink, W.C.; Buch, N. Local Calibration of Flexible Pavement Performance Models in Michigan. *Can. J. Civ. Eng.* **2016**, *43*, 986–997. [CrossRef]
29. Button, J.W.; Brock, J.D.; Decker, D.S.; Donnelly, D.E.; Harman, T.P.; Horan, R.D.; Huber, G.A.; King, G.N.; Newcomb, D.E.; Paul, H.R.; et al. *Perpetual Bituminous Pavements*; Transportation Research Board: Washington, DC, USA, 2001.
30. Schapery, R.A. Deformation and Fracture Characterization of Inelastic Composite Materials Using Potentials. *Polym. Eng. Sci.* **1987**, *27*, 63–76. [CrossRef]
31. Transportation Officials. *AASHTO Guide for Design of Pavement Structures*; American Association of State Highway and Transportation Officials: Washington, DC, USA, 2004; pp. 63–70.

32. Astm C29/C Bulk Density ("Unit Weight") and Voids in Aggregate. *ASTM Int.* **1997**, *4*, 2–5.
33. Eslaminia, M.; Thirunavukkarasu, S.; Guddati, M.N.; Kim, Y.R. Accelerated Pavement Performance Modeling Using Layered Viscoelastic Analysis. In *7th RILEM International Conference on Cracking in Pavements: Mechanisms, Modeling, Testing, Detection and Prevention Case Histories*; RILEM Bookseries; Springer: Berlin/Heidelberg, Germany, 2012; Volume 4, pp. 497–506. [CrossRef]
34. Federal Highway Administration. National Performance Management Measures—Assessing Pavement Condition for the National Highway Performance Program and Bridge Condition for the National Highway Performance Program. *Fed. Regist.* **2017**, *2017-00550*, 5886–5970.
35. Queensland Transport. Pavement Rehabilitation Manual, Pavement, Materials, Geotechnical Branch, State of Queensland, April 2012. 2012. Available online: https://www.tmr.qld.gov.au/business-industry/Technical-standards-publications/Pavement-Rehabilitation-Manual.aspx (accessed on 21 January 2023).
36. *AASHTO: TP-79*; Determining the Dynamic Modulus and Flow Number for Hot Mix Asphalt (HMA) Using the Asphalt Mixture Performance Tester (AMPT). American Association of State Highway and Transportation Officials: Washington, DC, USA, 2013.
37. *AASHTO T 19M/T 19-14*; Standard Method of Test for Bulk Density ("Unit Weight") and Voids in Aggregate. American Association of State Highway and Transportation Officials: Washington, DC, USA, 2015; pp. 2064–2072.

polymers

MDPI

Review

Review of the Application of Microwave Heating Technology in Asphalt Pavement Self-Healing and De-icing

Letao Zhang, Zihan Zhang, Weixiao Yu * and Yinghao Miao *

National Center for Materials Service Safety, University of Science and Technology Beijing, Beijing 100083, China
* Correspondence: yuweixiao@xs.ustb.edu.cn (W.Y.); miaoyinghao@ustb.edu.cn (Y.M.)

Abstract: In the past decades, a large amount of research was conducted to investigate the application prospect of microwave heating technology in improving the efficiency of asphalt pavement self-healing and de-icing. This paper reviewed the achievements in this area. Firstly, the properties of asphalt concrete after microwave heating were summarized, including microwave sensitivity and heating uniformity. Then, the evaluation indicators and influence factors of the self-healing properties of the asphalt mixtures heated by microwave were reviewed. Finally, the application of microwave heating in asphalt pavement de-icing was explored. In addition, asphalt pavement aging due to microwave heating was also reviewed. It was found that microwave heating technology has good prospects in promoting asphalt pavement self-healing and de-icing. There are also some problems that should be studied in depth, such as the cost-effectiveness of microwave-sensitive additives (MSAs), the performance of the pavement with MSAs, mechanism-based self-healing performance indicators, and the aging of asphalt pavements under cycling microwave heating.

Keywords: asphalt pavement; microwave heating; pavement de-icing; self-healing

Citation: Zhang, L.; Zhang, Z.; Yu, W.; Miao, Y. Review of the Application of Microwave Heating Technology in Asphalt Pavement Self-Healing and De-icing. *Polymers* **2023**, *15*, 1696. https://doi.org/10.3390/polym15071696

Academic Editor: Xavier Colom

Received: 16 January 2023
Revised: 25 March 2023
Accepted: 27 March 2023
Published: 29 March 2023

1. Introduction

Asphalt pavements have the advantages of high smoothness and comfortableness and they are the main form of road pavements. Asphalt mixture is a composite material formed using aggregates, fillers, asphalt binders, and voids. Asphalt binders are often used to bind the aggregates and their viscosity depends on external temperature. They behave as a Newtonian fluid at high temperatures, while their behavior is non-Newtonian at low temperatures. Under the long-term influence of climatic conditions and vehicle loads, asphalt pavements will inevitably be damaged, with cracks, potholes, and ruts appearing. In addition, icy roads caused by freezing rain and snowy weather bring potential safety hazards to traffic. Traditional pavement maintenance and de-icing methods have the disadvantages of inefficiency and serious environmental pollution. Researchers have been developing new technologies of pavement maintenance and de-icing, such as induction heating, Joule heating, and microwave heating.

The microwave is actually an electromagnetic wave with a frequency of 300 MHz to 300 GHz and a wavelength of 1 mm to 1 m. Energy can be transferred through space or medium in the form of electromagnetic waves [1]. Microwave heating is a process in which electromagnetic energy is converted into heat. Microwave heating technology possesses many advantages, such as high efficiency, low pollution, energy conservation, and uniform heating, as well as easy control. It has been widely used in cooking, drying, pasteurizing, and other fields [2]. The research in the field of pavement engineering using microwave heating can be traced back to 1974. Bosisio et al. [3] found that microwave radiation with 2450 MHz can effectively penetrate the asphalt concrete layer up to 12 cm. They also used microwave equipment with 1.6 kW to heat asphalt concrete slabs and concluded that actual asphalt pavement cracking can be well repaired using microwave heating. In 1989, Osborne et al. [4] found that the ice layer is facilely separated from the pavement surface under microwave action. Microwave heating has attracted researchers' attention for its potential in asphalt pavement maintenance and de-icing.

At present, the study on asphalt pavements using microwave heating mainly involves self-healing and de-icing. The principle of self-healing is that asphalt binders can flow and the cracks can be filled when the temperature of asphalt pavements is high. The principle of de-icing is that the microwave can directly heat asphalt concrete pavements through the ice layer and weaken the bond between the ice and the pavement surface. The influence factors of self-healing and de-icing include the properties of asphalt binders and aggregates, temperature, and heating modes. The qualities of self-healing and de-icing mainly depend on the microwave absorption ability of pavement materials. For enhancing the efficiency and effect of asphalt pavement self-healing and de-icing, researchers have carried out investigations from the aspects of microwave-sensitive additives (MSAs) and microwave heating devices [5–7]. There have also been a few studies on the aging of asphalt binders and mixtures heated by the microwave and the influence of MSA on pavement performance.

This paper reviewed the investigation into asphalt pavement self-healing and de-icing using microwave heating. The influences of different MSAs and heating conditions on the efficiency of microwave heating, the homogeneity of heating, and the self-healing performance of asphalt concrete were comparably analyzed. The achievements in de-icing by microwave heating were summarized. The research on the potential aging of asphalt under microwave heating was also reviewed. Finally, the issues that still need to be improved were proposed accordingly.

2. Microwave Heating Properties of the Asphalt Mixtures

The mechanism of microwave heating is the dielectric loss of materials under the microwave field, including polarized relaxation loss and conductive loss. When asphalt mixture is exposed to microwave radiation, heat is generated through the conversion of the energy of the electromagnetic field. In the conventional heating methods, energy is transferred from the surface of the materials by convection, conduction, and radiation. In contrast, microwave heating is achieved by molecular excitation inside the material without relying on the temperature gradient [8]. Figure 1 describes the microwave-absorbing principle of materials.

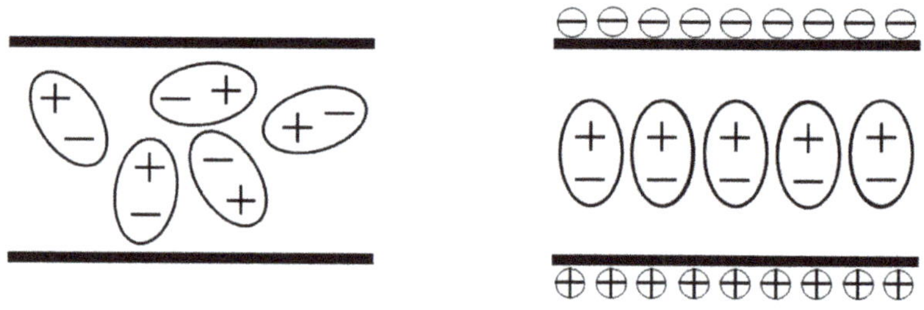

Figure 1. The microwave-absorbing principle of materials.

2.1. Microwave Sensitivity

The microwave sensitivity of a material determines its microwave absorption ability and microwave heating efficiency (MHE). It is commonly represented by dielectric properties. The complex permittivity (ε) and complex permeability (μ) are two indispensable parameters characterizing the performance of materials reflecting and absorbing microwaves, which are defined as Equation (1) [8].

$$\varepsilon = \varepsilon' - j\varepsilon'', \mu = \mu' - j\mu'' \quad (1)$$

where ε' and μ' are the real parts, indicating the extent of polarization or magnetization of microwave-absorbing materials under the action of electric or magnetic fields, respectively, and ε'' and μ'' are the imaginary parts, representing the loss magnitude caused by the rearrangement of electric- or magnetic-coupling moments of microwave-absorbing materials under the action of external electric or magnetic fields, respectively. The greater ε'' and μ'', the better the ability of the material to absorb microwaves. Therefore, the microwave heating properties can be improved by increasing the ε'' and μ'' of the asphalt mixture.

In addition, the matching attenuation features of materials are also two important conditions for achieving efficient microwave absorption. The former refers to the proportion of the incident microwaves entering materials, which depends on the input wave impedance of the interface between materials and free space. Only when materials match the wave impedance in free space, can the incident microwave get into the material to a large extent. The attenuation characteristics of materials refer to the rapid absorption and attenuation capacity of the microwave entering the materials, which requires that the materials have large electromagnetic loss. This depends on the physical performance of the material itself. The reflection loss (RL) represents the amount of microwave energy reflected by materials. It is another indicator used to evaluate the microwave absorption capacity (MAC) of materials, which can be calculated by Equations (2) and (3) [4]. When the reflection loss is negative, the smaller its value, and the stronger the MAC of materials. The reflection loss values of some representative materials used in asphalt mixtures are summarized in Table 1.

$$R_L = 20\lg\left|\frac{Zn_{in} - Zn_0}{Zn_{in} + Zn_0}\right| \tag{2}$$

$$Zn_{in} = Zn_0\sqrt{\mu_r/\varepsilon_r}\tan\mathrm{h}\left[j\left(\frac{2\pi fd}{c}\right)\sqrt{\mu_r\varepsilon_r}\right] \tag{3}$$

where Zn_{in} is the input impedance of materials, Zn_0 is the intrinsic impedance at free space, ε_r and μ_r indicate the relative complex permittivity and permeability of materials, respectively, f is the electromagnetic frequency, d is the thickness of materials, and c is the speed of light in free space.

Table 1. The reflection loss values of representative materials.

Materials	Electromagnetic Frequency (GHz)	Thickness (mm)	RL (db)
Asphalt carbon-coated graphene/magnetic $NiFe_2O_4$-modified multi-wall carbon nanotube composites [9]	4.6	3.2	−45.9
Asphalt carbon-coated reduced graphene oxide/magnetic $CoFe_2O_4$ hollow-particle-modified multi-wall carbon nanotube composites [10]	11.6	1.6	−46.8
Mg-Al layered double hydroxides (LDHs) [11]	15.71	8	−4.79
LDHs:Fe_3O_4 = 1:1	11.71	10	−6.88
LDHs:Fe_3O_4 = 2:1	11.88	10	−5.25
LDHs:Fe_3O_4 = 1:2	11.28	8	−10.73
SiC [12]	16.12	28	−12.53
SiC:Fe_3O_4 = 1:1	11.21	30	−18.93
SiC:Fe_3O_4 = 2:1	11.26	29.5	−15.82
SiC:Fe_3O_4 = 1:2	17.92	26	−22.18
LDHs [13]	15.49	8	−5.21
SiC attached LDHs	17.5	10	−13.65

Table 1. *Cont.*

Materials	Electromagnetic Frequency (GHz)	Thickness (mm)	RL (db)
SiC:Fe_3O_4 is 3:1 [14]	2.45	25	−28
Limestone filler [15]	16.1	-	−6.8
Manganese dioxide powder	11.6	-	−18.83
Carbon powder	2.36	-	−33.53
Ferrite powder	12.3	-	−41.68
Limestone [16]	15.88	-	−2.67
Ferrite	3.89	-	−10.62
Ferrite	13.67	-	−30.28
Fine SiC [17]	13.68	2	−22.34
Fine SiC	2.45	10	−13.55
Coarse SiC	8.08	10	−15.27
Coarse SiC	2.45	10	−10.51
Asphalt mixture added with natural magnetite power in grade of 80 [18]	2.9	30	−38
Asphalt mixture added with natural magnetite power in grade of 70	3	30	−27
Asphalt mixture added with natural magnetite power in grade of 60	3.15	30	−25

Asphalt binders and aggregates are the major components of the asphalt mixtures, while their microwave-absorbing ability is poor. For improving the microwave-absorbing efficiency of the asphalt mixture, researchers developed microwave-absorbing and magnetite-bearing aggregates. Other attempts have included the addition of graphite, carbonyl iron powders (CIPs), carbon nanotubes, and steel wool, as well as ferrite particles. Liu et al. [19] added activated carbon powder (ACP) into asphalt mixture for enhancing the MHE. It was found that the addition of ACP significantly improves the heating efficiency of asphalt mastic and mixture. Zhu et al. [20] studied the heating effect of the asphalt mixture with ferrite and concluded that the MHE of Ni-Zn ferrite powders is 3.91 times as much as that of mineral powder (MP). Wang et al. [21] pointed out that magnetite aggregate has better MHE compared with basalt aggregate. Zhang et al. [22] explored the MHE of ceramics prepared from low-grade pyrite cinder. The results showed that the MHE of the ceramics was far higher than that of limestone. Trigos et al. [23] studied the MHE of various aggregates and graded them. The results showed that the MHE of blast furnace slag, classified as having very high susceptibility, is 24 times higher than that of quartzite, classified as having very low susceptibility. Deng et al. [15] and Cao [24] found that the MHE of basalt is 1.46–5.58 times as much as that of limestone. Li et al. [25] conducted a microwave heating test of asphalt mastics and found that as the volume ratio of the slag filler to asphalt binder increases, the average heating rate of asphalt mastics gradually improves. Li et al. [26] synthesized SiC to wrap LDHs (SwL) at different temperatures and studied the effect of different combinations of SBS and SwL on the microwave absorption of modified asphalt. The results showed that the microwave absorption of SBS/SwL-200-modified asphalt is best. The heating rates of some other typical asphalt mixtures are summarized in Table 2.

The research results mentioned above indicate that the higher the oxide content of iron in the aggregate, the faster the heating rate. The MHE of the asphalt mixture can be promoted by the addition of microwave-sensitive powders. However, the experimental methods of microwave heating for aggregates and fillers are not standardized and unified, which means that the results from various sources cannot be compared. In addition, the cost of MSAs is often ignored in the studies.

Table 2. The heating rate of different asphalt mixtures.

Asphalt Mixtures	Aggregates	Heating Rate (°C/s)
Normal asphalt mixture [27–30]	Basalt	0.252–0.76
Normal asphalt mixture [14,17,31]	Limestone	0.2–0.548
Normal asphalt mixture [31]	Dolomite limestone	0.618
Normal asphalt mixture [31]	Granite	0.757
Normal asphalt mixture [6]	Andesite	0.355
Asphalt mixture with steel slag aggregates [27]	Basalt	0.623
Asphalt mixture with SiC and Fe_3O_4 powder [14]	Limestone	0.244–0.367
Asphalt mixture with SiC aggregates [17]	Limestone	0.458–0.476
Asphalt mixture with steel fiber and graphite [28]	Basalt	0.9–1.02
Asphalt mixture with graphite powder and magnetite powder [29]	Basalt	0.372
Asphalt mixture with aggregates coated by magnetic Fe_3O_4 films [30]	Basalt	0.888–0.9
Asphalt mixture with nano-graphite [31]	Limestone	0.579–0.815
Asphalt mixture with nano-graphite [31]	Dolomite limestone	0.658–0.92
Asphalt mixture with nano-graphite [31]	Granite	0.831–1.184
Asphalt mixture with steel fiber [6]	Andesite	0.804
Asphalt mixture with steel slag aggregate [6]	Andesite	0.696

2.2. *Heating Uniformity*

In the microwave heating process, the differences in the microwave absorption capacity and heat transfer performance of different materials may lead to uneven temperature distribution in asphalt pavements and large internal temperature gradient [32]. In addition, microwave heating relies on the action of the high-frequency alternating electromagnetic field to realize the transformation of electric energy and material heat energy. In the actual heating process, the electromagnetic field strength is usually distributed unevenly, which can also result in uneven temperature distribution [33].

Many studies were carried out from the aspects of microwave heating modes, microwave heating devices, and MSAs to improve heating uniformity. Zhu et al. [34] simulated intermittent and continuous microwave heating and found that intermittent microwave heating could produce more uniform temperature distribution compared with continuous microwave heating. Sun et al. [35] found that a 0.5 min intermission can bring about a more uniform temperature distribution than no intermission, 1 min intermission, and 2 min intermission. Sun et al. [36] investigated the energy distribution of the electromagnetic field in asphalt mixtures based on the Poynting theory and established an optimized model of the electromagnetic field and structure by building a relationship between the electric field and magnetic field. The results showed that the uniformity of energy distribution can be improved by adjusting the radiation electric field. Sun [37] designed four kinds of structures of horn antennas (as shown in Figure 2) to enhance the uniformity of the electromagnetic field. The results showed that the electromagnetic field is evenly distributed when the length/width ratio of the antenna aperture is close to that of the feed waveguide. The homogeneity of energy distribution can be improved by adjusting the radiation electric field. Lou et al. [16] found that replacing limestone filler with ferrite filler is an effective way to improve the microwave heating uniformity of the asphalt mixture with steel slag. This is because the asphalt binders containing ferrite filler have high thermal conductivity. Fakhri et al. [38] found that adding copper slag filler into the asphalt mixtures containing steel shavings or recycled tire steel fibers can improve their temperature and heating uniformity.

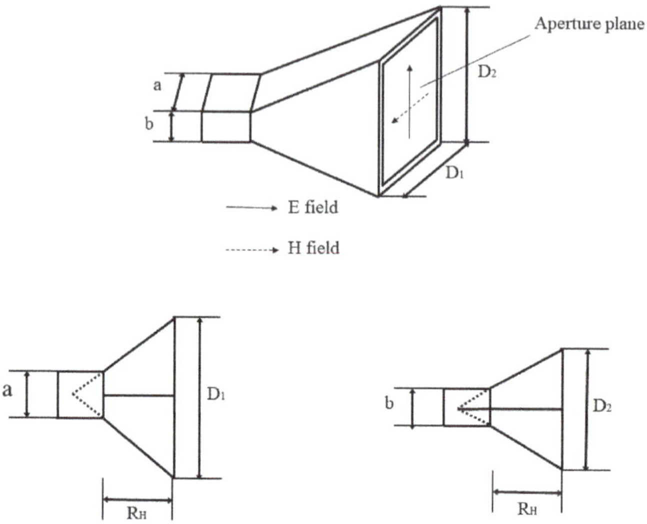

Figure 2. Structure of horn antenna [37].

In summary, researchers have acquired rich achievements in promoting the temperature uniformity of microwave heating for asphalt pavements. However, there is little research on the heat transfer mechanism in the process of microwave heating. Additionally, few researchers have carried out the simulation test from the mesoscopic perspective.

3. Self-Healing Properties of Asphalt Concrete under Microwave Heating

The heating technologies for accelerating the healing of the asphalt mixture mainly include induction heating, infrared heating, hot air heating, and microwave heating, as shown in Figure 3. In the process of induction heating, the heat energy diffuses into the asphalt mixture and the temperature of the asphalt pavement is increased via the Joule principle. Infrared heating is efficient, but continuous heating causes asphalt pavements to set on fire. Hot air heating has little influence on the aging of the asphalt mixture, but it will take more time. Under microwave heating, the temperature of the asphalt mixture is increased by the orientation change in polar molecules caused by the alternating magnetic field.

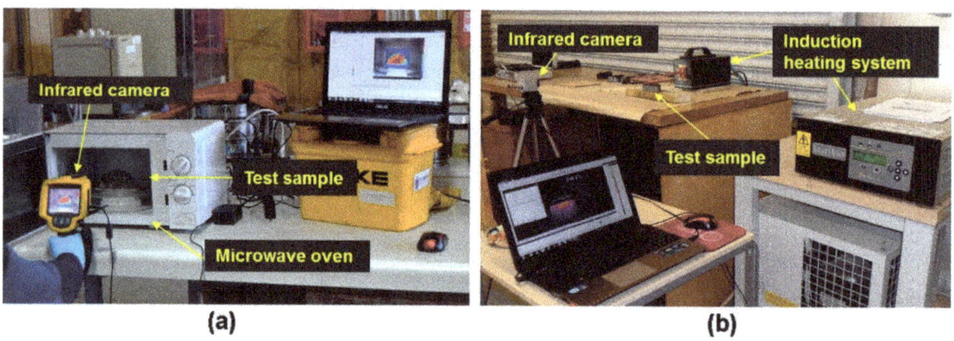

Figure 3. Heating devices: (**a**) microwave heating; (**b**) induction heating [7].

The self-healing behavior of asphalt concrete includes the self-healing of the asphalt binder and the self-healing of the adhesive interface between the binder and the aggregate.

As early as 1939, the self-healing phenomenon of asphalt concrete was observed in the Ghrib inclined wall dam in Algeria [39]. In 1967, Bazin et al. [40] discovered the self-healing properties of the asphalt mixtures. From the macroscopic perspective, the self-healing of the asphalt mixtures is the reverse process of cracking, which can be described as crack closure and strength recovery. From the microscopic perspective, it is the spontaneous interface infiltration and molecular diffusion of asphalt molecules on the upper and lower surfaces of the cracks.

Al-Ohaly et al. [41] and Gallego et al. [42] discovered the potential of microwave heating to improve the self-healing of the asphalt mixture. Microwave heating can reduce the viscosity of the asphalt binder and accelerate its capillary flow and molecular diffusion rate by increasing the temperature, therefore improving the self-healing of the asphalt mixture. Many researchers have carried out studies on the mechanism, test methods, evaluation indexes, and influence factors of asphalt mixture self-healing.

3.1. Self-Healing Mechanism

The surface energy theory as well as the capillary flow theory are the major theories to explain the self-healing mechanism of asphalt concrete [43,44]. Additionally, the molecular diffusion theory and the phase field theory are commonly used to describe the self-healing principle of the asphalt mixture. The definition of surface energy is the work conducted by external substances on the object under a certain temperature and pressure. It demonstrates the self-healing principle of the asphalt mixture from the aspect of the fracture mechanics and considers that the decrease in surface energy at the crack interface promotes the healing of asphalt concrete [45,46]. The application of molecular diffusion theory can be traced back to the self-healing of polymer materials [47]. In 1981, Wool et al. [48] proposed that the crack-healing of polymer materials undergoes five stages: rearrangement, surface approach, wetting, diffusion, and molecular random distribution. Afterwards, researchers adopted the molecular diffusion theory to study the healing mechanism of asphalt concrete. Little et al. [49] proposed that the self-healing of asphalt concrete is mainly attributed to the diffusion of the asphalt binder. As the asphalt molecules at the interface diffuse to the crack, the asphalt concrete strength gradually recovers. The phase field theory uses dynamic differential equations to express diffusion, ordering potential, and thermodynamic drive. Loeber et al. [50] observed that the asphalt binder has multiphase properties. On the basis of the phase field theory, the healing of the asphalt binder can be interpreted as the process of multiphase rearrangement to form a single phase [51]. The self-healing explanation on the basis of the model of capillary flow theory assumes that the flow of the asphalt binder in the micro cracks under the driving of capillary action contributes to the closure of the cracks. Álvaro García et al. [52,53] proposed a semi-empirical model to explain the healing of asphalt mastic and concrete according to the capillary flow theory. They considered that the capillary flow in cracks is the main cause of healing. When the binder shows as a Newtonian fluid at a high temperature and the junction of the crack makes contact, a pressure difference is generated at the contact point, which can be considered as the driving force of the capillary flow of the binder. In summary, researchers have comprehensively explained the self-healing mechanism of asphalt binders and asphalt mixture.

3.2. Evaluation Indicators of Self-Healing Properties

For evaluating the self-healing capacity of the asphalt binder and asphalt concrete, many researchers have conducted the three-point bending test [5], semi-circular bending test [54], DSR test [55], and ultrasonic technology [56] by the way of failure, healing, and failure. The self-healing indicators are mainly calculated based on dissipated energy, fatigue life, and dynamic shear modulus. Lou et al. [57] proposed the healing efficiency level (*HEL*) to evaluate the self-healing properties of asphalt concrete, as shown in Equation (4).

$$HLE(\%) = \frac{S_e}{S_w} \times 100\% \tag{4}$$

where S_e is the effective healing area of the sample in cm^2; S_w is the whole surface area of the sample. The effective healing area is defined as the area where the surface temperature threshold is between the softening point of the asphalt binder and 90 °C.

Wang et al. [55] defined the healing index (*HI*, %) as in Equation (5) to estimate the healing potential of asphalt concrete.

$$HI = \frac{G_2 - G_1}{G_0 - G_1} \tag{5}$$

where G_0 refers to the initial G^* value, and G_1 and G_2 represent the complex shear modulus before and after the rest, respectively.

Liu et al. [58] proposed the fatigue life extension ratio to evaluate the healing performance of asphalt concrete. It is defined as the ratio of the obtained extra fatigue life after damage due to healing and due to its original fatigue life. Li et al. [25] defined the *HI* from the perspective of dissipated energy, as shown in Equation (6).

$$HI = \frac{W_{Second}}{W_{First}} = \frac{\sum_{i=1}^{n_{Second}} DE_i}{\sum_{i=1}^{n_{First}} DE_i} = \frac{\sum_{i=1}^{n_{Second}} \pi \varepsilon_i^2 G_i^* sin\delta_i}{\sum_{i=1}^{n_{First}} \pi \varepsilon_i^2 G_i^* sin\delta_i} \tag{6}$$

where W_{First} and W_{Second} are the accumulative dissipated energy of the first and second fatigue loading, respectively, DE_i is the dissipated energy of the *i*-th loading cycle, and ε_i, δ_i, and G_i^* are the shear strain, phase angle, and complex modulus of the *i*-th loading cycle, respectively.

Shan et al. defined the HI by the ratio of the area between the healing curve and the initial curve to the area below the initial curve [59]. It can be calculated by the following equation.

$$HI = \frac{A_d}{A_{before}} \tag{7}$$

where A_d is the area between the healing curve and the initial curve, and A_{before} is the area below the initial curve.

In summary, the self-healing performance of the asphalt mixtures can be quantitatively evaluated by the above indicators. However, it is difficult to compare the results between different studies because there is no relevant standard for the self-healing test and evaluation of the asphalt mixtures. In addition, the self-healing and modulus recovery after rest are different. The existing evaluation indicators are unable to discriminate them by tests because the asphalt mixture is a complex composite material, which needs to be solved in future research.

3.3. Factors Influencing Self-Healing Properties

The influence factors of the self-healing performance of asphalt concrete under microwave heating can be categorized into external factors and internal factors. The external factors mainly refer to temperature, microwave heating mode, and time. The internal factors mainly involve material properties [60–62]. Wang et al. [28] and Fakhri et al. [38] studied the influence of microwave heating mode on asphalt concrete healing. The results indicated that intermittent microwave heating has a better healing effect compared with continuous heating. This is because the effective healing time is longer under intermittent heating. Norambuena-Contreras et al. [63] studied the self-healing properties of fiber-reinforced asphalt concrete under different microwave heating time using the three-point bending test and observed the crack size before and after healing. The results, shown in Figures 4 and 5, showed that the longer the heating time, the better the healing level. Forty seconds is the optimum heating time among the considered cases. The cracks can close under a long microwave heating time, which explains the cause of the self-healing of asphalt concrete.

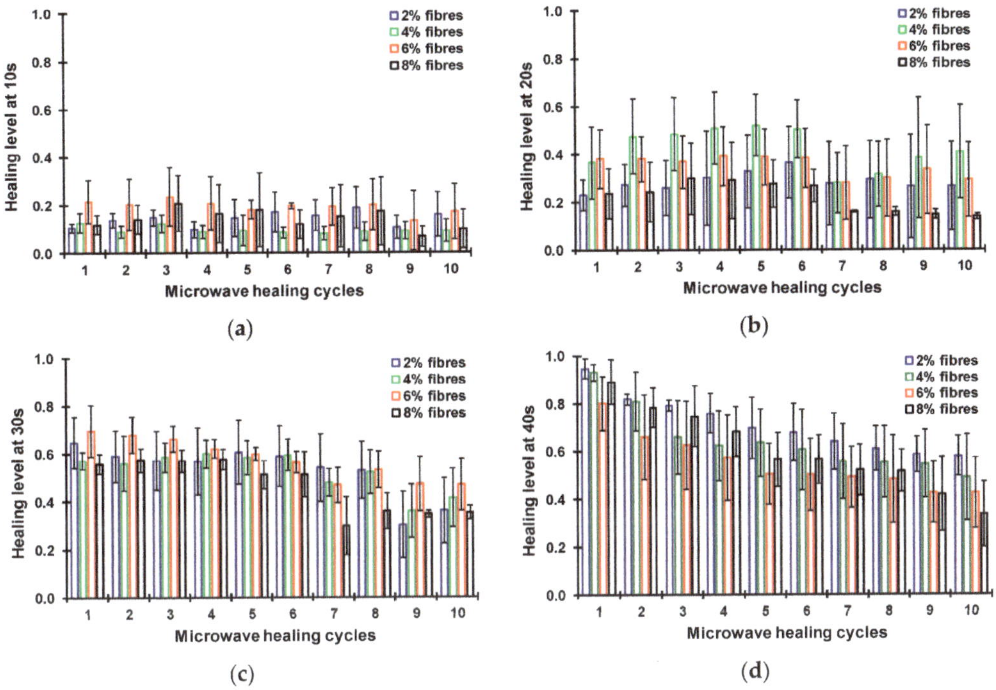

Figure 4. Healing levels of the asphalt mixtures under different microwave heating times: (**a**) 10 s; (**b**) 20 s; (**c**) 30 s; (**d**) 40 s. [63].

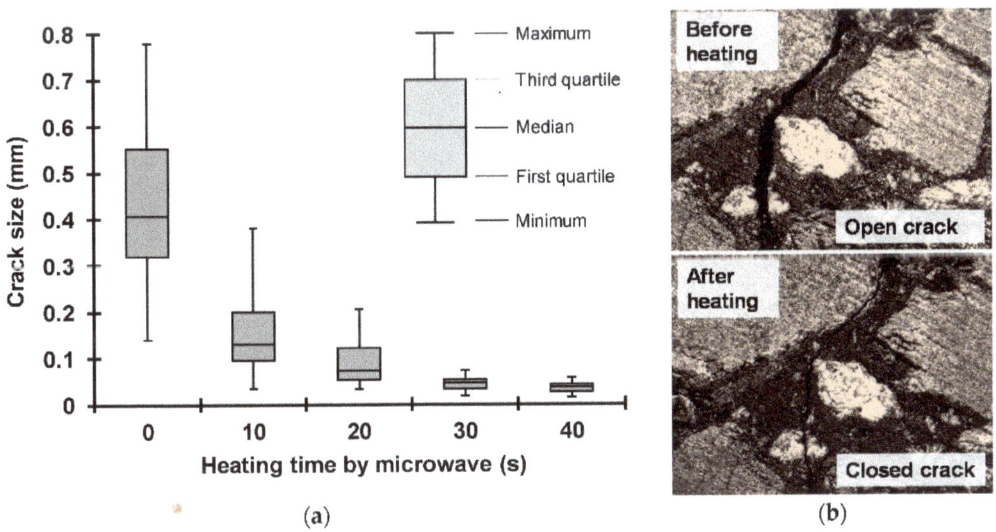

Figure 5. The crack size of the asphalt mixture samples: (**a**) Box plot representation of the crack size; (**b**) Images of a cracked specimen [63].

The material type also has an effect on the healing performance. Gonzalez et al. [64] studied the effect of RAP and metal fibers on the crack-healing of asphalt concrete under microwave heating. The results showed that the healing level is enhanced by adding metal fibers, but the opposite results occurred when adding RAP. In addition, asphalt concrete containing metal fibers and up to 30% RAP has the potential to repair cracks under microwave heating. Zhu et al. [65] evaluated the fracture-healing properties of AC-13 mixtures with 70# asphalt binder and SBS-modified asphalt binder under microwave heating via the SCB test. It was found that the SBS asphalt mixture has better healing performance at high temperatures (over 80 °C).

This can be interpreted as the special healing mechanisms of AC-13 mixture with SBS-modified asphalt, that is, the combined effect of the flow diffusion of the asphalt binder and the elastic recovery of SBS segments. Deng et al. [15] developed an asphalt mixture incorporating manganese dioxide powder (MDP) and steel fiber (SF), and investigated its heating–healing capacity through comparing with the mixtures only with MDP or SF. The results showed that the developed mixture exhibits better microwave healing performance. Wang et al. [66] found that the asphalt mixture containing carbon fiber has good healing performance under microwave heating. Two percent IM8-modified asphalt mixture has a better healing effect compared with 4% IM8-modified asphalt mixture. The reason for this is that the high content of carbon fiber may limit the Newtonian flow of the asphalt binder in the healing process. Phan et al. [67] analyzed the applicability of steel slag to promote the self-healing of the asphalt mixtures under microwave heating. They found that asphalt mixtures with 30% steel slag aggregate and 2% steel wool fibers have good healing effectiveness. Lou et al. [16] studied the self-healing performance of steel slag asphalt concrete containing ferrite fillers under microwave heating. It was found that an appropriate addition of ferrite can improve the average temperature and healing ratio of asphalt concrete. The optimal proportion of replacing limestone fillers with ferrite is 20% by volume among the considered cases. Adding excessive ferrite has a negative effect because abundant ferrite may lead to the agglomeration issue and obstruct the heat transfer process. The hardening of the asphalt binder also has adverse effects on the flow and diffusion process. Lou et al. [57] replaced the 4.75–9.5 mm aggregate of the asphalt mixture with hot braised steel slag (HBSS) in a certain volume percentage and studied its microwave self-healing performance via the SCB test. The results showed that the addition of HBSS promotes the healing of the asphalt mixture. However, excessive HBSS causes an overheating problem and reduces the effective healing area under microwave heating. González et al. [68] studied the crack-healing capability of normal asphalt mixtures and asphalt mixtures containing steel fiber, metal shavings, and silicon carbide via three-point bending tests. It was found that the additives (i.e., steel fibers, metal shavings, and silicon carbide) have little effect on crack-healing ability. The asphalt mixture without additives also showed good crack-healing ability. This is because the aggregates contain metals, as shown in Figure 6. In summary, asphalt binder is a temperature-sensitive and viscoelastic material. MSAs have good thermal conductivity performance, which enhances the diffusion rate of asphalt molecules and improves the healing ability of the asphalt mixture.

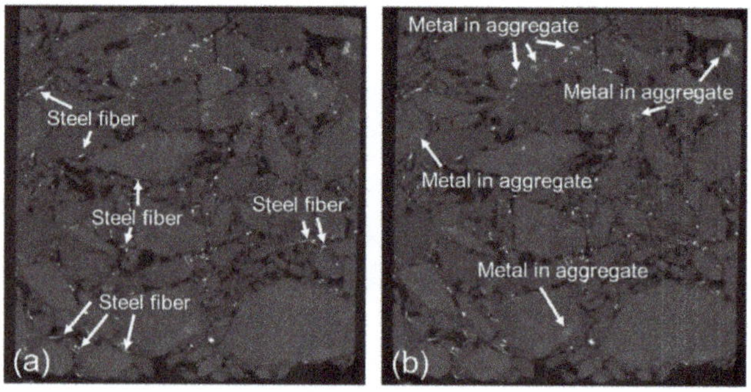

Figure 6. CT scan image of the asphalt mixture: (**a**) Steel fiber; (**b**) Mental contained in aggregate [68].

Some researchers also introduced microcapsules containing rejuvenator into asphalt concrete to promote self-healing. When the crack extends to the position of a microcapsule, it breaks and releases the rejuvenator, which accelerates the healing of asphalt concrete. Wan et al. [69] synthesized calcium alginate/nano-Fe_3O_4 composite capsules that can actively and rapidly release the rejuvenator under microwave heating. Kargari et al. [54] studied the self-healing effect of the asphalt mixture containing palm oil capsules under microwave heating. They found that the HI values of aged and non-aged asphalt mixtures heated by microwaves increase by 32% and 7%, respectively, when the added palm oil capsule accounts for 0.7% of the asphalt mixtures by mass. The reason for this is that microwave heating can control the release of rejuvenators in capsules. Thus, the released rejuvenator improves the healing effect of the asphalt mixtures.

In summary, most researchers explain the self-healing phenomenon of the asphalt mixture using macroscopic experimental study, but pay little attention to the influence mechanism from a multiscale perspective.

4. Asphalt Pavement De-icing Using Microwave Heating

Icy roads increase traffic accidents and pavement maintenance time and cost, reduce driving speed [70–74], and raise fuel consumption and exhaust emissions [75]. Pavement de-icing is significant to ensure traffic safety and efficiency. Traditional pavement de-icing methods mainly include the manual method, mechanical method [76], and chemical method [77]. The manual method is inefficient and costly. The mechanical method may cause damage to the pavement [78]. The chemical method usually damages vegetation, water sources, and atmosphere [79], and reduces pavement durability [80]. For overcoming the disadvantages of traditional methods, researchers have explored the applications of microwave heating [81], induction heating [82], Joule heating [83], solar energy [84], and environmentally friendly de-icing agent [85] in asphalt pavement de-icing. Microwave de-icing, developed in recent years, has great potential and application prospects. Figure 7 depicts the working mechanism of de-icing using a microwave heating vehicle.

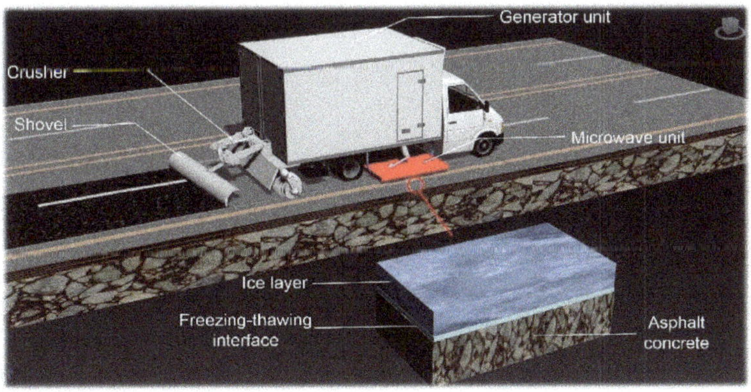

Figure 7. Schematic diagram of de-icing using microwave heating vehicle [86].

4.1. Microwave De-icing Mechanism and Efficiency Evaluation

The freezing adhesion between the ice and the pavement is the major resistance of de-icing. The adhesion strength can be defined as in Equations (8) and (9). The conditions of producing freezing adhesion mainly include the temperature of 0 °C or below, certain water content, and freezing time. Temperature is the most important factor. The higher the temperature, the lower the horizontal adhesion strength of the interface between ice and pavement [87] and the crushing strength of ice [88]. Because ice has a low dielectric constant and poor microwave-absorbing ability, it is difficult to heat directly using microwaves [89]. However, the microwave can penetrate the ice layer and be absorbed by pavement materials.

When the pavement temperature rises, the heat is transferred to the ice layer, which reduces the adhesion strength between the ice and the pavement [90]. Thus, it is easy to use mechanical devices to break and clean the ice layer [91].

$$c_\sigma = \sigma = F_1/S \tag{8}$$

$$c_q = \tau = F_2/S \tag{9}$$

where c_σ is the normal freezing adhesion coefficient, c_q is the tangential freezing adhesion coefficient, F_1 is the normal pull-out force, F_2 is the tangential pull-out force, and S is the area of the frozen interface.

Some indicators were developed for quantifying microwave de-icing efficiency. Tang et al. [92] proposed that the surface power density of the asphalt mixture can be used to evaluate de-icing efficiency. Jiao et al. [93] pointed out that the time for the ice–pavement interface temperature to reach 0 °C can be used as an evaluation indicator of microwave de-icing efficiency. Li et al. [94] found that the ice–pavement adhesion fails when the temperature reaches 3 °C. They believed that the time for the ice–pavement interface temperature to reach 3 °C is more accurate to evaluate de-icing efficiency. Gao et al. [95] evaluated the de-icing efficiency via the time for the ice to fall off the asphalt mixture under a horizontal force. Liu et al. [19] proposed that the ice melting speed (IMS), which is defined as in Equation (10), can also reflect the de-icing efficiency to a certain extent.

$$IMS = \frac{m_1 - m_2}{t} \tag{10}$$

where IMS is the ice melting speed, g/s; m_1 is the mass of the sample and ice before heating, g; m_2 is the mass of the sample and ice after heating, g; and t is the heating time, s.

4.2. Microwave De-icing Characteristics

The characteristics of microwave de-icing involve multiple factors, such as microwave frequency, microwave electric field strength, pavement materials, ambient temperature, ice thickness, and the microwave heating method [96]. As early as 1986, Monson began the research into microwave de-icing characteristics [97]. Some additives and alternatives, such as anthracite [98] and taconite [99,100], were included into the asphalt pavement to improve the de-icing capacity. Zhao et al. [101] developed a kind of asphalt mixture that can improve de-icing efficiency, in which magnetic powder completely or partially replaces limestone powder and magnetic metallurgical slag partially replaces natural coarse aggregate. Zhao et al. [102] also prepared a kind of asphalt concrete suitable for microwave absorption by completely or partially replacing calcareous mineral powder with magnetic powder and partially replacing natural aggregate with magnetic sand or silicon carbide sand. It is found that the heating rate of the new asphalt concrete is several times that of traditional asphalt concrete. Wang et al. [103] proposed a preparation method of microwave-absorbing asphalt mixture by adding hydroxy iron powder of 3~7% for improving de-icing efficiency. The results showed that the surface temperature of the material rises rapidly after microwave heating. Wang et al. [104] selected hydroxy iron power, ferroferric oxide, alumina, and expanded graphite as microwave-sensitive coating materials and carried out an ice melting test. It was found that ferroferric oxide and expanded graphite have better microwave-absorbing and ice-melting abilities. Jiao et al. [96] studied the effects of ambient temperature on microwave de-icing efficiency. They found that the lower the environmental temperature, the lower the de-icing efficiency. Some researchers focused on an improvement in microwave heating facilities. Guan et al. [105] designed a schematic of a microwave de-icer. Yang et al. [106] designed a horn antenna for a microwave de-icing device to improve heating uniformity and analyzed its heating characteristics and voltage standing wave ratio. The results showed that the microwave de-icing effect is the best when the slope angle of the horn antenna is 15°. Tang et al. [92], Jiao et al. [93], and Ding et al. [107] investigated the effects of microwave frequency on de-icing efficiency via simulation and laboratory tests. The results showed that the de-icing efficiency of 5.8 GHz microwaves is more than four times that of 2.45 GHz microwaves. The reason for this is that the microwave with a frequency of 5.8 GHz has a shallower penetration depth and higher power density, which can quickly increase the pavement temperature and improve the de-icing efficiency. Table 3 lists the de-icing efficiency of the asphalt mixtures containing different MSAs using 2.45 GHz microwaves. As can be seen from the table, the higher the environmental temperature (ET), the greater the improvement in the de-icing efficiency. The thicker the ice thickness (IT), the lower the de-icing efficiency.

Table 3. De-icing efficiency of the asphalt mixtures containing different MSAs using 2.45 GHz microwaves.

MSAs	Volume/Mass Fraction of MSAs	IT (mm)	ET (°C)	Efficiency Improvement
Ferrite [93]	10%	10	−10	3.1 times
		10	−15	2.8 times
		10	−19	2.9 times
Magnetite [21]	80%	10	−5	8.6 times
		10	−10	8.1 times
		10	−15	6.3 times
Steel slag [86]	80%	-	−5	3.1 times
		-	−20	2.6 times
MHCs [22]	100%	30	−10	5.8 times
		20	−10	6.1 times
		15	−10	6.6 times
		10	−10	8.9 times
		5	−10	9.2 times
		10	−20	7.8 times
2# steel wool fibers [95]	1%	-	−5	7 times
		-	−10	5.5 times
0# steel wool fibers	0.7%	-	−5	4.6 times
		-	−10	4.4 times
000# steel wool fibers	0.3%	-	−5	3.9 times
		-	−10	3.1 times
ACP [19]	100%	50	−15	2.5 times
Carbon fiber [108]	0.45%	-	−10	2.7 times

In summary, most investigations focus on the addition of MSAs for improving de-icing efficiency, while there are fewer studies on microwave frequency.

5. Asphalt Aging in Microwave Heating Process

Both the applications of microwave for self-healing and de-icing are based on heating the asphalt pavement, which may lead to asphalt pavement aging. It is necessary to study the asphalt aging caused by microwave heating. Flores et al. [109] compared asphalt aging induced by microwave heating and infrared radiant heating through the tests of penetration, softening point, and rheological properties. The results showed that microwave heating has less of an effect on the asphalt binder. Wu et al. [110] found that microwave heating has no perceptible negative effect on asphalt aging compared with TFOT by analyzing the indicators of penetration, ductility, and softening point. Fernandez et al. [111] quantified the effects of microwave heating and long-term aging on the rheological and chemical performance of recovered asphalt binders using a frequency sweep test and Fourier transform infrared spectrometry (FTIR) analysis. The results showed that microwave heating has little influence on the aging of the rheological properties of asphalt binders. With the increase in microwave heating and long-term aging cycles, the carbonyl and sulfoxide indices increase in both phases. Lou and Sha et al. [112,113] evaluated the physical and rheological properties and infrared spectra of asphalt binders obtained from steel slag asphalt mixtures under different microwave heating cycles via the tests of penetration, softening point, DSR, and FTIR spectroscopy. The results indicated that the penetration value as well as the softening point get worse during the first 10 microwave heating cycles. Compared with the aging methods of RTFOT and PAV, microwave heating has no significant effect on the deterioration of the physical properties of asphalt binders. In summary, several researchers have pointed out that microwave heating has little effect on asphalt aging. However, it is necessary to completely evaluate the aging of asphalt pavements under cycling microwave heating.

6. Summary

This paper reviewed asphalt pavement self-healing and de-icing using microwave heating. Some findings are summarized as follows.

(1) Microwave sensitivity and heating uniformity are two important aspects of microwave heating properties for asphalt pavements. They determine the microwave absorption ability and heating efficiency of the asphalt mixtures and have an important effect on the self-healing quality and de-icing efficiency. Adding MSAs (such as graphite powder, magnetite powder, steel slag aggregates, and steel fiber) into asphalt mixtures can improve microwave sensitivity and heating uniformity.

(2) Microwave heating promotes the self-healing of asphalt pavements by reducing the viscosity of asphalt binders and accelerating the capillary flow and molecular diffusion rate. Several indicators have been established for quantitatively evaluating the self-healing performance of asphalt mixtures, but the results between different studies are incomparable. The reason for this is that there is no corresponding standard for the self-healing test and assessment of the asphalt mixtures. In addition, most indicators are based on performance tests rather than the self-healing mechanism of asphalt pavements.

(3) Microwave heating mode and time as well as material properties have an important influence on the self-healing performance of asphalt pavements. The healing level of asphalt pavements gradually increases with the increase in the microwave heating time. The addition of MSAs also contributes to self-healing. This is mainly because MSAs can improve microwave-absorbing ability. However, the effect of MSAs on pavement performance should be studied in depth.

(4) Pavement materials and microwave frequency are two important factors affecting de-icing efficiency. Most investigations focus on the addition of MSAs for improving de-icing efficiency, while there are fewer studies on microwave frequency. The influence of microwave frequency should be further studied. In addition, many studies on de-icing characteristics are based on laboratory simulation experiments. Field tests should be conducted to verify the indoor test results.

7. Outlooks

Extensive research has been conducted in the past to study the application of microwave heating to asphalt pavement self-healing and de-icing. The following recommendations for future studies are summarized.

(1) The cost of MSAs and their effect on pavement performance are often ignored in studies. It is suggested to study the feasibility of MSA application in actual projects.

(2) There is no corresponding standard for the self-healing test and assessment of the asphalt mixtures. It is necessary to develop a unified test and assessment standard of asphalt mixture self-healing.

(3) Microwave heating technology has good prospects in promoting asphalt pavement self-healing and de-icing. However, microwave heating can also result in asphalt aging. Although several researchers have pointed out that microwave heating has little effect on asphalt aging, it is also necessary to completely evaluate the aging of asphalt pavements under cycling microwave heating.

Author Contributions: Conceptualization, Y.M., W.Y. and L.Z.; methodology, Y.M., W.Y. and Z.Z.; validation, W.Y. and Y.M.; investigation, Y.M. and Z.Z.; data curation, W.Y., L.Z. and Y.M.; writing—original draft preparation, L.Z. and Z.Z.; writing—review and editing, Y.M., L.Z. and W.Y.; visualization, W.Y. and Z.Z.; supervision, Y.M.; project administration, Y.M.; funding acquisition, Y.M. All authors have read and agreed to the published version of the manuscript.

Funding: This work was funded by the National Key R&D Program of China (grant number: 2019YFE0117600).

Institutional Review Board Statement: Not applicable.

Data Availability Statement: Not applicable.

Conflicts of Interest: The authors declare no conflict of interest.

References

1. Metaxas, A.A.; Meredith, R.J. *Industrial Microwave Heating;* IET: Stevenage, UK, 1983.
2. Belanger, J.M.R.; Pare, J.R.J.; Poon, O.; Fairbridge, C.; Ng, S.; Mutyala, S.; Hawkins, R. Remarks on various applications of microwave energy. *J. Microw. Power Electromagn. Energy* **2008**, *42*, 24–44. [CrossRef] [PubMed]
3. Bosisio, R.; Spooner, J.; Granger, J. Asphalt road maintenance with a mobile microwave power unit. *J. Microw. Power* **1974**, *9*, 381–386. [CrossRef]
4. Osborne, T.L.; Hutcheson, W.R. Asphalt Compounds and Method for Asphalt Reconditioning Using Microwave Radiation. US4849020A, 18 July 1989.
5. Yalcin, E. Effects of microwave and induction heating on the mechanical and self-healing characteristics of the asphalt mixtures containing waste metal. *Constr. Build. Mater.* **2021**, *286*, 122965. [CrossRef]
6. Sun, Y.; Wu, S.; Liu, Q.; Zeng, W.; Chen, Z.; Ye, Q.; Pan, P. Self-healing performance of asphalt mixtures through heating fibers or aggregate. *Constr. Build. Mater.* **2017**, *150*, 673–680. [CrossRef]
7. Norambuena-Contreras, J.; Garcia, A. Self-healing of asphalt mixture by microwave and induction heating. *Mater. Des.* **2016**, *106*, 404–414. [CrossRef]
8. Liu, W. *Study on Enhancement Mechanisim and Healing Evaluation of Microwave Absorption of Asphalt Mixture;* Southeast University: Nanjing, China, 2018.
9. Zhang, K.; Gao, X.; Zhang, Q.; Li, T.; Chen, H.; Chen, X. Synthesis, characterization and electromagnetic wave absorption properties of asphalt carbon coated graphene/magnetic $NiFe_2O_4$ modified multi-wall carbon nanotube composites. *J. Alloys Compd.* **2017**, *721*, 268–275. [CrossRef]
10. Zhang, K.; Gao, X.; Zhang, Q.; Li, T.; Chen, H.; Chen, X. Preparation and microwave absorption properties of asphalt carbon coated reduced graphene oxide/magnetic $CoFe_2O_4$ hollow particles modified multi-wall carbon nanotube composites. *J. Alloys Compd.* **2017**, *723*, 912–921. [CrossRef]
11. Li, C.; Wu, S.; Chen, Z.; Shu, B.; Li, Y.; Xiao, Y.; Liu, Q. Synthesis of Fe_3O_4-decorated Mg-Al layered double hydroxides magnetic nanosheets to improve anti-ultraviolet aging and microwave absorption properties used in asphalt materials. *Constr. Build. Mater.* **2019**, *220*, 320–328. [CrossRef]
12. Liu, X.; Wei, Z.; Zhao, Y.; Yan, D. Improving the Electromagnetic Properties of Bitumen Using SiC—Fe_3O_4 Composites. *J. Mater. Civ. Eng.* **2021**, *33*, 04021326. [CrossRef]
13. Li, C.; Wu, S.; Shu, B.; Li, Y.; Chen, Z. Microwave absorption and anti-aging properties of modified bitumen contained SiC attached layered double hydroxides. *Constr. Build. Mater.* **2019**, *227*, 116714. [CrossRef]
14. Liu, X.; Zhao, Y.; Wei, Z.; Zhang, D. Microwave absorption enhancement of asphalt concrete with SiC-Fe_3O_4 mixtures modifier. *Constr. Build. Mater.* **2020**, *254*, 119209. [CrossRef]
15. Deng, Y.; Ma, J.; Lu, T.; Sun, D. Enhanced heating-healing performance of asphalt concrete modified with heterogenous microwave sensitive admixtures. *Constr. Build. Mater.* **2021**, *299*, 123949. [CrossRef]
16. Lou, B.; Sha, A.; Barbieri, D.M.; Liu, Z.; Zhang, F.; Jiang, W. Improved microwave heating uniformity and self-healing properties of steel slag asphalt containing ferrite filler. *Mater. Struct./Mater. Et Constr.* **2021**, *54*, 1–14. [CrossRef]
17. Liu, X.; Yan, D.; Zhao, Y.; Wei, Z. Research on SiC sand as fine aggregate to reinforce thermal performance of asphalt concrete. SiC. *Zhongnan Daxue Xuebao (Ziran Kexue Ban)/J. Cent. South Univ. (Sci. Technol.)* **2021**, *52*, 2459–2469. [CrossRef]
18. Guan, B.; Yang, T.; Xiong, R.; Wang, Y. Electromagnetic wave absorbing properties of asphalt mixture with natural magnetite powder. *Jianzhu Cailiao Xuebao/J. Build. Mater.* **2016**, *19*, 198–203. [CrossRef]
19. Liu, Z.; Yang, X.; Wang, Y.; Luo, S. Engineering properties and microwave heating induced ice-melting performance of asphalt mixture with activated carbon powder filler. *Constr. Build. Mater.* **2019**, *197*, 50–62. [CrossRef]
20. Zhu, X.; Ye, F.; Cai, Y.; Birgisson, B.; Lee, K. Self-healing properties of ferrite-filled open-graded friction course (OGFC) asphalt mixture after moisture damage. *J. Clean. Prod.* **2019**, *232*, 518–530. [CrossRef]
21. Wang, Z.; Wang, H.; An, D.; Ai, T.; Zhao, P. Laboratory investigation on deicing characteristics of asphalt mixtures using magnetite aggregate as microwave-absorbing materials. *Constr. Build. Mater.* **2016**, *124*, 589–597. [CrossRef]

22. Zhang, J.; Yan, Y.; Hu, Z.; Fan, X.; Zheng, Y. Utilization of low-grade pyrite cinder for synthesis of microwave heating ceramics and their microwave deicing performance in dense-graded asphalt mixtures. *J. Clean. Prod.* **2018**, *170*, 486–495. [CrossRef]
23. Trigos, L.; Gallego, J.; Escavy, J.I. Heating potential of aggregates in asphalt mixtures exposed to microwaves radiation. *Constr. Build. Mater.* **2020**, *230*, 117035. [CrossRef]
24. Cao, J. *Study on Self Healing Performance of Basalt Fiber Asphalt Mixture Based on Microwave Heating*; Yangzhou University: Yangzhou, China, 2020.
25. Li, C.; Wu, S.; Chen, Z.; Tao, G.; Xiao, Y. Enhanced heat release and self-healing properties of steel slag filler based asphalt materials under microwave irradiation. *Constr. Build. Mater.* **2018**, *193*, 32–41. [CrossRef]
26. Li, C.; Zeng, G.; Zhou, M.; Fang, Y.; Chen, Z.; Xu, Y.; Ding, S.; Yuan, M.; Li, H.; Wu, S. Controllable synthesis of SiC wrapped LDHs to reinforce microwave absorption and exothermic properties of styrene-butadiene-styrene (SBS) polymer modified asphalt. *Mater. Res. Express* **2021**, *8*, 035501. [CrossRef]
27. Wang, H.; Zhang, Y.; Zhang, Y.; Feng, S.; Lu, G.; Cao, L. Laboratory and Numerical Investigation of Microwave Heating Properties of Asphalt Mixture. *Materials* **2019**, *12*, 146. [CrossRef] [PubMed]
28. Wang, H.; Yang, J.; Lu, G.; Liu, X. Accelerated healing in asphalt concrete via laboratory microwave heating. *J. Test. Eval.* **2020**, *48*. [CrossRef]
29. Wu, S.; Yang, J.; Yang, R.; Zhu, J.; Liu, S. Preparation and properties of microwave-absorbing asphalt mixtures containing graphite and magnetite powder. *J. Test. Eval.* **2021**, *49*, 573–589. [CrossRef]
30. Liu, W.; Wang, S.; Gu, X. Improving microwave heating efficiency of asphalt concrete by increasing surface magnetic loss of aggregates. *Road Mater. Pavement Des.* **2020**, *21*, 950–964. [CrossRef]
31. Khavandi Khiavi, A.; Asadi, M. Effect of specific heat capacity of aggregates and nano-graphite on self-healing of hot mix asphalt under microwave radiation. *Constr. Build. Mater.* **2022**, *328*, 127091. [CrossRef]
32. Liu, Z. *Investigation on Properties of Steel Wool Fiber Reinforced Asphalt Mixture Based on Induction Heating*; Chang'an University: Xi'an, China, 2018.
33. Chen, L. *The Study of Asphalt's Electric and Thermal Field Based on Microwave Heating and Its Control*; Anhui Polytechnic University: Wuhu, China, 2016.
34. Zhu, S.-Q.; Shi, J.-F.; Sun, T.-S.; McGean, T.J. Heat transfer model and numerical simulation for microwave hot in-place recycling of asphalt pavements. In Proceedings of the 1st International Symposium on Transportation and Development Innovative Best Practices 2008, TDIBP 2008, Beijing, China, 24–26 April 2008; pp. 462–467.
35. Sun, T.; Shi, J.; Zhang, Z.; Zheng, J. Temperature control strategies for microwave hot inplace recycling of asphalt pavements. In Proceedings of the 2009 International Workshop on Intelligent Systems and Applications, ISA 2009, Wuhan, China, 23–24 May 2009.
36. Sun, T.; Zhang, Z.; Tang, L. Research on electromagnetic field optimization for microwave hot recycling of asphalt mixtures. In Proceedings of the 2011 International Conference on Material Science and Engineering Technology, ICMSET 2011, Zhengzhou, China, 11–13 November 2011; pp. 575–579.
37. Sun, T. Optimized antenna for asphalt mixture recycling based on microwave heating. *J. Microw. Power Electromagn. Energy* **2015**, *49*, 245–255. [CrossRef]
38. Fakhri, M.; Bahmai, B.B.; Javadi, S.; Sharafi, M. An evaluation of the mechanical and self-healing properties of warm mix asphalt containing scrap metal additives. *J. Clean. Prod.* **2020**, *253*, 119963. [CrossRef]
39. Hao, J. Important topics in development of asphalt concrete technology in China. *J. Hydraul. Eng.* **2008**, *39*, 1213–1219.
40. Bazin, P.; Saunier, J. Deformability, fatigue and healing properties of asphalt mixes. In Proceedings of the International Conference on the Structural Design of Asphalt Pavements, Ann Arbor, MI, USA; 1967; pp. 553–569. Available online: https://trid.trb.org/view/100855 (accessed on 25 March 2023).
41. Al-Ohaly, A.A.; Terrel, R.L. *Effect of Microwave Heating on Adhesion and Moisture Damage of Asphalt Mixtures*; National Research Council, Transportation Research Board: Washington, DC, USA, 1988; Available online: https://trid.trb.org/view/295675 (accessed on 25 March 2023).
42. Gallego, J.; del Val, M.A.; Contreras, V.; Páez, A. Heating asphalt mixtures with microwaves to promote self-healing. *Constr. Build. Mater.* **2013**, *42*, 1–4. [CrossRef]
43. Li, Y.; Hao, P.; Zhang, M. Fabrication, characterization and assessment of the capsules containing rejuvenator for improving the self-healing performance of asphalt materials: A review. *J. Clean. Prod.* **2021**, *287*, 125079. [CrossRef]
44. Anupam, B.R.; Sahoo, U.C.; Chandrappa, A.K. A methodological review on self-healing asphalt pavements. *Constr. Build. Mater.* **2022**, *321*, 126395. [CrossRef]
45. Schapery, R.A. On the mechanics of crack closing and bonding in linear viscoelastic media. *Int. J. Fract.* **1989**, *39*, 163–189. [CrossRef]
46. Lytton, R.L.; Uzan, J.; Fernando, E.G.; Roque, R.; Hiltunen, D.; Stoffels, S.M. *Development and Validation of Performance Prediction Models and Specifications for Asphalt Binders and Paving Mixes*; Strategic Highway Research Program: Washington, DC, USA, 1993; Volume 357.
47. De Gennes, P.-G. Reptation of a polymer chain in the presence of fixed obstacles. *J. Chem. Phys.* **1971**, *55*, 572–579. [CrossRef]
48. Wool, R.P.; O'Connor, K.M. A theory crack healing in polymers. *J. Appl. Phys.* **1981**, *52*, 5953–5963. [CrossRef]

49. Little, D.N.; Bhasin, A. Exploring Mechanism of healing in asphalt mixtures and quantifying its impact. *Self-Heal. Mater.* **2008**, *100*, 205–218.
50. Loeber, L.; Sutton, O.; Morel, J.; Valleton, J.M.; Muller, G. New direct observations of asphalts and asphalt binders by scanning electron microscopy and atomic force microscopy. *J. Microsc.* **1996**, *182*, 32–39. [CrossRef]
51. Kringos, N.; Scarpas, A.; Pauli, T.; Robertson, R. A thermodynamic approach to healing in bitumen. In *Advanced Testing and Characterization of Bituminous Materials, Two Volume Set*; CRC Press: Boca Raton, FL, USA, 2009; pp. 139–148.
52. García, Á. Self-healing of open cracks in asphalt mastic. *Fuel* **2012**, *93*, 264–272. [CrossRef]
53. Garcia, A.; Bueno, M.; Norambuena-Contreras, J.; Partl, M.N. Induction healing of dense asphalt concrete. *Constr. Build. Mater.* **2013**, *49*, 1–7. [CrossRef]
54. Kargari, A.; Arabani, M.; Mirabdolazimi, S.M. Effect of palm oil capsules on the self-healing properties of aged and unaged asphalt mixtures gained by resting period and microwave heating. *Constr. Build. Mater.* **2022**, *316*, 125901. [CrossRef]
55. Wang, R.; Xiong, Y.; Ma, X.; Guo, Y.; Yue, M.; Yue, J. Investigating the differences between steel slag and natural limestone in asphalt mixes in terms of microscopic mechanism, fatigue behavior and microwave-induced healing performance. *Constr. Build. Mater.* **2022**, *328*, 127107. [CrossRef]
56. Franesqui, M.A.; Yepes, J.; Gallego, J. Ultrasound Monitoring and Microwave Self-healing of Top-Down Cracks in Asphalt Pavements. In *Lecture Notes in Civil Engineering*; Springer: Berlin/Heidelberg, Germany, 2020; Volume 76, pp. 263–273.
57. Lou, B.; Sha, A.; Li, Y.; Wang, W.; Liu, Z.; Jiang, W.; Cui, X. Effect of metallic-waste aggregates on microwave self-healing performances of asphalt mixtures. *Constr. Build. Mater.* **2020**, *246*, 118510. [CrossRef]
58. Liu, Q.; Schlangen, E.; Van De Ven, M.; Van Bochove, G.; Van Montfort, J. Evaluation of the induction healing effect of porous asphalt concrete through four point bending fatigue test. *Constr. Build. Mater.* **2012**, *29*, 403–409. [CrossRef]
59. Shan, L.; Tan, Y.; Richard Kim, Y. Establishment of a universal healing evaluation index for asphalt binder. *Constr. Build. Mater.* **2013**, *48*, 74–79. [CrossRef]
60. Daniel, J.S.; Kim, Y.R. Laboratory Evaluation of Fatigue Damage and Healing of Asphalt Mixtures. *J. Mater. Civ. Eng.* **2001**, *13*, 434–440. [CrossRef]
61. Kim, Y.-R.; Little, D.N.; Lytton, R.L. Fatigue and Healing Characterization of Asphalt Mixtures. *J. Mater. Civ. Eng.* **2003**, *15*, 75–83. [CrossRef]
62. Norambuena-Contreras, J.; Gonzalez, A.; Concha, J.L.; Gonzalez-Torre, I.; Schlangen, E. Effect of metallic waste addition on the electrical, thermophysical and microwave crack-healing properties of asphalt mixtures. *Constr. Build. Mater.* **2018**, *187*, 1039–1050. [CrossRef]
63. Norambuena-Contreras, J.; Gonzalez-Torre, I. Influence of the Microwave Heating Time on the Self-Healing Properties of Asphalt Mixtures. *Appl. Sci.* **2017**, *7*, 1076. [CrossRef]
64. Gonzalez, A.; Norambuena-Contreras, J.; Storey, L.; Schlangen, E. Effect of RAP and fibers addition on asphalt mixtures with self-healing properties gained by microwave radiation heating. *Constr. Build. Mater.* **2018**, *159*, 164–174. [CrossRef]
65. Zhu, H.; Yuan, H.; Liu, Y.; Fan, S.; Ding, Y. Evaluation of Self-Healing Performance of Asphalt Concrete for Macrocracks via Microwave Heating. *J. Mater. Civ. Eng.* **2020**, *32*, 04020248. [CrossRef]
66. Wang, Z.; Dai, Q.; Porter, D.; You, Z. Investigation of microwave healing performance of electrically conductive carbon fiber modified asphalt mixture beams. *Constr. Build. Mater.* **2016**, *126*, 1012–1019. [CrossRef]
67. Phan, T.M.; Park, D.-W.; Le, T.H.M. Crack healing performance of hot mix asphalt containing steel slag by microwaves heating. *Constr. Build. Mater.* **2018**, *180*, 503–511. [CrossRef]
68. Gonzalez, A.; Valderrama, J.; Norambuena-Contreras, J. Microwave crack healing on conventional and modified asphalt mixtures with different additives: An experimental approach. *Road Mater. Pavement Des.* **2019**, *20*, S149–S162. [CrossRef]
69. Wan, P.; Liu, Q.; Wu, S.; Zhao, Z.; Chen, S.; Zou, Y.; Rao, W.; Yu, X. A novel microwave induced oil release pattern of calcium alginate/ nano-Fe3O4 composite capsules for asphalt self-healing. *J. Clean. Prod.* **2021**, *297*, 126721. [CrossRef]
70. Kuemmel, D.; Hanbali, R. Accident Analysis of Ice Control Operations. 1992. Available online: https://epublications.marquette.edu/transportation_trc-ice/2/ (accessed on 25 March 2023).
71. Liu, J.-B.; Du, C.-G. Study on prediction model of friction coefficient of ICY road surface. In Proceedings of the 19th COTA International Conference of Transportation Professionals: Transportation in China—Connecting the World, CICTP 2019, Nanjing, China, 6–8 July 2019; pp. 3389–3396.
72. Öberg, G. Friction and Journey Speed on Roads with Various Winter Road Maintenance. 1981. Available online: https://trid.trb.org/view/187066 (accessed on 25 March 2023).
73. Malin, F.; Norros, I.; Innamaa, S. Accident risk of road and weather conditions on different road types. *Accid. Anal. Prev.* **2019**, *122*, 181–188. [CrossRef]
74. Tan, T.; Xing, C.; Tan, Y.; Gong, X. Safety aspects on icy asphalt pavement in cold region through field investigations. *Cold Reg. Sci. Technol.* **2019**, *161*, 21–31. [CrossRef]
75. Jelisejevs, B. Alternative methods of de-icing on highways. *Mot. Riga* **2001**, *3*, 31–34.
76. Hu, Z.D.; Du, S.R.; Shen, B.C.; Wang, L.H. Mechanical Property Analysis on Cutting Tool of the Ice and Snow Removing Machine Based on ANSYS. *Appl. Mech. Mater* **2015**, *779*, 74–79. [CrossRef]
77. Shi, X.; Jungwirth, S.; Akin, M.; Wright, R.; Fay, L.; Veneziano, D.A.; Zhang, Y.; Gong, J.; Ye, Z. Evaluating snow and ice control chemicals for environmentally sustainable highway maintenance operations. *J. Transp. Eng.* **2014**, *140*, 05014005. [CrossRef]

78. Nixon, W.A. *Improved Cutting Edges for Ice Removal*; Strategic Highway Research Program: Washington, DC, USA, 1993.
79. Guo, H.; Liu, H. Research progress and prospect of testing and application technology of highway snow-melting agent at home and abroad. In Proceedings of the 2011 2nd International Conference on Mechanic Automation and Control Engineering, MACE 2011, Inner Mongolia, China, 15–17 July 2011; pp. 6807–6811.
80. Zhang, K.; Luo, Y.; Li, Z.; Zhao, Y.; Zhao, Y. Evaluation of Performance Deterioration Characteristics of Asphalt Mixture in Corrosion Environment Formed by Snow-Melting Agents. *J. Mater. Civ. Eng.* **2022**, *34*, 04021481. [CrossRef]
81. Chen, H.; Xu, J.; Wu, Y.; Liu, J.; Huang, H. Surface during microwave deicing of airport pavement. *Materials* **2020**, *13*, 3557. [CrossRef]
82. Liu, K.; Xu, P.; Wang, F.; Jin, C.; Huang, M.; Dai, D.; Fu, C. Deicing efficiency analysis and economic-environment assessment of a novel induction heating asphalt pavement. *J. Clean. Prod.* **2020**, *273*, 123123. [CrossRef]
83. Pan, P.; Wu, S.; Xiao, F.; Pang, L.; Xiao, Y. Conductive asphalt concrete: A review on structure design, performance, and practical applications. *J. Intell. Mater. Syst. Struct.* **2015**, *26*, 755–769. [CrossRef]
84. Wu, C.; Geng, H.; Tan, S.; Lv, J.; Wang, H.; He, Z.; Wang, J. Highly efficient solar anti-icing/deicing: Via a hierarchical structured surface. *Mater. Horiz.* **2020**, *7*, 2097–2104. [CrossRef]
85. Jiang, X.; Li, G.; Wu, Z. Deicing and corrosive performances of calcium acetate deicer made from bamboo-Vinegar. *World Acad. Sci. Eng. Technol.* **2010**, *41*, 506–511.
86. Gao, J.; Sha, A.; Wang, Z.; Tong, Z.; Liu, Z. Utilization of steel slag as aggregate in asphalt mixtures for microwave deicing. *Journal of Cleaner Production* **2017**, *152*, 429–442. [CrossRef]
87. Xiang, H.; Xia, H.; Chen, Y.; Wu, Y.; Chen, H.; Yan, M. Pavement anti-icing coating based on a functional composite of NaCl microcapsules. *Constr. Build. Mater.* **2021**, *307*, 125010. [CrossRef]
88. Chatterji, S. Aspects of the freezing process in a porous material–water system: Part 1. Freezing and the properties of water and ice. *Cem. Concr. Res.* **1999**, *29*, 627–630. [CrossRef]
89. Meredith, R.J. *Engineers' Handbook of Industrial Microwave Heating*; IET: Stevenage, UK, 1998.
90. Li, X.; Xu, Y.G.; Liu, F.L. Application of microwave heating for ice removal on streets. *Harbin Gongye Daxue Xuebao/J. Harbin Inst. Technol.* **2003**, *35*, 1342–1343.
91. Gao, J.; Zhang, Z.; Han, Z.; Sha, A.; Wang, Z.; Jiang, W. A review of electromagnetic wave absorbing materials used in microwave deicing pavement. *Mater. Rep.* **2016**, *30*, 87–95.
92. Tang, X.W.; Jiao, S.J.; Gao, Z.Y.; Xu, X.L. Study of 5.8 GHz magnetron in microwave deicing. *J. Electromagn. Waves Appl.* **2008**, *22*, 1351–1360. [CrossRef]
93. Jiao, S.-J.; Tang, X.-W.; Gao, Z.-Y.; Wang, Q. Study of key technology on microwave deicing efficiency. *Zhongguo Gonglu Xuebao/China J. Highw. Transp.* **2008**, *21*, 121–126.
94. Li, B.; Sun, Y.; Liu, Q.; Fang, H.; Wu, S.; Tang, J.; Ye, Q. Ice melting properties of steel slag asphalt concrete with microwave heating. In Proceedings of the 17th IUMRS International Conference in Asia, IUMRS-ICA 2016, Qingdao, China, 20–24 October 2016.
95. Gao, J.; Guo, H.; Wang, X.; Wang, P.; Wei, Y.; Wang, Z.; Huang, Y.; Yang, B. Microwave deicing for asphalt mixture containing steel wool fibers. *J. Clean. Prod.* **2019**, *206*, 1110–1122. [CrossRef]
96. Jiao, S.-J.; Tang, X.-W.; Gao, Z.-Y.; Wang, Q. Influence of environmental temperature on road de-icing efficiency using microwave. *Chang. Daxue Xuebao (Ziran Kexue Ban)/J. Chang. Univ. (Nat. Sci. Ed.)* **2008**, *28*, 85–88.
97. Kulash, D. *Strategic Highway Research Program*; American Association of State Highway and Transportation Officials: Washington, DC, USA, 1991.
98. Long, W.H. Asphaltic Compositions and Uses Therefor. US6193793B1, 27 February 2001.
99. Hopstock, D. Microwave-Absorbing Road Construction and Repair Material. *Final Report to NRRI on Idea Evaluation Subcontract*. 2003.
100. Hopstock, D.M.; Zanko, L.M. Minnesota Taconite as a Microwave-Absorbing Road Aggregate Material for Deicing and Pothole Patching Applications. 2005. Available online: https://www.cts.umn.edu/publications/report/minnesota-taconite-as-a-microwave-absorbing-road-aggregate-material-for-deicing-and-pothole-patching-applications (accessed on 25 March 2023).
101. Zhao, P.; Ai, T.; Wang, Z. A Composition of Slag Asphalt Concrete Pavement Material for Microwave Heating. CN101774786B, 16 January 2013.
102. Zhao, P.; Wang, Z.; Ai, T. A Microwave Absorbing Asphalt Concrete Pavement Material Composition. CN101736671B, 29 June 2011.
103. Wang, X.; Wei, Q.; Zhou, P.; Cai, L.; Yuan, C.; Feng, G.; Zhang, S.; Wang, H.; Tang, Y.; Li, L.; et al. A Microwave Absorbing Asphalt Pavement Mixture and Its Preparation Method. CN104310860A, 28 January 2015.
104. Wang, J.; Wang, X.; Ding, L.; Fu, L. Microwave sensitive coating materials and equipment for snow removal. *J. Chang. Univ. (Nat. Sci. Ed.)* **2018**, *38*, 49–57.
105. Guan, M.-H.; Xu, Y.-G.; Lu, T.-J.; Xu, C.-H. Application of microwave heating on removing ice on streets. *Beifang Jiaotong Daxue Xuebao/J. North. Jiaotong Univ.* **2003**, *27*, 79.
106. Yang, M.-H.; Zhao, Q.; Li, H.-F.; Sun, X.; Ju, H.-J. Design of sloped horn antenna for microwave deicing of asphalt pavements. *Qiangjiguang Yu Lizishu/High Power Laser Part. Beams* **2007**, *19*, 1883–1886.
107. Ding, L.T.; Wang, X.C.; Zhang, W.G.; Wang, S.; Zhao, J.; Li, Y.Q. Microwave Deicing Efficiency: Study on the Difference between Microwave Frequencies and Road Structure Materials. *Appl. Sci.* **2018**, *8*, 2360. [CrossRef]

108. Wang, Z.; Gao, J.; Ai, T.; Zhao, P. Laboratory investigation on microwave deicing function of micro surfacing asphalt mixtures reinforced by carbon fiber. *J. Test. Eval.* **2014**, *42*, 1–10. [CrossRef]
109. Flores, G.; Gallego, J.; Giuliani, F.; Autelitano, F. Aging of asphalt binder in hot pavement rehabilitation. *Constr. Build. Mater.* **2018**, *187*, 214–219. [CrossRef]
110. Wu, S.; Yang, J.; Yang, R.; Zhu, J.; Liu, S. Investigation on Microwave Heating Technology for Rutting Maintenance in Asphalt Pavement. *J. Test. Eval.* **2018**, *48*, 2998–3011. [CrossRef]
111. Fernandez, M.; Canon, G.; Leischner, S.; Rochlani, M.; Norambuena-Contreras, J.; Gonzalez, A. Effects of microwave heating and long-term aging on the rheological and chemical properties of recovered bitumen. *Materials* **2021**, *14*, 7787. [CrossRef]
112. Lou, B.; Sha, A.; Barbieri, D.M.; Zhang, X.; Chen, H.; Hoff, I. Evaluation of microwave aging impact on asphalt mixtures. *Road Mater. Pavement Des.* **2022**, *24*, 730–743. [CrossRef]
113. Sha, A.; Lou, B.; Barbieri, D.M.; Hoff, I. Microwave Heating as an Innovative Road Maintenance Technology: Aging Effect on Binder and Feasibility Evaluation. *Materials* **2022**, *15*, 316. [CrossRef] [PubMed]

polymers

MDPI

Article

Emission of Volatile Organic Compounds in Crumb Rubber Modified Bitumen and Its Inhibition by Using Montmorillonite Nanoclay

Gang Liu [1], Shuaiyin Fang [1], Yong Wang [2], Jinjun Liu [3,*], Yangshi Liang [1], Tingwei Cao [1] and Quantao Liu [1,*]

1 State Key Laboratory of Silicate Materials for Architectures, Wuhan University of Technology, Wuhan 430070, China
2 College of Water and Architectural Engineering, Shihezi University, Shihezi 832003, China
3 School of Materials Science and Engineering, Wuhan University of Technology, Wuhan 430070, China
* Correspondence: whutm@whut.edu.cn (J.L.); liuqt@whut.edu.cn (Q.L.)

Abstract: Bitumen emits a large amount of volatile organic compounds (VOCs) during the production and construction of asphalt mixture, which can cause both environmental hazards and health risks. In this study, a setup was designed to collect the VOCs released by base and crumb rubber-modified bitumen (CRMB) binders and their composition was characterized by thermal desorption-gas chromatography-mass spectrometry (TD-GC-MS). Next, organic montmorillonite (Mt) nanoclay was added into CRMB binder and its inhibiting effect on the VOCs emission of the binder was investigated. Finally, the VOCs emission models for the CRMB and Mt-modified CRMB (Mt-CRMB) binders were established according to reasonable assumptions. The results indicated that the VOCs emission of CRMB binder was 3.2 times larger than that of the base binder. Due to its intercalated structure, the nanoclay can reduce the VOCs emission of CRMB binder by 30.6%. Especially, its inhibition effects on alkanes, olefins, and aromatic hydrocarbons were more significant. After finite element verification, the established model based on the Fick's second law can describe the emission behavior of CRMB and Mt-CRMB binders well. Overall, the Mt nanoclay can be used as an effective modifier to inhibit the VOCs emission of CRMB binder.

Keywords: VOCs; crumb rubber; bitumen; organic montmorillonite; inhibition

Citation: Liu, G.; Fang, S.; Wang, Y.; Liu, J.; Liang, Y.; Cao, T.; Liu, Q. Emission of Volatile Organic Compounds in Crumb Rubber Modified Bitumen and Its Inhibition by Using Montmorillonite Nanoclay. *Polymers* **2023**, *15*, 1513. https://doi.org/10.3390/polym15061513

Academic Editors: Xavier Colom and Klaus Werner Stöckelhuber

Received: 25 January 2023
Revised: 21 February 2023
Accepted: 15 March 2023
Published: 18 March 2023

1. Introduction

Every year, over 1.6 billion new tires and around 1 billion waste tires are generated [1]. Approximately 18 million tons of waste tires in China are produced yearly, but their recycling rate is less than 40% [2]. The accumulation of a large number of waste tires occupies land resources, and it has the risk of auto-ignition, which can produce a large amount of harmful smoke [3]. At present, a broad application of the waste tires is to produce crumb rubber (CR) powder to modify bitumen binder. Many studies have shown that asphalt mixture with crumb rubber-modified bitumen (CRMB) binder can effectively improve the resistance to the high-temperature deformation, fatigue performance, and rheological performance [4–7].

Bitumen can emit large amounts of volatile organic compounds (VOCs) during the production of asphalt mixture and the construction of asphalt pavement [8]. The CRMB mixture needs a higher production temperature, and its VOCs emission is larger than the traditional hot-mix asphalt [9]. Bitumen is composed of saturate, aromatic, resin, and asphaltene. Among them, aromatic is the most harmful and contains benzene, naphthalene, anthracene, phenanthrene, pyridine, acridine, carbazole, phenol, and other volatile substances [10]. With long-term exposure to such an environment, the harmful substances will enter the human body through skin contact and the respiratory tract, seriously harming the health of humans [11,12]. The influence of bitumen fumes, including VOCs, is attracting more and more attention. The International Agency for Research on Cancer listed bitumen fumes as a suspected carcinogen in 1987 [13]. In 2015, China implemented new regulations (GB31570-2015), which set the emission standard of asphalt fumes at 20 mg/m^3 [14,15]. Therefore, the emission problem of bitumen VOCs is one of the important factors limiting the further expansion of bitumen applications.

Lots of research has been carried out on how to inhibit the emission of bitumen VOCs. Temperature is an important factor affecting the VOCs emission of bitumen [16]. Compared with the traditional hot-mix asphalt (HMA), the warm-mix asphalt (WMA) technology can reduce the construction temperature by 20–40 °C, down to 100–140 °C [17]. The WMA containing the CR powder can effectively reduce polycyclic aromatic hydrocarbon emissions [18]. However, the warm-mix agents decreased the low-temperature cracking and moisture damage resistances of the asphalt mixture [19]. The nano-sized calcium carbonate and styrene butadiene styrene copolymer (SBS) can be used to absorb the VOCs emission of bitumen [20]. However, SBS polymer is susceptible to aging and degradation at high temperatures, which will reduce the inhibition effect of SBS on bitumen VOCs [21,22]. In addition, some inorganic materials such as zeolite, layered double hydroxides, and expanded graphite have been proven to be used as asphalt VOCs inhibitors, but the compatibility between the inorganic materials and asphalt is poor, tending to cause segregation problems at high temperatures [23–25].

Montmorillonite (Mt) is a widely available and inexpensive mineral. Its unit cell has a crystal structure comprising a 2:1 layered silicate, two silicon-oxygen tetrahedron layers, and a sandwiched layer of aluminum-oxygen octahedrons [26,27]. The surface of its silicate lamellar layer is negatively charged and is composed of accumulation by electrostatic interaction between layers [28]. Montmorillonite has unique one-dimensional layered nanostructures and cation exchange properties, which makes it possible to modify and expand the application field [29,30]. Modification by organic compounds can significantly improve its utilization rate. It has been reported that organic Mt can improve the anti-aging performance of bitumen, and the anti-aging performance was better than SBS-modified bitumen when the content exceeded 3% [31]. In addition, the physical and chemical stability of montmorillonite sheet can further enhance the flame-retardant performance of polymers [32]. Moreover, 5 wt.% Mt containing alkyl surfactants reduced VOCs emissions of base bitumen by more than 50% through interlayer restriction of volatilization of the light components of the bitumen [33]. Organic Mt with a particle size greater than 200 mesh size significantly reduced the types and quantities of SBS-modified bitumen VOCs [34]. The research above indicates that Mt can not only effectively improve the performance of bitumen, but also has great potential in inhibiting VOCs emission. However, using Mt to inhibit the VOCs emission of CRMB has not been reported.

This study aimed to characterize the VOCs emission of CRMB binder and evaluate the effect of Mt nanoclay on its inhibition. Firstly, a homemade setup was used to collect the VOCs of base and CRMB binders during the heating process, and their composition was characterized. Next, nanoclay was used to modify the CRMB binder, and its inhibition effect on the VOCs emission of the CRMB binder was evaluated. Finally, based on the Fick's second law, the VOCs emission models for the CRMB and Mt-modified CRMB (Mt-CRMB) binders were established.

2. Experimental

2.1. Materials

Base bitumen (IRPC-70) with a penetration grade of 70 was imported from Thailand and used in this study. The basic properties of the base bitumen are presented in Table 1.

Table 1. Basic properties of base bitumen.

Properties	Values	Standard
Penetration at 25 °C (0.1 mm)	66	ASTM D5
Softening point (°C)	48	ASTM D36
Ductility at 15 °C (cm)	181	ASTM D113
Viscosity at 60 °C (Pa·s)	166	ASTM D4402
Viscosity at 135 °C (Pa·s)	0.54	ASTM D4402

One waste tire from a passenger car was chosen to produce the CR powder. Its preparation process was as follows: first, the electric grinder was adopted to grind the tread part of the waste tire to produce the CR powder; next, the magnet was used to remove iron impurities; then, the CR powder was sieved and collected with a typical particle size between 40 and 60 mesh number (i.e., 250–425 μm) to be used in this study; finally, it was washed by the water to remove other impurities (e.g., dust) and dried at 60 °C in the oven for 12 h.

The organic Mt produced by Zhejiang Fenghong Clay Chemical Co. Ltd. (Huzhou, China) was used to modify the CRMB binder, and distearylammonium chloride was used as the surfactant with the content of 9%. The physical properties of montmorillonite are presented in Table 2.

Table 2. Physical properties of montmorillonite.

Properties	Values
Basal spacing (nm)	1.3
Density (g/cm^3)	1.7538
Appearance	White powder
Weight content of surfactant (%)	9

2.2. *Modified Bitumen Preparation*

During the preparation of the CRMB or Mt-CRMB binders, 250 g of base binder was melted and poured into a cylindrical container on the heating plate. Then, 45 g of CR powder (18% by weight) was added to the base binder, dispersing with a low-speed mixer (JJ-1, Shanghai Loikaw Instrument) at 200 rpm for 10 min, at a temperature of 150 °C. For the Mt-CRMB binder, 10 g of Mt (4% by weight) was simultaneously added with the CR powder into the base binder. Next, the temperature was increased to 180 °C, and the binder was sheared with a high-speed shearing mixer (ESR-500, Shanghai ELE Mechanical and Electrical Equipment, Shanghai, China) at 4500 ± 500 rpm for 1 h. All samples were cooled to room temperature and stored for further testing. More information about the bitumen binder modification process in this study can be found in the previous research [33].

2.3. *Test Methods*

2.3.1. VOCs Collection Methods

The VOCs sample collection device is shown in Figure 1. Two VOCs sampling methods were used in this study. Method 1 was used to compare the VOCs characteristics of CR powder, base, and CRMB binders, by opening or closing the clamps, and Method 2 was not only used to characterize the VOCs emission of the binders at different heating times, but also could be used to calculate the VOCs emission rates in order to understand the entire VOCs emission behavior. The specific collection processes for these two methods are outlined below.

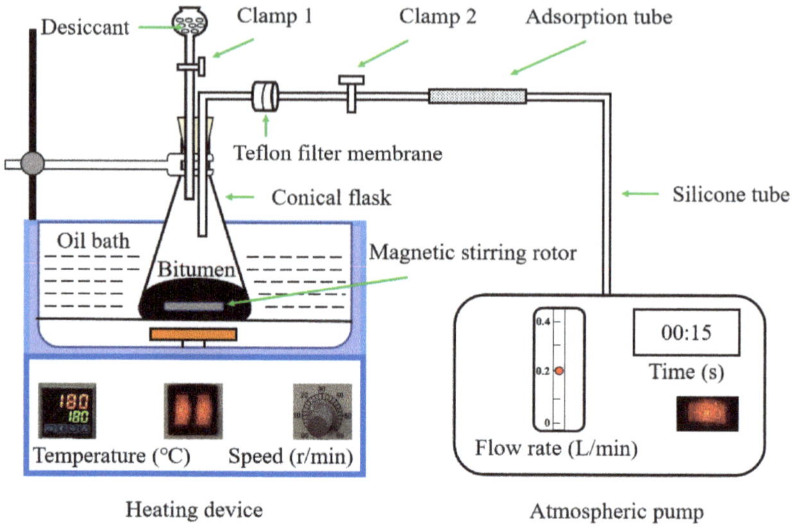

Figure 1. VOCs sample collection device.

Method 1: First, 50 g of binder or the CR powder sample was put into the conical flask, and Clamps 1 and 2 were closed (see Figure 1). When the temperature of the heating device rose to 180 °C, the magnetic stirring switch was turned on and the temperature was held for 10 min. Then, the two clamps were opened, and the VOCs were adsorbed into the ATD (automatic tube dispenser) by using an atmospheric pump at a gas flow rate of 0.2 L/min for 15 s.

Method 2: The CRMB binder or Mt-CRMB binder with the weight of 50 g was poured into the conical flask (see Figure 1). As shown in Figure 2, the VOCs at 180 °C were collected at five sampling points (i.e., 10, 30, 60, 90, and 120 min) to simulate the VOCs release characteristics of the asphalt mixture during the process of mixing, transporting, and paving. The collection time for each point was 15 s, with a flow rate of 0.2 L/min using the pump. Before each collection, the clamps were closed for 10 min to preserve enough VOCs to be absorbed by the ATD.

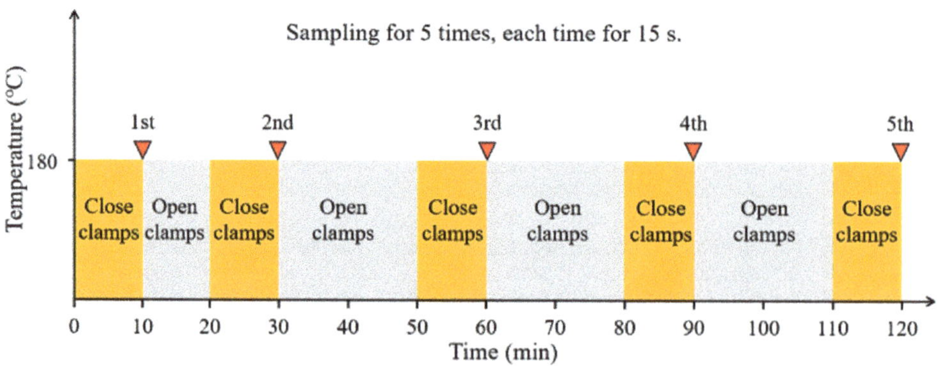

Figure 2. Operation process for Method 2 (sampling five times).

2.3.2. VOCs Characterization

Atomx P&T-Agilent 7890B-5977B thermal desorption-gas chromatography-mass spectrometry (TD-GC-MS, Agilent, Santa Clara, CA, USA) was used to characterize the VOCs emission of the binder. The gas temporarily adsorbed in the automatic tube dispenser

(ATD, PerkinElmer, Waltham, MA, USA) was desorbed in the thermal desorption system and then transferred to the ultra-inert capillary column (30 m × 250 mm × 0.25 mm) using high-purity helium (≥99.999%) at a flow rate of 40 mL/min. The GC-MS system was kept at 30 °C for 3 min and then heated up to 200 °C at 11 °C/min for 3 min. Finally, it was kept at 280 °C for 5 min. The ion-source temperature was maintained at 230 °C, and the MS detector (Agilent, Santa Clara, CA, USA) was operated in full-scan electron ionization (EI) mode, where data over 35–400 m/z were acquired [35,36].

2.3.3. X-ray Diffraction (XRD) Test

A D/Max-RB X-ray diffractometer (Rigaku, Tokyo, Japan) was used to analyze the structural changes of the organic montmorillonite in the CRMB binder, with the conditions: Cu Kα rays (λ = 1.54184 Å), the tube voltage of 40 kV, the tube current of 40 Ma, the step size of 0.02°, the scanning range of 1–10° (2θ), and the scanning rate of 5°/min.

3. Results and Discussion

3.1. VOCs Composition of Base and CRMB Binders

Figure 3 shows the GC-MS chromatograms of VOCs emitted by the base binder, CRMB binder, and pure CR powder. Based on the NIST database, there were 123 organic compounds detected in the base binder VOCs. Commonly, the bitumen VOCs can be classified into alkanes (ALK), olefins (OLE), hydrocarbon derivatives (HYD), aromatic hydrocarbons (ARH), and aromatic hydrocarbon derivatives (ARHD). Among these compounds, there were 49 ALKs, 31 OLEs, 19 HYDs, 24 ARHs, and 0 ARHD. It was indicated that most compounds in the base binder VOCs belonged to the ALK and OLE.

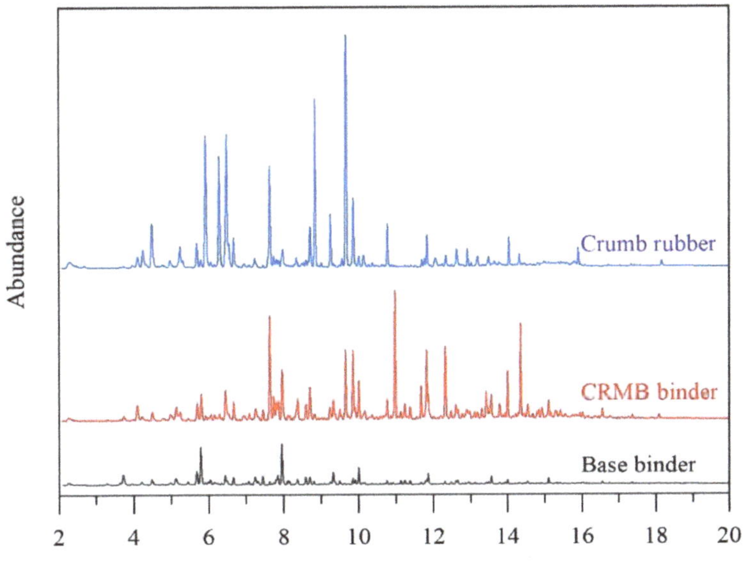

Figure 3. VOCs GC-MS chromatograms of CR powder, base, and CRMB binders.

For the CR powder, 116 VOCs were detected, containing 17 ALKs, 29 OLEs, 62 HYDs, 6 ARHs, and 2 ARHDs. It was indicated that the hydrocarbon derivatives dominated the VOCs of CR powder. Meanwhile, there were two aromatic hydrocarbon derivatives.

The GC-MS results indicated that there were 144 compounds in the CRMB binder VOCs. Among them, there were 42 ALKs, 43 OLEs, 28 HYDs, 29 ARHs, and 2 ARHDs. Compared with the base binder VOCs, all the species numbers increased, except in the ALE. Maybe some alkane compounds in the binder were absorbed by the rubber, and the ALE volatilization decreased. Two aromatic hydrocarbon derivatives were detected, possibly originating from the CR powder with the same VOCs.

In terms of ionic peak intensities, the CR power, base binder, and CRMB binder VOCs were significantly different. For the CR powder VOCs, the abundances of methyl acrolein, methyl vinyl ketone, benzene, 2-ethenyl-2-butenal, and methyl isobutyl ketone were larger. For the base binder VOCs, the abundances of n-hexane and heptane were larger. For the CRMB VOCs, the abundances of benzene, 4-ethenyl-cyclohexene, and D-limonene were larger.

Figure 4 shows the distribution of VOCs species for the CR powder, base, and CRMB binders. Three areas were defined: Area 1 meant the overlap of 21 VOCs among the CR powder, base, and CRMB binders, Area 2 meant 51 joint VOCs between the base and CRMB binders, and Area 3 represented 14 joint compounds between the CR powder and the CRMB binder. It was noted that 51 VOCs in the base binder and 83 VOCs in the CR powder were not detected in the CRMB binder. However, there were 58 new VOCs in the CRMB binder.

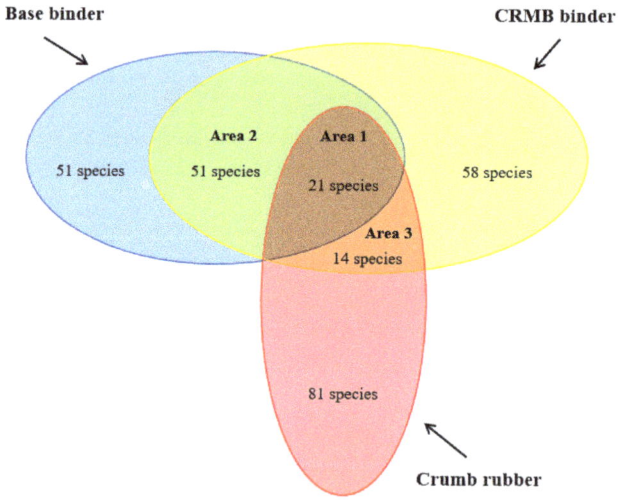

Figure 4. VOCs species overlap for crumb rubber, base, and CRMB binders.

Figure 5 shows the peak area changes of 21 compounds in Area 1 based on the GC-MS chromatograms of the CRMB binder. It indicated that except for pentane and n-hexane, the peak areas of the other 19 compounds increased compared with the base binder. Among them, the peak areas of benzene, cyclohexene, toluene, m-xylene, p-xylene, and propyl benzene increased by 28 times, 10 times, 10 times, 14 times, 17 times, and 11 times, respectively. It can be seen that because the waste CR powder itself contains these compounds, adding the CR powder to the base binder will have a direct impact on the emissions of these compounds in VOCs.

The CR powder mainly introduced 14 VOCs in Area 3. Among them, ethanol, methyl isobutyl ketone, ethyl acetate, 1-pentanol, and D-limonene are often used as essential solvents in the rubber tire. The 4-ethenyl-cyclohexene and styrene are critical raw materials for synthesizing rubber. Benzothiazole was used as the rubber vulcanization accelerant, while 2-methyl-furan, heptanal, octanal, nonanal, (E)-6-dodecane, and 2-ethenyl-2-butenal were used as the intermediates in the rubber synthesis.

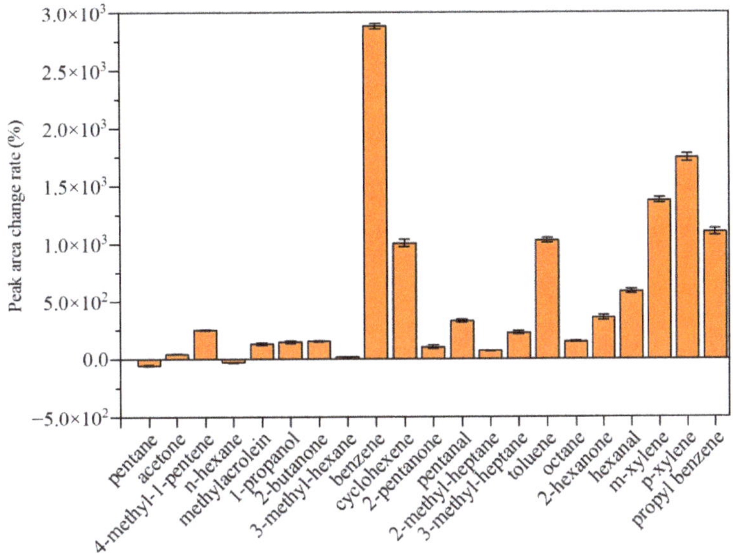

Figure 5. Peak area changes of 21 compounds in Area 1 based on the GC-MS chromatograms of the CRMB binder.

Based on the previous research [37–39], 23 normally used VOCs were selected as the fingerprint compounds for the CRMB binder (see Table 3). Among them, there were 12 species from Area 1, 5 from Area 2, and 6 from Area 3.

Table 3. The 23 VOCs fingerprint compounds from 3 areas and their standard curves for the GC-MS.

Area Code	No.	Retention Time (min)	Compounds	Standard Curve
1	1	3.743	Pentane	y = 6443x
	2	4.499	Acetone	y = 34,348x
	4	5.807	n-Hexane	y = 32,620x
	5	5.924	Methacrolein	y = 43,341x
	8	7.451	3-methyl-hexane	y = 57,274x
	9	7.632	Benzene	y = 116,197x
	10	7.958	Heptane	y = 82,427x
	11	9.326	2-methyl-heptane	y = 58,003x
	13	9.854	Toluene	y = 137,374x
	14	10.006	Octane	y = 160,012x
	16	11.680	m-Xylene	y = 74,470x
	17	11.825	p-Xylene	y = 137,036x
2	3	5.143	2-methyl-pentane	y = 41,944x
	7	6.453	Butanal	y = 33,394x
	19	14.016	Mesitylene	y = 138,233x
	21	15.115	Undecane	y = 124,317x
	22	17.369	Naphthalene	y = 466,879x
3	6	6.288	2-methyl-furan	y = 58,003x
	12	9.656	Methyl Isobutyl Ketone	y = 146,885x
	15	10.974	4-ethenyl-cyclohexene	y = 70,311x
	18	12.331	Styrene	y = 92,293x
	20	14.367	D-Limonene	y = 121,411x
	23	18.080	Benzothiazole	y = 24,376x

The external method was used to obtain the standard curve of the fingerprint compound. First, a standard solution was prepared by mixing 23 fingerprint compounds in the ethanol solvent. Each compound had 5, 10, 25, 50, and 100 ppm concentrations. For other non-fingerprint compounds, the standard curve of toluene was used [9]. After the GC-MS test, the standard curve describing the relationship between the peak area and mass of each compound was obtained as follows:

$$m_i = \frac{S_i}{k_i} \tag{1}$$

where, m_i is the mass of fingerprint compounds, ng, S_i is the peak area of fingerprint compounds, and k_i is the slope of the standard curve for fingerprint compounds, ng^{-1}.

The total amount of bitumen VOCs emission was calculated as follows:

$$\mathrm{m} = \sum_{i=1}^{\mathrm{n}} m_i + \frac{S_t - \sum S_i}{k_{tol}} \tag{2}$$

where, m is the total amount of bitumen VOCs emission, ng, n is the number of fingerprint compounds, S_t is the total peak of all VOCs, and k_{tol} is the slope of the standard curve measured by the standard toluene solution prepared in this study, ng^{-1}.

Figure 6 shows the VOCs emission amount of each fingerprint compound. The total base binder VOCs amount was 1306 ng/g. However, this amount for the CRMB binder VOCs was 4180 ng/g, three times that of the base binder. It was indicated that the addition of CR significantly increased the total emission of bitumen VOCs. Two lightweight compounds (No. 1 pentane and No. 4 n-hexane) in the CRMB binder VOCs had a lower emission amount, decreasing by 57.47% and 31.79%, respectively. The reason was that the CR powder absorbed part of the lightweight compounds in the base bitumen, thereby reducing their emissions. However, most fingerprint compounds had an increased emission. The emission amount of benzene, toluene, m-xylene, p-xylene, and mesitylene in the CRMB binder VOCs increased by 34, 11, 20, 15, and 9 times, respectively. It indicated that the CR powder significantly influenced the ARHs emission.

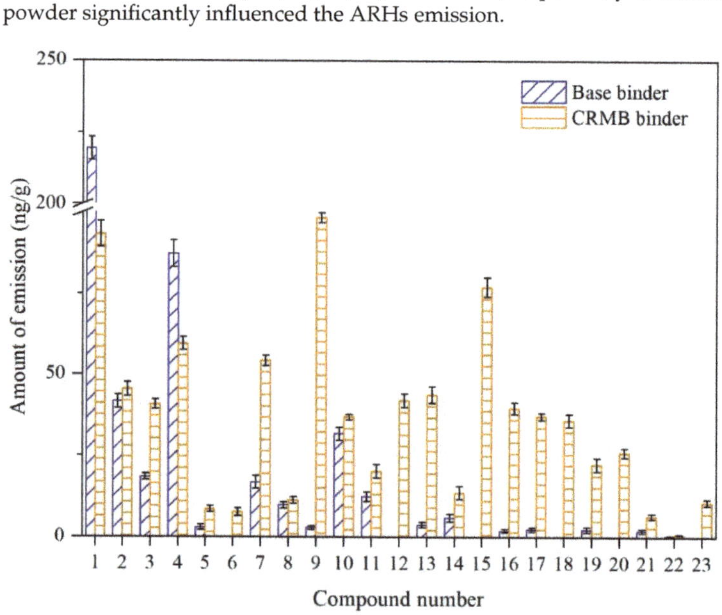

Figure 6. Emission comparison of each VOC fingerprint compound between the base and CRMB binders.

3.2. Inhibition Effect of Mt on the VOCs Emission of CRMB Binder

Figure 7 shows the VOCs GC-MS chromatograms of the CRMB and Mt-CRMB binders. The results indicated that Mt-CRMB VOCs had 108 compounds, 25% less than the CRMB VOCs, which had 144 compounds. It can be easily observed that most fingerprint compounds of the Mt-CRMB VOCs had a smaller peak than that of the CRMB VOCs. Calculated by Equation (2), the total emission of the Mt-CRMB binder was 2901 ng/g, 30.6% less than that of the CRMB binder.

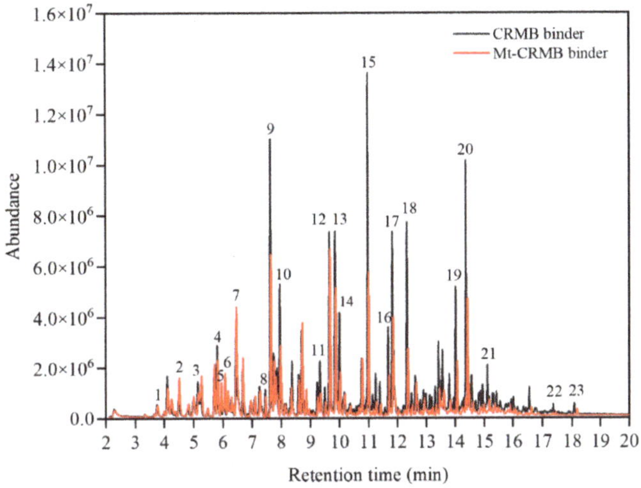

Figure 7. VOCs GC-MS chromatograms of CRMB and Mt-CRMB binders.

Figure 8 compares the emission amount of each VOC fingerprint compound between CRMB and Mt-CRMB binders. Due to the addition of Mt, 19 compounds had a decreased emission. Among them, the emission of compounds Nos. 3, 9, 10, 11, 14, 15, 16, 17, 18, 19, 20, 21, 22, and 23 was significantly reduced by more than 30%. These compounds belonged to the ALK, OLE, and ARH. However, the inhibition effect of Mt on the compounds of No. 2 (acetone), No. 5 (methacrolein), No. 6 (2-methyl-furan), and No. 7 (butanal) was not good.

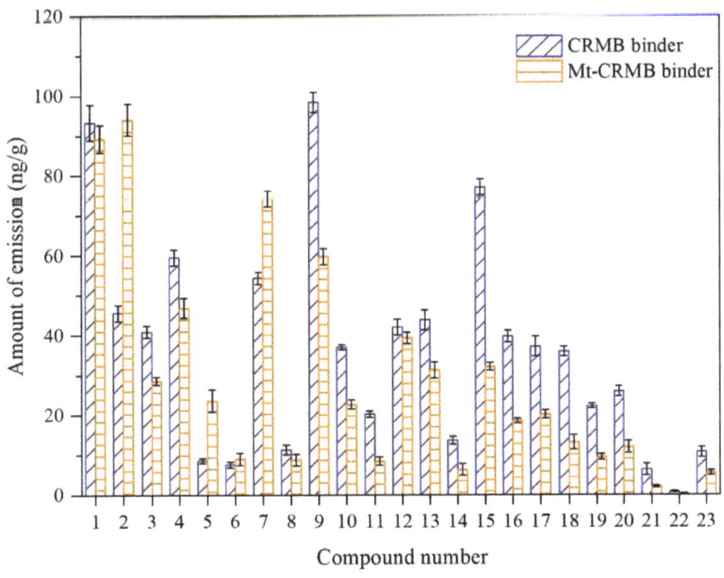

Figure 8. Emission amount comparison of each VOC fingerprint compound between CRMB and Mt-CRMB binders.

Figure 9 shows the VOCs GC-MS chromatograms of the CRMB and Mt-CRMB binders at different retention times, following sampling Method 2 (see Figure 2). The characteristic peak abundance of the CRMB and Mt-CRMB VOCs changed with the extension of the heating time. At Time 2 (i.e., 30 min), the peak abundance reached the maximum. It meant that the CRMB and Mt-CRMB binders at 180 °C had the most VOCs emission after heating for half an hour.

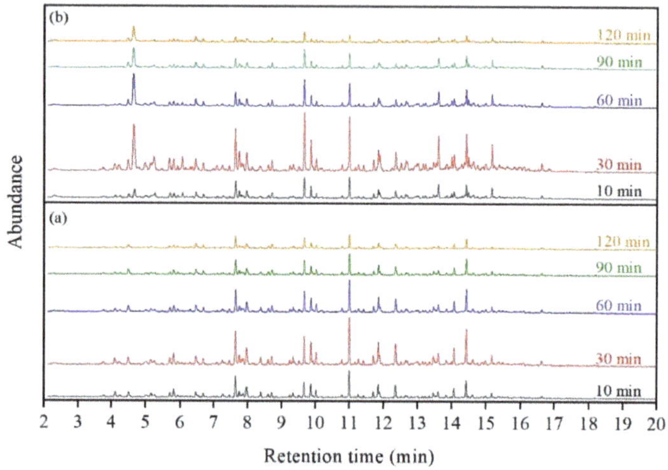

Figure 9. VOCs GC-MS chromatograms of (**a**) CRMB and (**b**) Mt-CRMB binders at different times.

Figure 10 shows the species numbers of CRMB and Mt-CRMB VOCs as a function of the heating time. As indicated, the species number for both binders had a similar change trend. At the same time, the Mt-CRMB VOCs had a lower number. Both had a very close maximum number at 30 min. With the extension of the heating time, the species number gradually decreased. At 120 min (i.e., the fifth sampling), the Mt-CRMB binder only had 79 VOCs species, reduced by 39% compared with the maximum value. However, this number for the CRMB binder only decreased by 27%. It indicated that adding Mt could effectively reduce the number of the VOCs of the CRMB binder.

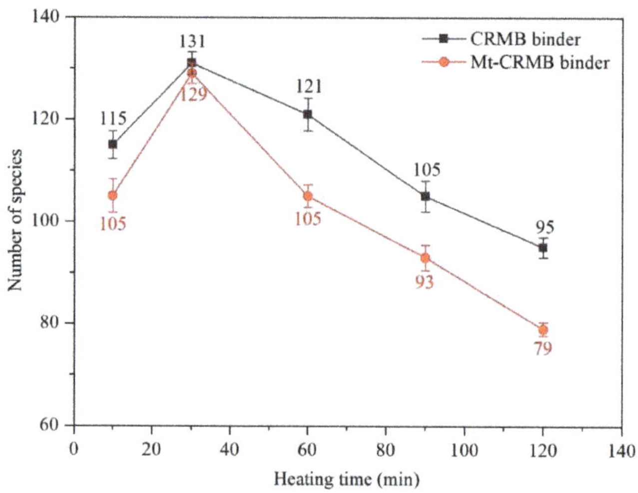

Figure 10. VOCs species number comparison between CRMB and Mt-CRMB binders at different times.

Figure 11 shows the structural change of Mt in the CRMB binder at different times. For the time at 0 min, there were two diffraction peaks. According to the Bragg equation, the basal spacing of Mt was 1.32 nm at 6.62° and 5.25 nm at 1.68°. It meant that parts of Mt remained the original structure, but some showed an intercalated structure with a basal spacing increased by almost three times. The reason was that some small bituminous molecules entered the interlayer of the Mt, forming an intercalation structure [40]. The structure limited the release of bitumen VOCs and reduced their escape paths.

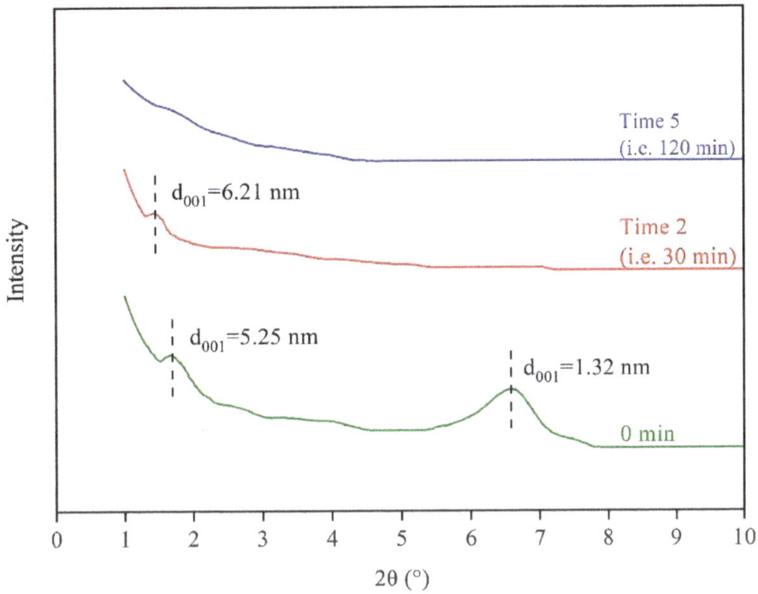

Figure 11. XRD spectrum of Mt-CRMB binder at different times.

For the time at 30 min, there was only one diffraction peak at 1.42°. The peak for the original Mt at 6.62° disappeared, probably due to the continuous stirring of the magnetic rotor improving the exfoliated nanoclay. According to the Bragg equation, the basal spacing of Mt increased from 5.25 nm to 6.21 nm. For the time at 120 min, there was no prominent diffraction peak. The reason was that most Mt layers were exfoliated due to the stirring function. This exfoliation structure improved the inhibition effect on the VOCs emission, with the smallest species number at 120 min.

3.3. *VOCs Emission Model of Bitumen*

3.3.1. Establishment of Model

Figure 12 shows the schematic diagram of VOCs emission from bitumen. Generally, the VOCs emission consists of internal molecular diffusion in the material, partition on the air–material interface, and external convective mass transfer in the ambient air [41]. Five assumptions were established in the model establishment: the initial concentration of VOCs was uniformly distributed, there was no chemical reaction during the VOCs diffusion, the concentration gradient of VOCs was considered to be the only mass transfer force, the equilibrium condition existed between the material surface and the air boundary layer, and the partition on the material–air interface and the external convective mass transfer resistance at the air–material interface can be ignored. Under the above assumptions, the bitumen VOCs emission can be regarded as one-dimensional, and its emission rate measured in the experiment equaled that at the upper surface of bitumen.

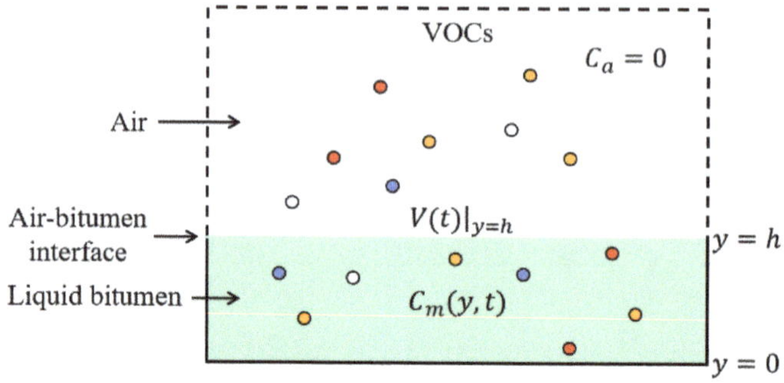

Figure 12. Schematic diagram of VOCs emission from bitumen into the air.

The VOCs diffusion in bitumen follows Fick's second law and can be described by the unsteady diffusion equation:

$$\frac{\partial C_m}{\partial t} = \frac{\partial}{\partial y}\left(D_m \frac{\partial C_m}{\partial y}\right),\ 0 \le y \le h \tag{3}$$

where, C_m is the concentration of each VOC in bitumen, μg/m^3, D_m is the diffusion coefficient of each VOC in bitumen, m^2/s, and h is the thickness of the bitumen sample, m.

The initial condition of VOCs is given as:

$$C_m|_{t=0} = C_0,\ 0 \le y \le h \tag{4}$$

where, C_0 is the initial concentration of each VOC inside the bitumen, μg/m^3, and t is the heating time, s.

Since the bitumen samples will not penetrate the glass substrate, the concentration gradient is 0 μg/m^4 (i.e., no mass flux) on the bottom boundary. It means that:

$$\left.\frac{\partial C_m}{\partial y}\right|_{y=0} = 0 \tag{5}$$

Since the operation process shown in Figure 2 was designed to simulate the open environment in the process of bitumen preparation, transportation, and spreading in the actual situation, the spread of VOCs emitted from bitumen to the external environment is infinite. Therefore, the assumption of zero concentration on the upper surface of the sample is valid in this case, as follows:

$$C_m|_{y=h} = 0 \tag{6}$$

When the time, t, is long enough, the model for the instantaneous emission rate of VOCs based on the Laplace transform can be expressed as [41]:

$$V(t) = \frac{2D_m C_0}{h} e^{\frac{-D_m \pi^2}{4h^2} t} \tag{7}$$

Equation (7) can also be expressed as:

$$lnV(t) = ln\left(\frac{2D_m C_0}{h}\right) - \frac{D_m \pi^2}{4h^2} t \tag{8}$$

$V(t)$ can be obtained from the experiment and calculated by the following Equation:

$$V(t) = \frac{m_e}{St_c} \tag{9}$$

where, m_e is the mass of each compound measured in the experiment, µg, S is the area of the upper surface of the sample, m^2, and t_c is the closed time before each sampling, min.

Two parameters, C_0 and D_m, can be obtained through the linear curve fitting based on Equation (8).

Figure 13 shows the total VOCs emission rates of CRMB and Mt-CRMB binders, and both total emission rates of the two binders sharply increased before 30 min and then gradually decreased. The VOCs emission rates of CRMB and Mt-CRMB binders were close at 30 min. However, with the extension of the heating time, their VOCs emission rates were significantly different. At 120 min, this value of the Mt-CRMB binder decreased by 41% compared with CRMB. For the VOCs collection device, a hysteresis of heat transfer existed at the initial state, leading to the uneven distribution of the VOCs concentration inside the binder. Therefore, there was an emission peak at 30 min. After this time, the internal concentration tended to be uniform, and the emission of bitumen VOCs met Fick's law. Therefore, the emission rates at 30, 60, 90, and 120 min were used to fit the model of Equation (7).

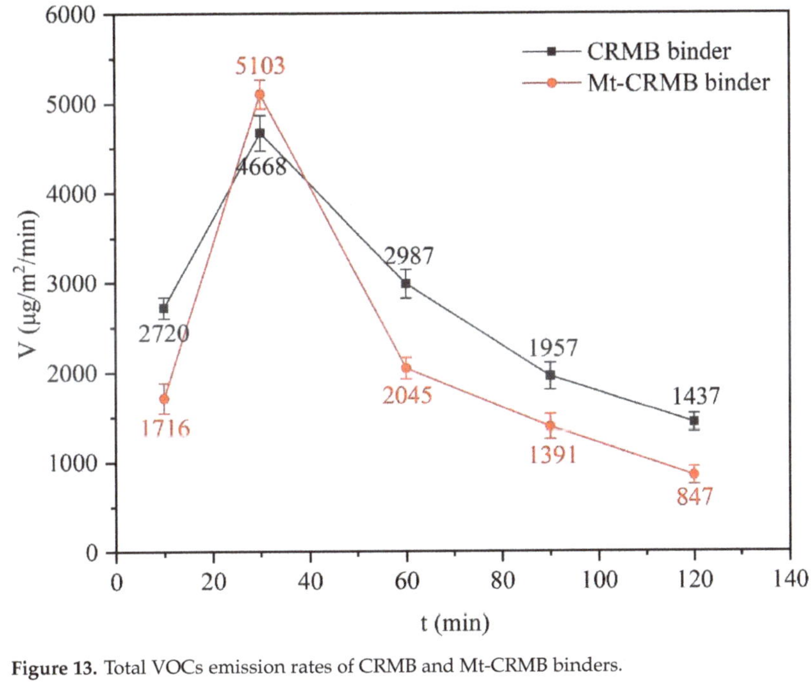

Figure 13. Total VOCs emission rates of CRMB and Mt-CRMB binders.

Table 4 presents the VOCs emission rate of each fingerprint compound. In comparison, the emission rates for all fingerprint compounds of the Mt-CRMB binder decreased at the same heating time, except for No. 12 and No. 21 compounds. It indicated that adding Mt significantly reduced the emission rate of most VOCs in the bitumen.

Table 5 presents the initial concentration, C_0, and the diffusion coefficient, D_m, in the emission model for 23 fingerprint compounds of CRMB and Mt-CRMB binders. As indicated, C_0 for most compounds after the Mt modification decreased, except for undecane. The acetone had the highest C_0 of 2.033×10^6 µg/m^3, and this value decreased by 27% in the Mt-CRMB binder. Meanwhile, C_0 of 2-methyl-furan, 2-methyl-heptane, octane, and mesitylene decreased by 53%, 58%, 51%, and 51%, respectively. For most compounds, D_m was relatively stable after the Mt modification.

Table 4. Emission rate, $V(t)$, of 23 fingerprint compounds for CRMB and Mt-CRMB binders at different times.

No.	Emission Rate (μg/m²/min)									
	Sampling Time for CRMB Binder					Sampling Time for Mt-CRMB Binder				
	1	2	3	4	5	1	2	3	4	5
1	117.00	160.73	99.80	63.50	45.32	46.52	111.93	44.42	35.73	33.12
2	92.65	142.55	129.52	111.15	72.33	53.28	139.05	88.20	75.12	53.47
3	33.50	44.68	27.95	17.85	12.03	13.85	43.35	15.48	8.27	5.10
4	54.98	79.20	45.92	29.37	21.05	20.52	52.52	21.47	15.63	11.23
5	12.22	18.32	15.58	12.38	10.80	8.72	16.17	10.40	8.68	7.23
6	4.78	9.32	8.23	6.72	5.93	3.08	8.73	4.68	4.05	2.95
7	26.68	46.28	38.48	29.18	27.80	21.62	44.07	26.00	20.27	19.37
8	8.75	14.13	8.77	5.78	4.12	3.85	11.00	4.37	3.02	1.70
9	64.10	98.18	64.57	45.12	35.60	34.90	83.05	28.52	19.17	10.42
10	28.97	39.13	23.53	15.65	12.27	13.60	28.93	12.50	9.17	6.93
11	13.32	23.43	14.48	9.28	6.30	4.52	13.05	5.32	3.40	1.88
12	29.72	52.68	38.22	27.48	21.58	26.30	72.45	33.83	23.50	14.33
13	26.67	43.55	27.17	16.47	10.82	16.42	39.88	15.24	8.85	4.08
14	7.73	13.32	8.43	5.48	3.93	3.28	8.88	3.92	2.72	1.83
15	48.85	83.82	56.92	37.22	25.50	24.37	64.00	27.27	17.98	8.83
16	22.37	37.70	24.22	14.97	9.92	10.92	26.78	10.33	6.83	3.35
17	21.93	37.07	24.28	15.47	10.50	9.48	22.83	9.25	6.32	3.27
18	18.38	30.13	18.88	11.17	7.05	8.00	18.42	6.93	4.47	2.08
19	13.30	23.73	15.60	10.08	6.88	6.08	13.60	6.32	4.52	2.52
20	14.37	29.52	18.43	12.68	8.70	7.38	19.47	9.02	6.47	3.52
21	5.28	10.48	6.65	4.23	2.97	7.60	18.62	8.65	6.03	3.18
22	0.37	0.67	0.48	0.30	0.22	0.25	0.50	0.30	0.23	0.15
23	5.77	9.18	6.60	4.50	3.47	2.07	3.65	3.28	3.53	2.07
Total SD	121.53	198.75	162.88	155.47	99.21	171.54	160.15	122.34	142.78	104.12

Table 5. The initial concentration, C_0, and the diffusion coefficient, D_m, in the emission model for 23 fingerprint compounds of CRMB and Mt-CRMB binders.

No.	CRMB Binder			Mt-CRMB Binder		
	C_0 (μg/m³)	D_m (m²/s)	R^2	C_0 (μg/m³)	D_m (m²/s)	R^2
1	1.355×10^6	2.163×10^{-8}	0.9914	1.001×10^6	7.439×10^{-9}	0.6893
2	2.033×10^6	1.607×10^{-8}	0.8316	1.479×10^6	1.086×10^{-8}	0.9538
3	3.780×10^5	3.182×10^{-8}	0.9978	2.135×10^5	2.229×10^{-8}	0.9540
4	6.198×10^5	2.266×10^{-8}	0.9829	3.144×10^5	1.673×10^{-8}	0.8896
5	3.058×10^5	1.302×10^{-8}	0.9865	1.801×10^5	8.930×10^{-9}	0.9151
6	1.734×10^5	1.592×10^{-8}	0.9849	8.160×10^4	7.839×10^{-9}	0.8928
7	7.445×10^5	1.247×10^{-8}	0.9155	4.265×10^5	9.344×10^{-9}	0.7946
8	1.198×10^5	2.919×10^{-8}	0.9917	7.308×10^4	2.089×10^{-8}	0.9513
9	9.257×10^5	2.718×10^{-8}	0.9784	4.840×10^5	3.094×10^{-8}	0.9399
10	3.250×10^5	1.974×10^{-8}	0.9657	1.888×10^5	1.490×10^{-8}	0.8808
11	1.959×10^5	2.978×10^{-8}	0.9966	8.314×10^4	2.230×10^{-8}	0.9693
12	5.721×10^5	2.509×10^{-8}	0.9932	5.173×10^5	1.524×10^{-8}	0.9660
13	3.651×10^5	3.385×10^{-8}	0.9983	2.769×10^5	3.849×10^{-8}	0.9711
14	1.144×10^5	2.194×10^{-8}	0.9925	5.576×10^4	2.024×10^{-8}	0.9378
15	7.753×10^5	3.009×10^{-8}	0.9993	4.629×10^5	3.344×10^{-8}	0.9767
16	3.260×10^5	3.513×10^{-8}	0.9988	1.906×10^5	2.255×10^{-8}	0.9672
17	3.305×10^5	3.288×10^{-8}	0.9988	1.661×10^5	2.125×10^{-8}	0.9648
18	2.492×10^5	3.679×10^{-8}	0.9991	1.293×10^5	2.453×10^{-8}	0.9708
19	2.125×10^5	2.848×10^{-8}	0.9990	1.047×10^5	2.114×10^{-8}	0.9680
20	2.579×10^5	2.888×10^{-8}	0.9950	1.489×10^5	1.946×10^{-8}	0.9687
21	9.030×10^4	2.148×10^{-8}	0.9949	1.406×10^5	2.981×10^{-8}	0.9746
22	6.363×10^3	2.087×10^{-8}	0.9948	4.483×10^3	1.874×10^{-8}	0.9700
23	9.552×10^4	1.669×10^{-8}	0.9929	5.750×10^4	9.606×10^{-9}	0.4464

3.3.2. The Model Validation

For validating the emission model of Equation (8), the Concentration Field of the PDEtoolbox in MATLAB software was used to construct the finite element model (FEM) for four typical components (toluene, styrene, benzothiazole, and methyl isobutyl ketone) in the CRMB and Mt-CRMB binders. In this FE model, a rectangle was selected as the model's shape and gridded. The boundary (left, right, and bottom) conditions were set with the diffusion concentration gradient equal to 0 $\mu g/m^4$. The parameters of C_0 and D_m in Table 5 were input into the software to obtain the diffusion model of VOCs inside bitumen. Figure 14 shows the two-dimensional cloud image of benzothiazole diffusion in the CRMB binder at different times. At the same time, the diffusion flux from the bottom to the surface gradually increased. As the volatilization time increased, the diffusion flux at the upper interface gradually decreased.

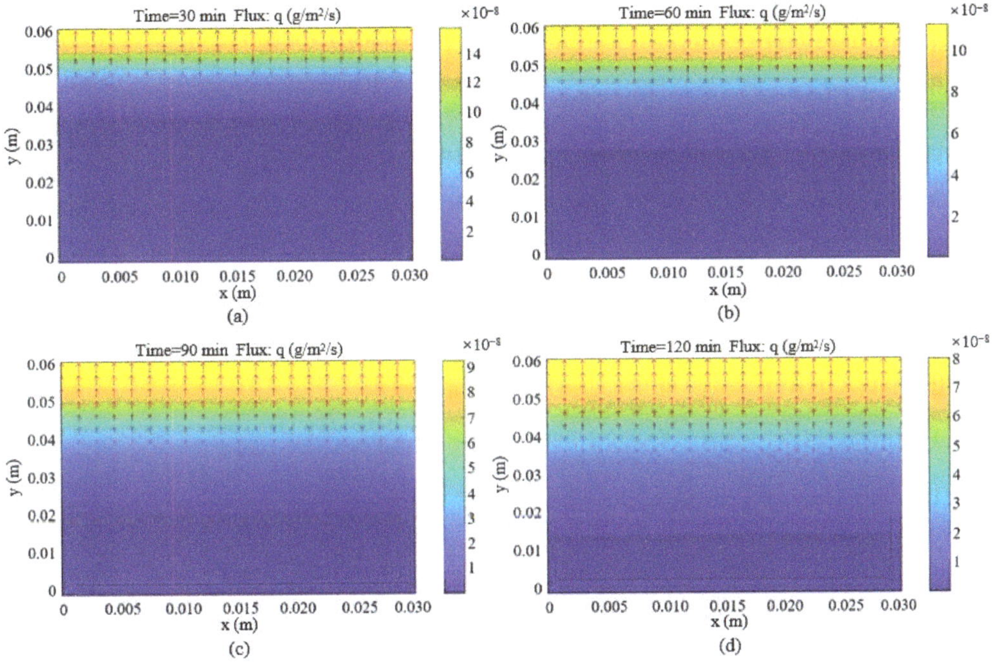

Figure 14. Two-dimensional cloud images of benzothiazole diffusion in CRMB binder at (**a**) 30 min, (**b**) 60 min, (**c**) 90 min, and (**d**) 120 min.

Figure 15 shows the emission rate of four typical compounds at the upper interface based on the FEM and the experiment. As indicated, the emission rate gradually decreased with the increase in the heating time. The emission curve of these compounds in the CRMB and Mt-CRMB binders agreed well with the experimental data. In comparison, the CRMB binder better fit the model and the experiment than the Mt-CRMB. It was observed that the data at 90 min and 120 min differed more from the curve than at 30 min and 60 min. The reason could be that the bitumen gradually aged, hindering the VOCs emission. Although only four fingerprint compounds were used for the verification, it was still proven that the analytical solutions derived from Equations (3)–(6) were reasonable and can be used to describe the emission behavior of bitumen very well. This model can provide a concise and efficient understanding of bitumen VOCs emission behavior and lays a solid foundation for further deepening the model (considering more influencing factors) in future studies.

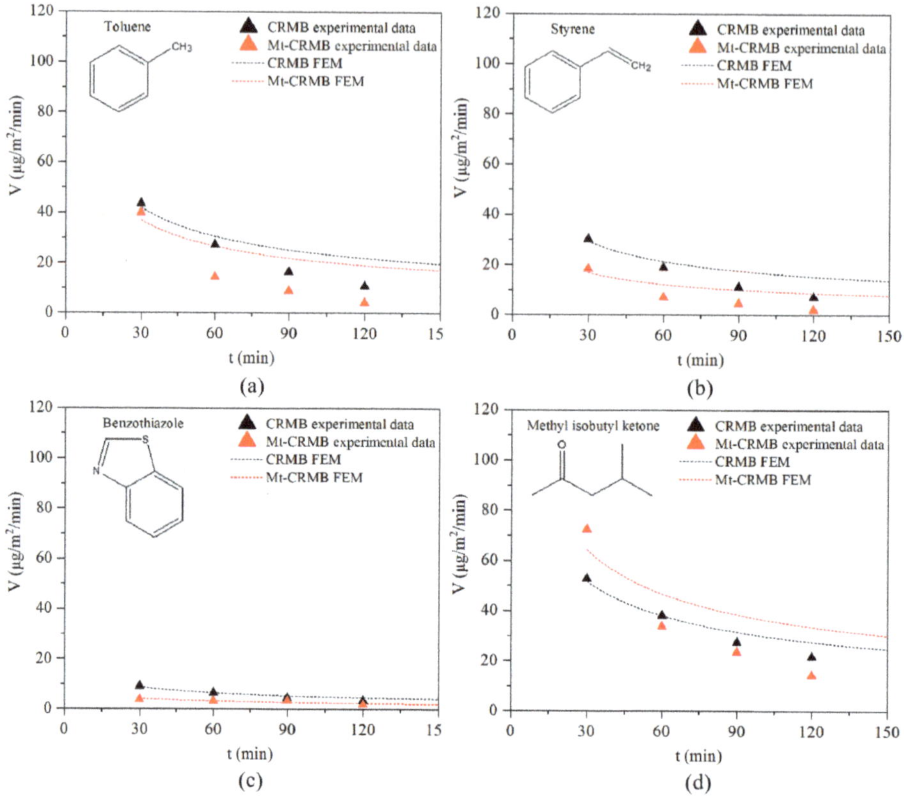

Figure 15. Emission rate of four typical compounds of (**a**) toluene, (**b**) styrene, (**c**) benzothiazole, and (**d**) methyl isobutyl ketone at the upper interface based on the FEM and the experiment.

4. Conclusions

The VOCs emission characteristics of base and CRMB binders during heating were characterized and the inhibiting effect of Mt on the VOCs emission of CRMB was investigated in this research. A VOCs emission model of bitumen was established to compare the VOCs emission behaviors of CRMB and Mt-CRMB. Based on the results and discussion, the following conclusions can be drawn:

(1) The addition of crumb rubber increased the VOCs species number from 123 to 144. The VOCs emission of the CRMB binder was 3.2 times larger than that of the base binder. Furthermore, the addition of crumb rubber significantly increased the ARHs of the base binder, and the emission of some ARHs such as benzene, toluene, m-xylene, and p-xylene emissions increased by more than 10 times. However, the emission of some lightweight compounds such as alkanes was reduced since they were absorbed by the rubber.

(2) Due to its special intercalated structure, the VOCs of Mt-CRMB included 108 compounds, 25% less than that of CRMB, and the Mt nanoclay can reduce the VOCs emission of CRMB binder by about 30%. Especially, its inhibition effects on alkanes, olefins, and aromatic hydrocarbons were more significant, and 14 fingerprint compounds belonging to the above types were reduced by more than 30%. With the extension of the heating time, the VOCs species number of CRMB and Mt-CRMB binders gradually decreased. At 120 min, the Mt-CRMB binder only emitted 79 species of VOCs, reduced by 39% compared with the maximum value.

(3) The VOCs emission rates of CRMB and Mt-CRMB binders were close at 30 min. However, with the extension of the heating time, their VOCs emission rates were signifi-

cantly different. The emission rate of the Mt-CRMB binder decreased by 41% compared with the CRMB binder at 120 min. Furthermore, the addition of Mt significantly reduced the emission rate of most VOCs in the bitumen. By calculation, the C_0 for most compounds after the Mt modification decreased, except for undecane, and for most compounds, the D_m was relatively stable after the Mt modification. Based on Fick's second law, the emission behavior of CRMB and Mt-CRMB binders can be well-described by the established model.

Overall, the Mt nanoclay can be used as an effective modifier to inhibit the VOCs emission of the CRMB binder, which is widely used in asphalt pavement.

Author Contributions: G.L., conceptualization, methodology, investigation, writing—original draft, project administration; S.F., methodology, supervision, formal analysis, writing—review and editing; Y.W., methodology, investigation; J.L., investigation, writing—review and editing; Y.L., investigation, writing—original draft; T.C., investigation, supervision; Q.L., supervision, writing—review and editing. All authors have read and agreed to the published version of the manuscript.

Funding: The research was supported by the Hainan Provincial Joint Project of Sanya Yazhou Bay Science and Technology City (No. 520LH016), the National Natural Science Foundation of China (Grant No. 52068061), and the Key R&D Program of Hubei Province (No. 20211G0098).

Institutional Review Board Statement: Not applicable.

Informed Consent Statement: Not applicable.

Data Availability Statement: Data are contained within the article.

Conflicts of Interest: The authors declare no conflict of interest.

References

1. Chen, G.; Farooq, M.Z.; Sun, B.; Lin, F.; Yan, B.; Rajput, G.; Chawla, M. Pollutants formation, distribution, and reaction mechanism during WT pyrolysis: A review. *J. Anal. Appl. Pyrolysis* **2021**, *157*, 105218. [CrossRef]
2. Zhang, J.; Chen, M.; Wu, S.; Zhou, X.; Zhao, G.; Zhao, Y.; Cheng, M. Evaluation of VOCs inhibited effects and rheological properties of asphalt with high-content waste rubber powder. *Constr. Build Mater.* **2021**, *300*, 124320. [CrossRef]
3. Tang, X.; Chen, Z.; Liu, J.; Chen, Z.; Xie, W.; Evrendilek, F.; Buyukada, M. Dynamic pyrolysis behaviors, products, and mechanisms of waste rubber and polyurethane bicycle tires. *J. Hazard. Mater.* **2021**, *402*, 123516. [CrossRef] [PubMed]
4. Ilyin, S.O.; Arinina, M.P.; Mamulat, Y.S.; Malkin, A.Y.; Kulichikhin, V.G. Rheological properties of road bitumens modified with polymer and solid nanosized additives. *Colloid J.* **2014**, *76*, 425–434. [CrossRef]
5. Sienkiewicz, M.; Janik, H.; Borzędowska-Labuda, K.; Kucińska-Lipka, J. Environmentally friendly polymer-rubber composites obtained from waste tyres: A review. *J. Clean. Prod.* **2017**, *147*, 560–571. [CrossRef]
6. Xiao, F.; Amirkhanian, S.N.; Shen, J.; Putman, B. Influences of crumb rubber size and type on reclaimed asphalt pavement (RAP) mixtures. *Constr. Build Mater.* **2009**, *23*, 1028–1034. [CrossRef]
7. Zhu, Y.; Xu, G.; Ma, T.; Fan, J.; Li, S. Performances of rubber asphalt with middle/high content of waste tire crumb rubber. *Constr. Build Mater.* **2022**, *335*, 127488. [CrossRef]
8. Chong, D.; Wang, Y.; Guo, H.; Lu, Y. Volatile Organic Compounds Generated in Asphalt Pavement Construction and Their Health Effects on Workers. *J. Constr. Eng. Manag.* **2014**, *140*, 04013051. [CrossRef]
9. Yang, X.; You, Z.; Perram, D.; Hand, D.; Ahmed, Z.; Wei, W.; Luo, S. Emission analysis of recycled tire rubber modified asphalt in hot and warm mix conditions. *J. Hazard. Mater.* **2019**, *365*, 942–951. [CrossRef]
10. Autelitano, F.; Bianchi, F.; Giuliani, F. Airborne emissions of asphalt/wax blends for warm mix asphalt production. *J. Clean. Prod.* **2017**, *164*, 749–756. [CrossRef]
11. Celebi, U.B.; Vardar, N. Investigation of VOC emissions from indoor and outdoor painting processes in shipyards. *Atmos. Environ.* **2008**, *42*, 5685–5695. [CrossRef]
12. Gągol, M.; Boczkaj, G.; Haponiuk, J.; Formela, K. Investigation of volatile low molecular weight compounds formed during continuous reclaiming of ground tire rubber. *Polym. Degrad. Stab.* **2015**, *119*, 113–120. [CrossRef]
13. Chen, S.; Wang, J.; Li, Q.; Zhang, W.; Yan, C. The Investigation of Volatile Organic Compounds (VOCs) Emissions in Environmentally Friendly Modified Asphalt. *Polymers* **2022**, *14*, 3459. [CrossRef]
14. Fostinelli, J.; Madeo, E.; Toraldo, E.; Sarnico, M.; Luzzana, G.; Tomasi, C.; De Palma, G. Environmental and biological monitoring of occupational exposure to polynuclear aromatic hydrocarbons during highway pavement construction in Italy. *Toxicol. Lett.* **2018**, *298*, 134–140. [CrossRef]
15. Schreiner, C.A. Review of mechanistic studies relevant to the potential carcinogenicity of asphalts. *Regul. Toxicol. Pharm.* **2011**, *59*, 270–284. [CrossRef] [PubMed]
16. Gasthauer, E.; Mazé, M.; Marchand, J.P.; Amouroux, J. Characterization of asphalt fume composition by GC/MS and effect of temperature. *Fuel* **2008**, *87*, 1428–1434. [CrossRef]

17. Rubio, M.C.; Martínez, G.; Baena, L.; Moreno, F. Warm mix asphalt: An overview. *J. Clean. Prod.* **2012**, *24*, 76–84. [CrossRef]
18. Xiu, M.; Wang, X.; Morawska, L.; Pass, D.; Beecroft, A.; Mueller, J.F.; Thai, P. Emissions of particulate matters, volatile organic compounds and polycyclic aromatic hydrocarbons from warm and hot asphalt mixes. *J. Clean. Prod.* **2020**, *275*, 123094. [CrossRef]
19. Cheraghian, G.; Cannone Falchetto, A.; You, Z.; Chen, S.; Kim, Y.S.; Westerhoff, J.; Moon, K.H.; Wistuba, M.P. Warm mix asphalt technology: An up to date review. *J. Clean. Prod.* **2020**, *268*, 122128. [CrossRef]
20. Yang, X.; Peng, X.; Zhang, X.; Qian, S. Experiments on the asphalt fume suppression agents and properties of asphalt concrete with fume suppression agent. *J. Chongqing Univ. (Nat. Sci. Ed.)* **2013**, *36*, 70–78.
21. Lei, M.; Wu, S.; Liu, G.; Amirkhanian, S. VOCs characteristics and their relation with rheological properties of base and modified bitumens at different temperatures. *Constr. Build Mater.* **2018**, *160*, 794–801. [CrossRef]
22. Liu, G.; Nielsen, E.; Komacka, J.; Greet, L.; van de Ven, M. Rheological and chemical evaluation on the ageing properties of SBS polymer modified bitumen: From the laboratory to the field. *Constr. Build Mater.* **2014**, *51*, 244–248. [CrossRef]
23. Cui, P.; Wu, S.; Xiao, Y.; Wan, M.; Cui, P. Inhibiting effect of Layered Double Hydroxides on the emissions of volatile organic compounds from bituminous materials. *J. Clean. Prod.* **2015**, *108*, 987–991. [CrossRef]
24. Huang, G.; He, Z.; Zhou, C.; Huang, T. Mechanism of fume suppression and performance on asphalt of expanded graphite for pavement under high temperature condition. *J. Wuhan Univ. Technol. Mater. Sci.* **2014**, *6*, 1229–1236. [CrossRef]
25. Sharma, A.; Lee, B.K. A novel nanocomposite of Ca(OH)(2)-incorporated zeolite as an additive to reduce atmospheric emissions of PM and VOCs during asphalt production. *Environ. Sci.-Nano* **2017**, *4*, 613–624. [CrossRef]
26. Alexandre, M.; Dubois, P. Polymer-layered silicate nanocomposites: Preparation, properties and uses of a new class of materials. *Mater. Sci. Eng. R Rep.* **2000**, *28*, 1–63. [CrossRef]
27. Slaný, M.; Jankovič, L'.; Madejová, J. Structural characterization of organo-montmorillonites prepared from a series of primary alkylamines salts: Mid-IR and near-IR study. *Appl. Clay Sci.* **2019**, *176*, 11–20. [CrossRef]
28. Brigatti, M.F.; Galan, E.; Theng, B.K.G. Chapter 2 Structures and Mineralogy of Clay Minerals. In *Developments in Clay Science*; Bergaya, F., Theng, B.K.G., Lagaly, G., Eds.; Elsevier: Amsterdam, The Netherlands, 2006; Volume 1, pp. 19–86.
29. de Paiva, L.B.; Morales, A.R.; Valenzuela Díaz, F.R. Organoclays: Properties, preparation and applications. *Appl. Clay Sci.* **2008**, *42*, 8–24. [CrossRef]
30. Yadykova, A.Y.; Ilyin, S.O. Bitumen improvement with bio-oil and natural or organomodified montmorillonite: Structure, rheology, and adhesion of composite asphalt binders. *Constr. Build Mater.* **2023**, *364*, 129919. [CrossRef]
31. Tan, Z.; Wang, J.; Shi, Z. Laboratory investigation on effects of organic montmorillonite on performance of crumb rubber modified asphalt. *J. Cent. South Univ.* **2020**, *27*, 3888–3898. [CrossRef]
32. Fox, J.B.; Ambuken, P.V.; Stretz, H.A.; Peascoe, R.A.; Payzant, E.A. Organo-montmorillonite barrier layers formed by combustion: Nanostructure and permeability. *Appl. Clay Sci.* **2010**, *49*, 213–223. [CrossRef]
33. Li, L.; Wu, S.; Liu, G.; Cao, T.; Amirkhanian, S. Effect of organo-montmorillonite nanoclay on VOCs inhibition of bitumen. *Constr. Build Mater.* **2017**, *146*, 429–435. [CrossRef]
34. Yang, X.; Shen, A.; Jiang, Y.; Meng, Y.; Wu, H. Properties and mechanism of flame retardance and smoke suppression in asphalt binder containing organic montmorillonite. *Constr. Build Mater.* **2021**, *302*, 124148. [CrossRef]
35. Li, N.; Jiang, Q.; Wang, F.; Cui, P.; Xie, J.; Li, J.; Wu, S.; Barbieri, D.M. Comparative Assessment of Asphalt Volatile Organic Compounds Emission from field to laboratory. *J. Clean. Prod.* **2021**, *278*, 123479. [CrossRef]
36. Li, N.; Jiang, Q.; Wang, F.; Xie, J.; Li, Y.; Li, J.; Wu, S. Emission behavior, environmental impact and priority-controlled pollutants assessment of volatile organic compounds (VOCs) during asphalt pavement construction based on laboratory experiment. *J. Hazard. Mater.* **2020**, *398*, 122904. [CrossRef]
37. Boczkaj, G.; Przyjazny, A.; Kamiński, M. Characteristics of volatile organic compounds emission profiles from hot road bitumens. *Chemosphere* **2014**, *107*, 23–30. [CrossRef]
38. Cui, P.; Wu, S.; Li, F.; Xiao, Y.; Zhang, H. Investigation on Using SBS and Active Carbon Filler to Reduce the VOC Emission from Bituminous Materials. *Materials* **2014**, *7*, 6130–6143. [CrossRef]
39. Xiao, Y.; Chang, X.; Dong, Q.; You, Z. Fingerprint Components and Quantitative Analysis of Volatile Organic Compounds of Asphalt Materials. *China J. Highw. Transp.* **2020**, *33*, 276–287.
40. Liu, G.; Wu, S.; Ven, M.; Yu, J.; Molenaar, A.J.A.C.S. Influence of sodium and organo-montmorillonites on the properties of bitumen. *Appl. Clay Sci.* **2010**, *49*, 69–73. [CrossRef]
41. Liu, Z.; Huang, Z.; Yan, Y.; Liu, T.; Lv, T.; Chen, Y.; Yang, J.; Die, Q.; Zhao, Y.; Huang, Q. Characterizing the emission behaviors of cumulative VOCs from automotive solvent-based paint sludge. *J. Environ. Manag.* **2022**, *317*, 115369. [CrossRef]

polymers

MDPI

Review

Applications of Synthetic, Natural, and Waste Fibers in Asphalt Mixtures: A Citation-Based Review

Adham Mohammed Alnadish [1,2,*], Narinderjit Singh Sawaran Singh [3,*] and Aawag Mohsen Alawag [4]

1 Department of Civil Engineering, Thamar University, Dhamar 87246, Yemen
2 Department of Transportation & Geotechnical Engineering, Balochistan Campus, National University of Sciences and Technology (NUST), Quetta 87300, Pakistan
3 Faculty of Data Science and Information Technology, INTI International University, Persiaran Perdana BBN Putra Nilai, Nilai 71800, Negeri Sembilan, Malaysia
4 Department of Civil Engineering, Universiti Teknologi PETRONAS, Seri Iskandar 32610, Perak, Malaysia
* Correspondence: adhmalnadish@gmail.com (A.M.A.); narinderjits.sawaran@newinti.edu.my (N.S.S.S.)

Abstract: The utilization of synthetic, natural, and waste fibers in asphalt mixtures is constantly increasing due to the capability of fibers to improve the mechanical performance of asphalt mixes. The combination of fibers in asphalt mixes contributes to ecological sustainability and cost benefits. The objective of this paper is to introduce a citation-based review on the incorporation of synthetic, natural, and waste fibers in bitumen, dense-graded asphalt mix, stone mastic asphalt, and porous asphalt mix. Additionally, this article aims to identify research gaps and provide recommendations for further work. The outputs of this article demonstrated that there has recently been a growing interest in the use of natural and waste fibers in asphalt mixtures. However, more future studies are needed to investigate the performance of fiber-modified stone mastic asphalt and porous asphalt mix in terms of resistance to aging and low-temperature cracking. Furthermore, the period of natural fibers' biodegradability in asphalt mixtures should be investigated.

Keywords: asphalt; natural fibers; synthetic fibers; waste fibers; dense asphalt mix; stone mastic asphalt; porous asphalt mix; biodegradability

Citation: Alnadish, A.M.; Singh, N.S.S.; Alawag, A.M. Applications of Synthetic, Natural, and Waste Fibers in Asphalt Mixtures: A Citation-Based Review. *Polymers* **2023**, *15*, 1004. https://doi.org/10.3390/polym15041004

Academic Editors: Wei Jiang, Quantao Liu, Jose Norambuena-Contreras and Yue Huang

Received: 26 January 2023
Revised: 5 February 2023
Accepted: 8 February 2023
Published: 17 February 2023

1. Introduction

Asphalt mixture refers to the combination of bitumen, coarse aggregates, fine aggregates, and filler. Asphalt mixtures can be divided into three categories: dense-graded asphalt mix, stone mastic asphalt (SMA), and porous asphalt mix (open-graded asphalt mix) [1,2]. The dense-graded asphalt mix is the most widespread due to its suitability for all pavement layers and traffic conditions and its reasonable cost. Stone mastic asphalt (SMA) consists of 6–7% bitumen, 70–80% coarse aggregates, 8–12% mineral filler, and 0.3–0.5% fibers. SMA is distinguished by its remarkable durability and high resistance to rutting as compared to other types. The main cons of SMA are its higher cost than the dense-graded asphalt by about 20–25%, longer mixing time, and delays in opening to traffic [3,4]. The porous asphalt mix is designed to be used as a drainage layer that allows runoff and rainfall to penetrate the surface layer to the drainage system. The porous asphalt mix is characterized by its high air void content, smooth surface, less tire splash, decreased runoff, and less energy in manufacturing, while its high cost and weakness in carrying heavy traffic loads are among its drawbacks [5,6].

On the other hand, permanent deformation, fatigue cracking, low-temperature cracking, and moisture damage are the major distresses that decrease the lifespan of an asphalt layer. Permanent deformation in an asphalt layer occurs due to densification, lateral flow, and subgrade consolidation. In addition, a heavy load, improper compaction, materials, and design for the asphalt layer cause permanent deformation [7–9]. Furthermore, the cyclic stress of traffic loads and structural weakness of an asphalt layer due to improper design or aging result in fatigue cracking, which starts from the bottom of the layer and propagates to the surface. Low-temperature cracking is caused by the shrinkage of the asphalt layer in cold regions [10,11]. Furthermore, moisture damage occurs by the loss of bonding between aggregates and bitumen due to the reaction between adhesion materials and water that penetrates the surface layer through cracks [12–14].

Accordingly, there is a growing global interest in improving the performance of asphalt layers by modifying bitumen or the asphalt mix to extend their lifespan. Fibers have been the most utilized materials in asphalt layers for many decades due to their capability in boosting the mechanical performance of asphalt layers. Fibers can be of natural origin, such as lignin, bamboo, sisal, kenaf, coconut, banana, and Jute, or manmade, such as polypropylene, polyester, polyacrylonitrile (PAN), carbon, aramid, basalt, glass, steel, and ceramic [15–18]. Moreover, fibers can be used as a bitumen modifier or mixture modifier [19]. It is reported that modifying bitumen with fibers enhances the physical and rheological properties of bitumen in terms of high-temperature performance (rutting) and intermediate-temperature (fatigue cracking) and low-temperature cracking [20–23]. Additionally, introducing fibers to an asphalt mix by a dry process (modifying mix) has been documented as an effective approach for improving the performance of the asphalt mix, reducing aging of the asphalt mix, and extending the service life of the asphalt layer [24–28]. The use of natural fibers in stone mastic asphalt (SMA) and porous asphalt mix is widely desirable due to their low cost, availability, and absorbency property, which reduces drain down in asphalt mixes [29–31]. Furthermore, it is documented that the mechanical properties of an asphalt mix integrated with reclaimed asphalt pavement are notably enhanced by adding fibers. In addition, utilizing fibers in an asphalt mix is cost effective due to the fibers' ability to decrease maintenance and extend the lifespan of the asphalt layer [32–35].

The objective of this article is to review and summarize the current state of utilizing natural, synthetic, and waste fibers in asphalt mixtures. Another aim is to identify research gaps and propose recommendations for future work.

2. Physical, Mechanical, and Thermal Properties of Fibers

The physical, mechanical, and thermal properties of fibers play a major role in determining the intended use of fibers. The fiber density is used to identify the suitable content in asphalt mixes. Moisture content refers to the absorbent property of fibers. The tensile strength of fibers indicates the stress that fibers can carry without failure. Young's modulus and elongation are used to describe the elastic behavior of fibers. The relationship between Young's modulus and elongation is linear; as Young's modulus increases, the elongation decreases. In other words, the higher the Young's modulus, the lower the elastic and ductile behavior of fibers. Thermal conductivity is the ability of fibers to transfer heat. Glass transition represents the temperature at which fibers change from the glassy state to the rubbery state. Glass transition has a major impact on the behavior of the physical properties of fibers [36–38].

Table 1 shows the general mechanical properties of natural and synthetic fibers. As can be seen in the table, natural fibers are characterized by low density and high absorption rate. Coconut and palm fibers are more elastic than the other fibers. However, the common utilization of natural fibers in stone mastic asphalt and porous asphalt mix is attributed to the high absorption rate. The absorbency of natural fibers decreases the drain down. Therefore, the use of natural fibers in dense asphalt mixes may increase the bitumen content [29–31]. On the other hand, synthetic fibers, i.e., polypropylene, polyester, and PAN, are characterized by low density, high elasticity, and low absorption rate, which make them suitable for dense asphalt mixes. Furthermore, the low melting point of polypropylene fibers indicates that polypropylene fibers are desirable for use as a bitumen modifier in hot mix asphalt. The high Young's modulus and low elongation of basalt fibers and glass fibers refer to the brittle behavior of these fibers, which may break during compaction [14]. Moreover, the prevalent use of carbon fibers and steel fibers as conductive additives for self-healing and microwave deicing of asphalt mixes is attributed to the high thermal conductivity of these fibers.

Table 1. Physical, mechanical, and thermal properties of fibers [39–45].

Fiber	Physical Properties		Mechanical Properties			Thermal Properties		
	Density (g/cm^3)	Moisture Content (%)	Tensile Strength (MPa)	Young's Modulus (GPa)	Elongation (%)	Thermal Conductivity (W/m.K)	Glass Transition (Tg) (°C)	Melting Point (Tm) (°C)
Natural fibers								
Bamboo [41]	0.6–1.1	8.9	140–230	11–17	-	-	-	-
Palm [41,44]	0.7–1.55	-	248	3.2	25	0.199	-	-
Coconut [40,44]	1.15–1.46	8	95–230	2.8–6	15–51.5	0.047	-	-
Sisal [40,44]	1.33–1.5	10–22	363–700	9–38	2–7	0.042	-	-
Kenaf [40]	1.40	9–12	223–930	14.5–53	1.5–2.7	-	-	-
Banana [39]	1.40	8.7–12	529–914	27–32	3	-	-	-
Jute [41]	1.30–1.40	12.60	393–773	13–26.50	1.2–1.5	-	-	-
Synthetic fibers								
Polypropylene [43,45]	0.91	-	500–700	3.5–6.8	21	0.12	−20 to −50	165
Polyester [43,45]	1.38	-	400–600	8.4–16	11-30	0.13	64	240
Basalt [43]	2.6–2.7	-	3100–4800	85–95	3.1	-	-	1450
Carbon [42]	1.8-1.9	-	1700–2600	140–200	0.8–1.5	8–70	-	3500
Glass [42]	2.5–2.56	-	1700–3500	27	2.5–3.2	0.04	-	1540
Aramid [41,44]	1.4	-	3000–3150	63–67	3.3–3.7	0.05	-	500
PAN [45]	1.17	-	200–400	20	27–48	-	97	330
Steel [43]	7.85	-	400–1200	200	3.5	50	-	800

3. Biodegradability of Natural Fibers

Despite the widespread use of naturals fibers due to their availability, low cost, and desirable mechanical properties, they are subjected to premature decomposition as compared to syntactic fibers due to the components of natural fibers such as cellulose, hemi cellulose, and lignin [46–49]. Table 2 summarizes the biodegradation periods of natural fibers. The degradation of natural fibers due to biodegradability or thermal degradation may negatively affect the performance of natural-fiber-modified asphalt mixes. This, in turn, leads to a decrease in the asphalt layer lifespan. Therefore, the biodegradability period of natural fibers may play a decisive role in the evaluation of asphalt mixes incorporating natural fibers.

Table 2. Degradation periods of natural fibers [46].

Fiber	Period (Days)
Bamboo	1–120
Oil palm	21–90
Sisal	21–90
Banana	28–90
Jute	21–35
Kenaf	30–180
Coconut	21–60
Cotton	21–28

4. Methodology

In this study, the literature review on the applications of fibers in asphalt mixtures was conducted by means of citation analysis. Citation analysis is used to identify the most influential research in a specific field through the number of citations. The higher the citations, the better the impact of the research [50,51]. Citation analysis was conducted using the software VOSviewer. The purpose of VOSviewer is to create a network visualization of the imported bibliometric data by creating connection links between citations and documents (clusters). The connection links ensure that the content of the documents are relevant. However, the size of cluster indicates the number of citation, while the link distance between clusters refers to the citations relatedness between documents. The short distance indicates the strong relatedness, while the long distance implies the less relatedness [52]. Moreover, the number of clusters depend on the resolution parameter. The higher the value of resolution parameter, the higher the number of clusters. In this study, the type of analysis in the VOSviewer tool was citation, while the unit of analysis was documents. Furthermore, the minimum number of document's citations was set to 0 to ensure that the number of citations does not affect the links between documents, while the resolution parameter was set to 1. In addition, the review was focused on the connected documents, while the unconnected, unrelated, and review documents were neglected. However, bibliometric data were obtained from the Dimensions AI database in the form of CSV. Dimensions AI is a free, accessible database that aims to reveal links between articles and their outputs. It is reported that the Dimensions AI database provides higher coverage than Scopus and Web of Science [53–55].

The keywords of "Polypropylene fibers AND asphalt", "Polyester fibers AND asphalt", "Basalt fibers AND asphalt ", "Glass fibers AND asphalt", "Steel fibers AND asphalt", "Aramid fibers AND asphalt", "Carbon fibers AND asphalt", "Polyvinyl alcohol fibers AND asphalt", "Ceramic fibers AND asphalt", "PAN fibers AND asphalt", "Cellulose fibers AND asphalt", "Bamboo fibers AND asphalt", "Lignin fibers AND asphalt", "Kenaf fibers AND asphalt", "Coconut fibers AND asphalt", "Palm fibers AND asphalt", "Sisal fibers AND asphalt", "Jute fibers AND asphalt", "Banana fibers AND asphalt", "PET fibers AND asphalt", "Tire textile fibers AND asphalt", and "Metallic fibers AND asphalt" were searched in the titles and abstracts of the Dimensions database. The publication year was set to the last five years (2018–2022) to ensure that the sources are recent. However, the publication year from 2000 to 2022 was also set to study the annual publication trend of utilizing fibers in asphalt mixtures.

5. Results and Analysis

5.1. Annual Publications

The annual publications for the applications of fibers in asphalt mixes are shown in Figure 1. It can be seen in the figure that the most utilized fibers in asphalt mixtures are basalt, polyester, carbon, steel, glass, polypropylene, lignin, and cellulose fibers. Moreover, the use of basalt fibers in the years 2018–2022 sharply increased by about 186% as compared to the utilization of basalt fibers from 2000 to 2017. In addition, bamboo, kenaf, coconut, palm banana, jute, aramid ceramic, PET, PAN, metallic, and tire textile fibers have been recently utilized in asphalt mixes. The annual publications of these fibers are located within the last five years. The increase in the use of natural and waste fibers in the last five years indicates the growing interest in this sustainable practice.

5.2. Applications of Synthetic Fibers in Asphalt Mixes

5.2.1. Polypropylene Fibers

Polypropylene is widely used as a bitumen modifier due to its low density and melting point of 165 °C. Figure 2 presents the network visualization of documents on utilizing polypropylene in asphalt mixtures. As shown in the figure, there are 3 clusters with 3 links. However, the number of citations for the documents of Park, Mohammed, Al-Badri, and Omaranian were 9, 6, 3, and 0, respectively. While the number of links were 1, 1, 1, and 3 respectively. Park et al. [56] investigated the addition of glass fiber coated with polypropylene into a hot asphalt mix incorporated with reclaimed asphalt pavement. As reported, 10–15 mm long glass fibers improved the tensile strength, moisture resistance, dynamic modulus, and rutting resistance. Mohammed et al. [57] studied amending bitumen with 2, 4, and 6% of polypropylene by bitumen weight. The authors concluded that as the content of polypropylene increases, the tensile strength ratio (TSR) of the warm mix improves. Omaranian et al. [58] reported that the integration of glass and polypropylene fibers at a content of 0.2% by aggregate weight improved the moisture resistance of hot mix asphalt and reduced the cons of freeze and thaw cycles on the performance of the asphalt mix. Al-Badri et al. [59] indicated that the combination of 0.3% steel fibers and 0.1% polypropylene by aggregates weight enhanced the stability and temperature sensitivity of hot mix asphalt.

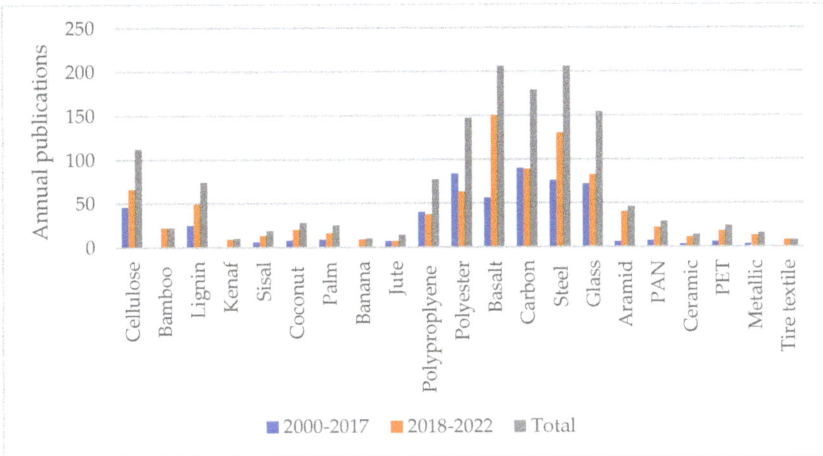

Figure 1. Annual publications.

Figure 2. Network visualization of polypropylene fibers.

5.2.2. Polyester Fibers

Polyester fibers are made with petroleum. The distinctive properties of polyester fibers, i.e., low density, high tensile strength, desirable melting point, and high elasticity, are the major reasons for their widespread use as a bitumen modifier and mixture modifier. Figure 3 shows the most influential documents on utilizing polyester fibers in asphalt mixes based on citation analysis. However, the bigger the size of node, the higher the number of citations, and the greater the impact of document. As seen in the figure, there are 9 clusters with 62 links. The documents of Qin, Kim, Hong, Zhang, and Zarie represents the significant clusters with high number of citations. However, the cluster of Qin incorporating 7 links and 101 citations, while the cluster of Kim containing 7 links and 49 citations. In addition, the document citations number of Hong, Zhang, and Zarie was 30, 26, and 19, respectively. Qin et al. [60] confirmed that modifying a binder with 6 mm long polyester at a content of 5% by bitumen weight enhanced the physical and rheological properties of the binder. Kim et al. [61] assessed hot mix asphalt reinforced by polyester fibers with a length of 6 mm and different contents, 0.5% and 1%, by mix volume. The performance tests showed that adding 1% polyester fibers enhanced the stability, tensile strength ratio (TSR), dynamic stability, and cracking resistance of the asphalt mix. Zhang et al. [62]

investigated the use of different fibers, i.e., lignin, polyester, and PAN fibers, in an open-graded asphalt mix. The authors stated that reinforcing the open-graded asphalt mix with polyester fibers at a dosage of 0.3% by mix weight showed the best performance in terms of stiffness modulus, rutting resistance, fatigue resistance, and drainage property in comparison with lignin and PAN fibers. In a similar study, Zhang et al. [63] observed that the integration of 50% steel slag aggregates and 0.45% polyester fibers in a permeable asphalt mix significantly improved the resistance of the mix to low-temperature cracking. Zarie et al. [64,65] evaluated the fracture energy of hot mix asphalt and warm mix asphalt at low and intermediate temperatures. The authors concluded that the addition of 8 mm long polyester fibers at a content of 0.25% by mix weight significantly improved the fracture energy of asphalt mixes. Alnadish et al. [66–68] noticed that reinforcing an asphalt mix incorporated with coarse steel slag aggregates with 0.3% polyester fibers enhanced moisture resistance, rutting resistance, cracking resistance, and aging resistance. Hong et al. [69] found that introducing 0.3% polyester fiber to an asphalt mix composed of 0.4% coal gangue powder notably boosted the resistance of the asphalt mix to cracking at low temperatures. In a comparative study conducted by Zhu et al. [70], polyester, basalt, and lignin fibers were separately added at a content of 0.3% by mix weight to an asphalt mix integrated with 40% reclaimed asphalt pavement. The researchers observed that the use of fibers improved the performance of the asphalt mix as compared to the unreinforced mix. Among the fibers, basalt fibers showed the best performance as compared to the reinforced asphalt mix with polyester and lignin fibers. Zhu et al. [71] noticed that reinforcing cold mix asphalt with 0.3% polyester fibers with a length of 6 mm showed better low-temperature performance than the control mix and integrated mix with 0.2% 6 mm long basalt fibers. Yu et al. [72] reported that adding 0.3% polyester fiber to cold asphalt mix increased the resistance of the mix to low temperatures by about 42% in comparison with the control mix.

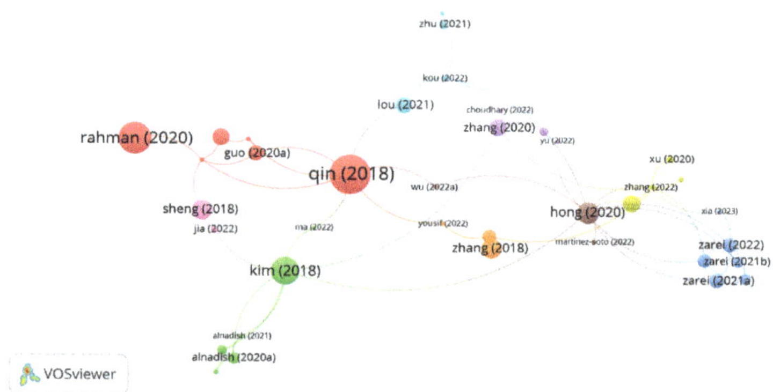

Figure 3. Network visualization of polyester fibers.

5.2.3. Basalt Fibers

Basalt fibers are produced from melted basalt rocks. Influential documents on using basalt fibers in asphalt mixes are shown in Figure 4. It can be seen in the figure that there are 15 clusters with 571 links. The documents of Qin, Xiang, Tanzadeh, Guo, Wang, and Li represents the significant studies with high number of citations. However, the document citations number of Qin, Xiang, Tanzadeh, Guo, Wang, and Li was 101, 67, 56, 46, 63, and 36, respectively. While the number of links was 48, 14, 14, 12, 31, and 26, respectively. Qin et al. [60] investigated the use of basalt fibers as binder modifier Qin et al. [60] investigated the use of basalt fibers as a binder modifier. In their study, 6, 9, and 15 mm long basalt fibers were added into base bitumen at different contents of 3, 5, 7, and 10% by bitumen weight. The findings of the study demonstrated that the amended bitumen with 6 mm basalt fibers at doses of 5% and 7% exhibited better performance than the modified binder with lignin and polyester fibers. The authors stated that adding basalt fibers at a content of 10% reduced the force distribution homogeneity and the bonding property between the binder and fibers, while introducing fibers at 3% did not provide a stable network. Tanzadeh et al. [73] investigated the use of 24 mm basalt fibers and 12 mm glass fibers in a porous asphalt mix. The study outputs demonstrated that the inclusion of glass fibers at a content of 0.2% showed better performance than basalt fibers in terms of tensile strength, TSR, Cantabro, and reducing drain down. In addition, the authors noticed that using fibers decreased the permeability of the porous asphalt mix. To overcome the negative effect of fibers in terms of decreasing permeability, the authors introduced nanosilica to the fiber-modified porous asphalt mix. Guo et al. [74] studied the performance of the low-temperature fracture energy of an asphalt mix reinforced separately with different fibers, i.e., basalt, glass, and steel, with various lengths of 6, 12, and 20 mm at a content of 0.5% by mix weight. The study results demonstrated that 6 mm long basalt fibers showed the highest ultimate stain and fracture energy followed by 20 mm long basalt fibers, 6 mm long glass fibers, and 6 mm long steel fibers. The impact of the glass fibers' length was not significant on the strain. Guo et al. [75] noticed that the addition of basalt fibers to SMA improved cracking resistance and skid resistance. Conversely, adding basalt fibers negatively affects the high-temperature performance and moisture sensitivity. the researchers stated that the combination of lignin fibers and basalt fibers in SMA showed the best performance regarding cracking resistance and rutting resistance. Wang et al. [76–78] conducted studies utilizing 6 mm long basalt fibers in hot mix asphalt, asphalt mix incorporated with steel slag, and modified bitumen. The authors concluded that introducing 0.3% basalt fibers by aggregate weight to an asphalt mix containing waste rubber enhanced Marshall stability and the properties of vibration absorption. In addition, the researchers found that hot mix asphalt with basalt fiber displayed improved tensile strength and resistance to freeze–thaw cycles. Furthermore, the authors noticed that the addition of 0.35% basalt fiber to the asphalt mix composed of steel slag aggregates showed the best performance as compared to polyester fibers and lignin with regard to integrity, moisture resistance, and reducing the negative impact of freezing–thawing. In another study conducted by Wang et al. [79], the performance of a porous asphalt mix containing 9 mm long basalt fibers was studied. The findings of the study showed that introducing basalt fibers at a content of 0.3% significantly improved the mechanical properties of the porous asphalt mix in terms of cracking at low temperatures, dynamic stability, Cantabro abrasion, and tensile strength. Xiang et al. [80] concluded that modifying the basalt fiber surface with a solution of silane coupling agent resulted in a rough surface, better compatibility, higher bonding of basalt-modified bitumen, and superior performance of the asphalt mix. According to Zhang et al. [81], the use of 0.3% basalt fibers improves the rheological properties of the binder, stiffness modulus, and creep behavior of the asphalt mix. Gong et al. [82] found that the use of nano TiO2/CaCO3 and basalt fibers enhanced the performance of the asphalt mix after freeze-thaw cycles regarding uniaxial compression static, stability, tensile strength, resilient modulus, and dynamic creep rate. Yoo et al. [83] demonstrated that adding 0.4% basalt fibers into a porous asphalt mix improves tensile strength, rutting resistance, and stability and reduces freezing–thawing effects. Li et al. [84] stated that the incorporation of 9 mm long basalt fibers at a content of 0.4% improved the low-temperature performance. Kong et al. [85] noticed that amending cold mix asphalt with 0.3% basalt fibers notably enhanced its strength and crack resistance. Alfalah et al. [86] indicated that reinforcing an asphalt mix with 0.3% basalt fibers improved the mix's resistance to rutting. However, the authors indicated that the addition of bitumen negatively affected the rutting performance of asphalt mix.

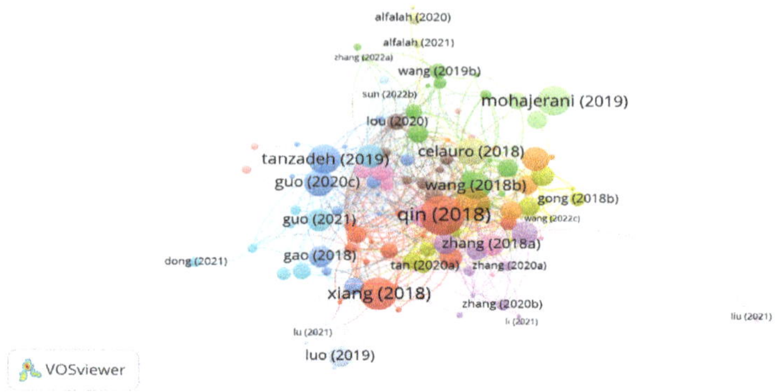

Figure 4. Network visualization of basalt fibers.

5.2.4. Glass Fibers

The influential documents on utilizing glass fibers in asphalt mixtures are shown in Figure 5. A bigger circle size in the network map indicates the stronger impact of the study. As shown in the figure, there are 9 clusters with 94 links. The documents of Tanzadeh, Zairi, Morea, Gupta, and Enieb represents the significant clusters with high number of citations. However, the author's document citations of Zairi, Tanzadeh, Shanbara, Gupta, Khater, Ramesh, Enieb, Morea, Khanghahi, Guo, Liu and Fu were 65, 56, 44, 19, 8, 4, 27, 44, 15, 46, 21 and 10, while the links were 10, 7, 4, 6, 2, 4, 10, 11, 6, 2, 3 and 2, respectively. Zairi [87] evaluated the low-temperature performance of an asphalt mix containing 12 mm long glass fibers at different proportions of 0.06, 0.12, and 0.18% by mix weight and various contents of reclaimed asphalt pavement, i.e., 25, 50, 75, and 100%. The authors concluded that inserting 0.12% glass fibers notably boosted the low-temperature cracking resistance of the asphalt mix. The authors claimed that introducing glass fibers can improve the performance of an asphalt mix incorporated with 100% reclaimed asphalt pavement. Tanzadeh et al. [73] investigated the use of 0.2% glass fibers, 0.2% basalt fibers, 2% nanosilica, and 4% SBS in a porous asphalt mix. The findings of the study showed that reinforcing the porous asphalt mix with basalt fibers led to the best performance in terms of decreasing drain down. However, nanosilica successfully reduced the sensitivity of the asphalt mix to aging. In similar content, Guo et al. [74] concluded that reinforcing asphalt mix with 6 mm long glass fibers at the content of 0.5% significantly improved the low-temperature performance of asphalt mix. Liu et al. [88] confirmed that treating glass fibers with 2 mol/L of etchant notably enhanced the fracture energy and the adhesion between glass fibers and emulsified asphalt. Shanbara et al. [89] investigated the impact of adding glass and hemp fiber into cold asphalt mix. The study findings elucidated that glass fibers with 14 mm length and 0.35% content by aggregate weight along with hemp fiber considerably enhanced cracking and rutting resistance. In a comparative study, Gupta [90] investigated the feasible incorporation of different fibers, i.e., 0.05% aramid fibers, 0.05% aramid pulp, 0.5% cellulose fibers, and 0.5% hybrid glass fibers (cellulose fibers + glass fibers) in a porous asphalt mix. The authors corroborated that the utilization of fibers in the porous asphalt mix successfully reduced drain down and improved indirect tensile strength (ITS). The integrated mixture with aramid pulp showed the best performance, followed by hybrid glass fibers, aramid fibers, and cellulose fibers. Khater et al. [91] studied the potential use of 12 mm long glass and lignin fibers in an asphalt mix. The study outputs demonstrated that the incorporation of glass and lignin fibers in the asphalt mix at a content of 0.3% by mix weight enhanced moisture resistance and low-temperature cracking resistance. Ramesh et al. [92] noticed that adding 0.3% nanoglass fiber by mix weight into warm mix asphalt composed of 70% reclaimed asphalt pavement improved fracture energy by about 1.8 times. Enieb et al. [93] evaluated the performance of an asphalt mix incorporated with 6 mm and 12 mm long glass fibers in terms of indirect tensile strength, moisture sensitivity, fatigue cracking, resilient modulus, and fracture energy. The authors indicated that adding 6 mm long glass fibers to the asphalt mix led to a positive impact in terms of improving the mechanical properties and aging resistance of the asphalt mix. According to Morea et al. [94], introducing 36 mm long glass fibers to hot mix asphalt at a dosage of 0.4% showed better fatigue resistance and permeant deformation resistance than the control mix, reinforced mix with 25 mm long polyester, and modified mix with 12 mm long glass fibers. Khanghahi et al. [95] noticed an increase in fracture energy at low temperatures for hot asphalt mix integrated with 6% gilsonite by bitumen weight and 12 mm long glass fibers at a content of 0.3% by mix weight. Mohammed et al. [96] concluded that modifying the binder with 12 mm long glass fibers at a proportion of 1% by bitumen volume led to better performance than the unmodified bitumen and modified bitumen with cellulose fibers in terms of cracking and rutting resistance. Fu et al. [97] evaluated an amended asphalt mix with glass fiber at the optimum content of 0.5% by mix weight. The outputs of the study emphasized that adding glass fiber enhanced the tensile strength and cracking resistance of the asphalt mix.

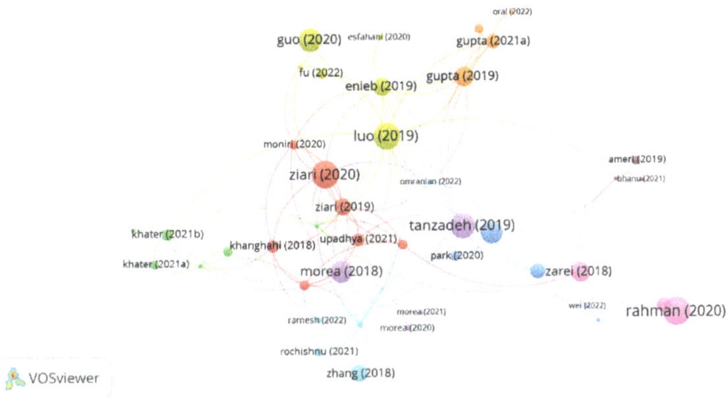

Figure 5. Network visualization of glass fibers.

5.2.5. Steel Fibers

According to the citation analysis, influential documents on the applications of steel fibers in asphalt mixes were identified using the VOSviewer tool. Figure 6 demonstrates the network connections between documents and citations. As seen in the figure, there are 9 clusters with 131 links. The significant clusters with high number of citations are represented by the studies of Liu, Phan, Gao, Tabaković, González, Yang, Jiao, Li, Dinh, Hosseinian, and Norambuena-Contreras. However, the citations for the document of Liu, Phan, Gao, Tabaković, González, Yang, Jiao, Li, Dinh, Hosseinian, and Norambuena-Contreras were 15, 53, 75, 11, 29, 7, 17, 15, 40, 19, and 45, respectively. While the number of links was 4, 6, 8, 4, 7, 3, 2, 4, 4, 4, and 12, respectively. Liu et al. [98] introduced steel fibers into the binder at various contents of 2, 3, and 4% by bitumen weight and different lengths, i.e., 1 mm–7 mm in 2 mm increments, to study healing efficiency. The findings of the study revealed that amending the binder with 5 mm long steel fibers at a content of 4% led to the best performance regarding healing efficiency and fatigue life recovery, while the dosage of 3% led to the worst performance. Phan et al. [99] studied the potential use of steel wool fibers and steel slag aggregates in an asphalt mix as self-healing additives using microwave heating. The authors concluded that the use of 2% steel wool fibers by binder weight and 30% of coarse steel slag aggregate significantly enhanced the healing efficiency of the asphalt mix. Similarly, Gao et al. [100] assessed microwave deicing of an asphalt mix incorporated with different sizes of steel wool fibers: 0.015–0.035, 0.050–0.070, and 0.075–0.125 mm. The authors pointed out that the ice thawing time for steel wool fibers (0.015–0.035) at a content of 0.3%, steel wool fibers (0.050–0.070) at a proportion of 0.6%, and steel wool fibers (0.075–0.125) at a dosage of 0.9% were 9.3, 11.3, and 14.8%, respectively. In another study conducted by Tabaković et al. [101], the self-healing property of a porous asphalt mix containing steel wool fibers 10 mm long and 0.09 mm in diameter was investigated. The results of the study indicated that the addition of steel wool fibers at the proportions of 10% and 15% by binder weight led to lower strength recovery than the amended bitumen with 5% steel wool fibers. González et al. [102] investigated microwave crack healing on an asphalt mix incorporated with different additives, i.e., 4% steel wool fibers, 4% metal shaving, and 5.3% silicon carbide, by bitumen volume. The findings of the study showed that there was a small reduction in the average healing ratio of cracks in the asphalt mix. The authors also stated that heating natural aggregates by microwave could heal cracks without additives. According to Yang et al. [103], utilizing steel fibers in an asphalt mix led to better fracture behavior, moisture resistance, crack healing ability, and less sensitivity to freeze–thaw cycles than the modified asphalt mix with steel wool fibers and carbon fibers. Furthermore, Jiao [104] investigated the low-temperature fracture resistance of an asphalt mix reinforced with steel fibers and basalt fibers. The authors noticed that the asphalt mix containing 6 mm long basalt fibers showed desirable ductal behavior, while the amended mix with 6 mm or 12 mm long steel fibers demonstrated brittle characteristics. Li et al. [105] observed that the healing crack reached 92.3% by modifying the asphalt mix with steel wool fibers at a dosage of 6% by bitumen volume. Additionally, after induction heating, the air void content in the modified asphalt mix was reduced by about 17%. Similarly, Dinh et al. [106] noticed that the combination of steel wool fibers and steel slag aggregates exhibited a higher healing rate than the mixtures consisting of steel wool fibers and granite aggregates. In addition, Dinh et al. [107] observed in another study that the use of reclaimed asphalt pavement in an asphalt mix reinforced with steel wool fibers decreased the efficiency of induction healing due to aging and oxidation processes. As a result, the authors suggested the use of a cooking oil/rejuvenator to enhance induction heating and the healing rate. Hosseinian et al. [108] investigated the moisture resistance and conductivity of an asphalt mix reinforced with different proportions of steel wool fibers in the range of 2–10% in 2% increments by binder volume. The authors concluded that introducing 6% steel wool fibers improved the indirect tensile strength and tensile strength ratio of the asphalt mix. Norambuena-Contreras et al. [109] indicated that adding steel wool fibers and metallic waste enhanced the conductivity of the asphalt mix but not steel shavings. However, a decrease in thermal conductivity was observed in the asphalt mix modified with metallic waste.

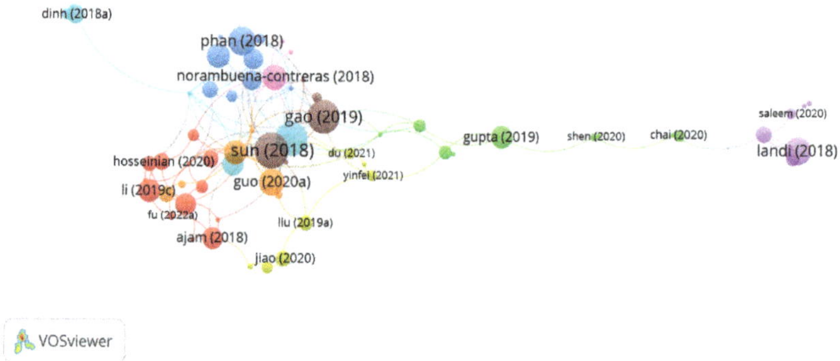

Figure 6. Network visualization of steel fibers.

5.2.6. Carbon Fibers

The components of carbon fibers are 90% PAN and 10% rayon. The network map of connections between documents and citations on using carbon fibers in asphalt mixes is displayed in Figure 7. As can be seen in the figure, there are 9 clusters with 82 links. The documents of Arabzadeh, Pirmohammad, Kim, Yoo, and Ullah represents the most significant clusters. However, the document citations of Arabzadeh, Pirmohammad, Kim,

Yoo, Zhang, Ismael, Gürer, and Ullah were 37, 56, 49, 32, 10, 4, and 12, respectively. Whereas the number of links was 9, 2, 4, 7, 4, and 3, respectively. Arabzadeh et al. [110] studied the use of carbon fibers as a conductive additive in an asphalt mix. The study outputs demonstrated that inserting 0.5% carbon fibers enhanced the conductive property of the asphalt mix. Pirmohammad et al. [111] evaluated the low-temperature fracture performance of an asphalt mix reinforced separately with 4, 8, and 12 mm long carbon and kenaf fibers at different dosages of 0.1, 0.2, and 0.3% by mix weight. The study results demonstrated that adding 8 mm kenaf fibers and 4 mm carbon fibers at a content of 0.3% effectively enhanced the fracture resistance of the asphalt mix at low temperatures. Kim et al. [61] stated that reinforcing an asphalt mix with 6 mm long carbon fibers at a content of 1% by mix volume led to better cracking resistance, permeant deformation resistance, and moisture resistance than the unreinforced asphalt mix. In a study conducted by Yoo et al. [112], the healing property of an asphalt mix modified by carbon fibers and graphite nanofibers was studied. The authors asserted that the incorporation of fibers at a proportion of 0.5% considerably improved the efficiency of healing. The researchers stated that the recovery improved by 40% as compared to the control mix. Zhang et al. [113,114] concluded that inserting 0.05% cured carbon fibers by mix weight to a porous asphalt mix significantly improved indirect tensile strength and rutting resistance. In addition, at the optimum fiber content of 0.15%, the infiltration and porosity of the asphalt mix decreased by about 20% and 7%, respectively. Moreover, the authors stated that with the increase in fiber dosage, the ductile behavior and cracking resistance of the asphalt mix improved. Ismael et al. [115] investigated the utilization of carbon and jute fibers in SMA. The outputs of the study demonstrated that adding 7.5 mm long carbon or jute fibers at a proportion of 0.5% by mix weight notably enhanced Marshall stability and rutting resistance and decreased drain down of the asphalt mix. According to Gürer et al. [116,117], adding 5 mm long carbon fibers with a length of 5 mm and content of 0.2% by mix mass into hot mix asphalt and SMA improved the conductive property of the asphalt mix. The researchers also stated that the gradation of the asphalt mix has a significant effect on the conductive property. The denser the gradation, the higher the conductivity. Schuster et al. [118] concluded that the integration of 1% carbon fiber and 16% steel wool fiber by bitumen weight significantly improved the healing rate. Similarity, Ullah et al. [119] indicated that the incorporation of 0.2–0.4% carbon fiber by mix weight in an asphalt mix remarkably enhanced the conductivity and performance of the asphalt mix.

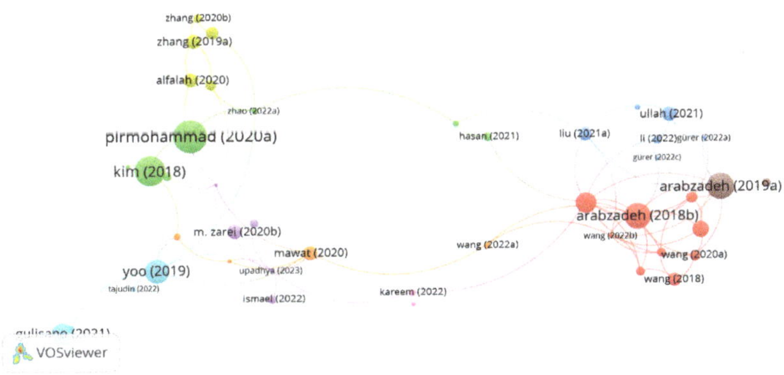

Figure 7. Network visualization of carbon fibers.

5.2.7. Aramid Fibers

Aramid fibers are produced from aromatic polyamides. The network map shown in Figure 8 demonstrates the citation analysis outputs. It is shown in the figure that there are 9 clusters with 85 links. However, the study's citations of Takaikaew, Klinsky, Slebi-Acevedo, Callomamani, Hajiloo, Gupta, Noorvand, Daniel, and Xing were 30, 62, 30, 3,

10, 6, 3, 2, and 11, respectively. While the number of links was 6, 18, 8, 2, 2, 16, 4, 5, and 3, respectively. Takaikaew et al. [120] studied the performance of hot mix asphalt reinforced with 19 mm long polyolefin-aramid fibers at a dosage of 0.05% by mix weight. The study findings illustrated that adding fibers significantly improved the resilient modulus, tensile strength, fatigue resistance, and rutting resistance. Similarly, Klinsky et al. [121] indicated that the addition of polypropylene-aramid enhanced the mechanical properties of an asphalt mix. According to the findings obtained by Slebi-Acevedo et al. [122], adding 19 mm long polyolefin-aramid or 4 mm long PAN at a content of 0.3% by mix weight improves the low-temperature performance of an asphalt mix regarding fracture energy and cracking resistance. Additionally, Callomamani et al. [123] found that introducing 0.065% aramid or 0.1% polyethylene terephthalate (PET) or 0.065% PAN fibers to hot asphalt mix boosted the rutting resistance, cracking resistance, and tensile strength of the asphalt mix. Hajiloo et al. [124] studied the performance of hot mix asphalt reinforced with different proportions of polyolefin-aramid fibers, i.e., 0.025, 0.05, and 0.1%, by mix weight. The researchers noticed that the higher the content of fibers, the better the fracture toughness. Gupta et al. [125] stated that reinforcing a porous asphalt mix with 12 mm long aramid fibers notably boosted the tensile strength, moisture resistance, and cracking resistance of the asphalt mix. Xing et al. [126] confirmed that modifying bitumen with 2 mm long aramid fibers at a proportion of 2% by bitumen weight improved rheological and physical properties at high and low temperatures. According to Noorvand [127], adding 19 mm long aramid fibers at a dosage of 0.05% by mix weight to hot mix asphalt incorporated with 20% reclaimed asphalt pavement enhanced fatigue and cracking resistance. Moreover, 19 mm long aramid fibers exhibited the best performance in comparison with 10 mm and 38 mm long aramid fibers. Additionally, Daniel et al. [128] found that introducing 19 mm long polyolefin-aramid fibers into warm mix asphalt improved stiffness modulus by 89% and fatigue life by 100%. Slebi-Acevedo et al. [129] concluded that the combination of a porous asphalt mix with 19 mm polyolefin-aramid fibers at a dosage of 0.05% by mix weight and hydrate lime enhanced the mechanical properties of the porous asphalt mix.

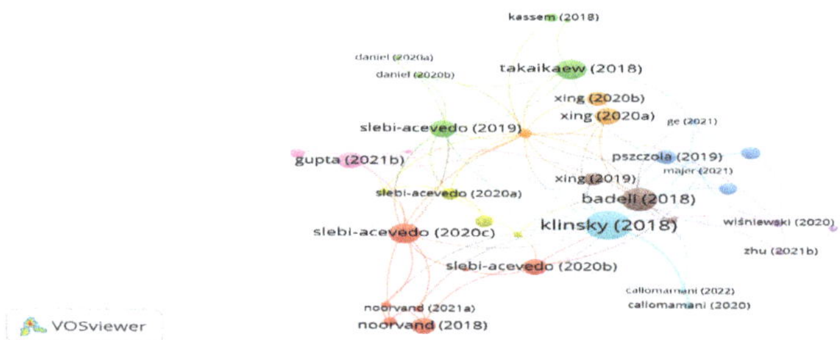

Figure 8. Network visualization of aramid fibers.

5.2.8. Polyacrylonitrile (PAN) Fibers

Significant studies using PAN fibers in asphalt mixes as reinforcement materials are demonstrated by network visualization shown in Figure 9. As seen in the figure, there are 4 clusters with 20 links. The clusters of Su, Slebi-Acevedo, Dalhat, and Wang exhibited the highest number of citations. However, the research's citations of Su, Slebi-Acevedo, Dalhat, and Wang were 25, 18, 15, and 20, respectively. Su et al. [130] investigated amending bitumen with 3% PAN fiber and modified PAN fiber with self-polymerization of dopamine and covalent grafting. The study results demonstrated that the amended bitumen with modified PAN fibers increased the adhesion property and moisture resistance by 8% as compared to the unmodified PAN fiber. Slebi-Acevedo et al. [131] affirmed that the addition of 6 mm long PAN fibers to a reclaimed asphalt mix at a content of 0.3% by mix weight enhanced the mechanical performance of the asphalt mix up to 50% without the use of rejuvenators. Dalhat et al. [132] noticed a significant enhancement in the dynamic modulus, rutting resistance, and cracking resistance of an asphalt binder modified with 1% PAN fiber and 1% SBS. According to Xing et al. [133], blending base bitumen with SBS and PAN fibers surface-modified by bionic coating and nanosilica improves the rheological properties and shear strength of bitumen. Wang et al. [134] indicated that reinforcing an asphalt mix with 6 mm long PAN at a dosage of 0.3% by mix weight boosted the fatigue resistance of the asphalt mix.

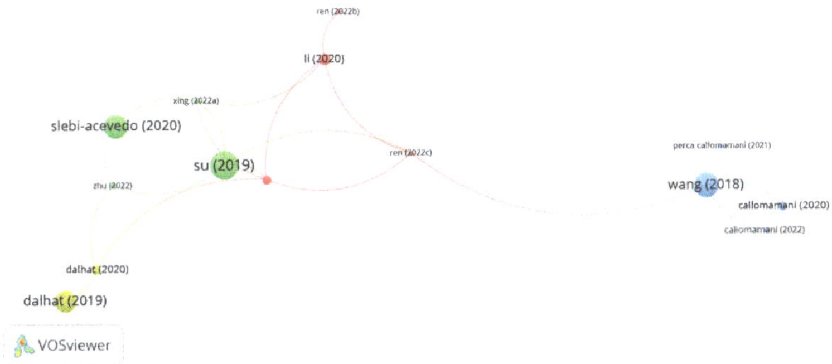

Figure 9. Network visualization of PAN fibers.

5.2.9. Ceramic Fibers

Recently, there is a growing use of ceramic fibers as bitumen fibers. Influential studies using ceramic fibers in asphalt mixes are shown in Figure 10. It is shown in the figure that the cluster of Arabani, Naseri, Wang, and Hamedi exhibited the highest number of citations. However, the author's document of Arabani, Naseri, Wang, and Hamedi was 35, 8, 5, and 3, respectively. Arabani et al. [135,136] confirmed that the combination of 20 mm long ceramic fibers at a dosage of 3% by bitumen weight into the base binder enhanced the high-temperature performance and decreased the performance of the binder at low temperatures. On the contrary, Naseri et al. [137] stated that modifying bitumen with 0.4% ceramic fibers by bitumen weight boosted the fracture energy and toughness of the binder, which, in turn, decreased cracking at low temperatures. Wang et al. [138] noticed the presumed increase in moisture resistance, rutting resistance, and resistance to low-temperature cracking of an asphalt mix reinforced with 2–4 mm long ceramic at a proportion of 0.4% by mix weight. Hamedi et al. [139] concluded that modifying bitumen with 3% ceramic fibers improved the resistance of SMA to rutting. In a study conducted by Liu et al. [140], an asphalt mix was modified with ceramic fibers at a dosage of 0.3% by mix mass to study the influence of adding ceramic fiber on the properties of the binder. The findings of the study illustrated that the addition of ceramic fibers improved the freeze–thaw splitting and moisture resistance of the asphalt mix.

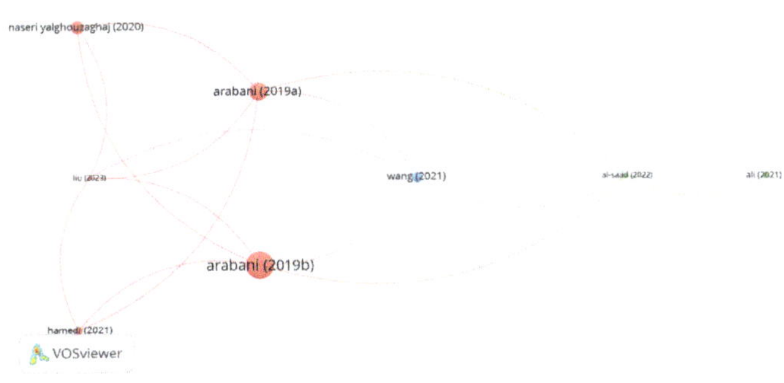

Figure 10. Network visualization of ceramic fibers.

5.3. Applications of Natural Fibers in Asphalt Mixes

5.3.1. Cellulose Fibers

Wood and paper industries and the byproducts of plants are the main sources of cellulose fibers. The outputs of the citation analysis are illustrated in Figure 11. A bigger size of the node implies a higher impact of the document, while the links describe the citation connections among the documents. It can be seen in the figure that there are 6 clusters with 22 links. In addition, the significant clusters with high number of citations are belong to the documents of Landi, Irfan, Fauzi, and Li. However, the citations for the document of Landi, Irfan, Fauzi, and Li were 50, 20, 5, and 13, respectively. Landi et al. [141,142] evaluated a porous asphalt mix incorporated with 0.3% cellulose fibers and 0.3% end-of-life tire fibers by mix weight. The study findings demonstrated that both fibers showed better fatigue resistance than the unreinforced asphalt mix. However, the asphalt mix incorporated with end-of-life tire fibers displayed higher fatigue resistance than the amended mix by about 70%. The researchers also found that the cumulative energy demand and global warming potential of reinforced porous asphalt with end-of-life tire decreased by 25% and 10%, respectively, in comparison with porous asphalt composed of cellulose fibers. Irfan et al. [143] found that adding 0.3% cellulose fibers to stone mastic asphalt significantly enhanced rutting and fatigue resistance. In similar content, Fauzi et al. [144] confirmed that introducing 0.3% cellulose fibers to stone mastic asphalt improved the stiffness modulus. According to Li et al. [33], aging has a negative impact on cellulose fibers; aging resulted in a decrease in the cracking resistance of an asphalt mix containing cellulose fibers.

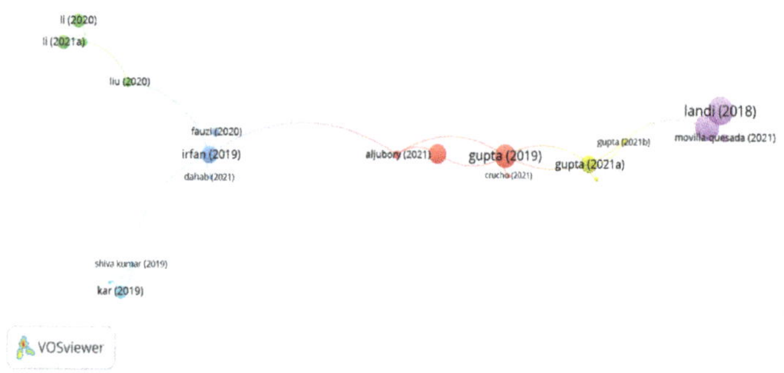

Figure 11. Network visualization of cellulose fibers.

5.3.2. Bamboo Fibers

Influential studies on the applications of bamboo fibers in asphalt mixes are shown in Figure 12. As seen in the figure, the document of Sheng, Liu, Xia, Jia, and Meng displayed bigger size of node than the other documents. However, the document citations of Sheng, Liu, Xia, Jia, and Meng were 26, 17, 11, 12, and 6, respectively. Sheng et al. [145] assessed the potential use of bamboo, lignin, and polyester fibers in a dense asphalt mix and stone mastic asphalt mix. The study concluded that adding 0.2% bamboo fibers to dense asphalt and 0.4% to stone mastic asphalt led to better performance than the control mix and modified mix with polyester and lignin fibers. Similarly, Liu et al. [146] investigated the performance of asphalt containing 0.4% lignin and 0.3% bamboo fibers. The findings of the study illustrated that both fibers improved the mechanical properties of the asphalt mix regarding low-temperature performance, high-temperature performance, and moisture sensitivity. However, the researchers stated that bamboo fibers notably boosted the thermal properties of the asphalt mix. Furthermore, the authors noticed that adding bamboo and lignin fibers enhanced the performance of asphalt mixes subjected to long-term aging. Xia et al. [147,148] evaluated the performance of SMA and hot mix asphalt incorporated with lignin and bamboo fibers. The authors noticed that the durability of SMA was better than dense hot mix asphalt. However, the positive effect of utilizing lignin and bamboo on freeze–thaw was almost the same. The researchers stated that the incorporation of bamboo fibers in the asphalt mix led to better aging performance than the modified mix with lignin fibers. Based on the findings obtained by Jia et al. [149], the inclusion of bamboo fibers in an asphalt mix at a proportion of 0.3% by mix weight remarkably improved the rutting and fatigue cracking resistance of the asphalt mix. Meng et al. [150] found that amending bitumen with bamboo fibers, SBS, and soybean bio-asphalt at a dosage of 3% by bitumen weight led to better performance than the modified bitumen with 5% SBS in terms of rheological properties.

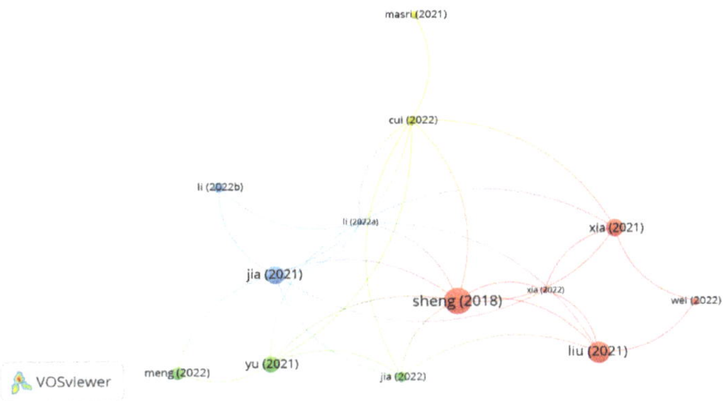

Figure 12. Network visualization of bamboo fibers.

5.3.3. Palm Fibers

The outputs of citation analysis by means of the VOSviewer tool for impactful studies using palm fibers as reinforcement materials are illustrated in Figure 13. It is seen in the figure that there are 3 clusters with 11 links. However, the citations number of the significant documents of Syammaun, Yaro and Tayh was 6, 7, and 2, respectively. Syammaun et al. [151] added palm oil fibers by a dry process at different proportions in the range of 1–5% in 1% increments by bitumen weight to investigate the influence of palm fibers on the resilient modulus of a porous asphalt mix. The authors noticed that adding palm fibers at 3% led to the best performance. Yaro et al. [152,153] concluded that the use of sequential mixing for palm oil fibers in stone mastic asphalt enhanced the moisture resistance, stiffness modulus, Cantabro, and drain down of SMA. It is reported that the optimum content of palm fiber was 0.3% by mix weight. Tayh et al. [154] indicated that modifying bitumen with 0.75% palm fibers by binder weight enhances the physical properties and rutting resistance of bitumen.

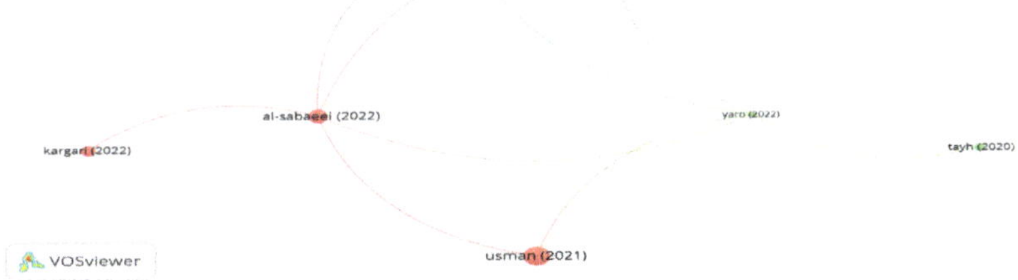

Figure 13. Network visualization of palm fibers.

5.3.4. Lignin Fibers

Wood pulp, cotton, hemp, jute, and paper industries are the sources of lignin. Figure 14 displays influential documents based on the conducted citation analysis. It can be seen in the figure, that there are 9 clusters with 74 links. However, the document of Qin, Chen, Wang, Wu, and Kou exhibited higher number of citations than the other documents. The document citations number of Qin, Chen, Wang, Wu, and Kou was 101, 38, 19, and 18, respectively. Qin et al. [60] investigated the rheological and physical properties of modified bitumen with lignin, polyester, and basalt fibers. The study findings demonstrated that the addition of basalt fibers at contents of 5% and 7% showed the best performance, followed by lignin fibers at a dosage of 3% by bitumen weight. Wang et al. [77] observed that introducing lignin fibers at a dosage of 0.23% by mix weight significantly improved the moisture resistance of an asphalt mix incorporated with steel slag aggregates. Chen et al. [155] studied the physical and rheological properties of modified binder with corn stalk, basalt, and lignin fibers. The findings of study demonstrated that modifying binder with stalk fibers showed the best performance, followed by basalt, and lignin fibers. Ma et al. [156] investigated the performance of modified bitumen with different types of fibers, i.e., basalt, lignin, polyester, and composite fibers, at proportions of 7, 7, 7, and 6% by binder weight. The researchers indicated that all fibers boosted the shear strength of the binder as compared to the unmodified binder. However, the amended binder with composite fibers showed superior performance, followed by basalt, lignin, and polyester. In studies conducted by Wu et al. [157,158], the cracking resistance and high-temperature performance of stone mastic asphalt reinforced with 0.3% lignin and 0.3% basalt fibers were assessed. The results of the studies illustrated that the use of fibers improves the performance of the asphalt mix in comparison with the control mix. Moreover, the authors concluded that basalt showed the best performance. According to Kou et al. [159,160], the appropriate contents of basalt fibers, polyester fibers, and lignin fibers as bitumen modifier are 2%, 3%, and 4% by bitumen weight, respectively. The authors confirmed that these contents remarkably enhanced performance in terms of permanent recovery and cracking resistance.

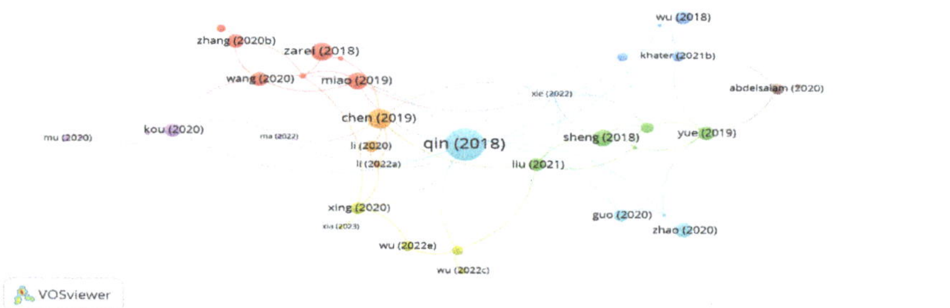

Figure 14. Network visualization of lignin fibers.

5.3.5. Coconut Fibers

Recently, the use of coconut fibers in asphalt mixes has increased. Significant studies utilizing coconut fibers in asphalt mixes are highlighted through a network map of documents and citations in Figure 15. The disappearance of links indicates that there were no linked citations among the documents. Khasawneh et al. [161] noticed that adding coconut fibers at a dosage of 0.4% by aggregate weight improved the stability of an asphalt mix. In addition, the researchers stated that the incorporation of coconut fibers has a significant effect on the volumetric properties of the asphalt mix. Maharaj et al. [162] found that adding 2.5 mm long coconut fibers as a bitumen modifier at a proportion of 6% by binder weight improved the rheological properties of bitumen in terms of deformation resistance and elastic behavior. In a study conducted by Parimita [163], the Marshall stability of stone mastic asphalt reinforced with 0.3% coconut and banana fibers were evaluated. The outputs of the study illustrated that adding coconut/ banana fibers significantly improved Marshall stability of asphalt mix as compared to the unreinforced mix. Norhidayah et al. [164] concluded that introducing 0.3% coconut fibers by mix weight into a porous asphalt mix composed of coconut shells as coarse aggregates reduced the drainage of the mix. In another study conducted by Haryati et al. [165], the potential integration of 10% coconut shell and 0.3% coconut fibers in a porous asphalt mix was assessed. The study findings demonstrated that the addition of coconut shells and fibers notably boosted the rutting resistance and stability of the asphalt mix.

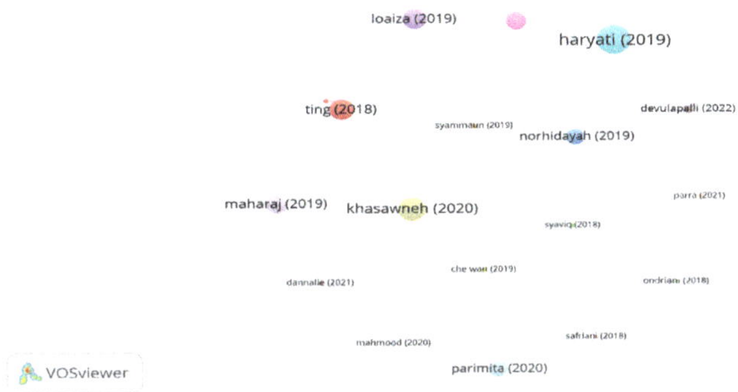

Figure 15. Network visualization of coconut fibers.

5.3.6. Sisal Fibers

The outputs of the citation analysis of influential documents on using sisal fibers are demonstrated in Figure 16. Kar et al. [166] evaluated the feasible use of sisal fibers and fly ash in dense asphalt mix and stone mastic asphalt in terms of Marshall stability. The results of the study indicated that adding 0.3% sisal fibers improved the Marshall stability of asphalt mixes. In a similar study, Kumar [167] observed that the stability and tensile strength of control stone mastic asphalt were almost comparable to stone mastic asphalt containing 0.3% sisal fibers. Singh et al. [168] concluded that using 0.4% sisal fibers, 0.3% coir fibers, and 0.3% rice straw fibers in SMA significantly improved the tensile strength and stability of stone mix, and drain down also decreased.

Figure 16. Network visualization of sisal fibers.

5.3.7. Kenaf Fibers

Kenaf fibers are one of the commonly used natural fibers in asphalt mixes. The research on reinforcing asphalt mixtures with kenaf fibers has been reviewed based on the impact of the studies. Figure 17 shows the poignant documents on the use of kenaf fibers. It can be seen in the figure that the number of clusters is 3, while the number of links is 9. However, the documents of Pirmohammad and Hainin showed the highest number of citations as compared to the other documents. Pirmohammad et al. [111] noticed that introducing 8 mm long kenaf fibers at a content of 0.3% into dense asphalt mix enhanced the low-temperature cracking resistance of the asphalt mix. In another study conducted by Pirmohammad et al. [169], the low-temperature performance of asphalt mixes modified with kenaf and goat wool fibers at the proportions of 0.1, 0.2, and 0.3% by mix weight and lengths of 4, 8, and 12 mm was evaluated. The study concluded that the addition of 8 mm long kenaf/4 mm long goat wool fibers significantly improved the fracture energy of the asphalt mix at low-temperature. Hainin et al. [170] confirmed that introducing 30 mm long kenaf fibers to dense asphalt mix enhances the tensile strength ratio and rutting resistance of the asphalt mix. Syafiqah et al. [171] noticed a significant enhancement in the resilient modulus, Marshall stability, and dynamic stability of stone mastic asphalt reinforced with 0.2% kenaf fibers. Masri et al. [172] stated that adding kenaf fibers to a porous asphalt mix at a dosage of 0.6% by mix weight enhanced the resilient modulus of the asphalt mix.

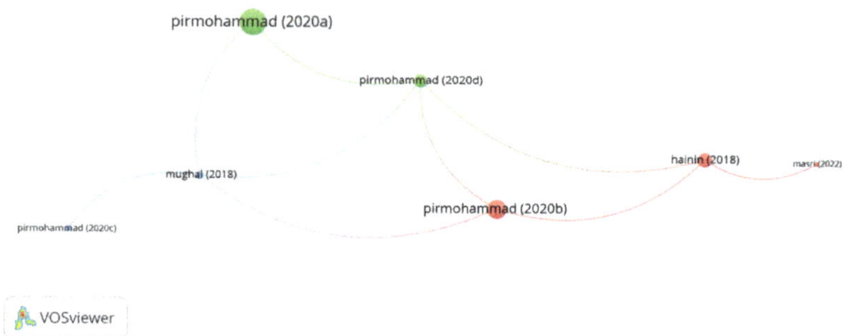

Figure 17. Network visualization of kenaf fibers.

5.3.8. Jute Fibers

Influential studies on reinforcing asphalt mixes with jute fibers are displayed in Figure 18. It can be seen in the figure that there are no links between documents. This is because there were no citations between the studies. Selvaraj et al. [173] investigated the interface behavior between coir geotextile, jute geotextile, and asphalt layer by means of numerical analysis. The outputs of study demonstrated that the use of coir geotextile showed better performance than the geotextile of jute in terms of delaying degradation and decreasing developed stress at surface of asphalt layer. According to Shanbara et al. [89,174], the use of 14 mm long jute fibers at a content of 0.35% by mix weight in cold mix asphalt significantly improved the rutting resistance and indirect tensile strength of the asphalt mix. Ismael et al. [115] demonstrated that modifying stone mastic asphalt with 7.5 mm long jute fibers at a dosage of 0.5% by mixture mass notably enhanced the rutting resistance and dynamic stability and decreased drain down of SMA. Gallo et al. [175] confirmed that the addition of 20 mm long jute fibers at the content of 0.2% into an asphalt mix increased air voids content, resilient modulus, and indirect tensile strength, while the tensile strength ratio slightly decreased.

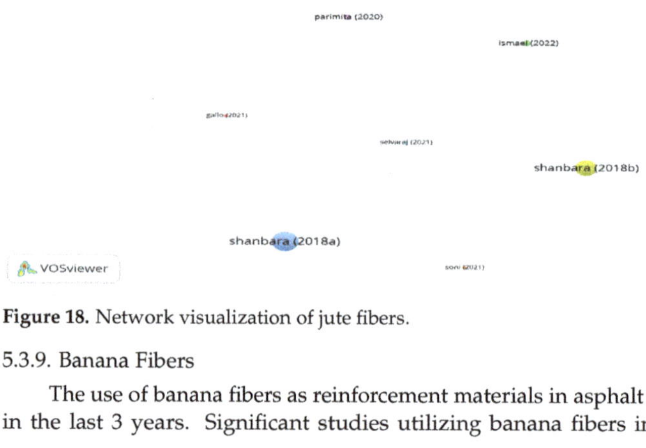

Figure 18. Network visualization of jute fibers.

5.3.9. Banana Fibers

The use of banana fibers as reinforcement materials in asphalt mixtures has increased in the last 3 years. Significant studies utilizing banana fibers in asphalt mixtures are highlighted in Figure 19. The significant studies were identified through citation analysis. Kumar et al. [167,176] concluded that amending SMA with 0.3% banana fibers by mix mass enhanced the rutting and cracking resistance and reduced drain down of SMA. Costa et al. [177] studied the performance of modified stone mastic asphalt at various lengths and proportions. The superior performance in terms of indirect tensile strength, resilient modulus, dynamic modulus, and drain down was observed by adding 20 mm long banana fibers at a proportion of 0.3% by SMA weight. In another study conducted by Costa et al. [178], the addition of 20 mm long banana fibers at a dosage of 0.3% by mix weight into SMA enhanced Marshall stability, modified Lottman, indirect tensile strength (ITS), and Cantabro.

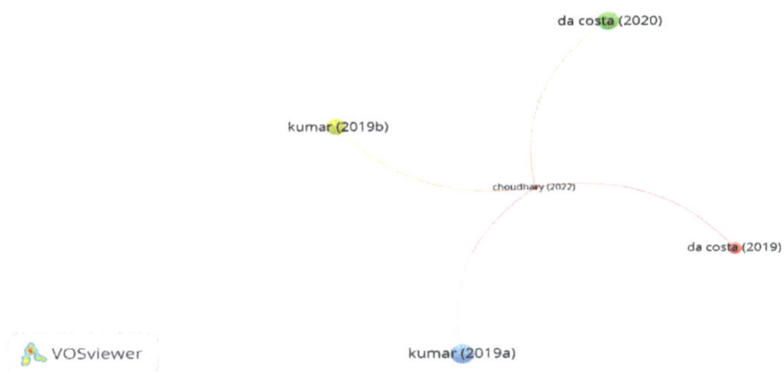

Figure 19. Network visualization of banana fibers.

5.4. Applications of Waste Fibers in Asphalt Mixes

5.4.1. Polyethylene Terephthalate (PET) Fibers

PET fibers are produced from waste plastic bottles. The global increase in the waste of plastic bottles prompted scientists and engineers to use these wastes in civil engineering applications [179]. Recently, the use of PET fibers in asphalt mixes remarkably increased, in particular in the last three years. Significant studies using PET in asphalt applications have been reviewed based on the citation analysis. Figure 20 demonstrates the network map of influential studies. It is seen in the figure that the number of clusters is 5, while the number of links is 17. However, the document citations number of Dehghan, Usman, Movilla-Quesada, Jegatheesan, and Babalghaith was 48, 7, 13, 0, and 1, respectively. Dehghan et al. [179] stated that adding 20 mm long PET fibers by a dry process at a dosage of 1% and crumb PET at a content of 2% by binder weight significantly improved fatigue cracking by 148% and 163%, respectively, as compared to the control mix. Usman et al. [180] investigated the potential use of 10 mm long PET fibers in hot mix asphalt. PET fibers were introduced to the mix at proportions of 0.3, 0.5, 0.7, and 1% by mix weight. The authors noticed that adding PET fibers at 0.5% showed the best performance regarding rutting resistance and resilient modulus. Movilla-Quesada et al. [181] concluded that amending bitumen with 30–50 mm long PET fibers at doses of 10% and 12% by bitumen weight enhanced the physical properties of the binder in terms of penetration, consistency, ductility, and viscosity. According to Jegatheesan et al. [182,183], modifying bitumen with 4–6 mm long PET fibers at a content of 10% by weight of bitumen improved the physical properties of the binder and the performance of the asphalt mix composed of 18% carbonized wood as fine aggregates in terms of Marshall stability, resilient modulus, and indirect tensile strength. Babalghaith et al. [184] indicated that adding 4 mm long black PET fibers to SMA at a content of 0.6% by mix weight significantly enhanced the stiffness modulus and indirect tensile strength of SMA.

Figure 20. Network visualization of PET fibers.

5.4.2. Tire Textile Fibers

Tire textile fibers represent 10–15% of end-of-life tire weight. End-of-life tires have become a major environmental challenge due to the annual constant increase in these wastes. As a result, utilizing tire textiles as reinforcement materials in asphalt mixes is a successful approach in terms of decreasing the quantities of end-of-life tires [185]. Figure 21 summarizes the influential studies utilizing tire textile fibers in asphalt mixes. As seen in the figure, the number of clusters is 3, while the links number is 9. The documents of Landi, Bocci, and Valdés-Vidal represents the significant clusters. However, the number of citations for the document of Landi, Bocci, Valdés-Vidal, and Calabi-Floody was 50, 15, 3, and 0, respectively. Landi et al. [141,142] confirmed that the addition of 0.3% end-of-life tires to a porous asphalt mix significantly improved the fatigue life of the asphalt mix. Bocci et al. [185] concluded that reinforcing hot mix asphalt with 1–2.5 mm long tire textile fibers at a content of 0.3% by mix weight enhanced the fatigue cracking resistance of the asphalt mix. However, the stiffness modulus and indirect tensile strength of the reinforced asphalt mix were almost similar to the control mix. Additionally, the authors stated that a filler with a content of 2% should be added due to the decrease in the compactability of the reinforced asphalt mix. Valdés-Vidal et al. [186] studied amending a binder with tire textile fibers at different percentages of 2–8% in 3% increments by bitumen weight. The researchers found that adding fibers at the percentages of 2% and 5% significantly enhanced the stiffness modulus, permeant deformation, and cracking resistance at intermediate and low temperatures of the asphalt mix. Furthermore, the authors noted that the addition of fibers requires a higher compaction level to achieve the required density. Calabi-Floody et al. [187] detected that adding 0.5% tire textile fibers as a bitumen modifier decreased the permanent deformation of the binder. In addition, the researchers confirmed that there was no negative impact on the performance of the binder at low temperatures.

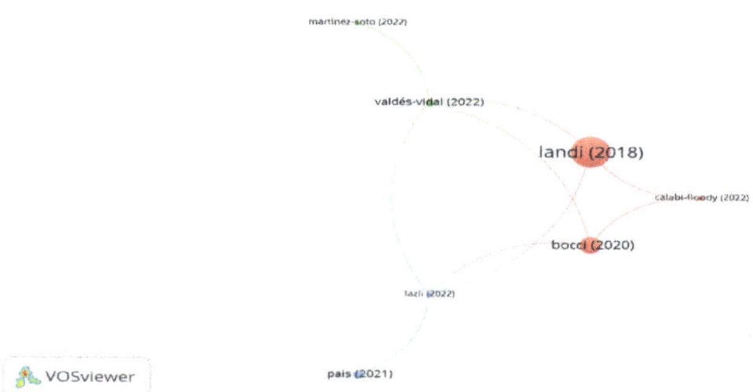

Figure 21. Network visualization of tire textile fibers.

5.4.3. Metallic Fibers

Metallic fibers are common waste materials produced from waste tires. Lately, the use of metallic fibers in asphalt mixes has increased as a contribution towards sustainable practices. In accordance with the outputs of citation analysis, the network map of the significant studies using metallic fibers in asphalt mixes is shown in Figure 22. As seen in the figure, the document of González and Norambuena-Contrera showed the highest number of citations. However, the citations for document of González and Norambuena-Contrera were 56 and 45, respectively. González et al. [188] assessed the performance of hot mix asphalt incorporating amended bitumen with different proportions of metallic fibers of 1.5, 2.5, and 3.5% by bitumen volume. The findings of the study demonstrated that the higher the dosage of fiber, the higher the air void content, clustering, and electrical conductivity. Additionally, the authors stated that fibers decreased the stiffness modulus and indirect tensile strength, while the tensile strength ratio was above 80%. In another study conducted by González et al. [189], the feasible use of reclaimed asphalt pavement (RAP) in an asphalt mix modified with metallic fibers by the wet process was assessed. The study findings illustrated that the integration of RAP increased the stiffness modulus and tensile strength of the mix, while metallic fibers improved cracking healing through microwave heating. Similarly, Norambuena-Contrera et al. [109] concluded that the incorporation of metallic fibers increased air voids and thermal conductivity, while the distribution of fibers was not uniform. Table 3 summarizes the findings of the review studies.

Figure 22. Network visualization of metallic fibers.

Table 3. Summary of findings.

Citations	Fiber	Content (wt%)	Length (mm)	Penetration (%)	Softening Point (%)	Viscosity (%)	Ductility (%)	Rutting (%)	Fatigue (%)	Shear Strength (%)
					Fiber-modified bitumen					
[60]	Polyester	5	6	393				90		↑
[60]	Basalt	7	6	394				285		↑
[160]	Basalt	3	6		12		11	90		
[96]	Glass	1 (vl%)	6	89	40	500		20		
[130]	PAN	3	6					8		
[133]	PAN	1	6							
[135]	Ceramic	3	20	50	20	33	690	55	35	
[154]	Carbon	2.75	0.15	335	16	400		8		
[126]	Aramid	2	2	10	33		≅	3		
[154]	Palm	0.75	<0.58	58	8	900		5		
[60]	Lignin	3	<3	77				329		310
[160]	Lignin	3	<3	≅			60	10		
[162]	Coconut	6	2.5					50		
[96]	cellulose	1 (vl%)		13	31	400		30		
					Fiber-modified dense asphalt mix					
Citations	Fiber	Content (wt%)	Length (mm)	Stability (%)	TSR (%)	Rutting (%)	Fatigue (%)	Low temperature (%)		Fracture energy (%)
[61]	polypropylene	0.5 (vl%)	6	13	50	40		60 (−10 °C)		
[64]	Polyester	0.25	8					37 (−18 °C)		23 (25 °C)
[94]	Polyester	0.4	25			15	≅			
[69]	Polyester	0.40	12							28 (0 °C)
[61]	Polyester	1 (vl%)	6	15	4	30		32 (−10 °C)		
[70]	Polyester	0.3	6		13	10	28			
[74]	Basalt	0.5	6					300 (−10 °C)		350 (−10 °C)
[84]	Basalt	0.4	9					40 (−20 °C)		
[70]	Basalt	0.3	6		12	23	33			
[74]	Glass	0.5	6					60 (−10 °C)		210 (−10 °C)
[87]	Glass	0.12	12					10 (−15 °C)		50 (0 °C)
[91]	Glass	0.3	12		12			19 (−10 °C)		
[94]	Glass	0.4	36			90	20 (10 °C)			
[111]	Carbon	0.3	4					20 (−10 °C)		
[61]	Carbon	1 (vl%)	6	≅	40	≅		15 (−10 °C)		
[74]	Steel	0.5	6					40 (−10 °C)		90 (−10 °C)
[122]	polypropylene-aramid	0.3	19					26 (−15 °C)		7 (15 °C)
[122]	PAN	0.3	4					22 (−15 °C)		13 (15 °C)
[123]	PAN	0.2	6		↓					79
[138]	Ceramic	0.4	2–4	18	6	23		10 (−10 °C)		
[145]	Bamboo	0.2	6	9	10	30		22 (−10 °C)		
[146]	Bamboo	0.35	<6	16	9	67		12 (−10 °C)		
[91]	Lignin	0.3	1.1		6			20 (−10 °C)		
[70]	Lignin	0.3	<5		320	15	19			
[146]	Lignin	0.4	<5	14	7	58		1.09 (−10 °C)		
[161]	Coconut	0.4	10	15						
[166]	Sisal	0.3	10	13	12					

Table 3. *Cont.*

Citations	Fiber	Content (wt%)	Length (mm)	Penetration (%)	Softening Point (%)	Viscosity (%)	Ductility (%)	Rutting (%)	Fatigue (%)	Shear Strength (%)
[111,169]	Kenaf	0.3	8						1.15 (−15 °C)	
[170]	Kenaf	0.3	30	3	13	50	↑			
[179]	PET	0.5	20			↑	↑			
[180]	PET	0.5	10							
[185]	Tire textile	0.3	1–2.5			↑	↑			
Fiber-modified stone mastic asphalt										
Citations	Fiber	Content (wt%)	Length (mm)	Stability (%)	Stiffness modulus (%)	ITS (%)	TSR (%)	Rutting (%)	Low temperature (%)	Drain down
[158]	Basalt	0.4	6					25	10 (−10 °C)	
[147]	Bamboo	0.4	<6	11				6	11 (−10 °C)	
[152]	Palm	0.3	5–25	18	30	23	12			↓
[167]	Coconut	0.3	0.1–1.5			38	88			
[167]	Sisal					35	86			
[166]	Sisal	0.3	10	16		16	9			↓
[171]	Kenaf	0.2	10	7	≅			↓		
[115]	Jute	0.5	7.5	19				1.5		↓
[115]	Carbon	0.5	7.5	30				2.1		↓
[178]	Banana	0.3	20		↓	50		2.2	↑	↓
Fiber-modified porous asphalt mix										
Citations	Fiber	Content (wt%)	Length (mm)	Cantabro (%)	Stiffness modulus (%)	Air voids/permeability	ITS/TSR	Rutting (%)	Fatigue (%)	Drain down
[62]	Polyester	0.3	6	500	↓			2	↑	↓
[73]	Basalt	0.2	24	18		↓	16/0.9			
[79]	Basalt	0.3	9	85			30/2	52	40	↓
[62]	Basalt	0.15	6	500	≅			200	↑	↓
[73]	Glass	0.2	12	227		↓	43/5			↓
[62]	PAN	0.15	6	490	↓			60	↑	↓
[125]	Aramid	0.05	12	7		↓	5/0.7			
[62]	Lignin	0.3	<6	95	↓			200	↑	↓
[164,165]	Coconut	0.3	-			↓		12		↓

↑ Increase of the parameter; ↓ decrease of the parameter; ≅ results similar to control.

5.5. Effect of Adding Fibers on the Volumetric Properties of Asphalt Mix

Based on the previous studies, Table 4 demonstrates the impact of fibers on the volumetric properties of an asphalt mix. As can be seen in the table, the addition of fibers decreases the specific gravity of the asphalt mix as compared to the control mix. The higher the content of fibers, the lower the density of the asphalt mix. This is attributed to the lightweight fibers. Furthermore, it is observed in the table that introducing fibers to an asphalt mix at the optimum bitumen content of the control mix notably increases the air void content (AV) due to the absorption property of some fibers. The increase in the absorption rate of fibers results in higher air void content. Thus, the addition of bitumen is required to meet the specification of air void content. Furthermore, the voids in mineral aggregates (VMA) refer to the volume of unabsorbed bitumen by aggregates and air void content in an asphalt mix. VMA is a critical property because it has a direct effect on the performance of an asphalt mix. The higher the VMA content, the thicker the film thickness, and the better the resistance to aging and moisture susceptibility. Moreover, it is displayed in Table 4 that the incorporation of fibers in an asphalt mix results in a higher VMA content as compared to an unreinforced mix. The higher the content of fibers, the higher the VMA. This is attributed to the slight reduction in the specific gravity of asphalt mixes integrated with fibers and the increase in bitumen content. Furthermore, voids filled with asphalt (VFA) refer to the voids filled with a binder in the compacted sample. VFA depends on VMA and AV, and as the percentage of VMA increases and the percentage of AV decreases, the percentage of VFA increases [67,86,124,146].

Table 4. Influence of adding fibers on the volumetric properties of asphalt mix.

Fiber	Fiber Content (%)	Bitumen Content (%)	AV (%)	VMA (%)	VFA (%)	Gmb (g/cm^3)
Control [86]	0	5.6	3.3	16.80	80.35	2.593
Basalt	0.15	5.80	3.70	17.60	78.98	2.573
	0.30	6.20	3.80	19.20	80.20	2.537
Glass	0.15	5.60	3.70	17.10	78.36	2.584
	0.30	6.20	3.40	18.50	81.62	2.560
Carbon	0.15	5.60	3.70	17.10	78.36	2.584
	0.30	6.40	3.60	19.30	81.35	2.541
Control [145]	0	4.85	4.30	15.00	71.70	-
Lignin	0.30	5.10	4.30	16.00	72.70	-
Polyester	0.20	4.90	4.20	15.20	72.40	-
Bamboo	0.30	5.30	4.00	15.90	74.80	-
Control [123]	0	5.50	4.00	14.90	73.80	2.337
PAN	0.30	5.50	6.43	-	-	2.270
	0.30	6.00	5.25	-	-	2.130
PET	0.30	5.50	6.81	-	-	2.300
	0.30	6.00	4.79	-	-	2.310
Control [67]	0	4.90	4.00	16.00	75.00	2.560
PVA	0.30	5.10	4.00	16.20	75.30	2.539
Acrylic	0.30	5.20	4.00	16.53	75.80	2.536
Polyester	0.30	5.20	4.00	16.13	75.20	2.541

6. Research Gaps and Recommendations

The use of natural and synthetic fibers in asphalt applications is increasing remarkably, in particular in the last five years. However, there are gaps in the research literature that need to be addressed. There is a lack of information that describes the performance of reinforced asphalt mixtures subjected to long-term aging. Additionally, the biodegradability of natural fibers and its impact on the performance of asphalt mixtures are not addressed. Furthermore, the period of natural fibers' decomposition in asphalt mixtures is not evaluated in the research literature. The effect of the adsorption of bitumen oil and light components by fibers on the aging resistance of asphalt mixes is not addressed well. The behavior and performance of elastic materials depend on the type of applied load stress/strain. Accordingly, the elastic and ductile behavior of a fiber-modified asphalt mix under different strains and stress levels is not addressed well. The low-temperature performance of SMA and porous asphalt mix incorporated with natural fibers is not highlighted well. Furthermore, the influence of coarser/finer gradation of dense asphalt mixtures on the performance of fiber-modified asphalt mixes needs to be assessed.

In accordance with the gaps mentioned above, the following recommendations are proposed.

- The performance of fiber-reinforced asphalt mixes subjected to long-term aging should be evaluated.
- The decomposition period of natural fibers in asphalt mixtures should be identified.
- The influence of load type stress/strain on the performance of reinforced asphalt mixes should be assessed.
- The low-temperature performance of fiber-modified SMA/porous asphalt mix should be highlighted.
- The gradation impact on the performance of asphalt mixes incorporated with fibers should be investigated.
- The integration of natural and synthetic fibers in asphalt mixtures should be evaluated.
- Effect of short-term aging and long-term aging on the thermal degradation of fiber-modified bitumen should be studied.
- Performance of asphalt mix incorporated natural fibers with treated surface should be highlighted.
- The addition of natural/synthetic fibers to polymer-modified bitumen should be assessed.

7. Conclusions

This paper aimed to review the most influential studies using synthetic, natural, and waste fibers in asphalt mixes. The conclusions can be stated as follows:

- Based on the annual publications, there is a growing interest in the incorporation of natural and waste fibers in asphalt mixes.
- Modifying bitumen with basalt, carbon, polyester, cellulose, lignin, PAN, polypropylene, palm, and coconut significantly enhanced the physical and rheological properties of the base binder.
- The most used dosage and length of fibers in asphalt mixes are 0.3% by aggregates/mix weight and 6 mm, while a proportion of 5% by bitumen weight is used for modifying bitumen.
- The incorporation of fibers in asphalt mixes remarkably improved the low-temperature and high-temperature performance of asphalt mixes.
- Introducing carbon and steel fibers enhances the conductive property of asphalt mixes.
- Utilizing natural fibers in SMA and porous asphalt mixes notably reduces drain down and improves the mechanical performance of asphalt mixes. However, adding fibers to porous asphalt mixes decreases the air void content and permeability.
- Adding fibers into asphalt mixes incorporated with reclaimed asphalt pavement significantly improves the performance of asphalt mixes.
- Introducing fibers to asphalt mixes has a significant impact on the volumetric properties of the mixes in terms of bitumen content, density, VMA, and VFA.

Author Contributions: Conceptualization, A.M.A. (Adham Mohammed Alnadish), N.S.S.S. and A.M.A. (Aawag Mohsen Alawag); methodology, A.M.A. (Adham Mohammed Alnadish); software, A.M.A. (Adham Mohammed Alnadish); validation, A.M.A. (Adham Mohammed Alnadish), N.S.S.S. and A.M.A. (Aawag Mohsen Alawag); formal analysis, A.M.A. (Adham Mohammed Alnadish); investigation, A.M.A. (Adham Mohammed Alnadish), N.S.S.S. and A.M.A. (Aawag Mohsen Alawag); resources, A.M.A. (Adham Mohammed Alnadish); data curation, A.M.A. (Adham Mohammed Alnadish); writing—original draft preparation, A.M.A. (Adham Mohammed Alnadish); writing—review and editing, A.M.A. (Adham Mohammed Alnadish); visualization, A.M.A. (Adham Mohammed Alnadish); supervision, N.S.S.S.; project administration, N.S.S.S.; funding acquisition, N.S.S.S. All authors have read and agreed to the published version of the manuscript.

Funding: This research received no external funding.

Institutional Review Board Statement: Not applicable.

Data Availability Statement: The presented data are available in the article.

Acknowledgments: The authors wish to extend their sincere thanks to the faculty of data science and information technology, INTI International University, Malaysia for providing the support to carry on this research.

Conflicts of Interest: The authors declare no conflict of interest.

References

1. Roberts, F.L.; Kandhal, P.S.; Brown, E.R.; Lee, D.Y.; Kennedy, T.W. *Hot Mix Asphalt Materials, Mixture Design and Construction*; NAPA Educational Foundation: Napa, CA, USA, 1991.
2. Read, J.; Whiteoak, D.; Hunter, R.N. *The Shell Bitumen Handbook*; Thomas Telford: London, UK, 2003.
3. Rebbechi, J. *Guide to Pavement Technology: Part 4B: Asphalt*; Austroads: Sydney, Australia, 2007. Available online: https://trid.trb.org/view/840820 (accessed on 5 December 2022).
4. Watson, D.E. Updated review of stone matrix asphalt and Superpave® projects. *Transp. Res. Rec.* **2003**, *1832*, 217–223. [CrossRef]
5. Moore, L.M.; Hicks, R.G.; Rogge, D.F. Design, construction, and maintenance guidelines for porous asphalt pavements. *Transp. Res. Rec.* **2001**, *1778*, 91–99. [CrossRef]
6. Hall, K.D.; Schwartz, C.W. Development of structural design guidelines for porous asphalt pavement. *Transp. Res. Rec.* **2018**, *2672*, 197–206. [CrossRef]
7. Kakar, M.R.; Hamzah, M.O.; Valentin, J. A review on moisture damages of hot and warm mix asphalt and related investigations. *J. Clean. Prod.* **2015**, *99*, 39–58. [CrossRef]
8. Alamnie, M.M.; Taddesse, E.; Hoff, I. Advances in Permanent Deformation Modeling of Asphalt Concrete—A Review. *Materials* **2022**, *15*, 3480. [CrossRef] [PubMed]
9. Alnadish, A.M.; Aman, M.Y.; Katman, H.Y.B.; Ibrahim, M.R. Characteristics of warm mix asphalt incorporating coarse steel slag aggregates. *Appl. Sci.* **2021**, *11*, 3708. [CrossRef]
10. Hajj, R.; Bhasin, A. The search for a measure of fatigue cracking in asphalt binders–a review of different approaches. *Int. J. Pavement Eng.* **2018**, *19*, 205–219. [CrossRef]
11. Mun, S.; Guddati, M.N.; Kim, Y.R. Fatigue cracking mechanisms in asphalt pavements with viscoelastic continuum damage finite-element program. *Transp. Res. Rec.* **2004**, *1896*, 96–106. [CrossRef]
12. Caro, S.; Masad, E.; Bhasin, A.; Little, D.N. Moisture susceptibility of asphalt mixtures, Part 1: Mechanisms. *Int. J. Pavement Eng.* **2008**, *9*, 81–98. [CrossRef]
13. Hamzah, M.O.; Kakar, M.R.; Hainin, M.R. An overview of moisture damage in asphalt mixtures. *J. Teknol.* **2015**, *73*. [CrossRef]
14. Slebi-Acevedo, C.J.; Lastra-González, P.; Pascual-Muñoz, P.; Castro-Fresno, D. Mechanical performance of fibers in hot mix asphalt: A review. *Constr. Build. Mater.* **2019**, *200*, 756–769. [CrossRef]
15. Taher, B.M.; Mohamed, R.K.; Mahrez, A. A review on fatigue and rutting performance of asphalt mixes. *Sci. Res. Essays* **2011**, *6*, 670–682.
16. Xie, T.; Zhao, K.; Wang, L. Reinforcement Effect of Different Fibers on Asphalt Mastic. *Materials* **2022**, *15*, 8304. [CrossRef] [PubMed]
17. Jia, B.; Xian, C.; Tsau, J.S.; Zuo, X.; Jia, W. Status and Outlook of Oil Field Chemistry-Assisted Analysis during the Energy Transition Period. *Energy Fuels* **2022**, *36*, 12917–12945. [CrossRef]
18. Morón, A.; Ferrández, D.; Saiz, P.; Vega, G.; Morón, C. Influence of Recycled Aggregates on the Mechanical Properties of Synthetic Fibers-Reinforced Masonry Mortars. *Infrastructures* **2021**, *6*, 84. [CrossRef]
19. Ota, A.; Beyer, R.; Hageroth, U.; Müller, A.; Tomasic, P.; Hermanutz, F.; Buchmeiser, M.R. Chitin/cellulose blend fibers prepared by wet and dry-wet spinning. *Polym. Adv. Technol.* **2021**, *32*, 335–342. [CrossRef]
20. Bellatrache, Y.; Ziyani, L.; Dony, A.; Taki, M.; Haddadi, S. Effects of the addition of date palm fibers on the physical, rheological and thermal properties of bitumen. *Constr. Build. Mater.* **2020**, *239*, 117808. [CrossRef]

21. Chen, J.S.; Lin, K.Y. Mechanism and behavior of bitumen strength reinforcement using fibers. *J. Mater. Sci.* **2005**, *40*, 87–95. [CrossRef]
22. Apostolidis, P.; Liu, X.; Daniel, G.C.; Erkens, S.; Scarpas, T. Effect of synthetic fibres on fracture performance of asphalt mortar. *Road Mater. Pavement Des.* **2020**, *21*, 1918–1931. [CrossRef]
23. Tan, I.A.; Wu, W.H.; Chan, R.A.; Lim, L.L. Effect of mercerization and acetylation on properties of coconut fiber and its influence on modified bitumen. *J. Civ. Eng. Sci. Technol.* **2014**, *5*, 17–22. [CrossRef]
24. Chen, H.; Xu, Q.; Chen, S.; Zhang, Z. Evaluation and design of fiber-reinforced asphalt mixtures. *Mater. Des.* **2009**, *30*, 2595–2603. [CrossRef]
25. Mohammed, M.; Parry, T.; Thom, N.; Grenfell, J. Microstructure and mechanical properties of fibre reinforced asphalt mixtures. *Constr. Build. Mater.* **2020**, *240*, 117932. [CrossRef]
26. Shanbara, H.K.; Ruddock, F.; Atherton, W. Improving the mechanical properties of cold mix asphalt mixtures reinforced by natural and synthetic fibers. In *Airfield and Highway Pavements*; ASCE Library: Reston, VL, USA, 2017; pp. 102–111.
27. Lee, S.J.; Rust, J.P.; Hamouda, H.; Kim, Y.R.; Borden, R.H. Fatigue cracking resistance of fiber-reinforced asphalt concrete. *Text. Res. J.* **2005**, *75*, 123–128. [CrossRef]
28. Stempihar, J.J.; Souliman, M.I.; Kaloush, K.E. Fiber-Reinforced Asphalt Concrete as Sustainable Paving Material for Airfields. *Transp. Res. Rec. J. Transp. Res. Board* **2012**, *2266*, 60–68. [CrossRef]
29. Vale, A.C.D.; Casagrande, M.D.T.; Soares, J.B. Behavior of natural fiber in stone matrix asphalt mixtures using two design methods. *J. Mater. Civ. Eng.* **2014**, *26*, 457–465. [CrossRef]
30. Jain, S.; Singh, H.; Chopra, T. Laboratory investigations and performance evaluation of stone matrix asphalt as a wearing course using three different fibers. *Int. J. Appl. Sci. Eng.* **2020**, *17*, 411–418.
31. Awanti, S.S.; Habbal, A.; Hiremath, P.N.; Tadibidi, S.; Hallale, S.N. Characterization of stone matrix asphalt with cellulose and coconut fiber. In *Advance in Civil Engineering and Building Materials*; Chang, S., al Bahar, S.K., Zhao, J., Eds.; Taylor & Francis Group: London, UK, 2012; pp. 639–642.
32. Andrew, B.; Buyondo, K.A.; Kasedde, H.; Kirabira, J.B.; Olupot, P.W.; Yusuf, A.A. Investigation on the use of reclaimed asphalt pavement along with steel fibers in concrete. *Case Stud. Constr. Mater.* **2022**, *17*, e01356. [CrossRef]
33. Li, N.; Zhan, H.; Yu, X.; Tang, W.; Xue, Q. Investigation of the aging behavior of cellulose fiber in reclaimed asphalt pavement. *Constr. Build. Mater.* **2020**, *271*, 121559. [CrossRef]
34. Abtahi, S.M.; Sheikhzadeh, M.; Hejazi, S.M. Fiber-reinforced asphalt-concrete—A review. *Constr. Build. Mater.* **2010**, *24*, 871–877. [CrossRef]
35. Mokhtari, A.; Nejad, F.M. Mechanistic approach for fiber and polymer modified SMA mixtures. *Constr. Build. Mater.* **2012**, *36*, 381–390. [CrossRef]
36. Elanchezhian, C.; Ramnath, B.V.; Ramakrishnan, G.; Rajendrakumar, M.; Naveenkumar, V.; Saravanakumar, M.K. Review on mechanical properties of natural fiber composites. *Mater. Today Proc.* **2018**, *5*, 1785–1790. [CrossRef]
37. Zeronian, S.H.; Xie, Q.; Buschle-Diller, G.; Holmes, S.; Inglesby, M.K. Relationships between the mechanical properties of synthetic fibers. *J. Text. Inst.* **1994**, *85*, 293–300. [CrossRef]
38. Kumar Kesharwani, A.; Yadav, R. A Short review on synthesis and applications of zinc based metal organic frameworks. *Int. J. Eng. Appl. Sci. Technol.* **2022**, *6*, 120–131. [CrossRef]
39. Nassar, M.M.A.; Arunachalam, R.; Alzebdeh, K.I. Machinability of natural fiber reinforced composites: A review. *Int. J. Adv. Manuf. Technol.* **2016**, *88*, 2985–3004. [CrossRef]
40. Lotfi, A.; Li, H.; Dao, D.V.; Prusty, G. Natural fiber–reinforced composites: A review on material, manufacturing, and machinability. *J. Thermoplast. Compos. Mater.* **2019**, *34*, 238–284. [CrossRef]
41. Ramamoorthy, S.K.; Skrifvars, M.; Persson, A. A Review of Natural Fibers Used in Biocomposites: Plant, Animal and Regenerated Cellulose Fibers. *Polym. Rev.* **2015**, *55*, 107–162. [CrossRef]
42. Dittenber, D.B.; GangaRao, H.V.S. Critical review of recent publications on use of natural composites in infrastructure. *Compos. Part A Appl. Sci. Manuf.* **2012**, *43*, 1419–1429. [CrossRef]
43. Reddy, B.B.K.; Tejaswini, K. Strength Properties of Polypropylene Fibre Reinforced Concrete. *Int. J. Eng. Res. Adv. Technol.* **2018**. [CrossRef]
44. Dashtizadeh, Z.; Abdan, K.; Jawaid, M.; Khan, M.A.; Behmanesh, M.; Dashtizadeh, M.; Cardona, F.; Ishak, M. Mechanical and thermal properties of natural fibre based hybrid composites: A review. *Pertanika J. Sci. Technol.* **2017**, *25*, 1103–1122.
45. Fred-Ahmadu, O.H.; Bhagwat, G.; Oluyoye, I.; Benson, N.U.; Ayejuyo, O.O.; Palanisami, T. Interaction of chemical contaminants with microplastics: Principles and perspectives. *Sci. Total. Environ.* **2019**, *706*, 135978. [CrossRef]
46. Patil, A.Y.; Banapurmath, N.R.; Sunal, S. Review on Period of Biodegradability for Natural Fibers Embedded Polylactic Acid Biocomposites. In *Biodegradation, Pollutants and Bioremediation Principles*; CRC Press: Boca Raton, FL, USA, 2021; pp. 234–271.
47. Jaffe, M.; Menczel, J. (Eds.) *Thermal Analysis of Textiles and Fibers*; Woodhead Publishing: Sawston, UK, 2020.
48. Zhu, K.; Tu, H.; Yang, P.; Qiu, C.; Zhang, D.; Lu, A.; Luo, L.; Chen, F.; Liu, X.; Chen, L.; et al. Mechanically Strong Chitin Fibers with Nanofibril Structure, Biocompatibility, and Biodegradability. *Chem. Mater.* **2019**, *31*, 2078–2087. [CrossRef]
49. Su, S.-K.; Wu, C.-S. Polyester biocomposites from recycled natural fibers: Characterization and biodegradability. *J. Appl. Polym. Sci.* **2010**, *119*, 1211–1219. [CrossRef]
50. Nicolaisen, J. Citation analysis. *Annu. Rev. Inf. Sci. Technol.* **2007**, *41*, 609–641. [CrossRef]

51. MacDonald, K.I.; Dressler, V. Using citation analysis to identify research fronts: A case study with the Internet of Things. *Sci. Technol. Libr.* **2018**, *37*, 171–186. [CrossRef]
52. Van Eck, N.J.; Waltman, L. Citation-based clustering of publications using CitNetExplorer and VOSviewer. *Scientometrics* **2017**, *111*, 1053–1070. [CrossRef]
53. Singh, V.K.; Singh, P.; Karmakar, M.; Leta, J.; Mayr, P. The journal coverage of Web of Science, Scopus and Dimensions: A comparative analysis. *Scientometrics* **2021**, *126*, 5113–5142. [CrossRef]
54. Thelwall, M. Dimensions: A competitor to Scopus and the Web of Science? *J. Informetr.* **2018**, *12*, 430–435. [CrossRef]
55. Martín-Martín, A.; Thelwall, M.; Orduna-Malea, E.; Delgado López-Cózar, E. Google Scholar, Microsoft Academic, Scopus, Dimensions, Web of Science, and OpenCitations' COCI: A multidisciplinary comparison of coverage via citations. *Scientometrics* **2021**, *126*, 871–906. [CrossRef]
56. Park, K.S.; Shoukat, T.; Yoo, P.J.; Lee, S.H. Strengthening of hybrid glass fiber reinforced recycled hot-mix asphalt mixtures. *Constr. Build. Mater.* **2020**, *258*, 118947. [CrossRef]
57. Mohammed, S.F.; Ismael, M.Q. Effect of Polypropylene Fibers on moisture Susceptibility of Warm Mix Asphalt. *Civ. Eng. J.* **2021**, *7*, 988–997. [CrossRef]
58. Omranian, S.R.; Van den bergh, W.; He, L.; Manthos, E. Incorporating 3D image analysis and response surface method to evaluate the effects of moisture damage on reinforced asphalt mixtures using glass and polypropylene fibers. *Constr. Build. Mater.* **2022**, *353*, 129177. [CrossRef]
59. Al-Bdairi, A.; Al-Taweel, H.M.; Noor, H.M. Improving the properties of asphalt mixture using fiber materials. *IOP Conf. Ser. Mater. Sci. Eng.* **2020**, *870*, 012092. [CrossRef]
60. Qin, X.; Shen, A.; Guo, Y.; Li, Z.; Lv, Z. Characterization of asphalt mastics reinforced with basalt fibers. *Constr. Build. Mater.* **2018**, *159*, 508–516. [CrossRef]
61. Kim, M.-J.; Kim, S.; Yoo, D.-Y.; Shin, H.-O. Enhancing mechanical properties of asphalt concrete using synthetic fibers. *Constr. Build. Mater.* **2018**, *178*, 233–243. [CrossRef]
62. Zhang, J.; Huang, W.; Zhang, Y.; Lv, Q.; Yan, C. Evaluating four typical fibers used for OGFC mixture modification regarding drainage, raveling, rutting and fatigue resistance. *Constr. Build. Mater.* **2020**, *253*, 119131. [CrossRef]
63. Zhang, T.; Wu, J.; Hong, R.; Ye, S.; Jin, A. Research on low-temperature performance of steel slag/polyester fiber permeable asphalt mixture. *Constr. Build. Mater.* **2022**, *334*, 127214. [CrossRef]
64. Zarei, M.; Kordani, A.A.; Zahedi, M. Evaluation of fracture behavior of modified Warm Mix Asphalt (WMA) under modes I and II at low and intermediate temperatures. *Theor. Appl. Fract. Mech.* **2021**, *114*, 103015. [CrossRef]
65. Zarei, M.; Kordani, A.A.; Ghamarimajd, Z.; Khajehzadeh, M.; Khanjari, M.; Zahedi, M. Evaluation of fracture resistance of asphalt concrete involving Calcium Lignosulfonate and Polyester fiber under freeze–thaw damage. *Theor. Appl. Fract. Mech.* **2022**, *117*, 103168. [CrossRef]
66. Alnadish, A.M.; Aman, M.Y.; Katman, H.Y.B.; Ibrahim, M.R. Laboratory Evaluation of Fiber-Modified Asphalt Mixtures Incorporating Steel Slag Aggregates. *Comput. Mater. Contin.* **2021**, *70*, 5967–5990.
67. Alnadish, A.M.; Aman, M.Y.; Katman, H.Y.B.; Ibrahim, M.R. Influence of the Long-Term Oven Aging on the Performance of the Reinforced Asphalt Mixtures. *Coatings* **2020**, *10*, 953. [CrossRef]
68. Alnadish, A.M.; Aman, M.Y.; Katman, H.Y.B.; Ibrahim, M.R. Laboratory assessment of the performance and elastic behavior of asphalt mixtures containing steel slag aggregate and synthetic fibers. *Int. J. Pavement Res. Technol.* **2020**, *14*, 473–481. [CrossRef]
69. Hong, R.-B.; Wu, J.-R.; Cai, H.-B. Low-temperature crack resistance of coal gangue powder and polyester fibre asphalt mixture. *Constr. Build. Mater.* **2020**, *238*, 117678. [CrossRef]
70. Zhu, Y.; Li, Y.; Si, C.; Shi, X.; Qiao, Y.; Li, H. Laboratory Evaluation on Performance of Fiber-Modified Asphalt Mixtures Containing High Percentage of RAP. *Adv. Civ. Eng.* **2020**, *2020*, 5713869. [CrossRef]
71. Zhu, S.; Xu, Z.; Qin, X.; Liao, M. Fiber-Reinforcing Effect in the Mechanical and Road Performance of Cement-Emulsified Asphalt Mixtures. *Materials* **2021**, *14*, 2779. [CrossRef] [PubMed]
72. Yu, X.; Han, Z.; Cai, Y.; Liu, L.; Sun, L. Study on Low-Temperature Index and Improvement Measures of Emulsified Asphalt Cold Recycled Mixture. *Materials* **2022**, *15*, 7867. [CrossRef] [PubMed]
73. Tanzadeh, R.; Tanzadeh, J.; Honarmand, M.; Tahami, S.A. Experimental study on the effect of basalt and glass fibers on behavior of open-graded friction course asphalt modified with nano-silica. *Constr. Build. Mater.* **2019**, *212*, 467–475. [CrossRef]
74. Guo, Q.; Wang, H.; Gao, Y.; Jiao, Y.; Liu, F.; Dong, Z. Investigation of the low-temperature properties and cracking resistance of fiber-reinforced asphalt concrete using the DIC technique. *Eng. Fract. Mech.* **2020**, *229*, 106951. [CrossRef]
75. Guo, S.; Cheng, P.; Yi, J.; Chang, L. Performance Evaluation of SMA Mixture Reinforced by Basalt Fiber and Composite Fiber. *CICTP* **2022**, *2022*, 2232–2241.
76. Wang, W.; Cheng, Y.; Chen, H.; Tan, G.; Lv, Z.; Bai, Y. Study on the Performances of Waste Crumb Rubber Modified Asphalt Mixture with Eco-Friendly Diatomite and Basalt Fiber. *Sustainability* **2019**, *11*, 5282. [CrossRef]
77. Wang, W.; Shen, A.; Jin, X.; Yang, J. Optimization and Performance Evaluation of Steel Slag Asphalt Mixture Modified with Fibers under Freeze–Thaw Cycles. *J. Mater. Civ. Eng.* **2022**, *35*, 04022419. [CrossRef]
78. Wang, W.; Cheng, Y.; Tan, G. Design Optimization of SBS-Modified Asphalt Mixture Reinforced with Eco-Friendly Basalt Fiber Based on Response Surface Methodology. *Materials* **2018**, *11*, 1311. [CrossRef]

79. Wang, S.; Kang, A.; Xiao, P.; Li, B.; Fu, W. Investigating the Effects of Chopped Basalt Fiber on the Performance of Porous Asphalt Mixture. *Adv. Mater. Sci. Eng.* **2019**, *2019*, 2323761. [CrossRef]
80. Xiang, Y.; Xie, Y.; Long, G. Effect of basalt fiber surface silane coupling agent coating on fiber-reinforced asphalt: From macro-mechanical performance to micro-interfacial mechanism. *Constr. Build. Mater.* **2018**, *179*, 107–116. [CrossRef]
81. Zhang, X.; Liu, J. Viscoelastic creep properties and mesostructure modeling of basalt fiber-reinforced asphalt concrete. *Constr. Build. Mater.* **2020**, *259*, 119680. [CrossRef]
82. Gong, Y.; Song, J.; Bi, H.; Tian, Z. Optimization Design of the Mix Ratio of a Nano-TiO_2/$CaCO_3$-Basalt Fiber Composite Modified Asphalt Mixture Based on Response Surface Methodology. *Appl. Sci.* **2020**, *10*, 4596. [CrossRef]
83. Cheng, Y.; Chai, C.; Zhang, Y.; Chen, Y.; Zhu, B. A New Eco-Friendly Porous Asphalt Mixture Modified by Crumb Rubber and Basalt Fiber. *Sustainability* **2019**, *11*, 5754. [CrossRef]
84. Li, Z.; Shen, A.; Wang, H.; Guo, Y.; Wu, H. Effect of basalt fiber on the low-temperature performance of an asphalt mixture in a heavily frozen area. *Constr. Build. Mater.* **2020**, *253*, 119080. [CrossRef]
85. Kong, L.; Lu, Z.; He, Z.; Shen, Z.; Xu, H.; Yang, K.; Yu, L. Characterization of crack resistance mechanism of fiber modified emulsified asphalt cold recycling mixture based on acoustic emission parameters. *Constr. Build. Mater.* **2022**, *327*, 126939. [CrossRef]
86. Alfalah, A.; Offenbacker, D.; Ali, A.; Mehta, Y.; Elshaer, M.; Decarlo, C. Evaluating the impact of fiber type and dosage rate on laboratory performance of Fiber-Reinforced asphalt mixtures. *Constr. Build. Mater.* **2021**, *310*, 125217. [CrossRef]
87. Ziari, H.; Aliha MR, M.; Moniri, A.; Saghafi, Y. Crack resistance of hot mix asphalt containing different percentages of reclaimed asphalt pavement and glass fiber. *Constr. Build. Mater.* **2020**, *230*, 117015. [CrossRef]
88. Liu, Y.; Zhang, Z.; Tan, L.; Xu, Y.; Wang, C.; Liu, P.; Yu, H.; Oeser, M. Laboratory evaluation of emulsified asphalt reinforced with glass fiber treated with different methods. *J. Clean. Prod.* **2020**, *274*, 123116. [CrossRef]
89. Shanbara, H.K.; Ruddock, F.; Atherton, W. A laboratory study of high-performance cold mix asphalt mixtures reinforced with natural and synthetic fibres. *Constr. Build. Mater.* **2018**, *172*, 166–175. [CrossRef]
90. Gupta, A.; Castro-Fresno, D.; Lastra-Gonzalez, P.; Rodriguez-Hernandez, J. Selection of fibers to improve porous asphalt mixtures using multi-criteria analysis. *Constr. Build. Mater.* **2021**, *266*, 121198. [CrossRef]
91. Khater, A.; Luo, D.; Abdelsalam, M.; Yue, Y.; Hou, Y.; Ghazy, M. Laboratory Evaluation of Asphalt Mixture Performance Using Composite Admixtures of Lignin and Glass Fibers. *Appl. Sci.* **2021**, *11*, 364. [CrossRef]
92. Ramesh, A.; Ramayya, V.V.; Reddy, G.S.; Ram, V.V. Investigations on fracture response of warm mix asphalt mixtures with Nano glass fibres and partially replaced RAP material. *Constr. Build. Mater.* **2022**, *317*, 126121. [CrossRef]
93. Enieb, M.; Diab, A.; Yang, X. Short- and long-term properties of glass fiber reinforced asphalt mixtures. *Int. J. Pavement Eng.* **2019**, *22*, 64–76. [CrossRef]
94. Morea, F.; Zerbino, R. Improvement of asphalt mixture performance with glass macro-fibers. *Constr. Build. Mater.* **2018**, *164*, 113–120. [CrossRef]
95. Khanghahi, S.H.; Tortum, A. Determination of the Optimum Conditions for Gilsonite and Glass Fiber in HMA under Mixed Mode I/III Loading in Fracture Tests. *J. Mater. Civ. Eng.* **2018**, *30*, 04018130. [CrossRef]
96. Mohammed, M.; Parry, T.; Grenfell, J. Influence of fibres on rheological properties and toughness of bituminous binder. *Constr. Build. Mater.* **2018**, *163*, 901–911. [CrossRef]
97. Fu, L.; Jiao, Y.; Chen, X. Reinforcement evaluation of different fibers on fracture resistance of asphalt mixture based on acoustic emission technique. *Constr. Build. Mater.* **2022**, *314*, 125606. [CrossRef]
98. Liu, Z.; Luo, S.; Wang, Y.; Chen, H. Induction Heating and Fatigue-Damage Induction Healing of Steel Fiber–Reinforced Asphalt Mixture. *J. Mater. Civ. Eng.* **2019**, *31*, 04019180. [CrossRef]
99. Phan, T.M.; Park, D.-W.; Le, T.H.M. Crack healing performance of hot mix asphalt containing steel slag by microwaves heating. *Constr. Build. Mater.* **2018**, *180*, 503–511. [CrossRef]
100. Gao, J.; Guo, H.; Wang, X.; Wang, P.; Wei, Y.; Wang, Z.; Huang, Y.; Yang, B. Microwave deicing for asphalt mixture containing steel wool fibers. *J. Clean. Prod.* **2019**, *206*, 1110–1122. [CrossRef]
101. Tabaković, A.; O'Prey, D.; McKenna, D.; Woodward, D. Microwave self-healing technology as airfield porous asphalt friction course repair and maintenance system. *Case Stud. Constr. Mater.* **2019**, *10*, e00233. [CrossRef]
102. González, A.; Valderrama, J.; Norambuena-Contreras, J. Microwave crack healing on conventional and modified asphalt mixtures with different additives: An experimental approach. *Road Mater. Pavement Des.* **2019**, *20*, S149–S162. [CrossRef]
103. Yang, H.; Ouyang, J.; Jiang, Z.; Ou, J. Effect of fiber reinforcement on self-healing ability of asphalt mixture induced by microwave heating. *Constr. Build. Mater.* **2023**, *362*, 129701. [CrossRef]
104. Jiao, Y.; Zhang, L.; Guo, Q.; Guo, M.; Zhang, Y. Acoustic Emission-Based Reinforcement Evaluation of Basalt and Steel Fibers on Low-Temperature Fracture Resistance of Asphalt Concrete. *J. Mater. Civ. Eng.* **2020**, *32*, 04020104. [CrossRef]
105. Li, H.; Yu, J.; Wu, S.; Liu, Q.; Wu, Y.; Xu, H.; Li, Y. Effect of moisture conditioning on mechanical and healing properties of inductive asphalt concrete. *Constr. Build. Mater.* **2020**, *241*, 118139. [CrossRef]
106. Dinh, B.H.; Park, D.-W.; Phan, T.M. Healing Performance of Granite and Steel Slag Asphalt Mixtures Modified with Steel Wool Fibers. *KSCE J. Civ. Eng.* **2018**, *22*, 2064–2072. [CrossRef]
107. Dinh, B.H.; Park, D.-W.; Le, T.H.M. Effect of rejuvenators on the crack healing performance of recycled asphalt pavement by induction heating. *Constr. Build. Mater.* **2018**, *164*, 246–254. [CrossRef]

108. Hosseinian, S.M.; Najafi Moghaddam Gilani, V.; Mehraban Joobani, P.; Arabani, M. Investigation of Moisture Sensitivity and Conductivity Properties of Inductive Asphalt Mixtures Containing Steel Wool Fiber. *Adv. Civ. Eng.* **2020**, *2020*, 8890814. [CrossRef]
109. Norambuena-Contreras, J.; Gonzalez, A.; Concha, J.L.; Gonzalez-Torre, I.; Schlangen, E. Effect of metallic waste addition on the electrical, thermophysical and microwave crack-healing properties of asphalt mixtures. *Constr. Build. Mater.* **2018**, *187*, 1039–1050. [CrossRef]
110. Arabzadeh, A.; Notani, M.A.; Zadeh, A.K.; Nahvi, A.; Sassani, A.; Ceylan, H. Electrically conductive asphalt concrete: An alternative for automating the winter maintenance operations of transportation infrastructure. *Compos. Part B Eng.* **2019**, *173*, 106985. [CrossRef]
111. Pirmohammad, S.; Shokorlou, Y.M.; Amani, B. Laboratory investigations on fracture toughness of asphalt concretes reinforced with carbon and kenaf fibers. *Eng. Fract. Mech.* **2020**, *226*, 106875. [CrossRef]
112. Yoo, D.-Y.; Kim, S.; Kim, M.-J.; Kim, D.; Shin, H.-O. Self-healing capability of asphalt concrete with carbon-based materials. *J. Mater. Res. Technol.* **2019**, *8*, 827–839. [CrossRef]
113. Zhang, K.; Liu, Y.; Nassiri, S.; Li, H.; Englund, K. Performance evaluation of porous asphalt mixture enhanced with high dosages of cured carbon fiber composite materials. *Constr. Build. Mater.* **2021**, *274*, 122066. [CrossRef]
114. Zhang, K.; Lim, J.; Nassiri, S.; Englund, K.; Li, H. Reuse of Carbon Fiber Composite Materials in Porous Hot Mix Asphalt to Enhance Strength and Durability. *Case Stud. Constr. Mater.* **2019**, *11*, e00260. [CrossRef]
115. Ismael, M.; Fattah, M.Y.; Jasim, A.F. Permanent Deformation Characterization of Stone Matrix Asphalt Reinforced by Different Types of Fibers. *J. Eng.* **2022**, *28*, 99–116. [CrossRef]
116. Gürer, C.; Gürgöze, H. Investigation the CF Based Conductive Asphalt Mixtures for Anti-icing. *J. Eng. Res.* **2022**. [CrossRef]
117. Gürer, C.; Akbulut, H.; Elmaci, A.; Korkmaz, B.E.; Düzağaç, S. An Investigation of the Electrical Conductivity of Different Stone Mastic Asphalt Mixtures. *Proc. Inst. Civ. Eng. Transp.* **2022**, *10*, 1–25. [CrossRef]
118. Schuster, L.; de Melo, J.V.S.; Del Carpio, J.A.V. Effects of the associated incorporation of steel wool and carbon nanotube on the healing capacity and mechanical performance of an asphalt mixture. *Int. J. Fatigue* **2023**, *168*, 107440. [CrossRef]
119. Ullah, S.; Yang, C.; Cao, L.; Wang, P.; Chai, Q.; Li, Y.; Wang, L.; Dong, Z.; Lushinga, N.; Zhang, B. Material design and performance improvement of conductive asphalt concrete incorporating carbon fiber and iron tailings. *Constr. Build. Mater.* **2021**, *303*, 124446. [CrossRef]
120. Takaikaew, T.; Tepsriha, P.; Horpibulsuk, S.; Hoy, M.; Kaloush, K.E.; Arulrajah, A. Performance of Fiber-Reinforced Asphalt Concretes with Various Asphalt Binders in Thailand. *J. Mater. Civ. Eng.* **2018**, *30*, 04018193. [CrossRef]
121. Klinsky, L.M.G.; Kaloush, K.E.; Faria, V.C.; Bardini, V.S.S. Performance characteristics of fiber modified hot mix asphalt. *Constr. Build. Mater.* **2018**, *176*, 747–752. [CrossRef]
122. Slebi-Acevedo, C.J.; Lastra-González, P.; Castro-Fresno, D.; Bueno, M. An experimental laboratory study of fiber-reinforced asphalt mortars with polyolefin-aramid and polyacrylonitrile fibers. *Constr. Build. Mater.* **2020**, *248*, 118622. [CrossRef]
123. Callomamani, L.A.P.; Hashemian, L.; Sha, K. Laboratory Investigation of the Performance Evaluation of Fiber-Modified Asphalt Mixes in Cold Regions. *Transp. Res. Rec.* **2020**, *2674*, 323–335. [CrossRef]
124. Hajiloo, H.R.; Karimi, H.R.; Aliha, M.R.M.; Farahani, H.Z.; Salehi, S.M.; Hajiloo, M.; Haghighatpour, P.J. Crack resistance of fiber-reinforced asphalt mixtures: Effect of test specimen and test condition. *Fatigue Fract. Eng. Mater. Struct.* **2022**, *45*, 921–937. [CrossRef]
125. Gupta, A.; Lastra-Gonzalez, P.; Castro-Fresno, D.; Rodriguez-Hernandez, J. Laboratory Characterization of Porous Asphalt Mixtures with Aramid Fibers. *Materials* **2021**, *14*, 1935. [CrossRef]
126. Xing, X.; Pei, J.; Shen, C.; Li, R.; Zhang, J.; Huang, J.; Hu, D. Performance and Reinforcement Mechanism of Modified Asphalt Binders with Nano-Particles, Whiskers, and Fibers. *Appl. Sci.* **2019**, *9*, 2995. [CrossRef]
127. Noorvand, H.; Brockman, S.C.; Mamlouk, M.; Kaloush, K. Effect of Aramid Fibers on Balanced Mix Design of Asphalt Concrete. *CivilEng* **2021**, *3*, 21–34. [CrossRef]
128. Daniel, C.G. Analysis of The Effect of Using Fiber Aramid-Polyolefin on The Strength, Stiffness, and Durability of Warm Mix Asphalt. *J. Tek. Sipil* **2020**, *27*, 9–16. [CrossRef]
129. Slebi-Acevedo, C.J.; Lastra-González, P.; Calzada-Pérez, M.A.; Castro-Fresno, D. Effect of Synthetic Fibers and Hydrated Lime in Porous Asphalt Mixture Using Multi-Criteria Decision-Making Techniques. *Materials* **2020**, *13*, 675. [CrossRef] [PubMed]
130. Su, Z.; Muhammad, Y.; Sahibzada, M.; Li, J.; Meng, F.; Wei, Y.; Zhao, Z.; Zhang, L. Preparation and properties of aminated graphene fiber incorporated modified asphalt. *Constr. Build. Mater.* **2019**, *229*, 116836. [CrossRef]
131. Slebi-Acevedo, C.J.; Lastra-González, P.; Castro-Fresno, D.; Vega-Zamanillo, Á. Experimental evaluation and recyclability potential of asphalt concrete mixtures with polyacrylonitrile fibers. *Constr. Build. Mater.* **2022**, *317*, 125829. [CrossRef]
132. Dalhat, M.A.; Al-Adham, K.; Wahhab, H.I.A.-A. Multiple Stress–Creep–Recovery Behavior and High-Temperature Performance of Styrene Butadiene Styrene and Polyacrylonitrile Fiber–Modified Asphalt Binders. *J. Mater. Civ. Eng.* **2019**, *31*, 04019087. [CrossRef]
133. Xing, S.; Muhammad, Y.; Chen, Y.; Li, Z.; Ren, D.; Zhao, Z.; Li, J. Preparation and performance evaluation of surface-modified polyacrylonitrile fiber and SBS composite modified asphalt binder based on bionic hierarchy. *Constr. Build. Mater.* **2022**, *326*, 126866. [CrossRef]
134. Wang, H.; Yang, Z.; Zhan, S.; Ding, L.; Jin, K. Fatigue Performance and Model of Polyacrylonitrile Fiber Reinforced Asphalt Mixture. *Appl. Sci.* **2018**, *8*, 1818. [CrossRef]

135. Arabani, M.; Shabani, A. Evaluation of the ceramic fiber modified asphalt binder. *Constr. Build. Mater.* **2019**, *205*, 377–386. [CrossRef]
136. Arabani, M.; Shabani, A.; Hamedi, G.H. Experimental Investigation of Effect of Ceramic Fibers on Mechanical Properties of Asphalt Mixtures. *J. Mater. Civ. Eng.* **2019**, *31*, 04019203. [CrossRef]
137. Naseri Yalghouzaghaj, M.; Sarkar, A.; Hamedi, G.H.; Hayati, P. Effect of Ceramic Fibers on the Thermal Cracking of Hot-Mix Asphalt. *J. Mater. Civ. Eng.* **2020**, *32*, 04020325. [CrossRef]
138. Wang, X.; Zhou, H.; Hu, X.; Shen, S.; Dong, B. Investigation of the Performance of Ceramic Fiber Modified Asphalt Mixture. *Adv. Civ. Eng.* **2021**, *2021*, 8833468. [CrossRef]
139. Hamedi, G.H.; Sakanlou, F.; Omari, B.; Azarhoosh, A. Laboratory Investigation of the Effect of Ceramic Fiber on Stone Matrix Asphalt Rutting Performance. *J. Mater. Civ. Eng.* **2021**, *33*, 04020431. [CrossRef]
140. Liu, F.; Pan, B.; Bian, J.; Zhou, C. Experimental investigation on the performance of the asphalt mixture with ceramic fiber. *J. Clean. Prod.* **2023**, *384*, 135585. [CrossRef]
141. Landi, D.; Marconi, M.; Bocci, E.; Germani, M. Comparative life cycle assessment of standard, cellulose-reinforced and end of life tires fiber-reinforced hot mix asphalt mixtures. *J. Clean. Prod.* **2020**, *248*, 119295. [CrossRef]
142. Landi, D.; Gigli, S.; Germani, M.; Marconi, M. Investigating the feasibility of a reuse scenario for textile fibres recovered from end-of-life tyres. *Waste Manag.* **2018**, *75*, 187–204. [CrossRef] [PubMed]
143. Irfan, M.; Ali, Y.; Ahmed, S.; Iqbal, S.; Wang, H. Rutting and Fatigue Properties of Cellulose Fiber-Added Stone Mastic Asphalt Concrete Mixtures. *Adv. Mater. Sci. Eng.* **2019**, *2019*, 5604197. [CrossRef]
144. Fauzi NA, N.M.; Masri, K.A.; Ramadhansyah, P.J.; Samsudin, M.S.; Ismail, A.; Arshad, A.K.; Shaffie, E.; Norhidayah, A.H.; Hainin, M.R. Volumetric Properties and Resilient Modulus of Stone Mastic Asphalt incorporating Cellulose Fiber. *IOP Conf. Ser. Mater. Sci. Eng.* **2020**, *712*, 012028. [CrossRef]
145. Sheng, Y.; Zhang, B.; Yan, Y.; Li, H.; Chen, Z.; Chen, H. Laboratory Investigation on the Use of Bamboo Fiber in Asphalt Mixtures for Enhanced Performance. *Arab. J. Sci. Eng.* **2018**, *44*, 4629–4638. [CrossRef]
146. Liu, K.; Li, T.; Wu, C.; Jiang, K.; Shi, X. Bamboo fiber has engineering properties and performance suitable as reinforcement for asphalt mixture. *Constr. Build. Mater.* **2021**, *290*, 123240. [CrossRef]
147. Xia, C.; Xu, M.; Li, Q.; Liu, K.; Jiang, K. Study on the Freeze–Thaw Cycle Durability of Bamboo Fiber Asphalt Mixture. *Int. J. Pavement Res. Technol.* **2022**, *11*, 1–14. [CrossRef]
148. Xia, C.; Wu, C.; Liu, K.; Jiang, K. Study on the Durability of Bamboo Fiber Asphalt Mixture. *Materials* **2021**, *14*, 1667. [CrossRef]
149. Jia, H.; Sheng, Y.; Lv, H.; Kim, Y.R.; Zhao, X.; Meng, J.; Xiong, R. Effects of bamboo fiber on the mechanical properties of asphalt mixtures. *Constr. Build. Mater.* **2021**, *289*, 123196. [CrossRef]
150. Meng, Y.; Qin, P.; Yaseen, M.; Chen, J.; Yan, T.; Gan, W.; Lei, J.; Li, J. Effect of tannic acid modified bamboo fiber on the performance of soybean bio-asphalt/styrene-butadiene-styrene modified asphalt. *Polym. Compos.* **2022**, *43*, 2288–2302. [CrossRef]
151. Syammaun, T.; Rani, H.A. Resilient modulus of porous asphalt using oil palm fiber. *IOP Conf. Ser. Mater. Sci. Eng.* **2018**, *403*, 012023. [CrossRef]
152. Yaro, N.S.A.; Bin Napiah, M.; Sutanto, M.H.; Usman, A.; Saeed, S.M. Performance Evaluation of Waste Palm Oil Fiber Reinforced Stone Matrix Asphalt Mixtures Using Traditional and Sequential Mixing Processes. *Case Stud. Constr. Mater.* **2021**, *15*, e00783. [CrossRef]
153. Yaro, N.S.A.; Sutanto, M.H.; Habib, N.Z.; Napiah, M.; Usman, A.; Jagaba, A.H.; Al-Sabaeei, A.M. Application and circular economy prospects of palm oil waste for eco-friendly asphalt pavement industry: A review. *J. Road Eng.* **2022**, *2*, 309–331. [CrossRef]
154. Tayh, S.A.; Yousif, R.A.; Banyhussan, Q.S. A comparative study of physical properties using various grades asphalt binder with different type of fibers. *J. Eng. Res.* **2020**, *17*, 34–40. [CrossRef]
155. Chen, Z.; Yi, J.; Chen, Z.; Feng, D. Properties of asphalt binder modified by corn stalk fiber. *Constr. Build. Mater.* **2019**, *212*, 225–235. [CrossRef]
156. Ma, F.; Jin, Y.; Fu, Z.; Dai, J.; Zhang, P.; Zhang, C.; Wen, Y. Influencing factors and evaluation methods of reinforcement effect of fiber-modified asphalt binder. *Polym. Compos.* **2022**, *43*, 8986–8999. [CrossRef]
157. Wu, B.; Meng, W.; Xia, J.; Xiao, P. Influence of Basalt Fibers on the Crack Resistance of Asphalt Mixtures and Mechanism Analysis. *Materials* **2022**, *15*, 744. [CrossRef]
158. Wu, B.; Wu, X.; Xiao, P.; Chen, C.; Xia, J.; Lou, K. Evaluation of the Long-Term Performances of SMA-13 Containing Different Fibers. *Appl. Sci.* **2021**, *11*, 5145. [CrossRef]
159. Kou, C.; Chen, Z.; Kang, A.; Zhang, M.; Wang, R. Rheological behaviors of asphalt binders reinforced by various fibers. *Constr. Build. Mater.* **2022**, *323*, 126626. [CrossRef]
160. Kou, C.; Wu, X.; Xiao, P.; Liu, Y.; Wu, Z. Physical, Rheological, and Morphological Properties of Asphalt Reinforced by Basalt Fiber and Lignin Fiber. *Materials* **2020**, *13*, 2520. [CrossRef] [PubMed]
161. Khasawneh, M.A.; Alyaseen, S.K. Analytic methods to evaluate bituminous mixtures enhanced with coir/coconut fiber for highway materials. *Mater. Today Proc.* **2020**, *33*, 1752–1757. [CrossRef]
162. Maharaj, R.; Ali, R.; Ramlochan, D.; Mohamed, N. Utilization of coir fibre as an asphalt modifier. *Prog. Rubber Plast. Recycl. Technol.* **2019**, *35*, 59–74. [CrossRef]

163. Parimita, P. Influence of Natural Fibers as Additive on Characteristics of Stone Mastic Asphalt. *IOP Conf. Ser. Mater. Sci. Eng.* **2020**, *970*, 012021. [CrossRef]
164. Norhidayah, A.H.; Haryati, Y.; Nordiana, M.; Idham, M.S.M.K.; Juraidah, A.; Ramadhansyah, P.J. Permeability coefficient of porous asphalt mixture containing coconut shells and fibres. *IOP Conf. Ser. Earth Environ. Sci.* **2019**, *244*, 012037. [CrossRef]
165. Haryati, Y.; Norhidayah, A.H.; Nordiana, M.; Juraidah, A.; Hayati, A.H.N.; Ramadhansyah, P.J.; Azman, M.K.; Haryati, A. Stability and rutting resistance of porous asphalt mixture incorporating coconut shells and fibres. *IOP Conf. Ser. Earth Environ. Sci.* **2019**, *244*, 012043. [CrossRef]
166. Kar, D.; Giri, J.P.; Panda, M. Performance Evaluation of Bituminous Paving Mixes Containing Sisal Fiber as an Additive. *Transp. Infrastruct. Geotechnol.* **2019**, *6*, 189–206. [CrossRef]
167. Kumar, N.L.N.K.; Ravitheja, A. Characteristics of stone matrix asphalt by using natural fibers as additives. *Mater. Today Proc.* **2019**, *19*, 397–402. [CrossRef]
168. Singh, S.; Khairandish, M.I.; Razahi, M.M.; Kumar, R.; Chohan, J.S.; Tiwary, A.; Sharma, S.; Li, C.; Ilyas, R.A.; Asyraf, M.R.M.; et al. Preference Index of Sustainable Natural Fibers in Stone Matrix Asphalt Mixture Using Waste Marble. *Materials* **2022**, *15*, 2729. [CrossRef]
169. Pirmohammad, S.; Shokorlou, Y.M.; Amani, B. Influence of natural fibers (kenaf and goat wool) on mixed mode I/II fracture strength of asphalt mixtures. *Constr. Build. Mater.* **2019**, *239*, 117850. [CrossRef]
170. Hainin, M.R.; Idham, M.K.; Yaro, N.S.A.; Hussein, S.O.A.E.; Warid, M.N.M.; Mohamed, A.; Naqibah, S.N.; Ramadhansyah, P.J. Performance of Hot Mix Asphalt Mixture Incorporating Kenaf Fibre. *IOP Conf. Ser. Earth Environ. Sci.* **2018**, *140*, 012092. [CrossRef]
171. Syafiqah, S.M.Z.N.; Masri, K.A.; Jasni, N.E.; Hasan, M. Performance of Stone Mastic Asphalt incorporating Kenaf fiber. *IOP Conf. Ser. Earth Environ. Sci.* **2021**, *641*, 012001.
172. Masri, K.A.; Katini, N.H.; Arshad, A.K.; Shahnewaz, S.; Ferdaus, R. Microstructure analysis of porous asphalt incorporating kenaf fiber in the pavement. *Mater. Today Proc.* **2022**, *57*, 1191–1195. [CrossRef]
173. Selvaraj, S.; Karpurapu, R. Numerical Analysis of Leutner Shear Tests on Interface Between Geosynthetic and Asphalt Layers. *Int. J. Geosynth. Ground Eng.* **2021**, *7*, 61. [CrossRef]
174. Shanbara, H.K.; Ruddock, F.; Atherton, W. Predicting the rutting behaviour of natural fibre-reinforced cold mix asphalt using the finite element method. *Constr. Build. Mater.* **2018**, *167*, 907–917. [CrossRef]
175. Gallo, P.; Valentin, J.; Mondschein, P. Asphalt Concrete for Binder Courses with Different Jute Fibre Content. In *IOP Conference Series: Materials Science and Engineering, 1 Nov 2021*; IOP Publishing: Bristol, UK, 2021.
176. Kumar, G.S.; Shankar, A.U.R.; Teja, B.V.S.R. Laboratory Evaluation of SMA Mixtures Made with Polymer-Modified Bitumen and Stabilizing Additives. *J. Mater. Civ. Eng.* **2019**, *31*, 04019026. [CrossRef]
177. da Costa, L.F.; de Barros, A.G.; de Figueirêdo Lopes Lucena, L.C.; de Figueirêdo Lopes Lucena, A.E. Asphalt mixture reinforced with banana fibres. *Road Mater. Pavement Des.* **2020**, *22*, 1881–1893. [CrossRef]
178. da Costa, L.F.; de Figueirêdo Lopes Lucena, L.C.; de Figueirêdo Lopes Lucena, A.E.; de Barros, A.G. Use of Banana Fibers in SMA Mixtures. *J. Mater. Civ. Eng.* **2019**, *32*, 04019341. [CrossRef]
179. Dehghan, Z.; Modarres, A. Evaluating the fatigue properties of hot mix asphalt reinforced by recycled PET fibers using 4-point bending test. *Constr. Build. Mater.* **2017**, *139*, 384–393. [CrossRef]
180. Usman, N.; Masirin, M.I.M. 20 Performance of asphalt concrete with plastic fibres. In *Use of Recycled Plastics in Eco-Efficient Concrete*; Woodhead Publishing: Sawston, UK, 2019; pp. 427–440.
181. Movilla-Quesada, D.; Raposeiras, A.C.; Olavarría, J. Effects of Recycled Polyethylene Terephthalate (PET) on Stiffness of Hot Asphalt Mixtures. *Adv. Civ. Eng.* **2019**, *2019*, 6969826. [CrossRef]
182. Jegatheesan, N.; Rengarasu, T.M.; Bandara, W.M.K.R.T.W. Mechanical properties of modified hot mix asphalt containing polyethylene terephthalate fibers as binder additive and carbonized wood particles as fine aggregate replacement. *Asian Transp. Stud.* **2020**, *6*, 100029. [CrossRef]
183. Jegatheesan, N.; Rengarasu, T.M.; Bandara, W.M.K.R.T.W. Modelling the Properties of Modified Hot Mix Asphalt Containing Polyethylene Terephthalate Fibers and Carbonized Wood Particles. *J. East. Asia Soc. Transp. Stud.* **2022**, *14*, 1692–1711.
184. Babalghaith, A.M.; Koting, S.; Ibrahim, M.R.; Mohd, N.S.; Rozali, S.; Muhamad, M.R.; Zubir MN, M.; El-Shafei, A.; Khairuddin, F.H.; Yusoff, N.I.M. Effect of Black PET Fiber as Additive on the Mechanical Properties of Stone Mastic Asphalt (SMA) Mixtures. In *Advances in Civil Engineering Materials*; Springer: Singapore, 2021; pp. 249–260.
185. Bocci, E.; Prosperi, E. Recycling of reclaimed fibers from end-of-life tires in hot mix asphalt. *J. Traffic Transp. Eng. (Engl. Ed.)* **2020**, *7*, 678–687. [CrossRef]
186. Valdés-Vidal, G.; Calabi-Floody, A.; Duarte-Nass, C.; Mignolet, C.; Díaz, C. Development of a New Additive Based on Textile Fibers of End-of-Life Tires (ELT) for Sustainable Asphalt Mixtures with Improved Mechanical Properties. *Polymers* **2022**, *14*, 3250. [CrossRef]
187. Calabi-Floody, A.; Mignolet-Garrido, C.; Valdés-Vidal, G. Evaluation of the effects of textile fibre derived from end-of-life tyres (TFELT) on the rheological behaviour of asphalt binders. *Constr. Build. Mater.* **2022**, *360*, 129583. [CrossRef]

188. González, A.; Norambuena-Contreras, J.; Poulikakos, L.; Varela, M.J.; Valderrama, J.; Flisch, A.; Arraigada, M. Evaluation of Asphalt Mixtures Containing Metallic Fibers from Recycled Tires to Promote Crack-Healing. *Materials* **2020**, *13*, 5731. [CrossRef]
189. González, A.; Norambuena-Contreras, J.; Storey, L.; Schlangen, E. Effect of RAP and fibers addition on asphalt mixtures with self-healing properties gained by microwave radiation heating. *Constr. Build. Mater.* **2018**, *159*, 164–174. [CrossRef]

polymers

MDPI

Article

Self-Healing Performance of Asphalt Concrete with Ca-Alginate Capsules under Low Service Temperature Conditions

Huoming Wang [1], Quantao Liu [2,*], Jie Wu [1], Pei Wan [2,*] and Feiyang Zhao [2]

[1] National Engineering and Research Center for Mountainous Highways, China Merchants Chongqing Communications Technology Research and Design Institute Co., Ltd., Chongqing 400067, China
[2] State Key Laboratory of Silicate Materials for Architectures, Wuhan University of Technology, Wuhan 430070, China
* Correspondence: liuqt@whut.edu.cn (Q.L.); wanpei@whut.edu.cn (P.W.)

Abstract: Calcium alginate capsules containing rejuvenators represent a promising method for asphalt concrete premaintenance, but their healing capacities under lower temperature conditions are still unknown. This paper investigated the healing performance of asphalt concrete containing calcium alginate capsules at low service temperatures. The Ca-alginate capsules were synthesized, and their morphology, compressive strength, thermal resistance, and relative oil content were evaluated. Besides, evaluations for the healing of asphalt concrete and the rejuvenator-release ratio of the capsules were determined via fracture-healing-refracture testing and Fourier-transform infrared spectrum experiments. Meanwhile, the glass transition temperature and rheological property of asphalt binder after compressive loading under different temperatures were explored via a differential scanning calorimeter and dynamic shear rheometer. The results showed that the capsules had good thermal resistance and mechanical strength. The capsules released less oil under −15, −10, and −5 °C than at 20 °C, and the healing ratios of the asphalt concrete with the capsules at −15, −10, and −5 °C were obviously lower than that at 20 °C. The released rejuvenator from the capsules could decrease the complex modulus and glass transition temperature of the asphalt binder. When compared with low service temperatures, the asphalt binder containing the capsules and serving at a high temperature has a better softening effect and low-temperature performance due to more oil being released.

Keywords: calcium alginate capsules; asphalt concrete; selfhealing level; low service temperature

Citation: Wang, H.; Liu, Q.; Wu, J.; Wan, P.; Zhao, F. Self-Healing Performance of Asphalt Concrete with Ca-Alginate Capsules under Low Service Temperature Conditions. *Polymers* **2022**, *15*, 199. https://doi.org/10.3390/polym15010199

Academic Editor: Andrea Pucci

Received: 3 December 2022
Revised: 22 December 2022
Accepted: 27 December 2022
Published: 30 December 2022

1. Introduction

Asphalt mixtures represent the leading surfacing material for roads and bridges due to excellent pavement performance. Different asphalt modification technologies were developed to enhance the road performance and service life of asphalt pavement [1]. Nonetheless, under the combined actions of cold temperature [2], ultraviolet radiation exposure [3], water erosion [4], and repetitive vehicle loading [5], microcracks will inevitably occur on asphalt concrete during its service life. Without effective treatment for the pavement, the microcrack will fatally evolve into macrocracking, which will lead to the failure of the concrete structure and, thus, shorten the lifetime of the asphalt pavement.

In order to keep asphalt pavement in a good serviceable condition, road agencies usually take maintenance measures regarding pavement. The current crack repair treatments can extend the service life of asphalt pavements to some extent, but these are conducted on pavements passively after the occurrence of structural damage, which requires considerable raw materials and imposes an additional burden on the environment [6–8]. Therefore, there is an urgent need for sustainable treatments in pavement maintenance in a carbon-neutral context.

Asphalt, as a natural healing material, can recover its original properties spontaneously by molecular diffusion during rest time or under elevated thermal conditions [9–11]. In this respect, the self-healing potential of asphalt can be considered as an eco-sustainable solution strategy to conserve natural resources and relieve environmental pressure. However, in practice, the self-healing capacity of asphalt is weak due to cryogenic conditions and inevitable aging. Therefore, improving the self-healing capacity of asphalt is necessary. In recent years, the rejuvenator encapsulation method, which aims to enhance asphalt healing ability, has drawn researchers' attention. It can enhance the microcrack repair ability of asphalt concrete and regenerate the aging asphalt binder in place due to the timely replenishment of the healing agent, which represents a hopeful premaintenance approach for asphalt pavements in the future [12,13].

The capsules with different asphalt rejuvenators are the common form of encapsulation to boost the auth-healing capability of asphalt [14,15]. The core-shell microcapsule (μm size) [16,17] and muti-chamber capsule (mm size) [18–20] are the primary storage mediums for asphalt rejuvenators. The microcapsules with a rejuvenator showed excellent healing enhancement for asphalt binders in fracture recovery tests [16,21]. However, asphalt concrete with microcapsules cannot effectively enhance the healing ratio of the crack due to limited rejuvenator content and an unsustainable release mode. The muti-chamber calcium alginate capsules showed obvious gradual rejuvenator release potential and provided a sustainable healing capability for asphalt concrete under external cyclic loading [22–24]. Hence, incorporating calcium alginate capsules containing a rejuvenator into asphalt concrete will be an ideal treatment strategy for sustainable asphalt pavement.

It is remarkable that the current self-healing experiments on asphalt mixtures mixed with Ca alginate capsules are performed at an idealized temperature condition. The healing capability of asphalt concrete containing capsules depends to a large extent on the capillary flow of the rejuvenator in the microcrack zone, which is closely related to environmental temperature. In order to obtain high healing levels, most of the relevant research tested the healing capacities of the specimens at a moderate test temperature (20 °C) [25–32]. The temperature for capsule activation (healing agent released from Ca alginate capsules via the external compressive loading) was set at 20 °C. Besides, the temperature of healing process was also set at 20 °C. In general, the test temperature does not match the ambient temperature of the actual road. The ambient temperature of asphalt pavement cannot always maintain 20 °C in the actual service environment. The fluctuation of the service temperature has an impact on the healing of asphalt concrete. In cold conditions, the capillary flow speed of the healing agent is slow around the microcrack zone, which decreases the healing rate of asphalt concrete.

Regarding the extent of information available to our knowledge, research concerning the low service temperature (low induction release temperature of the rejuvenator and low healing temperature of the test asphalt mixture specimen) of Calcium alginate capsules embedded in bituminous concrete has yet not been conducted. Even though cracks usually occur in asphalt concrete during a cold period, there are few studies that focus on the crack-healing ratio of bituminous concrete with capsules under low-temperature service conditions. Therefore, this paper focused on the self-healing performance of asphalt concrete with Ca alginate capsules under simulative low-temperature service conditions. Firstly, the Ca alginate capsules were synthesized based on the hole-coagulation bath technique. Secondly, the performance experiments were performed to assess the main properties of the prepared capsules. Thirdly, the healing ratios of the bituminous concrete mixed with capsules under simulative low service conditions were evaluated by the three-point-bending (3PB) testing and cyclic loading–healing testing. Besides, the rejuvenator release ratios of the capsules within the bituminous concrete were determined by Fourier transform infrared spectroscopy (FTIR). Finally, the glass transition temperature and rheological property of the asphalt binder after loadings under different temperatures were explored via a differential scanning calorimeter and a dynamic shear rheometer, respectively.

2. Materials and Experimental Methods

2.1. Raw Materials

The synthetic ingredients of the Ca alginate capsules include sodium alginate (SA), calcium chloride ($CaCl_2$), and Tween-80, which were provided by Shanghai Sinopharm Chemical Reagent Co., Ltd. (Shanghai, China), and the sunflower oil (healing agent) was provided by Arowana Group Co., Ltd. (Shanghai, China). In this study, the density, penetration, and softening point of virgin asphalt (#70) were 1.034 g/cm^3, 68 (0.1 mm, 25 °C), and 48.4 °C respectively. The density, viscosity, and flash point of the sunflower oil were 0.935 g/cm^{-3}, 0.285 pa·s, and 230 °C, respectively. Sunflower oil can supply light component for aged asphalt and help it restore its original property. Besides, sunflower oil has a distinct peak at 1745 cm^{-1}, while virgin asphalt has no absorption peak between 1700–1750 cm^{-1} [23,31,33]. Therefore, according to FTIR spectra, this characteristic peak can be used to evaluate the healing agent release ratio of the capsules in bituminous concrete after cyclic loading.

2.2. Fabrication of Ca alginate Capsules

The Ca alginate capsules were synthesized at room temperature, and the specific process is shown in Figure 1. The process is divided into four steps: (1) add SA powder to room temperature water and stir for 5 min to form a 2.25 wt% homogeneous SA solution. (2) Add Tween-80 and rejuvenator to the SA solution and shear at 5000 r/min for 15 min to obtain a homogeneous sodium alginate-oil (SA-O) emulsion. The fixed water–oil ratio and surfactant dosage were 1:10 and 5% (oil mass), respectively. (3) Pour the finished emulsion into a custom funnel and instill it into a $CaCl_2$ solution (3.0 wt%); this is then kept for 12 h to ensure the complete reaction of the alginate chain with Ca^{2+}. (4) Remove the wet capsule, rinse it with water, and leave it in a tray at a moderate temperature for 48 h to obtain a dry capsule.

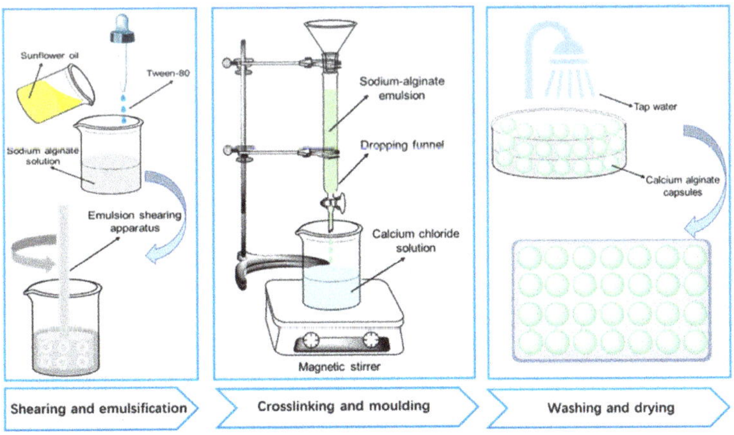

Figure 1. Preparation procedure of Calcium alginate capsules.

2.3. Characterization of Ca Alginate Capsules

The characterization tests were performed on the prepared capsules to determine their basic properties. (1) The microstructure inside the capsules was obtained by scanning electron microscopy (Zeiss, Gemini 300, Jena, Germany). The sample preparation process was as follows: (a) the upper and lower parts of the capsules were cut with a blade to obtain a thin section with a thickness of 1 mm; (b) the oil adhered to the surface of the thin section was removed by oil-absorbing paper; (c) the capsule slices were dispersed on conductive carbon adhesive tapes; (d) the sample was sprayed with gold for 30 s. (2) The compression strength of the Ca alginate capsules was determined by the uniaxial compression test (electric compression test machine, ZQ-990, Wuhan, China). Given that the temperature has impaction on the mechanical strength [27,34,35], hence three types of capsules were kept at −15 °C (4 h), 25 °C (4 h) and 160 °C (2 h), respectively, before the test. The capsules were compressively loaded at loading rate with 0.5 mm/min until the occurrence of yield point. In this work, the yield strength was selected as the mechanical strength of capsules. (3) A simultaneous thermal analyzer (STA449F3) with a heating rate of 10 °C/min was used to determine the thermal sensitivity of the capsules. The whole test was conducted under nitrogen atmosphere protection with a heating range of 40–1000 °C. The relative revertant content of capsules was calculated through the remained masses of the blank capsule and rejuvenator.

2.4. Preparation of Asphalt Concrete Mixed with Ca alginate Capsules

A dense asphalt mixture [24,36] was selected in this work, and the relevant gradation (AC-13) is presented in Figure 2. The Ca alginate capsules were placed into the bituminous concrete at the end of the mixing procedure, and the weight of the capsules was 0.5% over the total mass of the mixtures. After the mixing procedure, standard rutting noncapsule and capsule asphalt concrete plates were prepared via rutting-slab-forming equipment. Finally, asphalt mix beams (98 mm × 45 mm × 50 mm) were cut from the bituminous concrete slabs, and a notch (10 mm × 4 mm) was cut in the middle zone.

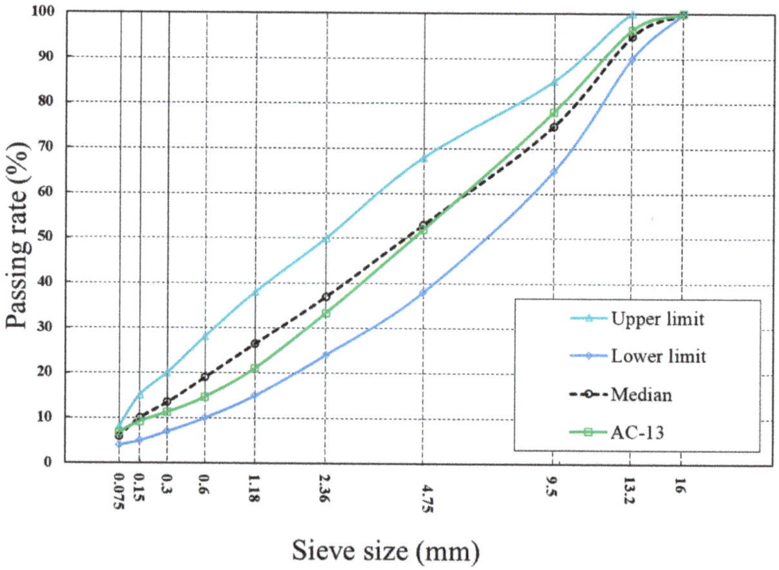

Figure 2. Gradation curve of the asphalt mixture. Reprinted with permission from Ref [36]. 2021, Elsevier.

2.5. Healing Evaluation of Bituminous Concrete with Ca Alginate Capsules

As illustrated in Figure 3, the fracture-cyclic loading-healing-refracture test was divided into four steps: (1) to obtain the initial bending strength, 3PB tests were performed sequentially on the asphalt mixture beams with and without capsules. The trial temperature was set as −20 °C, and the load conduction rate was set as 0.5 mm/min. (2) The cracked beam was placed into a steel mold, and the ballast pressure (0.7 MPa) was uniformly dispersed by placing steel plates on the beam. The loading cycles were set to 0, 16,000, 32,000, 48,000, and 64,000 to simulate actual vehicle tire loading on the capsules within an asphalt road for a period of 1 year, 2 years, and 4 years, respectively [37]. The test temperatures were set to −15, −10, −5, and 20 °C, respectively. The low-temperature service conditions (−15 °C, −10 °C, and −5 °C) were selected by a combination of average temperature data from the cold regions of China and the temperature limit range of the laboratory test apparatus. A service temperature of 20 °C was chosen to maintain consistency with the healing temperatures mentioned in the existing literature and to serve as an ambient temperature reference group for comparative studies of the healing of asphalt concrete under low-temperature service conditions. External loading was used to cause the capsule to release the encapsulated rejuvenator. (3) After the completion of cyclic loading, the beams in the steel molds were put in thermostats at −15, −10, −5, and 20 °C, respectively, and left to rest for 96 h to recover their strength. (4) At the end of the healing period, the 3PB test (as described in step 1) was conducted on the healed beams again.

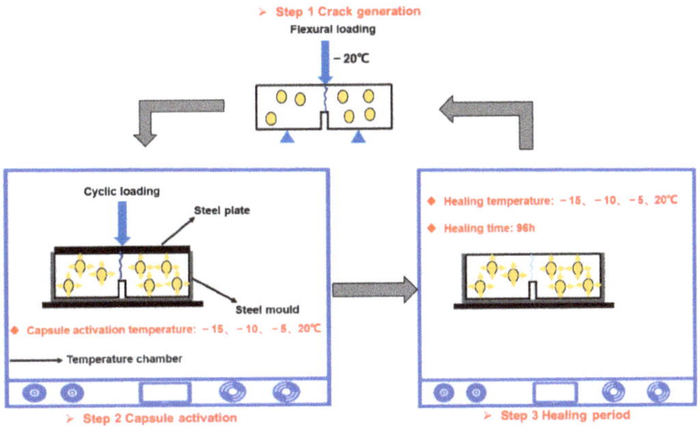

Figure 3. The fracture-healing-refracture test procedure for a bituminous concrete beam.

The healing rate of above tested bituminous concrete was determined by the strength restoration rate (HI_S). As shown in Equation (1), the HI_S was defined as the specific value between the beam's flexural strength after the rest period (F_2) and the initial flexural strength (F_1). In this trial, three specimens were tested in each group of tests to obtain the average healing rate.

$$HI_S = \frac{F_2}{F_1} \quad (1)$$

2.6. Quantification of the Healing Agent Released from the Capsules at Different Service Temperatures

In this work, the chemical evaluation of the healing agents released by the capsules in the asphalt mixtures was performed by FTIR spectroscopy (Thermo Fisher Scientific, Nicolet 6700, Waltham, CT, USA). Sunflower oil has a peak at 1745 cm^{-1} with a peak area lined with the oil content in asphalt, while asphalt has no peak in this characteristic range [23,33]. Figure 4a shows the infrared spectra of the oil and asphalt used in this study. Thus, an area of 1745 cm^{-1} can be used to determine the release ratio of the healing agent inside the capsule after compression loading.

The samples of asphalt containing rejuvenator (0, 2, 4, 6, and 8% of asphalt mass) were prepared by mixing oil with asphalt and stirring at 120 °C for 40 min. The prepared asphalt samples were subjected to FTIR tests to obtain the relationship between the 1745 cm^{-1} area and the sunflower oil content in the asphalt binder. Rao et al. found that the peak index in equation (2) could be used to determine the rejuvenator content in the asphalt binder [38].

Figure 4b presents the correlation curve between $I_{1745cm}{}^{-1}$ and the oil content of asphalt. In this work, Origin software was used to find the relationship between the peak area index and the healing agent content of the asphalt, and the linear fitting equation of the peak area index and oil content was derived.

$$I_{1745cm^{-1}} = \frac{\text{The peak area of } 1745cm^{-1}}{\Sigma\text{Area of spectral bands between 2000 and 600 } cm^{-1}} \quad (2)$$

After the cyclic loading-healing procedure, the trial beams were put in an oven and heated at 70 °C for 30 min, and the capsules inside the asphalt mix were removed by hand.

The loose asphalt mixtures were dissolved in trichloroethylene for 48 h, and then the extracted supernatant was put in a fuming cabinet for 24 h to allow the solvent to evaporate. We added 0.1 g of asphalt into a centrifuge tube and dissolved the asphalt with 2 mL CS_2. The prepared asphalt oil was poured onto KBr wafers and dried to form a layer of asphalt. The FTIR experiments were conducted within a mid-infrared wave number (400~4000 cm^{-1}). The resolution and total scan period were set as 4 cm^{-1} and 64, respectively.

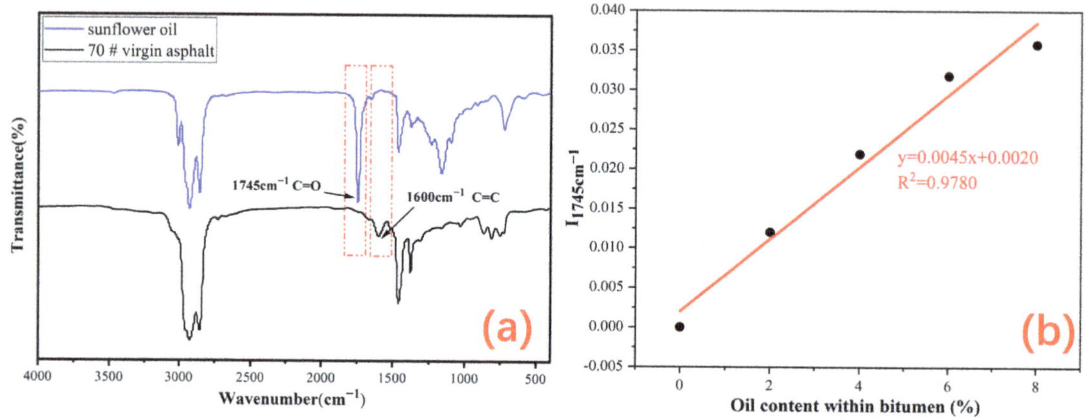

Figure 4. (**a**) Infrared spectra of virgin asphalt and sunflower oil and (**b**) the standard curve between $I_{1745cm^{-1}}$ and the rejuvenator content within the virgin asphalt.

2.7. Differential Scanning Calorimetry (DSC) Test of the Extracted Asphalt Binders

It has been confirmed that the addition of the rejuvenator into the asphalt binder decreases its glass transition temperature and reduces the modulus of asphalt at a low temperature [39–43]. Hence, the DSC test was performed on different extracted asphalt binders after 64,000 cycles of loading under different test temperatures (−15, −10, −5, and 20 °C) via a TA instrument (TA-DSC2500, New Castle, PA, USA) to determine their glass transition temperatures (Tg). The specific test conditions were as follows: (1) temperature range: −50~160 °C; (2) heating ratio: 10 °C/min; (3) protect atmosphere: N_2 (flow rate: 50 mL/min). In this work, according to the midpoint method (as presented in Figure 5), the Tg of asphalt binder was obtained.

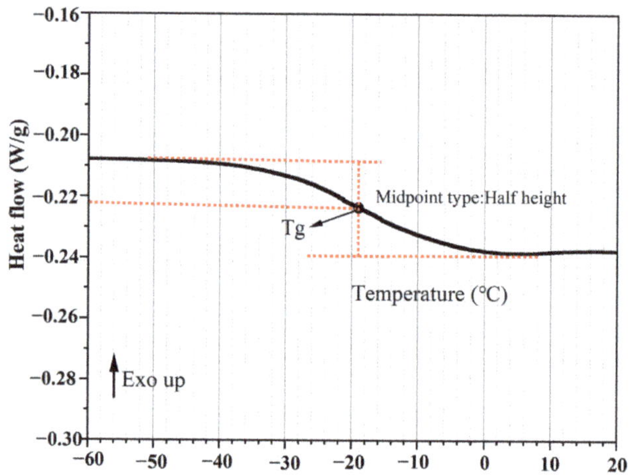

Figure 5. The curve of the glass transition temperature of the asphalt binder.

2.8. Rheological Property Testing of the Extracted Asphalt Binders

In order to characterize the rheological performance of the extracted asphalt binders after compressive loading under different temperatures, temperature scan tests were performed by a dynamic shear rheometer (DSR) (Smart Pavement 102) provided by Anton Paar Instruments, Ltd. After 64,000 cycles of compressive loading under different test temperature (−15, −10, −5, and 20 °C), the extracted asphalt binder was selected as test samples. The temperature of the DSR test was from −15 °C to 20 °C. The strain was 0.5%, and the frequency was 10 rad/s. The diameter of the rotor was 8 mm.

3. Results and Discussion

3.1. Basic Property of the Polymer Capsules

Figure 6 presents the morphology and interior structure of the calcium alginate capsules. As can be observed from Figure 6a, the capsules showed near sphere shape, and the average diameter was 1.7 mm. Hence, the capsules can be put into bituminous mixtures as part of the fine aggregates. Besides, as can be observed from Figure 6b–d, the capsule presented a multicavity inter structure. The healing agent droplets were stored in disjunctive cavities with different sizes and shapes. The unique storage mode may make the capsule release its inner healing agent gradually under the action of external loading.

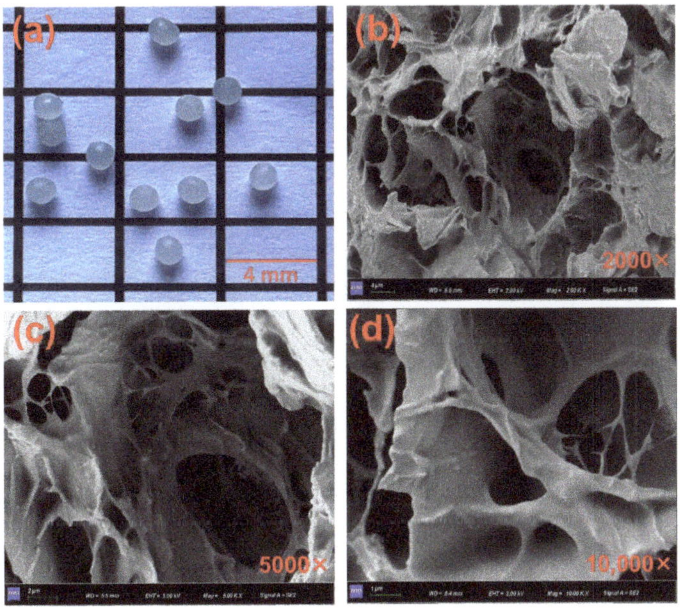

Figure 6. (**a**) Appearance of calcium alginate capsules and (**b–d**) the interior structure of the capsules with different magnification times.

The mass losses of the rejuvenator, the blank Ca alginate capsules (without the healing agent) and the Ca alginate capsules containing the healing agent are presented in Figure 7. As shown in Figure 7a, the healing agent (sunflower oil) starts to volatilize at 345 °C and is completely volatilized at 498 °C. Hence, the remaining mass of the oil capsules and blank capsules in this temperature interval (345~498 °C) can be used to obtain the capsules' actual healing agent content. Figure 7b shows the retained mass of the blank Ca alginate capsules, and their residual mass at 345 °C and 498 °C were 50.1% and 40.9%, respectively. Figure 7c presents the retained mass of the Ca alginate capsules with the healing agent. From 100 °C to 180 °C, the residual mass of the capsule decreased stably due to the vaporization of the moisture and the destruction of the minor glycosidic bonds that exist in the alginate chains. It is noteworthy that the mass loss of the capsules with the rejuvenator was 4.3% at 200 °C (above the fabrication temperature of asphalt mixtures); the capsules have favorable thermal stability at the asphalt mixture production temperature. Besides, the residual masses of the capsules at 345 °C and 498 °C were 86.2% and 17.1%, respectively. Hence, the relative healing agent content of the capsules can be calculated based on formulas (3) and (4). The relative rejuvenator content of the capsules was 58.5%.

$$\varepsilon + (\alpha_1 - \alpha_2)(1 - \varepsilon) = \beta_1 - \beta_2 \tag{3}$$

$$\varepsilon = \frac{\beta_1 - \beta_2 + \alpha_2 - \alpha_1}{1 + \alpha_2 - \alpha_1} \tag{4}$$

where ε is the relative healing agent content in the capsule (%), α_1 and α_2 are the retained mass of the blank capsules without the healing agent at 345 °C and 498 °C, respectively, and β_1 and β_2 are the retained mass of the capsules with the healing agent at 345 °C and 498 °C, respectively.

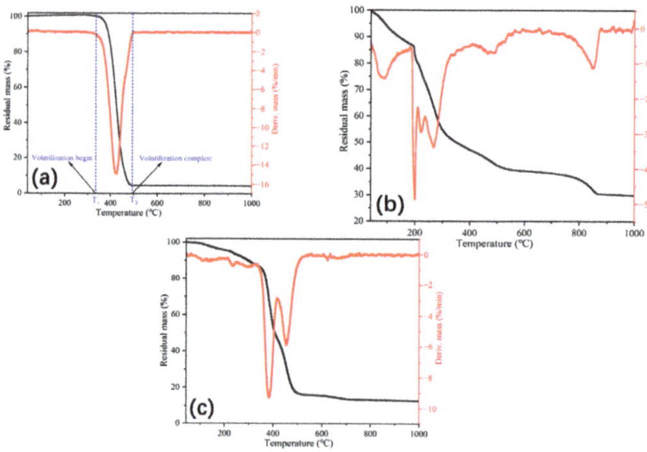

Figure 7. The mass loss of different materials. (**a**) Healing agent, (**b**) blank Ca alginate capsule without the healing agent, and (**c**) Ca alginate capsule containing the healing agent.

The capsules (mm size) are designed to be added to bituminous mixtures as part of the fine aggregates. Therefore, the Ca alginate capsules must withstand the mechanical action created in the production period of the asphalt concrete. Practice has shown that the mechanical strength of the capsules used for bituminous mixtures must be greater than 10 N [33,34]. Figure 8 presents the compressive strength of the capsules after specific temperature treatment. The mechanical strength of capsules after different treatments (0, 25, and 160 °C) were 14.9 N, 12.6 N, and 10.2 N, respectively. Lowering the temperature reduces the ability of the calcium alginate molecule chains to move, thus reducing the deformability of the calcium alginate capsules; hence, the capsule has higher strength at lower temperatures. The calcium alginate capsules softened with the increase in temperature, decreasing the compressive strength slightly [27]. The strength of the capsules under 160 °C was still above 10 N, which implied that the capsule could survive the production of asphalt concrete.

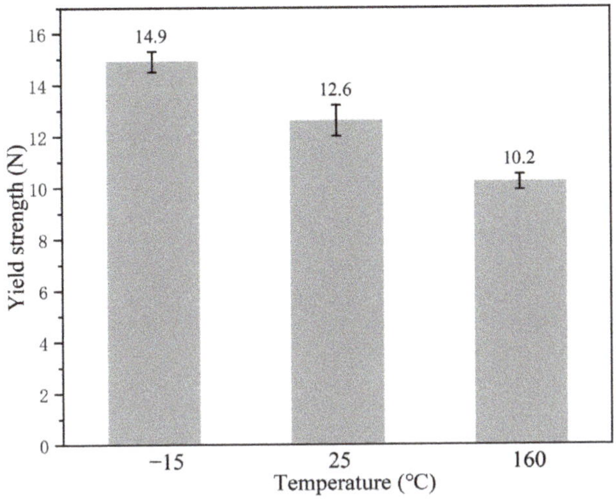

Figure 8. Compressive strength of the Ca alginate capsules.

3.2. Oil Release Assessment of the Capsules in Asphalt Concrete after Cyclic Loading at Different Service Temperatures

The oil released by the capsules can soften the asphalt and reduce its viscosity, which enhances the flow ability of asphalt and, thus, improves its crack repair efficiency. Therefore, the healing rate of asphalt concrete is closely dependent on the oil release rate of the capsules presented in the bituminous mixtures. Figure 9 shows the oil release ratio of the capsules in the specimen beams after cyclic loading at different service temperatures. When no external loading was applied on the asphalt concrete beams, the oil release ratios of the capsules after the constant healing period at −15, −10, −5, and 20 °C were 4.9, 5.1, 4.8, and 5.4%, respectively. It could be seen that there were no differences between the oil release ratios, which implied that the service temperature had no influence on the release of the oil in the capsules within the bituminous mixtures without the action of external loading. The premature leakage of oil from the asphalt concrete during the production period was the result of a combination of thermal and mechanical action. Under specific service temperatures, the capsules' oil release ratio rose with the loading cycles, which implied that the oil release of the capsule needed to be triggered by external loading and that the capsule could gradually let out the encapsulated oil under cyclic loading. The capsule's own gradual release pattern is based on the unique multi-chamber. The oil was stored in the separated chambers, and the capsule gradually released the healing agent by elastic contraction–expansion under cyclic loading [13,27].

After loading with specific cycles, the oil release ratio of the capsules at 20 °C was obviously higher than that of the capsules at low temperatures (−15, −10, and −5 °C). The oil release ratio of capsules improved with the increase in service temperature (loading applied temperature and healing temperature). For instance, after compression loading with 64,000 cycles, the capsule oil release ratios at 15, −10, −5, and 20 °C reached 18.2, 21.9, 25.3, and 57.1%, respectively. This interesting phenomenon was consistent with the result reported by Al-Mansoori [44]. The stiffness of the bituminous mixture beams improved with the reduction in test temperature, with the deformation resistance becoming stronger, which made the capsules in the test beams subject to mild compaction. Besides, the strength of the capsules increased with a reduction in temperature, which was proven by the mechanical strength results of the capsules under different temperatures, as shown in Figure 8. Under constant loading cycles, the capsule became stiffer with the reduction in temperature and, thus, more difficult to deform, which made the capsule release less of the rejuvenator.

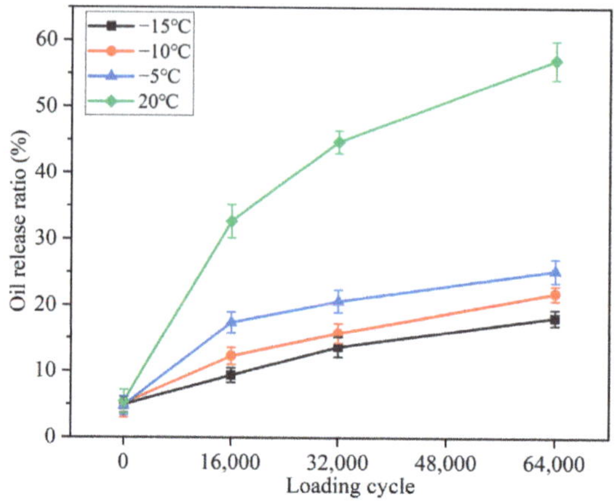

Figure 9. Oil release ratios of Ca alginate capsules within asphalt concrete after loading at different service temperatures.

It was clear that the release of the rejuvenator from the capsules was triggered by the external loading. For pavement areas that cannot be reached by traffic loads, microwave heating can be used to actively induce the release of the healing agent in the capsule at the right time to accelerate the crack healing of asphalt concrete [36].

3.3. Selfhealing Performance of Asphalt Concrete Beams under Different Service Temperatures

Figures 10 and 11 present the healing ratios of the bituminous mixture beams with and without the capsules after cyclic loading at different service temperatures. As can be seen from Figure 10, the fractured specimen beams without capsules could not obtain any strength recovery after different cycles of loading under −15, −10, and −5 °C, respectively. The reason is that the asphalt was rigid at low temperatures, and the asphalt around the crack zone failed to flow and diffuse; hence, the fractured beams were hard to heal. The weak healing ability of the noncapsule bituminous mixture at a low temperature was also confirmed by the literature [44,45]. When the test temperature was 20 °C, the noncapsule beams had partial strength restoration stemming from the inherent healing property of asphalt at moderate temperatures. Besides, the selfhealing ratios of the noncapsule beams improved slightly with the increase in the loading cycles. The healing ratios of the non-capsule beams were 33.9, 36.1, 37.4, and 40.2%, respectively, after 0, 16,000, 32,000, and 64,000 cycles of loading. The cyclic loading caused the gradual compaction of the aggregate inside the asphalt mixture beam, which caused a slight width reduction between the two surfaces and improved the strength restoration rate of the nonencapsulated beam.

It can be observed from Figure 10 that with zero cyclic loadings applied to the asphalt concrete containing the capsules, the level of healing was 0% at −15, −10, and −5 °C, respectively. Although there was a tiny amount of rejuvenator in asphalt mixture beams, the viscosity of the asphalt was still high (hard to flow). When cyclic loading was performed on the beams with the capsules, they all gained partial strength recovery at low temperatures. The healing ratios all increased with the loading cycles but were still in a low range. When compared to the healing ratios of the non-capsule asphalt beams, it could be inferred that the released healing agent may aid in the recovery of strength at low temperatures after cyclic loading. Due to the limitation of the low temperature, the viscosity reduction effect for the rejuvenator (sunflower oil) was limited. Besides, when the cyclic loading was fixed, the healing ratio of the asphalt concrete beams with capsules all improved with the increase in service temperature. For example, after 64,000 cycles of loading, the healing rates of the asphalt concrete beams with capsules were 9.3, 11.8, and 16.6% at −15, −10, and −5 °C, respectively. When the loading cycles equaled 64000, the oil release rates of the capsules at

−15, −10, and −5 °C were 18.2, 21.9, and 25.3%, respectively. The oil released and service temperature jointly affected the healing capability of the asphalt concrete beams with the capsules, and a higher oil release ratio and service temperature gave the beams higher healing ratios.

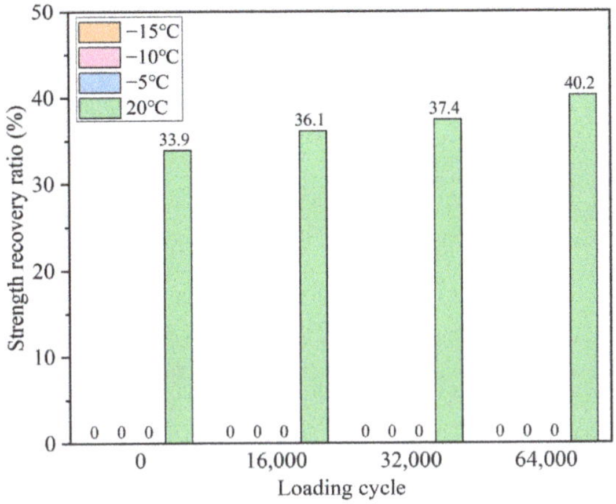

Figure 10. Healing ratio of non-capsule asphalt concrete beams after compression loading at different service temperatures.

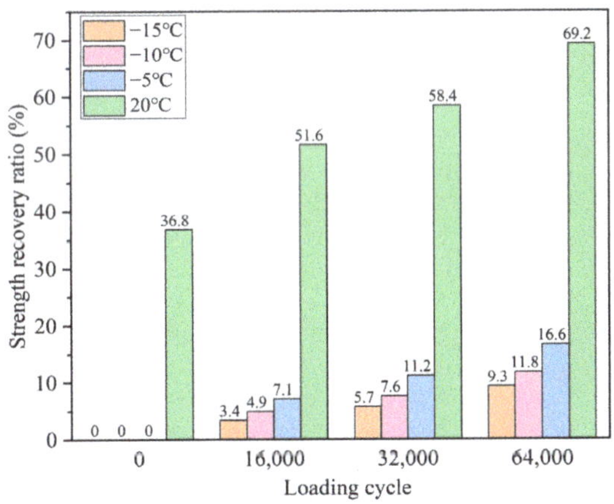

Figure 11. Healing ratio of capsule asphalt concrete beams after compression loading at different service temperatures.

When the trial temperature was 20 °C, the healing rates of the capsule beams were all significantly higher than those of the beams at low temperatures. The healing ratio of capsule beams improved significantly with the increase in loading cycles. After 0, 16,000, 32,000, and 64,000 cycles of loading, the strength recovery rates of the capsule concrete beams were 36.8, 51.6, 58.4, and 69.2%, respectively, which were higher than those of the noncapsule concrete beams. The improvement in the level of healing was attributed to the collective action of more of the rejuvenator being released and an enhanced healing temperature. In the previous literature [22,23,27,29,33,46,47], when the test temperature was 20 °C, the asphalt concrete with the calcium alginate capsules all obtained higher healing ratios than plain asphalt concrete (without capsules).

3.4. Analysis of the Glass Transition Temperature (T_g) of Asphalt with Oil

According to the mechanical state within various temperature ranges, the asphalt binder (as a temperature-sensitive material) can be classified into three states: the glassy state, the viscoelastic state, and the Newtonian liquid state [48,49]. The transition from the glassy to the viscoelastic state is defined as the glass transition, where the temperature at this point is considered the glass transition temperature (Tg). From the point of view of molecular motion, the glass transition of macromolecules corresponds to the critical states of the onset and freezing of chain segment motions. When the temperature exceeds the T_g, the molecular chain segments in the asphalt begin to show signs of activity, and the molecular chain conformation changes by rotating within the single bond around the main chain, which results in a rapid increase in deformation and a dramatic decrease in the modulus [39,50].

The T_g of the asphalt binders from the mixture beams after loading with 64,000 cycles at different service conditions are presented in Figure 12. The T_g of the asphalt binders with oil was apparently reduced when compared with the asphalt binder without oil. This implied that the release of oil from the capsules could improve the possibility of molecular motions and, thus, decrease the T_g of the asphalt. Besides, the healing agent release ratio increased with the rise in the simulated service temperature, and thus, the T_g of the asphalt decreased due to the rejuvenation effect of the healing agent, which is consistent with other literature [41,51]. Hence, when compared with a mild temperature (20 °C), the regeneration effect of the healing agent on the asphalt binder under low temperatures (−15, −10, and −5 °C) is less pronounced.

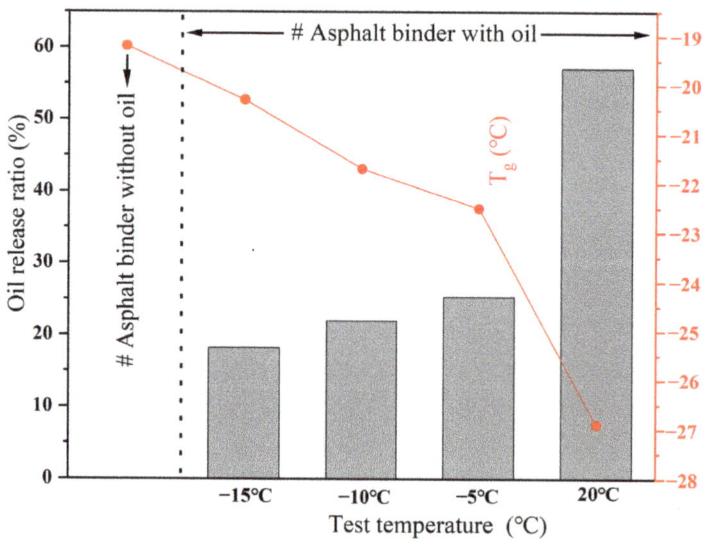

Figure 12. The glass transition temperatures of asphalt after loading with 64,000 cycles under different service temperatures.

3.5. Rheological Property of the Asphalt Binder Containing Oil at Simulated Low-Service Temperatures

Figures 13 and 14 present the complex modulus (G*) and phase angel (δ) of the asphalt binder after 64,000 cycles of compressive loading at different temperatures (capsule activation temperature and healing temperature), respectively. The G* and δ of the asphalt with and without oil all decreased and increased, respectively, with temperature due to the viscoelastic transition of the asphalt binder. When compared to the asphalt without oil, the complex modulus of the asphalt with oil apparently decreased at specific test temperatures. In contrast, the phase angel of the asphalt with oil apparently increased at specific test temperatures. This indicated that the release of oil could change the rheological property of the asphalt binder. When the sunflower oil consists of soft molecules flowing into the asphalt, it could increase the flexible chain ratio of the asphalt and, thus, decrease the complex modulus. Hence, the asphalt with added oil can has better flow ability than the asphalt without oil under the same service temperature, thus having a lower G* and higher δ.

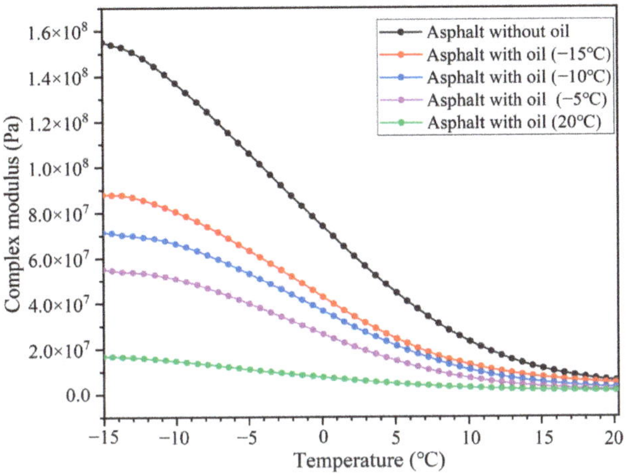

Figure 13. Complex modulus of asphalt binder after loading with 64,000 cycles under different service temperature conditions.

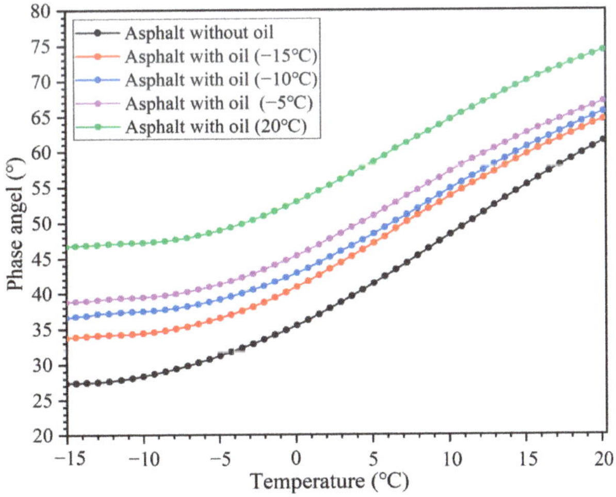

Figure 14. Phase angel of asphalt binder after loading with 64,000 cycles under different service temperature conditions.

Besides, with a rise in test temperature (from −15 to 20 °C), the G* of the asphalt binder decreased, while the phase angel of the asphalt binder increased. The oil release ratios of the capsules after loading with 64,000 cycles at −15, −10, −5, and 20 °C were 18.2, 21.9, 25.3, and 57.1%, respectively. A higher oil release ratio led to a lower G* and higher δ in the asphalt binder, which indicated that the asphalt binder had better flow ability when servicing at 20 °C than the asphalt binder servicing at low temperatures (−15, −10, and −5 °C). Therefore, when compared with low service temperatures, the asphalt binder within the asphalt concrete containing capsules and serving at a high temperature will have a better softening effect and low-temperature performance due to more oil being released.

4. Conclusions

In this work, Ca alginate capsules containing oil were prepared. Evaluation tests were conducted on the capsules to determine their morphological structure, thermal stability, and mechanical resistance. The glass transition temperature and rheological properties of the asphalt binder with and without oil were explored. Besides, the healing release ratio and the level of healing of the capsules in asphalt concrete after loadings under a stimulated low temperature were evaluated. The conclusions are as follows:

(1) The Ca alginate capsules with a mutichamber structure own good thermal stability and compression strength and stratify the production process requirements of asphalt mixtures in a laboratory;
(2) The oil release ratio of the capsules in the asphalt mixture beams increased with the loading cycles. Nevertheless, after loading with specific cycles, the capsules under low-temperature conditions (−15, −10, and −5 °C) had a lower oil release ratio than the capsules at 20 °C;
(3) The release of the oil into the asphalt reduced its complex modulus and increased its phase angle at low service temperatures. Besides, the oil released reduced the glass transition temperature of the asphalt binder and thus enhanced the probability of molecule motion within the asphalt under low temperatures;
(4) The fractured asphalt mixture beams without capsules did not retain any strength recovery at −15, −10, and −5 °C, owing to their reduced inherent healing ability under low temperatures. The incorporation of sunflower oil capsules into the asphalt concrete improved its healing level slightly under low temperatures due to the viscosity reduction effect of the oil. Besides, the healing ratios of the asphalt concrete beams with the capsules under 20 °C were higher than the beam specimens at −15, −10, and −5 °C, respectively.

The asphalt mixture beams with the sunflower oil capsules obtained partial strength recovery (<20%) under a low temperature, and the low healing ability of the asphalt fails to heal the cracks within the asphalt concrete under low-temperature conditions. As asphalt cracking is prone to appear under low temperatures, it is essential to improve the healing ratio of cracked asphalt concrete under low temperatures. Considering the temperature-sensitive properties of asphalt and the viscosity reduction effect of the rejuvenator, the application of the joint action of thermally induced healing and rejuvenator-induced healing may be a promising external healing enhancement strategy for asphalt pavement in cold regions. Microwave absorbing materials can be introduced into asphalt concrete to stimulate heat generation through external microwave heating to promote the flow and diffusion of the rejuvenator, which is expected to solve the problem of the poor healing performance of asphalt concrete under low temperature conditions.

Author Contributions: H.W.: Conceptualization, Methodology, Investigation, Writing—Original Draft. Q.L.: Project Administration, Supervision, Formal analysis, Writing—Review & Editing. J.W.: Methodology, Investigation. P.W.: Investigation, Writing—review & editing. F.Z.: Investigation, Writing—Original Draft. All authors have read and agreed to the published version of the manuscript.

Funding: The work presented in this paper was funded by the National Natural Science Foundation of China (No. 51978547) and Fund of National Engineering and Research Center for Mountainous Highways (The project No. is GSGZJ-2020-02).

Institutional Review Board Statement: Not applicable.

Informed Consent Statement: Not applicable.

Data Availability Statement: Data is contained within the article.

Conflicts of Interest: The authors declare no conflict of interest.

References

1. Hassan, N.A.; Airey, G.; Jaya, R.P.; Mashros, N.; Aziz, M.A. A Review of Crumb Rubber Modification in Dry Mixed Rubberised Asphalt Mixtures. *J. Teknol.* **2014**, *70*, 127–134.
2. Tang, J.; Fu, Y.; Ma, T.; Zheng, B.; Zhang, Y.; Huang, X. Investigation on low-temperature cracking characteristics of asphalt mixtures: A virtual thermal stress restrained specimen test approach. *Constr. Build. Mater.* **2022**, *347*, 128541. [CrossRef]
3. Li, Y.; Feng, J.; Wu, S.; Chen, A.; Kuang, D.; Bai, T.; Gao, Y.; Zhang, J.; Li, L.; Wan, L.; et al. Review of ultraviolet ageing mechanisms and anti-ageing methods for asphalt binders. *J. Road Eng.* **2022**, *2*, 137–155. [CrossRef]
4. Zou, Y.; Amirkhanian, S.; Xu, S.; Li, Y.; Wang, Y.; Zhang, J. Effect of different aqueous solutions on physicochemical properties of asphalt binder. *Constr. Build. Mater.* **2021**, *286*, 122810. [CrossRef]
5. Li, J.; Yu, J.; Wu, S.; Xie, J. The Mechanical Resistance of Asphalt Mixture with Steel Slag to Deformation and Skid Degradation Based on Laboratory Accelerated Heavy Loading Test. *Materials* **2022**, *15*, 911. [CrossRef]
6. Ma, F.; Dong, W.; Fu, Z.; Wang, R.; Huang, Y.; Liu, J. Life cycle assessment of greenhouse gas emissions from asphalt pavement maintenance: A case study in China. *J. Clean. Prod.* **2021**, *288*, 125595. [CrossRef]
7. Lizasoain-Arteaga, E.; Indacoechea-Vega, I.; Pascual-Muñoz, P.; Castro-Fresno, D. Environmental impact assessment of induction-healed asphalt mixtures. *J. Clean. Prod.* **2019**, *208*, 1546–1556. [CrossRef]
8. Nalbandian, K.M.; Carpio, M.; González, Á. Assessment of the sustainability of asphalt pavement maintenance using the microwave heating self-healing technique. *J. Clean. Prod.* **2022**, *365*, 132859. [CrossRef]
9. Sun, D.; Sun, G.; Zhu, X.; Ye, F.; Xu, J. Intrinsic temperature sensitive self-healing character of asphalt binders based on molecular dynamics simulations. *Fuel* **2018**, *211*, 609–620. [CrossRef]
10. Sun, G.; Hu, M.; Sun, D.; Deng, Y.; Ma, J.; Lu, T. Temperature induced self-healing capability transition phenomenon of bitumens. *Fuel* **2020**, *263*, 116698. [CrossRef]
11. García, Á. Self-healing of open cracks in asphalt mastic. *Fuel* **2012**, *93*, 264–272. [CrossRef]
12. Anupam, B.; Sahoo, U.C.; Chandrappa, A.K. A methodological review on self-healing asphalt pavements. *Constr. Build. Mater.* **2022**, *321*, 126395. [CrossRef]
13. Wan, P.; Wu, S.; Liu, Q.; Zou, Y.; Zhao, Z.; Chen, S. Recent advances in calcium alginate hydrogels encapsulating rejuvenator for asphalt self-healing. *J. Road Eng.* **2022**, *2*, 181–220. [CrossRef]
14. Concha, J.L.; Arteaga-Pérez, L.E.; Gonzalez-Torre, I.; Liu, Q.; Norambuena-Contreras, J. Biopolymeric Capsules Containing Different Oils as Rejuvenating Agents for Asphalt Self-Healing: A Novel Multivariate Approach. *Polymers* **2022**, *14*, 5418. [CrossRef]
15. Xu, S.; Liu, X.; Tabaković, A.; Lin, P.; Zhang, Y.; Nahar, S.; Lommerts, B.J.; Schlangen, E. The role of rejuvenators in embedded damage healing for asphalt pavement. *Mater. Des.* **2021**, *202*, 109564. [CrossRef]
16. Wang, X.; Guo, G.; Zou, F.; Zhao, H.; Li, Y. Enhancing self-healing properties of microcrack on aged asphalt incorporating with microcapsules encapsulating rejuvenator. *Constr. Build. Mater.* **2022**, *344*, 128123. [CrossRef]
17. Yang, P.; Gao, X.; Wang, S.; Su, J.-F.; Wang, L.-Q. Novel waterproof bituminous coating using self-healing microcapsules containing ultraviolet light curing agent. *Constr. Build. Mater.* **2022**, *329*, 127189. [CrossRef]
18. Norambuena-Contreras, J.; Arteaga-Perez, L.E.; Guadarrama-Lezama, A.Y.; Briones, R.; Vivanco, J.F.; Gonzalez-Torre, I. Microencapsulated Bio-Based Rejuvenators for the Self-Healing of Bituminous Materials. *Materials* **2020**, *13*, 1446. [CrossRef]
19. Norambuena-Contreras, J.; Concha, J.L.; Arteaga-Pérez, L.E.; Gonzalez-Torre, I. Synthesis and Characterisation of Alginate-Based Capsules Containing Waste Cooking Oil for Asphalt Self-Healing. *Appl. Sci.* **2022**, *12*, 2739. [CrossRef]
20. Abadeen, A.Z.U.; Hussain, A.; Kumar, V.S.; Murali, G.; Vatin, N.I.; Riaz, H. Comprehensive Self-Healing Evaluation of Asphalt Concrete Containing Encapsulated Rejuvenator. *Materials* **2022**, *15*, 3672. [CrossRef]
21. Yang, S.; Ji, J.; Tao, H.; Muhammad, Y.; Huang, J.; Wang, S.; Wei, Y.; Li, J. Fabrication of urea formaldehyde-epoxy resin microcapsules for the preparation of high self-healing ability containing SBS modified asphalt. *Polym. Compos.* **2021**, *42*, 4128–4137. [CrossRef]
22. Grangeiro de Barros, A.; Figueirêdo Lopes Lucena, L.C.d.; García Hernandez, Á. Addition of Encapsulated Soybean Oil and Waste Cooking Oil in Asphalt Mixtures: Effects on Mechanical Properties and Self-Healing of Fatigue Damage. *J. Mater. Civ. Eng.* **2022**, *34*, 04022002. [CrossRef]
23. Garcia-Hernández, A.; Salih, S.; Ruiz-Riancho, I.; Norambuena-Contreras, J.; Hudson-Griffiths, R.; Gomez-Meijide, B. Self-healing of reflective cracks in asphalt mixtures by the action of encapsulated agents. *Constr. Build. Mater.* **2020**, *252*, 118929. [CrossRef]
24. Wan, P.; Wu, S.; Liu, Q.; Wang, H.; Zhao, F.; Wu, J.; Niu, Y.; Ye, Q. Sustained-release calcium alginate/diatomite capsules for sustainable self-healing asphalt concrete. *J. Clean. Prod.* **2022**, *372*, 133639. [CrossRef]
25. Micaelo, R.; Freire, A.C.; Pereira, G. Asphalt self-healing with encapsulated rejuvenators: Effect of calcium-alginate capsules on stiffness, fatigue and rutting properties. *Mater. Struct.* **2020**, *53*, 20. [CrossRef]
26. Al-Mansoori, T.; Micaelo, R.; Artamendi, I.; Norambuena-Contreras, J.; Garcia, A. Microcapsules for self-healing of asphalt mixture without compromising mechanical performance. *Constr. Build. Mater.* **2017**, *155*, 1091–1100. [CrossRef]
27. Bao, S.; Liu, Q.; Li, H.; Zhang, L.; Barbieri, D.M. Investigation of the Release and Self-Healing Properties of Calcium Alginate Capsules in Asphalt Concrete under Cyclic Compression Loading. *J. Mater. Civ. Eng.* **2021**, *33*, 04020401. [CrossRef]
28. Norambuena-Contreras, J.; Yalcin, E.; Hudson-Griffiths, R.; García, A. Mechanical and Self-Healing Properties of Stone Mastic Asphalt Containing Encapsulated Rejuvenators. *J. Mater. Civ. Eng.* **2019**, *31*, 04019052. [CrossRef]

29. Ruiz-Riancho, N.; Saadoon, T.; Garcia, A.; Grossegger, D.; Hudson-Griffiths, R. Optimisation of self-healing properties for asphalts containing encapsulated oil to mitigate reflective cracking and maximize skid and rutting resistance. *Constr. Build. Mater.* **2021**, *300*, 123879. [CrossRef]
30. Xu, S.; Tabaković, A.; Liu, X.; Schlangen, E. Calcium alginate capsules encapsulating rejuvenator as healing system for asphalt mastic. *Constr. Build. Mater.* **2018**, *169*, 379–387. [CrossRef]
31. Yamaç, E.; Yilmaz, M.; Yalçın, E.; Kök, B.V.; Norambuena-Contreras, J.; Garcia, A. Self-healing of asphalt mastic using capsules containing waste oils. *Constr. Build. Mater.* **2021**, *270*, 121417. [CrossRef]
32. Zhang, L.; Liu, Q.; Li, H.; Norambuena-Contreras, J.; Wu, S.; Bao, S.; Shu, B. Synthesis and characterization of multi-cavity Ca-alginate capsules used for self-healing in asphalt mixtures. *Constr. Build. Mater.* **2019**, *211*, 298–307. [CrossRef]
33. Yu, X.; Liu, Q.; Wan, P.; Song, J.; Wang, H.; Zhao, F.; Wang, Y.; Wu, J. Effect of Ageing on Self-Healing Properties of Asphalt Concrete Containing Calcium Alginate/Attapulgite Composite Capsules. *Materials* **2022**, *15*, 1414. [CrossRef] [PubMed]
34. Ruiz-Riancho, N.; Garcia, A.; Grossegger, D.; Saadoon, T.; Hudson-Griffiths, R. Properties of Ca-alginate capsules to maximise asphalt self-healing properties. *Constr. Build. Mater.* **2021**, *284*, 122728. [CrossRef]
35. Concha, J.L.; Arteaga-Pérez, L.E.; Alpizar-Reyes, E.; Segura, C.; Gonzalez-Torre, I.; Kanellopoulos, A.; Norambuena-Contreras, J. Effect of rejuvenating oil type on the synthesis and properties of alginate-based polynuclear capsules for asphalt self-healing. *Road Mater. Pavement Des.* **2022**, *23*, 1–26. [CrossRef]
36. Wan, P.; Wu, S.; Liu, Q.; Xu, H.; Wang, H.; Peng, Z.; Rao, W.; Zou, Y.; Zhao, Z.; Chen, S. Self-healing properties of asphalt concrete containing responsive calcium alginate/nano-Fe_3O_4 composite capsules via microwave irradiation. *Constr. Build. Mater.* **2021**, *310*, 125258. [CrossRef]
37. Bao, S.; Liu, Q.; Rao, W.; Yu, X.; Zhang, L. Synthesis and characterization of calcium alginate-attapulgite composite capsules for long term asphalt self-healing. *Constr. Build. Mater.* **2020**, *265*, 120779. [CrossRef]
38. Rao, W.; Liu, Q.; Yu, X.; Wan, P.; Wang, H.; Song, J.; Ye, Q. Efficient preparation and characterization of calcium alginate-attapulgite composite capsules for asphalt self-healing. *Constr. Build. Mater.* **2021**, *299*, 123931. [CrossRef]
39. Xu, J.; Fan, Z.; Lin, J.; Yang, X.; Wang, D.; Oeser, M. Predicting the low-temperature performance of asphalt binder based on rheological model. *Constr. Build. Mater.* **2021**, *302*, 124401. [CrossRef]
40. Sun, G.; Li, B.; Sun, D.; Zhang, J.; Wang, C.; Zhu, X. Roles of aging and bio-oil regeneration on self-healing evolution behavior of asphalts within wide temperature range. *J. Clean. Prod.* **2021**, *329*, 129712. [CrossRef]
41. Zhang, L.; Tan, Y.; Hussain, B. Relationship between glass transition temperature and low temperature properties of oil modified binders. *Constr. Build. Mater.* **2016**, *104*, 92–98.
42. García, A.; Schlangen, E.; van de Ven, M.; Sierra-Beltrán, G. Preparation of capsules containing rejuvenators for their use in asphalt concrete. *J. Hazard. Mater.* **2010**, *184*, 603–611. [CrossRef] [PubMed]
43. Norambuena-Contreras, J.; Yalcin, E.; Garcia, A.; Al-Mansoori, T.; Yilmaz, M.; Hudson-Griffiths, R. Effect of mixing and ageing on the mechanical and self-healing properties of asphalt mixtures containing polymeric capsules. *Constr. Build. Mater.* **2018**, *175*, 254–266. [CrossRef]
44. Al-Mansoori, T.; Norambuena-Contreras, J.; Garcia, A. Effect of capsule addition and healing temperature on the self-healing potential of asphalt mixtures. *Mater. Struct.* **2018**, *51*, 1–12. [CrossRef]
45. Sun, G.; Sun, D.; Guarin, A.; Ma, J.; Chen, F.; Ghafooriroozbahany, E. Low temperature self-healing character of asphalt mixtures under different fatigue damage degrees. *Constr. Build. Mater.* **2019**, *223*, 870–882. [CrossRef]
46. Norambuena-Contreras, J.; Liu, Q.; Zhang, L.; Wu, S.; Yalcin, E.; Garcia, A. Influence of encapsulated sunflower oil on the mechanical and self-healing properties of dense-graded asphalt mixtures. *Mater. Struct.* **2019**, *52*, 1–13. [CrossRef]
47. Zhang, L.; Hoff, I.; Zhang, X.; Yang, C. Investigation of the self-healing and rejuvenating properties of aged asphalt mixture containing multi-cavity Ca-alginate capsules. *Constr. Build. Mater.* **2022**, *361*, 129685. [CrossRef]
48. Anderson, D.; Christensen, D.; Bahia, H.; Dongre, R.; Sharma, M.; Antle, C.; Button, J. *Strategic Highway Research Program*; Report No. SHRP-A-369; National Research Council: Ottawa, ON, Canada, 1994.
49. Marateanu, M.; Anderson, D. Time-temperature dependency of asphalt binders—An improved model (with discussion). *J. Assoc. Asph. Paving Technol.* **1996**, *65*, 408–448.
50. Kang, Y.; Zhou, D.; Wu, Q.; Liang, R.; Shangguan, S.; Liao, Z.; Wei, N. Molecular dynamics study on the glass forming process of asphalt. *Constr. Build. Mater.* **2019**, *214*, 430–440. [CrossRef]
51. Zhang, L.; Hussain, B.; Tan, Y. Effect of bio-based and refined waste oil modifiers on low temperature performance of asphalt binders. *Constr. Build. Mater.* **2015**, *86*, 95–100.

polymers

MDPI

Article

Investigations on Adhesion Characteristics between High-Content Rubberized Asphalt and Aggregates

Xiaofeng Wang [1], Jianan Liu [2], Zhenjun Wang [2,3,*], Haosen Jing [2] and Bo Yang [1]

[1] Henan Provincial Communications Planning & Design Institute, Zhengzhou 450052, China
[2] School of Materials Science and Engineering, Chang'an University, Xi'an 710061, China
[3] Engineering Research Center of Pavement Materials, Ministry of Education of China, Chang'an University, Xi'an 710064, China
* Correspondence: zjwang@chd.edu.cn; Tel.: +86-02982337245

Abstract: The use of waste tires to prepare rubberized asphalt has been a hot trend in recent years, and the characteristics of adhesion between rubberized asphalt and aggregates are important factors affecting the performance of asphalt pavement. However, there is a lack of uniform results on the adhesion characteristics of rubberized asphalt. Therefore, crumb-rubber-modified asphalt (CRMA) with 15%, 20%, and 25% rubber contents was prepared in this work, and the basic rheological parameters and cohesive energy of the rubberized asphalt were characterized by DSR. The adhesion properties between rubberized asphalt and aggregates were characterized based on macroscopic binder bond strength (BBS), surface free energy (SFE) theory, and nanoscale atomic force microscopy (AFM) tests. The results show that crumb rubber (CR) can improve the high-temperature elastic properties of asphalt; secondly, CR can have a negative impact on the maximum tensile strength of asphalt and aggregates. CR can improve the SFE parameter of asphalt. The work of adhesion of rubberized asphalt and limestone is the highest, followed by basalt and, finally, granite. Finally, CR can cause the catanaphase in asphalt to gradually break down and become smaller, and the adhesion of rubberized asphalt can be reduced. Overall, CR can reduce the adhesion performance of asphalt, and this work provides a reference for the application of rubberized asphalt.

Keywords: rubberized asphalt; adhesion characteristics; binder bond strength (BBS); surface free energy (SFE); atomic force microscope (AFM)

Citation: Wang, X.; Liu, J.; Wang, Z.; Jing, H.; Yang, B. Investigations on Adhesion Characteristics between High-Content Rubberized Asphalt and Aggregates. *Polymers* **2022**, *14*, 5474. https://doi.org/10.3390/polym14245474

Academic Editors: Wei Jiang, Quantao Liu, Jose Norambuena-Contreras and Yue Huang

Received: 2 December 2022
Accepted: 12 December 2022
Published: 14 December 2022

1. Introduction

With the continuous increase in car ownership, the disposal of waste tires has become a pressing issue for scholars from all over the world [1]. The crumb rubber (CR) prepared by recycling and reprocessing waste tires can be used as a modifier in asphalt binders [2,3]. CR powder can not only improve various properties of asphalt binders, but also significantly improve the environmental problems caused by waste tires [3]. As a typical solid waste product, CR powder improves the sustainability of road development. CR powder has great potential in the research of road materials and related fields because of its huge output, excellent performance, and environmental protection advantages [4–7].

The properties of CR-powder-modified asphalt are affected by many factors, including the particle size of the CR powder, the type of CR powder molding, the CR content, the type of blending, the pretreatment process, etc. Xiao Feipeng et al. focused on the plasma treatment of CR powder, and the internal de-crosslinking process of the CR powder improved the compatibility between the CR powder and asphalt [8,9]. The compatibility of CR powder and asphalt can also be enhanced by using CR powder desulfurized by microwaves, or by adding waste oil containing more light components in the process of CR powder modification of asphalt, and the rheological properties of crumb-rubber-modified asphalt (CRMA) can be improved [10,11]. The particle size of CR powder can affect the rheological properties of CRMA, and larger CR powder particles can help to enhance the fatigue performance of CRMA [12]. Some scholars have used graphene/carbon black composite materials and CR powder to create composite-modified asphalt, and the results show that the rutting resistance and healing properties of this composite-modified asphalt were improved [13]. In addition, the blending compatibility of CRMA and waste plastics and the aging resistance of CRMA have also been research hotspots in recent years [14,15].

Asphalt pavement is repeatedly affected by traffic loads and the environment during its use. Especially in a moisture-immersed state, the adhesion between the asphalt and aggregate may fail. The aggregate can fall off and the performance of the asphalt pavement will be seriously deteriorated [16–18]. This situation not only increases the cost of road maintenance, but also causes security risks [19]. However, the results of research on the moisture damage resistance of CRMA are inconsistent. Through static contact angle studies, Zahid Hossain et al. found that the incorporation of CR powder can improve the surface energy of the binder, improve the viscosity of the binder, and reduce the penetration value of the binder, showing a positive effect on resistance to moisture damage [20]. M.N. Partl et al. used CR powder to prepare an open-graded asphalt mixture and conducted a coaxial shear test (CAST), and the study found that compared with traditional porous or semi-porous asphalt mixtures, the moisture sensitivity of the CRMA mixture was reduced [21]. However, there are also different viewpoints. Quan Lv et al. studied the pull-off tensile strength of CRMA, polymer-modified asphalt, and matrix asphalt and basalt slabs on a large scale through binder bond strength (BBS) tests, on the basis of controlling the asphalt film thickness [22–25]. It was found that the polymer and CR powder adversely affected the pull-off tensile strength of the asphalt aggregates.

In addition to the common water boiling test, water immersion test, surface free energy theory, and BBS test, atomic force microscopy (AFM) for nanoscale research has also developed rapidly in recent years. The microstructural characteristics of asphalt surfaces show multinomial heterogeneity at the nanoscale, and the nanoscale properties of asphalt have always been the focus of academic research [26–28]. AFM is a powerful tool for evaluating the microstructure of asphalt. In 1996, L. Loeber et al. first used AFM to discover the bee structure of the asphalt surface [29]. The bee structure is also called catanaphase in the later research classification, which is temperature-reversible [28]. The chemical composition of the catanaphase was initially thought to be mainly asphaltenes, but increasing evidence suggests that interactions between wax crystals and other chemical constituents in asphalt lead to the formation of the catanaphase [30]. AFM is often used to evaluate asphalt's modification effects, degree of aging, and adhesion [31–33].

The loss of adhesive bonds and the fracture of cohesive bonds under the action of water are the main causes of moisture damage [34]. Common adhesion theories include chemical reaction, surface energy, molecular orientation, and mechanical adhesion [35]. When the deformation exceeds the influence of mechanical interlocking and surface molecular orientation, cohesive bond failure occurs [36]. Conventional adhesive strength tests such as the pneumatic adhesion tensile testing instrument (PATTI) and BBS have certain limitations and have been continuously improved [23]. The compression pull-off test has been developed and proven to be excellent [37]. Considering the complexity of adhesion and the diversity of test methods, the adhesion mechanism of crumb-rubber-modified asphalt needs to be further explored.

In this work, the rheological parameters of asphalt binder and base asphalt were characterized for three CR powder contents. The pull-off tensile strength between asphalt with di and limestone, basalt, and granite slabs was tested, and the effect of crumb rubber powder on the BBS was analyzed. Secondly, the SFE parameters of asphalt binders with different crumb rubber powder contents were studied, and the work of adhesion between different asphalt binders and three aggregates was calculated. In addition, the changes in the catanaphase and the adhesion force were used to analyze the influence of the modification of the CR powder on the adhesion and its mechanism of action via AFM. Finally, the effects of CR powder on the adhesion between the binders and aggregates were compared and analyzed from the perspectives of macroscopic strength, surface free energy theory, and nanomechanical properties. The results of this research can help to understand the effect of CR powder on the moisture damage resistance of the mixture, so as to ensure the long-term durable use of the pavement.

2. Materials and Methods

2.1. Materials

The base asphalt was used in this work, and its properties were tested according to the "Standard Test Methods of Asphalt and Bituminous Mixtures for Highway Engineering" (JTG E20-2011) [38], as shown in Table 1. The CR powder used in this work was 80 mesh, its relative density was 1.128, its water content was 0.49%, and its metal content and sieve residue were 0.02% and 4.27%, respectively. Distilled water, ethylene glycol, and glycerol were used to measure the contact angles of asphalt with different rubber powder contents and aggregates, and the surface energies of the three liquids are shown in Table 2 [39]. The detailed parameters can be found in Section 2.3.3. The chemical composition of the aggregates is shown in Table 3.

Table 1. Properties of the original asphalt.

Properties	Test Results	Test Methods
Penetration (25 °C, 100 g, 5 s; 0.1 mm)	92	T0604
Ductility (15 °C, 5 cm/min; cm)	>100	T0605
Softening point (°C)	46.5	T0606
Density (g/cm^3)	1.023	T0603
Solubility (%)	99.71	T0607
Flash point (°C)	295	T0611

Table 2. Surface free energy parameters of the test liquids.

Type	γ_d (mJ/m^2)	γ_p (mJ/m^2)	γ (mJ/m^2)
Distilled water	21.8	51.0	72.8
Ethylene glycol	29.3	19.0	48.3
Glycerol	34.0	30.0	64.0

Table 3. Chemical components of the aggregates.

Type	SiO_2	Al_2O_3	Fe_2O_3	CaO	MgO	TiO_2	Na_2O	K_2O	P_2O_5	MnO	Ignition Loss
Limestone	19.31	9.5	13.2	24.17	4.34	1.96	1.07	1.28	0.94	0.29	23.62
Basalt	47.32	15.0	16.36	8.21	3.76	1.44	2.13	0.44	0.21	0.17	1.94
Granite	72.05	12.81	2.13	0.75	0.11	0.07	2.81	4.63	0.04	0.02	2.88

2.2. Preparation of Rubberized Asphalt Binder

The asphalt was heated and melted, and then CR powder with a mass of 15%, 20%, or 25% of the asphalt was added. The asphalt was then sheared at 4000 rpm for 60 min at 180 °C, followed by low-speed stirring for 30 min (800 rpm) [23]. The base asphalt was named 90#, and the three rubberized asphalts were named CR-15, CR-20, and CR-25.

2.3. Methodology

2.3.1. Dynamic Shear Rheometer (DSR) Test

The Anton Paar SmartPave 102 DSR was used to test the rheological parameters of the different asphalts. The test adopted a temperature sweep; the temperature range was from 46 °C to 82 °C. The test was performed once at an interval of 6 °C; the frequency was 10 rad/s, and the strain was controlled to 1.5% to ensure that the asphalt's rheological behavior was within the linear viscoelastic (LVE) range [40,41]. Three replicate experiments were performed on the same sample to eliminate accidental errors.

2.3.2. Binder Bond Strength (BBS) Test

The BBS test is conducted based on AASHTO TP-91 [42], and a American Defelsko Positest AT-A adhesion tester was used to evaluate the binder bond strength of different kinds of asphalt. The diameter of the stub was 20 mm and the tensile strength loading rate was 0.7 MPa/s. The thickness of the asphalt film was maintained by the crumb rubber gasket at 0.8 mm [23]. At this time, in addition to adhesion failure, cohesive ductile damage still interfered. The aggregates' base materials were limestone, basalt, or granite, as shown in Figure 1. Prior to the test, each specimen was subjected to 48 h of moisture conditioning in a 40 °C water bath. In water conditioning, 15 h of conditioning can affect the bond strength and failure mode. As a result, most samples exhibited an adhesive failure. The peak tensile strength was recorded to quantitatively evaluate the adhesion properties between the different asphalts and aggregates [25]. Pull-off tensile strength (POTS) is the maximum tensile strength of the stub pulling away from the aggregates in the BBS test.

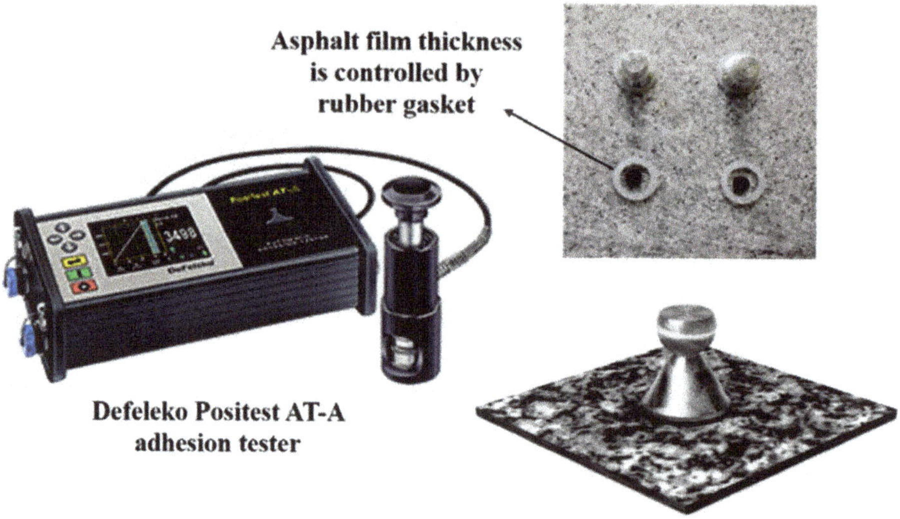

Figure 1. The process of the BBS test.

2.3.3. Surface Free Energy (SFE) Theory

The surface energy consists of two parts: the dispersion component and the polar component. The expression is shown in Equation (1).

$$\gamma = \gamma^d + \gamma^p \tag{1}$$

where γ is the surface free energy (mJ/m^2), γ^d is the dispersive component (mJ/m^2), and γ^p is the polarity component (mJ/m^2).

$$W_a = 2\gamma \tag{2}$$

The cohesion energy or cohesive bond energy (W_a) is defined as the value of the energy needed to create two new surfaces with unit areas [43,44]. A higher value of cohesion energy implies a higher level of energy needed for propagating a crack and fracturing the material into two new surfaces [45].

The surface energy of asphalt can be calculated using Equations (3)–(5), which can be obtained from Young's equation and surface energy theory [46].

$$\gamma_l \cos\theta = \gamma_s - \gamma_{sl} \tag{3}$$

$$\gamma_{sl} = \gamma_s + \gamma_l - 2\sqrt{\gamma_s{}^d \gamma_l{}^d} - 2\sqrt{\gamma_s{}^p \gamma_l{}^p} \tag{4}$$

$$\frac{1+\cos\theta}{2}\frac{\gamma_l}{\sqrt{\gamma_l{}^d}} = \sqrt{\gamma_s{}^p}\sqrt{\frac{\gamma_l{}^p}{\gamma_l{}^d}} + \sqrt{\gamma_s{}^d} \tag{5}$$

where γ_s, γ_l, and γ_{sl} are the surface free energy of the solid, liquid, and solid–liquid phases, respectively (mJ/m^2); $\gamma_l{}^d$ and $\gamma_l{}^p$ express the dispersion component and polar component of the surface energy of the liquid phase, respectively (mJ/m^2); $\gamma_s{}^p$ and $\gamma_s{}^d$ are the dispersion component and polar component of the solid (asphalt) phase, respectively (mJ/m^2); θ is the angle connecting the solid–liquid interface; and γ_b is the surface free energy of the aggregate (mJ/m^2).

In Equation (5), $\frac{(1+\cos\theta)\gamma_l}{2\sqrt{\gamma_l{}^d}}$ can be regarded as the y coordinate and $\sqrt{\frac{\gamma_l{}^p}{\gamma_l{}^d}}$ can be regarded as the x coordinate. The measured contact angle and the surface energy data of the three liquids can be substituted into the x and y coordinates for linear fitting. The square of the slope of the fitting line is the polar component of the solid surface energy. The square of the fitting intercept is the dispersion component of the surface energy. The sum of the two is the total surface energy.

The work of adhesion is used to evaluate the difficulty of water penetrating the asphalt film into the binder–aggregate interface of the asphalt mixture, and the work of adhesion for asphalt–aggregate systems can be calculated as shown in Equation (6):

$$W_{as} = \gamma_l(1+\cos\theta) \tag{6}$$

where W_{as} is the work of adhesion between the asphalt and the solid (limestone) (mJ/m^2).

In the actual measurement, the actual heating temperature, drop height, and droplet size of the asphalt are difficult to control, so Equation (7) can be used to calculate the asphalt–aggregate adhesion work.

$$W_{as} = 2\left(\sqrt{\gamma_s{}^d \gamma_a{}^d} + \sqrt{\gamma_s{}^p \gamma_a{}^p}\right) \tag{7}$$

where $\gamma_s{}^d$ and $\gamma_s{}^p$ express the dispersion component and polar component of the surface energy of the solid (limestone) phase, respectively (mJ/m^2), while $\gamma_a{}^p$ and $\gamma_a{}^d$ are the dispersion component and polar component of the asphalt, respectively (mJ/m^2).

2.3.4. Contact Angle Test of Asphalt Samples

The surface free energy of asphalts with different rubber powder contents was tested by using a German DataPhysics dynamic surface tensiometer, as shown in Figure 2. A glass slide with a flat asphalt film was formed by the Wilhelmy hanging plate method, and the contact angle of the asphalt was measured [39]. The surface free energy of the asphalt was quantified using two liquids with known surface energies that are insoluble in asphalt and do not chemically react with asphalt. The test was conducted at a temperature of 25 °C, and each set of experiments was run in parallel with three times to rule out accidental errors.

2.3.5. Contact Angle Test of Aggregates

The contact angle of the aggregates was tested based on the static contact angle, and three probe solutions of distilled water, ethylene glycol, and glycerol were also used. The surface free energy parameters of the aggregates were calculated according to the abovementioned surface energy theory. Before the aggregate contact angle test, 200-mesh, 400-mesh, and 1000-mesh sandpapers were used to preliminarily grind one side of the aggregate slices to avoid contact angle lag caused by the rough surface of the aggregates. In this test, a JC000D1 contact angle tester was used for testing, and the experimental temperature was 25 °C. After the contact angle test and calculation, the SFE parameters of the three aggregates were obtained.

Figure 2. Dynamic surface tensiometer.

2.3.6. Atomic Force Microscopy (AFM)

Microscopic images and nanomechanical characterizations of the four asphalt surfaces were obtained using a Bruker Dimension Icon Atomic Force Microscope (AFM), as shown in Figure 3a. The selected probe cantilever was a TAP300-G with a thickness of 4 μm, a width of 30 μm, a length of 125 μm, a nominal spring constant of 40 N/m, and a nominal resonance frequency of 300 kHz. The probe was uncoated, and the probe tip was made of monolithic silicon. The scanning frequency of the probe was set to 1.0 Hz, and different kinds of asphalt surfaces were scanned in tapping mode to obtain 20 μm × 20 μm topographic images and force curves of the asphalt surfaces [47]. Typically, three primary microstructures developed on the asphalt surface at around room temperature after annealing of asphalt from its melting temperature. The wrinkled areas were named the catanaphase (bee structures), the islands around the wrinkled domains were called the periphase, and the paraphase was the smoother phase neighboring the periphase, as shown in Figure 3b. The AFM images were analyzed by using the software Nanoscope Analysis 1.9 to quantitatively calculate the roughness of the samples. The adhesion force was determined from the measured force curve, as shown in Figure 3c.

2.3.7. Cohesive and Adhesive

As shown in Figure 4, the cohesion energy or cohesive bond energy is defined as the value of the energy needed to create two new surfaces with unit areas [43]. The amount of energy required for deboning the binder–aggregate interface of the asphalt is called the adhesion energy (or adhesive bond energy) [43,44].

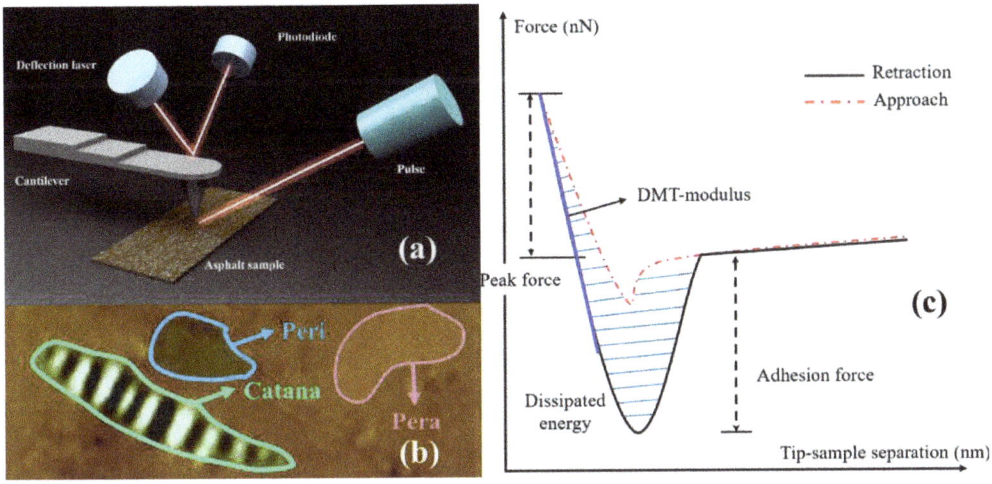

Figure 3. Operational principles of AFM: (**a**) operational principle of AFM; (**b**) schematic diagram of the three phases; (**c**) schematic force curve.

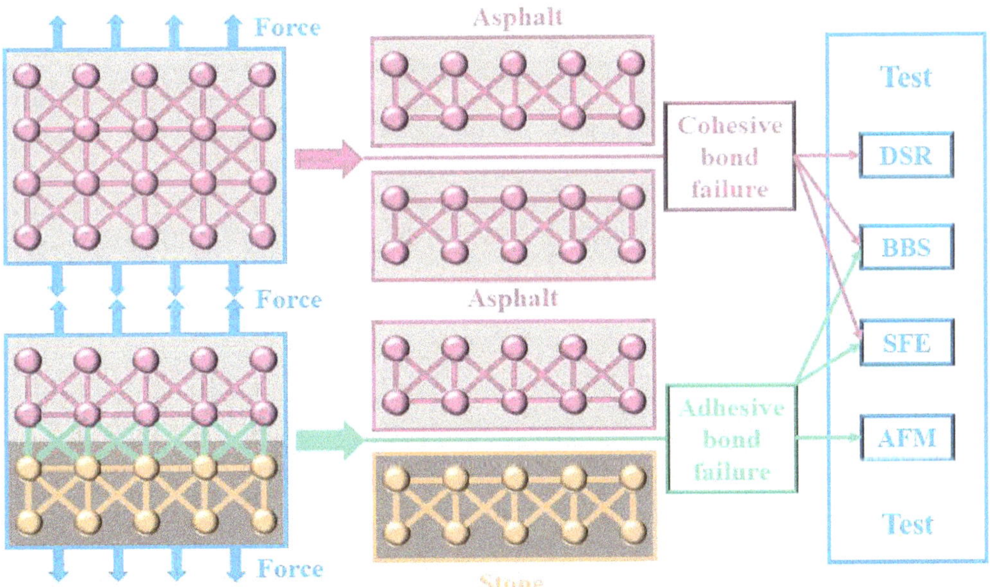

Figure 4. Cohesive and adhesive bond failure.

3. Results and Discussion

3.1. Analyses of Rheological Parameters

The rheological parameters of the four different binders under temperature sweep are shown in Figure 5. From Figure 5a, it can be seen that the phase angle of the base asphalt is relatively high, at 80–90° in the tested temperature range, while the phase angle of the rubberized asphalt is relatively low. The phase angle of CR-15 is within 60–80°, and it is greatly affected by temperature changes. The phase angle of CR-20 and CR-25 does not change much over the tested temperature range, within 50–60°. Unlike the base asphalt and CR-15, the phase angle of CR-20 and CR-25 decreases slightly within 76–82 °C. In general, the crumb rubber powder can reduce the phase angle of the binder.

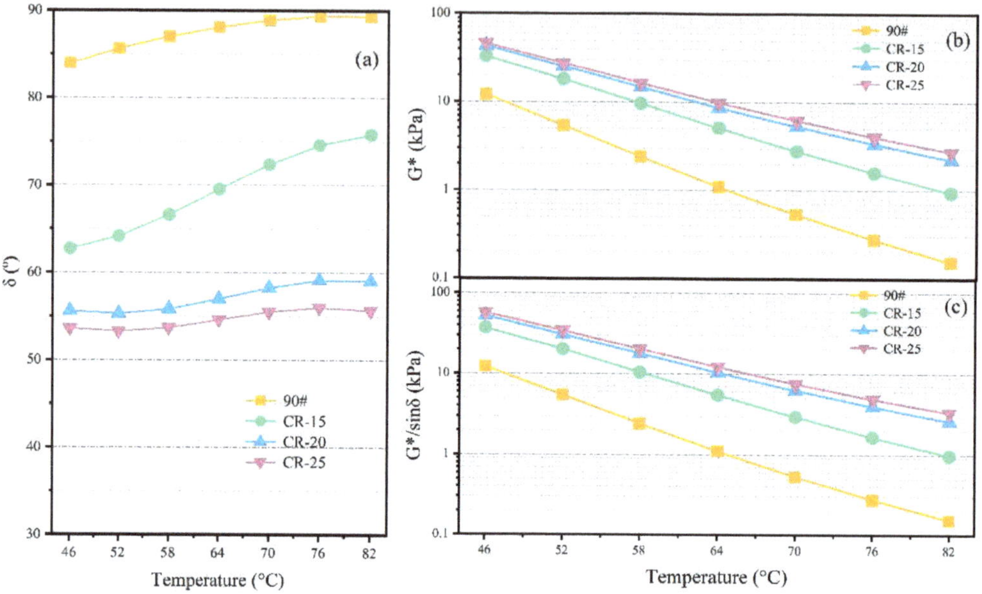

Figure 5. Results of the DSR test: **(a)** δ, **(b)** G*, and **(c)** G*/sin δ.

Figure 5b shows the variation in the complex shear modulus for the different binders. It can be intuitively found from the figure that the complex shear modulus of the rubberized asphalt is several times that of the base asphalt at the same test temperature, and the gap is further expanded at high temperatures. Similar to the change in the phase angle, the complex shear moduli of CR-20 and CR-25 are similar, which indicates that 20% CR powder content can achieve a relatively stable CRMA system. The addition of CR powder can improve the complex shear modulus.

Figure 5c shows the variation in the rutting factor of different binders. The rutting factor can reflect the ability of the asphalt binder to resist permanent deformation at high temperatures. The variation in the rutting factor of the four binders is essentially the same as the variation trend of the complex shear modulus. In general, the rubberized asphalt improves the high-temperature rheological properties of the binder.

3.2. Pull-Off Tensile Strength (POTS) Analyses

The results of the POTS between the different rubberized asphalts and limestone, basalt, and granite are shown in Figure 6. It can be clearly seen from the figure that the CR powder can have a negative impact on the adhesion between the asphalt binder and the aggregate. Comparing the adhesion between the three kinds of rubberized asphalt and limestone, it can be found that the POTS of CR-15, CR-20, and CR-25 decreases by 42.1%, 52.6%, and 56.1%, respectively, compared with the base asphalt.

The addition of more CR powder has a greater negative impact on the adhesion performance of the asphalt binder, and similar results can also be found in SBS-modified asphalt [25]. The reason for these results is that additives such as CR powder can have a negative impact on the homogeneity of the asphalt. The CR powder is not inherently sticky, so it does not improve adhesion in the rubberized asphalt–binder system during POTS test. In addition, the CR powder has certain volume characteristics that are blended in physical form in the rubberized asphalt. The CR powder occupies a certain contact area at the interface between the rubberized asphalt and the aggregate, and the adhesion force brought by this part of the contact area is lower than the adhesion force between the asphalt and the aggregate, so the adhesion performance between the rubberized asphalt and the aggregate deteriorates [23].

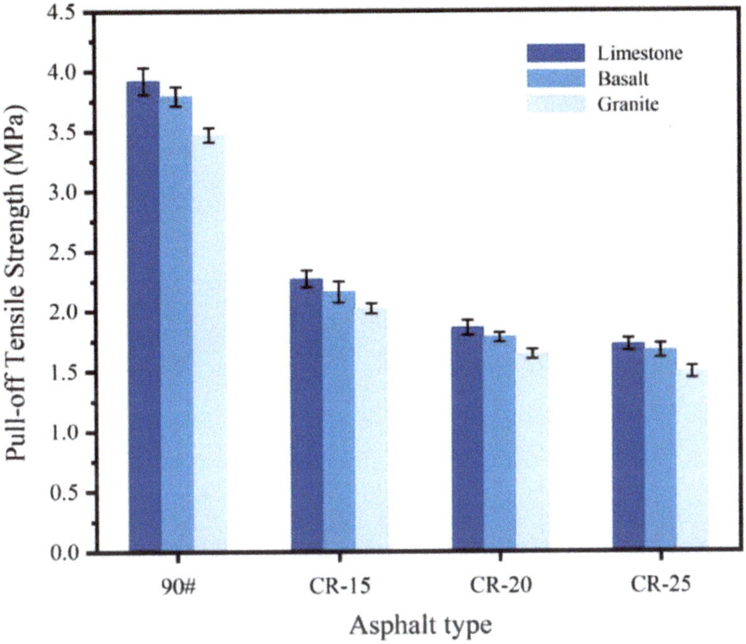

Figure 6. BBS test results of different rubberized asphalts.

There are also obvious differences in the adhesion between rubberized asphalt and different aggregates. It can be seen from the Figure 6 that the adhesion between rubberized asphalt and limestone is the best, followed by basalt, while the adhesion with granite is relatively poor. Taking CR-20 as an example, its POTS with granite is 1.64 MPa, and its POTS with basalt is 8.5% higher than that with granite, reaching 1.78 MPa, while its POTS with limestone is 13.4% higher than that with granite, reaching 1.86 MPa. The above results are caused by the differences in the properties of the different lithological aggregates, including the differences in the surface texture and composition of the aggregates [16].

3.3. *Surface Free Energy (SFE) Analyses*

Figure 7 shows the calculation results of the SFE parameters of the four different asphalts. The polar components of the four asphalts are all much smaller than the dispersion components. Overall, the polar component decreases with the increase in dosage. The dispersive component and total SFE of the asphalts with the higher crumb rubber powder contents are higher. The total surface energy of the four asphalts is between 23 mJ/m^2 and 27 mJ/m^2. Surface energy theory states that the surface energy of a substance in a stable state is low. As shown in Equation (2), asphalt with a higher surface energy has higher cohesive energy. Based on the SFE parameters of the base asphalt and the three kinds of rubberized asphalt, 90# $\approx$ CR-15 < CR-20 $\approx$ CR-25.

The polar components of these four asphalts are 1.66, 1.667, 0.321, and 0.0968, respectively, and the change rule is a decreasing trend. Some scholars believe that the rubberized asphalt has a negative impact on the smooth glass slide, and samples with rough surfaces can easily lead to inaccurate test results [25]. In this work, the dynamic contact angle test method was used to continuously test the asphalt slides to avoid the uncertainty of a single static contact angle. With the increase in rubber powder content, the polar component decreases, the dispersion component increases, and the cohesive energy increases. Therefore, it is recommended to use the dynamic contact angle to test the SFE parameters of rubberized asphalt.

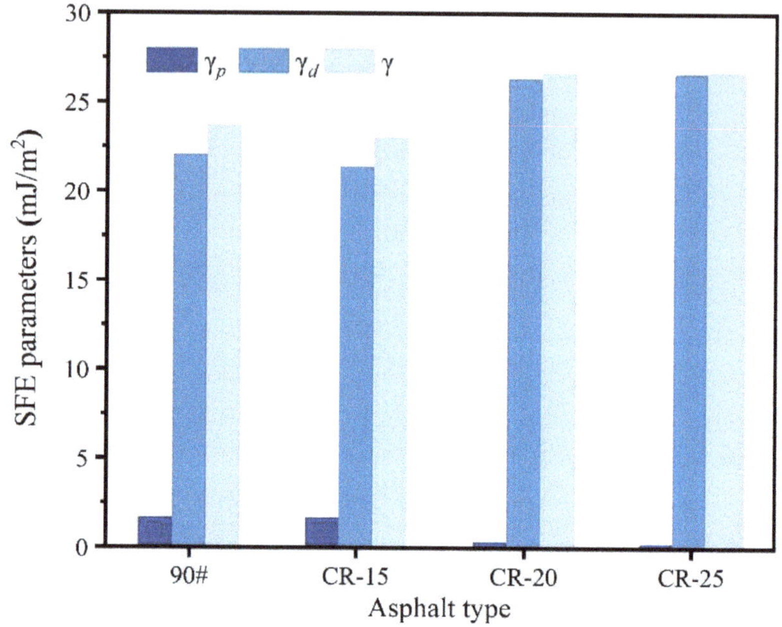

Figure 7. SFE parameters of different rubberized asphalts.

From the SFE parameters of three kinds of aggregates, as shown in Table 4, it can be seen that the surface energy of limestone is greater than that of basalt, while the surface energy of granite is the lowest. The maximum polar component of limestone is 21.37 mJ/m^2, and the minimum polar component of granite is 16.91 mJ/m^2. Figure 8 shows the calculation results of the work of adhesion between the different asphalts and the three aggregates. The four kinds of asphalt have the strongest adhesion to limestone, followed by basalt, and the worst adhesion to granite. Taking the base asphalt as an example, the work of adhesion of base asphalt to limestone, basalt, and granite is 51.031 mJ/m^2, 49.291 mJ/m^2, and 44.917 mJ/m^2, respectively. As an acidic substance, the adhesion of asphalt to the aggregate is greatly affected by the acidity and alkalinity of the aggregate. The adhesion between alkaline aggregates and asphalt is better than that of acidic aggregates [48].

Table 4. Surface free energy parameters of the three aggregates.

Type	γ_d (mJ/m^2)	γ_p (mJ/m^2)	γ (mJ/m^2)
Limestone	17.43	21.41	38.64
Basalt	16.22	19.71	35.93
Granite	13.28	16.71	30.17

In addition, from the perspective of asphalt, the addition of CR powder can reduce the work of adhesion between the rubberized asphalt and the aggregate, thereby weakening the adhesion performance. Taking the work of adhesion between the different asphalts and basalt as an example, the work of adhesion between the base asphalt and basalt was 49.29 mJ/m^2, while the work of adhesion between CR-15, CR-20, and CR-25 and basalt was 48.75 mJ/m^2, 46.38 mJ/m^2, and 44.25 mJ/m^2, respectively. The adhesion of CR-15, CR-20, and CR-25 to basalt was 1.10%, 5.91%, and 10.22% lower than that of base asphalt, respectively. The reason for this is also because the crumb rubber powder, which does not have adhesive properties, occupies a certain area of the adhesive interface between the asphalt and the aggregate.

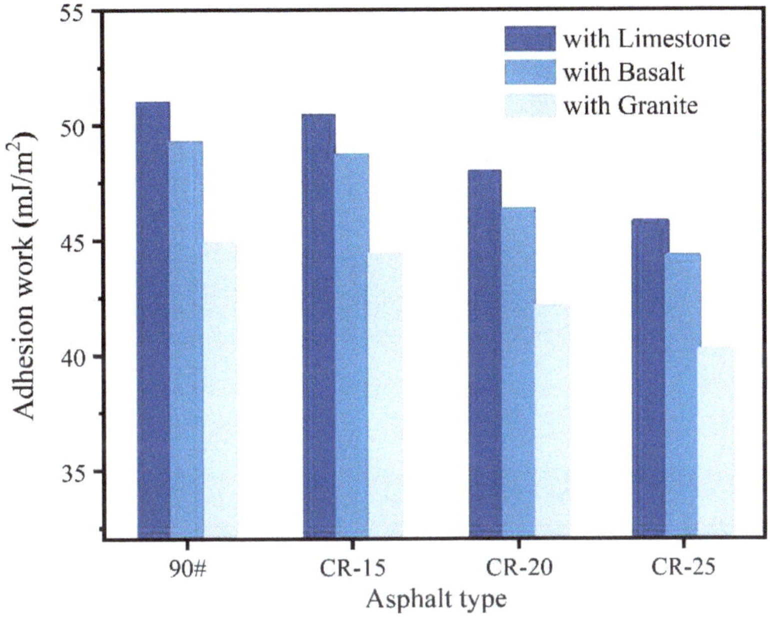

Figure 8. Work of adhesion of different rubberized asphalts.

3.4. Micromorphological Analyses

The modification with crumb rubber particles absorbed the light components of the asphalt and swelled, forming a uniform interconnection network in the asphalt system [49]. Rubber powder modification has a mainly physical effect [50]. The asphalt–rubber interaction stages can be divided into three steps, as shown in Figure 9 [51,52].

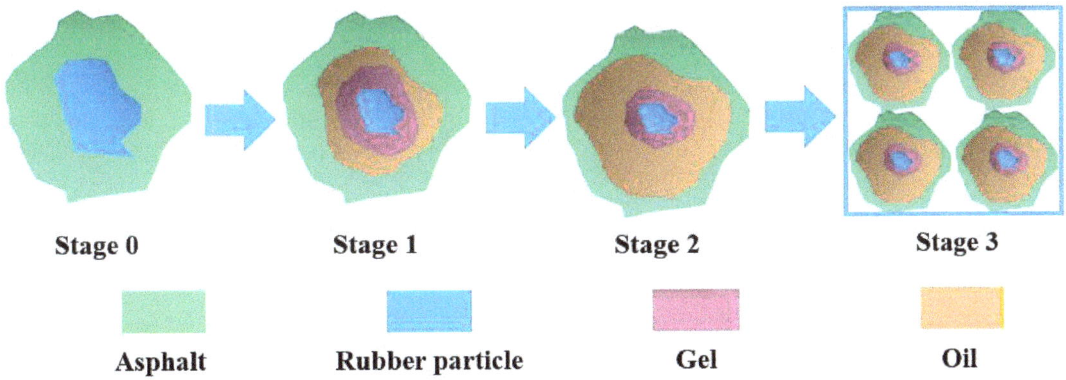

Figure 9. The asphalt–rubber interaction stages.

Stage 0—initial configuration: Rubber particles are immersed in the fluid asphalt.

Stage 1—swelling phase: Rubber particles start swelling by absorbing the light fractions of bitumen and form a gel layer adjacent to the bitumen–rubber interface.

Stage 2—post-swelling and beginning of degradation: The swelling of the rubber particles continues. Meanwhile, chemical degradation takes place through the breakup of the crosslinked network and polymer chains. Swollen rubber particles are split into smaller ones due to the destruction of the network structure.

Stage 3—degradation and complete dissolution: The degradation of the rubber particles continues progressing until they are completely dissolved into the bitumen matrix, which produces a homogeneous binder.

Regarding the microstructure of rubberized asphalt, some scholars believe that the dispersion of CR powder after absorbing the light components hinders the aggregation of asphaltenes, resulting in a reduction in the catanaphase and difficulty in identification [53]. However, there is evidence that an increase in the oil content can actually reduce the catanaphase [54]. Some studies suggest that asphaltenes do not play a decisive role in the formation of the catanaphase [28,55]. Thus, the reduction in light oil adsorption is related to the relevance of the reduced catanaphase is questionable. The investigation of the effect of crumb rubber on the microstructure using atomic force microscopy–infrared spectroscopy (AFM-IR) indicated that the main chemical change takes place in the paraphase [27], but that chemical change is not the main mechanism of rubber modification. Therefore, the nanomorphological changes in crumb-rubber-modified asphalt still tend to be physical changes caused by the unbalanced stress between phases.

Figure 10 shows the AFM images of different crumb rubber powder dosages. The catanaphase has obvious characteristic changes. The CR powder can be clearly seen in the three-dimensional image with the CR powder dosage of 25%. The change in the apparent structure may be more due to the change in the interfacial tension caused by the floating and agglomeration of the micro-rubber powder [56]. When the dosage is 15% and 20%, the catanaphase is broken; the details can be seen in Figure 10.

The common roughness indices are S_a, S_q, and S_Z, which are shown in Table 5 [57].

Table 5. Three-dimensional (3D) roughness parameters used in this study.

Parameter	Describe	Formula
S_a	Roughness average	$S_a = \frac{1}{A}\iint\limits_{A} \lvert Z(x,y)\rvert dxdy$
S_q	Root-mean-square roughness	$S_a = \sqrt{\frac{1}{A}\iint\limits_{A} \lvert Z(x,y)\rvert dxdy}$
S_Z	Maximum height of the roughness	$S_Z = S_p - S_V$
S_p	Maximum roughness peak height	Z_{max}
S_V	Maximum roughness valley depth	Z_{min}

As shown in Figure 11, it can be seen that the three indicators are consistent for the roughness changes with different dosages of CR powder. The smaller the roughness value, the smaller the difference between phases and the more stable the microstructural properties [58]. As shown in Figure 11, the roughness value is the highest and the microscopic morphology is the most unstable when the rubber powder content is 25%. The roughness value increases continuously with the increase in the rubber powder content. This shows that the content of rubber powder particles will continuously destroy the apparent morphology of the asphalt and reduce the stability of its microscopic properties. The roughness values did not change significantly at low rubber powder contents.

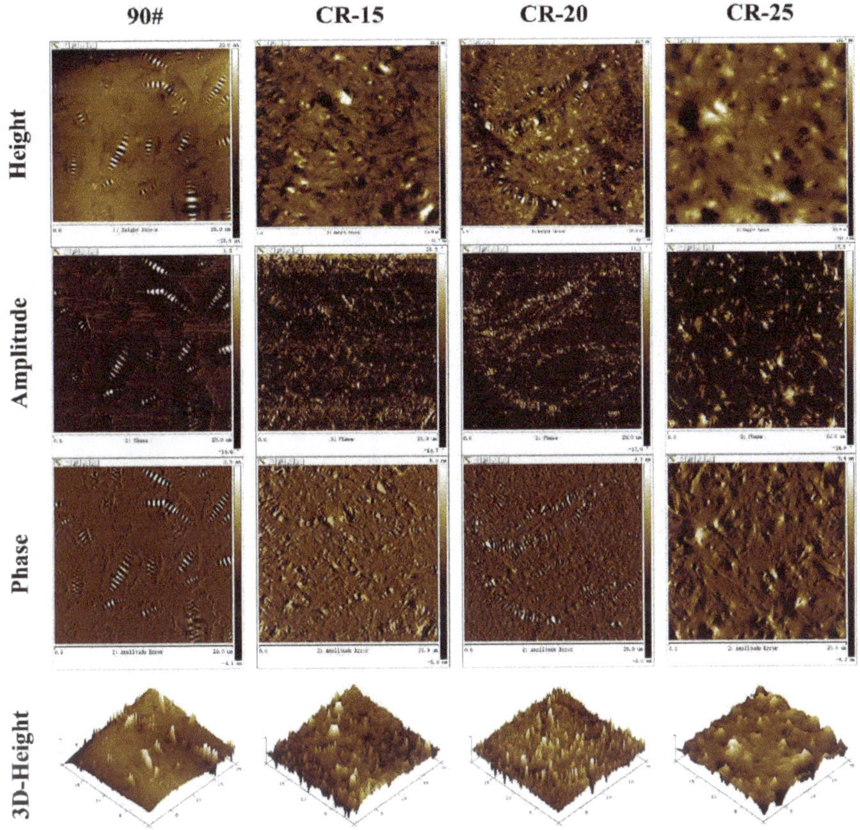

Figure 10. AFM diagrams of different rubberized asphalts.

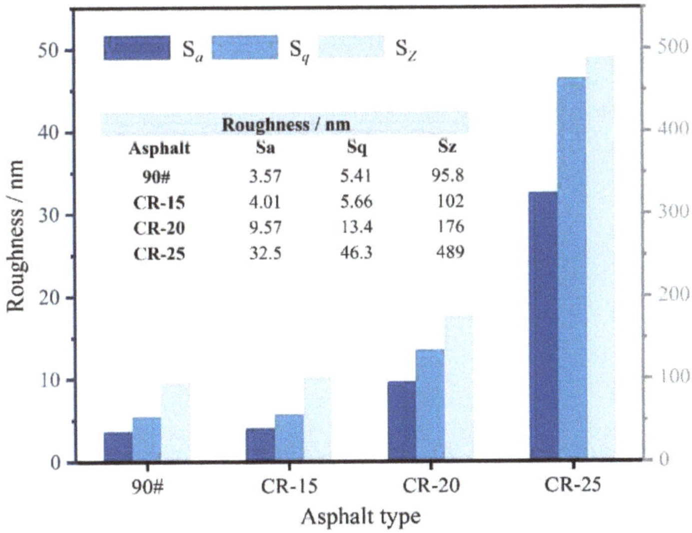

Figure 11. Roughness at different crumb rubber dosages.

The schematic diagram of the scatter of morphological changes in the catanaphase is shown in Figure 12. The area is the area of the catanaphase, and the aspect is the ratio of the major and minor axes of the catanaphase. With the increase in the amount of CR powder, the slender catanaphase of the base asphalt becomes more dispersed when the amount of CR powder is 15%, and the length of the catanaphase also decreases. When the dosage is 20%, the original large catanaphase is broken into several smaller sections of catanaphase by the crumb rubber powder. When the dosage reaches 25%, the catanaphase effectively does not exist, and even if the catanaphase exists, it is blurred and counted in order to have a certain contrast, and it can actually be considered to be non-existent.

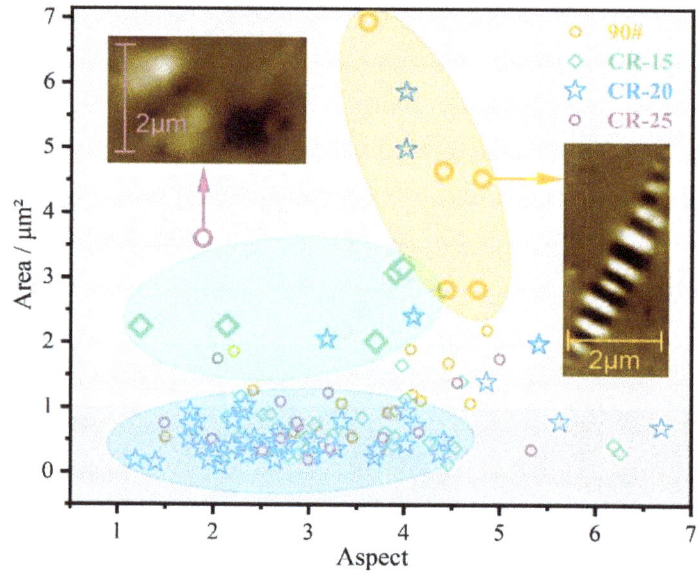

Figure 12. Morphological transformation of the catanaphase.

As shown in Figure 13, the adhesive force shows a trend of first increasing and then decreasing with the increase in the CR powder content. Because the probe itself will be affected by van der Waals forces, the rubber forms a gel structure to improve the cohesion by absorbing the light components. The cohesion of the rubber-modified asphalt interferes with the test results, such that the adhesion of the rubber-modified asphalt is greater than that of the original asphalt. However, the comparison rule of rubber-modified asphalt is consistent with the previous test, indicating that atomic force microscopy is more suitable for the comparison of two-phase systems. When the dosage is higher, the CR microparticles can aggregate on the surface of the asphalt, destroying the surface tension. The stability of microscopic properties decreases, leading to a decrease in adhesion. In this process, the catanaphase is squeezed and broken by the continuously aggregated microgel powder particles until it completely disappears. The change trend of the catanaphase change coefficient Tb (as shown in Equation (8)) is consistent with the change in the adhesive force, indicating that the microstructure is strongly related to the performance, and the change in the catanaphase can be used to evaluate the modification effect of the CR powder.

$$T_b = n \times \overline{A_s}^{-3} \tag{8}$$

where n is the number of small catanaphase, and $\overline{A_s}$ is the average aspect of the catanaphase.

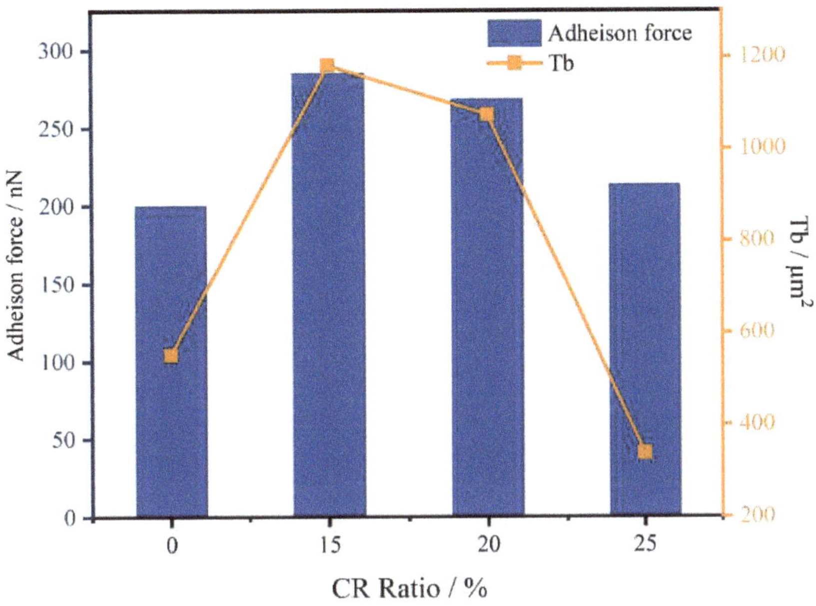

Figure 13. Adhesion force and catanaphase shape parameters.

4. Conclusions

In this work, the adhesion characteristics of the original asphalt and three kinds of rubberized asphalt were studied. The rheological properties of the different asphalts were characterized by DSR, while the adhesion characteristics of the rubberized asphalts were analyzed from three perspectives by BBS tests, contact angle tests, and AFM. The following conclusions can be drawn:

1. The incorporation of CR can improve the complex shear modulus and reduce the phase angle of the asphalt. CR can significantly improve the stiffness, modulus, and cohesive energy of asphalt, thereby improving the high-temperature elastic properties of asphalt, which is also an important prerequisite for its wide application.
2. According to the BBS test results, CR has a negative effect on the pull-off tensile strength of asphalt and aggregates. CR itself does not have adhesive performance and can occupy part of the contact area at the interface between the asphalt and the aggregate, resulting in a deterioration in the adhesive properties of the rubberized asphalt.
3. CR can improve the SFE parameters of asphalt, and the changes in the total surface energy and dispersion components are significantly affected by the changes in CR content. The work of adhesion between asphalt and limestone is the highest, followed by basalt and, finally, granite, due to the differences in chemical composition between the different aggregates.
4. The large catanaphase of asphalt with higher CR contents is continuously broken with the increase in dosage. The roughness value increases with the increase in the dosage. CR-25 had the highest roughness value and the worst microscopic properties. The roughness values did not change significantly at low rubber powder contents.
5. The adhesion deteriorates with the increase in CR content. AFM is more suitable for the comparison of two-phase systems. The change factor of the catanaphase is consistent with the change trend of the adhesion force, and the microscopic morphology has a strong correlation with the change in the adhesion performance.

5. Further Study

In this work, a variety of test methods were used to evaluate the polymerization energy and adhesion energy of modified asphalt with different rubber powder contents, and the relationships between different test indices and cohesive energy and adhesion energy were explored. However, this work did not consider the effect of aggregate texture on adhesion performance, and it also did not explore the role of moisture. The evaluation of the adhesion properties of rubber powder still needs further verification.

Author Contributions: Conceptualization, B.Y. and H.J.; software, H.J.; investigation, H.J.; data curation, H.J. and J.L.; writing—original draft preparation, H.J. and J.L.; writing—review and editing, H.J. and Z.W.; supervision, X.W. and Z.W.; funding acquisition, X.W. and Z.W. All authors have read and agreed to the published version of the manuscript.

Funding: This work was supported by the Central Plains Thousand People Plan of Henan Province (No. 204200510004), the Science and Technology Development Project of Shaanxi (No. 2022GY-417), the Project of Xinjiang Production and Construction Corps (No. 2021DB005), and the Shaanxi Transportation Technology Project (No. 21-53K). The authors also thank the reviewers for their valuable comments and suggestions concerning our manuscript.

Institutional Review Board Statement: Not applicable.

Data Availability Statement: All data are contained within the article.

Conflicts of Interest: The authors declare no conflict of interest.

References

1. Yuan, D.; Jiang, W.; Xiao, J.; Tong, Z.; Jia, M.; Shan, J.; Ogbon, A.W. Assessment of the Aging Process of Finished Product–Modified Asphalt Binder and Its Aging Mechanism. *J. Mater. Civ. Eng.* **2022**, *34*, 04022174. [CrossRef]
2. Xu, P.; Gao, J.; Pei, J.; Chen, Z.; Zhang, J.; Li, R. Research on highly dissolved rubber asphalt prepared using a composite waste engine oil addition and microwave desulfurization method. *Constr. Build. Mater.* **2021**, *282*, 122641. [CrossRef]
3. Ameli, A.; Maher, J.; Mosavi, A.; Nabipour, N.; Babagoli, R.; Norouzi, N. Performance evaluation of binders and Stone Matrix Asphalt (SMA) mixtures modified by Ground Tire Rubber (GTR), waste Polyethylene Terephthalate (PET) and Anti Stripping Agents (ASAs). *Constr. Build. Mater.* **2020**, *251*, 118932. [CrossRef]
4. Liu, J.; Zhang, T.; Guo, H.; Wang, Z.; Wang, X. Evaluation of self-healing properties of asphalt mixture containing steel slag under microwave heating: Mechanical, thermal transfer and voids microstructural characteristics. *J. Clean. Prod.* **2022**, *342*, 130932. [CrossRef]
5. Liu, J.; Wang, Z.; Guo, H.; Yan, F. Thermal transfer characteristics of asphalt mixtures containing hot poured steel slag through microwave heating. *J. Clean. Prod.* **2021**, *308*, 127225. [CrossRef]
6. Yuan, D.; Jiang, W.; Sha, A.; Xiao, J.; Shan, J.; Wang, D. Energy output and pavement performance of road thermoelectric generator system. *Renew. Energy* **2022**, *201*, 22–33. [CrossRef]
7. Liu, J.; Wang, Z.; Li, M.; Wang, X.; Wang, Z.; Zhang, T. Microwave heating uniformity, road performance and internal void characteristics of steel slag asphalt mixtures. *Constr. Build. Mater.* **2022**, *353*, 129155. [CrossRef]
8. Li, J.; Xiao, F.; Amirkhanian, S.N. High temperature rheological characteristics of plasma-treated crumb rubber modified binders. *Constr. Build. Mater.* **2020**, *236*, 117614. [CrossRef]
9. Mohamed, A.S.; Wang, W.; Weng, H.; Fang, Y.; Xiao, F. Impacts of colloidal and interfacial interactions on bonding strength of rubberized RAP asphalt binder. *Constr. Build. Mater.* **2022**, *349*, 128768. [CrossRef]
10. Li, B.; Zhu, X.; Zhang, X.; Yang, X.; Su, X. Surface area and microstructure of microwave activated crumb rubber modifier and its influence on high temperature properties of crumb rubber modifier binders. *Mater. Express* **2020**, *10*, 272–277. [CrossRef]
11. Zhou, T.; Zhou, J.; Li, Q.; Li, B. Aging Properties and Mechanism of Microwave-Activated Crumb Rubber Modified Asphalt Binder. *Front. Mater.* **2020**, *7*, 603938. [CrossRef]
12. Lei, Y.; Wei, Z.; Wang, H.; You, Z.; Yang, X.; Chen, Y. Effect of crumb rubber size on the performance of rubberized asphalt with bio-oil pretreatment. *Constr. Build. Mater.* **2021**, *285*, 122864. [CrossRef]
13. Liu, J.; Hao, P.; Dou, Z.; Wang, J.; Ma, L. Rheological, healing and microstructural properties of unmodified and crumb rubber modified asphalt incorporated with graphene/carbon black composite. *Constr. Build. Mater.* **2021**, *305*, 124512. [CrossRef]
14. Ma, Y.; Wang, S.; Zhou, H.; Hu, W.; Polaczyk, P.; Zhang, M.; Huang, B. Compatibility and rheological characterization of asphalt modified with recycled rubber-plastic blends. *Constr. Build. Mater.* **2021**, *270*, 121416. [CrossRef]
15. Li, D.; Leng, Z.; Zou, F.; Yu, H. Effects of rubber absorption on the aging resistance of hot and warm asphalt rubber binders prepared with waste tire rubber. *J. Clean. Prod.* **2021**, *303*, 127082. [CrossRef]
16. Liu, J.; Wang, Z.; Zhao, X.; Yu, C.; Zhou, X. Quantitative evaluations on influences of aggregate surface texture on interfacial adhesion using 3D printing aggregate. *Constr. Build. Mater.* **2022**, *328*, 127022. [CrossRef]

17. Tayebali, A.A.; Kusam, A.; Bacchi, C. An Innovative Method for Interpretation of Asphalt Boil Test. *J. Test. Eval.* **2018**, *46*, 1622–1635. [CrossRef]
18. Yu, D.; Jing, H.; Liu, J. Effects of Freeze-Thaw Cycles on the Internal Voids Structure of Asphalt Mixtures. *Materials* **2022**, *15*, 3560. [CrossRef]
19. Nazirizad, M.; Kavussi, A.; Abdi, A. Evaluation of the effects of anti-stripping agents on the performance of asphalt mixtures. *Constr. Build. Mater.* **2015**, *84*, 348–353. [CrossRef]
20. Hossain, Z.; Bairgi, B.; Belshe, M. Investigation of moisture damage resistance of GTR-modified asphalt binder by static contact angle measurements. *Constr. Build. Mater.* **2015**, *95*, 45–53. [CrossRef]
21. Partl, M.N.; Pasquini, E.; Canestrari, F.; Virgili, A. Analysis of water and thermal sensitivity of open graded asphalt rubber mixtures. *Constr. Build. Mater.* **2010**, *24*, 283–291. [CrossRef]
22. Lv, Q.; Huang, W.; Zheng, M.; Hao, G.; Yan, C.; Sun, L. Investigating the asphalt binder/mastic bonding healing behavior using bitumen bonding strength test and X-ray Computed Tomography scan. *Constr. Build. Mater.* **2020**, *257*, 119504. [CrossRef]
23. Lv, Q.; Lu, J.; Tang, X.; Hu, Y.; Yan, C. Evaluation of the moisture resistance of rubberized asphalt using BBS/UTM bonding test, TSR and HWT test. *Constr. Build. Mater.* **2022**, *340*, 127831. [CrossRef]
24. Zhou, L.; Huang, W.; Xiao, F.; Lv, Q. Shear adhesion evaluation of various modified asphalt binders by an innovative testing method. *Constr. Build. Mater.* **2018**, *183*, 253–263. [CrossRef]
25. Zhou, L.; Huang, W.; Zhang, Y.; Lv, Q.; Sun, L. Mechanical evaluation and mechanism analysis of the stripping resistance and healing performance of modified asphalt-basalt aggregate combinations. *Constr. Build. Mater.* **2021**, *273*, 121922. [CrossRef]
26. Xie, S.; Yi, J.; Zhao, H.; Feng, D.; Sun, Z. Micromorphology and Micromechanical Properties Evolution of Bitumen and Bitumen Fractions Using Atomic Force Microscopy Considering Temperature Effect. *Energy Fuels* **2021**, *35*, 17434–17445. [CrossRef]
27. Rodriguez-Fernandez, I.; Baheri, F.T.; Cavalli, M.C.; Poulikakos, L.D.; Bueno, M. Microstructure analysis and mechanical performance of crumb rubber modified asphalt concrete using the dry process. *Constr. Build. Mater.* **2020**, *259*, 119662. [CrossRef]
28. Fischer, H.R.; Dillingh, E.C.; Hermse, C.G.M. On the microstructure of bituminous binders. *Road Mater. Pavement Des.* **2014**, *15*, 1–15. [CrossRef]
29. Loeber, L.; Sutton, O.; Morel, J.; Valleton, J.-M.; Muller, G. New direct observations of asphalts and asphalt binders by scanning electron microscopy and atomic force microscopy. *J. Microsc.* **1996**, *182*, 32–39. [CrossRef]
30. Xing, C.; Liu, L.; Cui, Y.; Ding, D. Analysis of base bitumen chemical composition and aging behaviors via atomic force microscopy-based infrared spectroscopy. *Fuel* **2020**, *264*, 116845. [CrossRef]
31. Pipintakos, G.; Hasheminejad, N.; Lommaert, C.; Bocharova, A.; Blom, J. Application of Atomic Force (AFM), Environmental Scanning Electron (ESEM) and Confocal Laser Scanning Microscopy (CLSM) in bitumen: A review of the ageing effect. *Micron* **2021**, *147*, 103083. [CrossRef] [PubMed]
32. Roja, K.L.; Aljarrah, M.F.; Sirin, O.; Al-Nuaimi, N.; Masad, E. Rheological, Thermal, and Chemical Evaluation of Asphalt Binders Modified Using Crumb Rubber and Warm-Mix Additive. *J. Mater. Civ. Eng.* **2022**, *34*, 04022049. [CrossRef]
33. Xu, J.; Sun, L.; Pei, J.; Xue, B.; Liu, T.; Li, R. Microstructural, chemical and rheological evaluation on oxidative aging effect of SBS polymer modified asphalt. *Constr. Build. Mater.* **2021**, *267*, 121028. [CrossRef]
34. Kiggundu, B.M.; Roberts, F.L. *Stripping in Hma Mixtures: State-of-The-Art and Critical Review of Test Methods*; NCAT Report 88-02; National Centre for Asphalt Technology, Auburn University: Auburn, AL, USA, 1988. Available online: https://trid.trb.org/view.aspx?id=485075 (accessed on 1 December 2022).
35. Bagampadde, U.; Isacsson, U.; Kiggundu, B.M. Classical and contemporary aspects of stripping in bituminous mixes. *Road Mater. Pavement Des.* **2004**, *5*, 7–43. [CrossRef]
36. Chaturabong, P.; Bahia, H.U. Effect of moisture on the cohesion of asphalt mastics and bonding with surface of aggregates. *Road Mater. Pavement Des.* **2018**, *19*, 741–753. [CrossRef]
37. Rahim, A.; Thom, N.; Airey, G. Development of compression pull-off test (CPOT) to assess bond strength of bitumen. *Constr. Build. Mater.* **2019**, *207*, 412–421. [CrossRef]
38. *JTG E20-2011*; Standard Test Methods of Bitumen and Bituminous Mixtures for Highway Engineering. Ministry of Transport of the People's Republic of China: Beijing, China, 2011.
39. Yang, J.; Zhang, Z.; Shi, J.; Yang, X.; Fang, Y. Comparative analysis of thermal aging behavior and comprehensive performance of high viscosity asphalt (HVA) from cohesion, adhesion and rheology perspectives. *Constr. Build. Mater.* **2022**, *317*, 125982. [CrossRef]
40. Liu, J.; Wang, Z.; Luo, R.; Bian, G.; Liang, Q.; Yan, F. Changes of components and rheological properties of bitumen under dynamic thermal aging. *Constr. Build. Mater.* **2021**, *303*, 124501. [CrossRef]
41. Yuan, D.; Jiang, W.; Hou, Y.; Xiao, J.; Ling, X.; Xing, C. Fractional derivative viscoelastic response of high-viscosity modified asphalt. *Constr. Build. Mater.* **2022**, *350*, 128915. [CrossRef]
42. *AASHTO T 361*; Standard Method of Test for Determining Asphalt Binder Bond Strength by Means of the Binder Bond Strength (BBS) Test. AASHTO: Washington, DC, USA, 2015.
43. Moraes, R.; Velasquez, R.; Bahia, H. Using bond strength and surface energy to estimate moisture resistance of asphalt-aggregate systems. *Constr. Build. Mater.* **2017**, *130*, 156–170. [CrossRef]
44. Bhasin, A.; Masad, E.; Little, D.; Lytton, R. Limits on Adhesive Bond Energy for Improved Resistance of Hot-Mix Asphalt to Moisture Damage. *Transp. Res. Rec.* **2006**, *1970*, 2–13. [CrossRef]

45. Bhasin, A.; Howson, J.; Masad, E.; Little, D.N.; Lytton, R.L. Effect of Modification Processes on Bond Energy of Asphalt Binders. *Transp. Res. Rec.* **2007**, *1998*, 29–37. [CrossRef]
46. Guo, P.; Feng, Y.; Wei, W.; He, L.; Tang, B. Adhesion of Warm-Mix Recycled Asphalt Aggregate Mixtures Based on Surface Free Energy Theory. *J. Mater. Civ. Eng.* **2019**, *31*, 04019209. [CrossRef]
47. Liu, J.; Qi, L.; Wang, X.; Li, M.; Wang, Z. Influence of aging induced by mutation in temperature on property and microstructure development of asphalt binders. *Constr. Build. Mater.* **2022**, *319*, 126083. [CrossRef]
48. Guo, F.; Pei, J.; Zhang, J.; Xue, B.; Sun, G.; Li, R. Study on the adhesion property between asphalt binder and aggregate: A state-of-the-art review. *Constr. Build. Mater.* **2020**, *256*, 119474. [CrossRef]
49. Xu, A.; Wang, X.; Xiong, R.; Chen, H.; Fang, J.; Kuang, D.; Wang, X.; Liu, Z. Experimental Investigation on Preparation Technology and Performance of Rubber Powder Modified Asphalt. Available online: https://www.webofscience.com/wos/alldb/full-record/CSCD:5979213 (accessed on 1 December 2022).
50. Niu, D.; Xie, X.; Zhang, Z.; Niu, Y.; Yang, Z. Influence of binary waste mixtures on road performance of asphalt and asphalt mixture. *J. Clean. Prod.* **2021**, *298*, 126842. [CrossRef]
51. Wang, S.; Cheng, D.; Xiao, F. Recent developments in the application of chemical approaches to rubberized asphalt. *Constr. Build. Mater.* **2017**, *131*, 101–113. [CrossRef]
52. Wang, H.; Apostolidis, P.; Zhu, J.; Liu, X.; Skarpas, A.; Erkens, S. The role of thermodynamics and kinetics in rubber–bitumen systems: A theoretical overview. *Int. J. Pavement Eng.* **2021**, *22*, 1785–1800. [CrossRef]
53. Yang, X.; Shen, A.; Li, B.; Wu, H.; Lyu, Z.; Wang, H.; Lyu, Z. Effect of microwave-activated crumb rubber on reaction mechanism, rheological properties, thermal stability, and released volatiles of asphalt binder. *J. Clean. Prod.* **2020**, *248*, 119230. [CrossRef]
54. Rafiq, W.; Napiah, M.; Habib, N.Z.; Sutanto, M.H.; Alaloul, W.S.; Khan, M.I.; Musarat, M.A.; Memon, A.M. Modeling and design optimization of reclaimed asphalt pavement containing crude palm oil using response surface methodology. *Constr. Build. Mater.* **2021**, *291*, 123288. [CrossRef]
55. Yu, X.; Burnham, N.A.; Tao, M. Surface microstructure of bitumen characterized by atomic force microscopy. *Adv. Colloid Interface Sci.* **2015**, *218*, 17–33. [CrossRef] [PubMed]
56. Ramm, A.; Downer, M.C.; Sakib, N.; Bhasin, A. Morphology and kinetics of asphalt binder microstructure at gas, liquid and solid interfaces. *J. Microsc.* **2019**, *276*, 109–117. [CrossRef] [PubMed]
57. Long, Z.; Guo, N.; Tang, X.; Ding, Y.; You, L.; Xu, F. Microstructural evolution of asphalt induced by chloride salt erosion. *Constr. Build. Mater.* **2022**, *343*, 128056. [CrossRef]
58. Nahar, S.N.; Schmets, A.J.M.; Scarpas, A.; Schitter, G. Temperature and thermal history dependence of the microstructure in bituminous materials. *Eur. Polym. J.* **2013**, *49*, 1964–1974. [CrossRef]

polymers

MDPI

Article

Biopolymeric Capsules Containing Different Oils as Rejuvenating Agents for Asphalt Self-Healing: A Novel Multivariate Approach

Jose L. Concha [1], Luis E. Arteaga-Pérez [2], Irene Gonzalez-Torre [1], Quantao Liu [3,*] and Jose Norambuena-Contreras [1,*]

[1] LabMAT, Department of Civil and Environmental Engineering, University of Bío-Bío, Concepción 4051381, Chile
[2] LPTC, Department of Wood Engineering, University of Bío-Bío, Concepción 4051381, Chile
[3] State Key Laboratory of Silicate Materials for Architectures, Wuhan University of Technology, Wuhan 430070, China
* Correspondence: liuqt@whut.edu.cn (Q.L.); jnorambuena@ubiobio.cl (J.N.-C.)

Abstract: This study evaluated the effect of two encapsulation methods (i.e., dropping funnel and syringe pump), two concentrations of the alginate-based encapsulating material (2%, and 3%), and three oils as bitumen rejuvenators (virgin sunflower oil, waste cooking oil, and virgin engine oil) on the morphological, physical, chemical, thermal, and mechanical properties of encapsulated rejuvenators for asphalt self-healing purposes. A general factorial design $2 \times 2 \times 3$ was proposed to design 12 different Ca-alginate capsules. Significant differences on the morphological, physical, and mechanical properties of the capsules were analysed by three-way ANOVA and Tukey HSD Post Hoc analyses. The effect of the type of oil on the self-healing capacity of cracked bitumen samples was also evaluated. The main results showed that the design parameters and their interactions significantly affected the morphological, physical, and mechanical properties of the capsules. Capsules synthesised via syringe pump method, with virgin cooking oil and 2% alginate was the most appropriate for asphalt self-healing purposes since its uniform morphology, encapsulation efficiency up to 80%, thermal degradation below 5% wt., and compressive strength above the reference asphalt compaction load of 10 N. Finally, the healing tests showed that virgin cooking oil can be potentially used as a rejuvenator to promote asphalt crack-healing.

Keywords: self-healing; rejuvenators; alginate; encapsulation; factorial design; post hoc analyses

Citation: Concha, J.L.; Arteaga-Pérez, L.E.; Gonzalez-Torre, I.; Liu, Q.; Norambuena-Contreras, J. Biopolymeric Capsules Containing Different Oils as Rejuvenating Agents for Asphalt Self-Healing: A Novel Multivariate Approach. *Polymers* **2022**, *14*, 5418. https://doi.org/10.3390/polym14245418

Academic Editor: Enzo Martinelli

Received: 7 November 2022
Accepted: 9 December 2022
Published: 11 December 2022

1. Introduction

Asphalt pavements are commonly used for the construction of roads around the world. When exposed to rest periods, asphalt pavement has the potential to restore its stiffness and strength by closing the micro-cracks that occur when the pavement is subjected to traffic loads [1]. This property is known as self-healing ability, and it is conferred by the bitumen fraction in the asphalt mixture (about 5%wt.) [2–4]. Nonetheless, the natural self-healing in bitumen is a slow process due to low ambient temperature and asphalt ageing, while the ageing of asphalt is catalysed by several factors during the service life of the asphalt pavement [5,6]. Consequently, the mechanical properties of the asphalt pavement are affected, becoming more prone to temperature cracking and fatigue cracking, shortening the durability of pavement [7].

Traditionally, cracking of asphalt pavement caused by asphalt ageing has been treated using petroleum-based rejuvenators, via superficial application [8]. They consist of softening additives with a high proportion of organic components that aid in restoring the asphaltene/maltene ratio and the thermomechanical and rheological properties of the aged bitumen [9]. Nonetheless, the superficial application of rejuvenators on a pavement layer of

40–80 mm thickness [10] penetrated only 5 to 10 mm of a dense surface layer and no more than 20 mm of a porous surface layer [11]. To overcome this inefficiency, the incorporation of encapsulated bitumen rejuvenators into asphalt pavement have been developed over the past years, turning this topic into an emerging field of study [12]. The healing mechanism of the encapsulated rejuvenators is based on the rejuvenation of the aged binder [13]. The process is as follows: once a microcrack appears, the encapsulating material is broken by the fracture energy at the tip of the crack [14], resulting in the activation of the capsule; then, the bitumen rejuvenator is released, diffusing and permeating into the aged asphalt binder along the cracked surface; finally, the bitumen is softened, accelerating the healing process and reconstituting the aged asphalt binder's chemical composition [15].

Different morphologies of encapsulated rejuvenators such as microcapsules [16], hollow fibres [17], saturated porous aggregates [18], and more recently, biobased spore microcapsules [19] have been developed for asphalt self-healing. In particular, spherical microcapsules with polynuclear internal structure have been widely studied, where different encapsulation methods, encapsulating materials, and bitumen rejuvenators have been proposed [12]. In reference to the encapsulation methods, dropping funnel [20] and syringe pressure pump [21] devices have commonly been used. In these methods, an oil-in-water emulsion (O/W), consisting of a disperse (oil droplets) and a continuous (encapsulating solution) phase, is extruded and precipitated in the form of droplets into a hardening solution (typically calcium chloride), resulting in the formation of capsules via ionic gelation principle. Among the encapsulating materials, sodium alginate biopolymer has lately become the most-used material because of its nontoxic, stable, and biodegradable properties [22]. Additionally, bitumen rejuvenators coming from (i) natural sources, e.g., sunflower [23] and rapeseed oil [24]; (ii) the automotive industry, e.g., virgin engine oil [25]; and the valorisation of waste from (iii) industrial sources such as pyrolytic oil from end-of-life tyres [26] and waste engine oil [20]; and (iv) natural sources such as waste vegetable oils [27] and bio-oils [21] have been successfully encapsulated for asphalt crack self-healing purposes.

The effectiveness of different encapsulated rejuvenators as bitumen healing agents has been proven in different bituminous materials with some promising results. Sun et al. [28] synthesised microcapsules based on an in situ polymerisation principle, using a light-oil as rejuvenator and melamine–urea–formaldehyde as the encapsulating agent. They concluded that bitumen with 5% wt. of microcapsules reached a strength–recovery ratio of up to 85%, enhancing the healing efficiency of asphalt binder. Xu et al. [29] synthesised capsules based on ionic gelation principle, using a commercial rejuvenator and polyethylene-alt-maleic–anhydride as the encapsulating agent. They showed that asphalt mortars with 2.6% wt. of capsules reached healing index up to 40%, indicating that the encapsulated asphalt rejuvenator can be released upon cracking and rejuvenate the aged material. Norambuena-Contreras et al. [30] synthesised capsules based on the ionic gelation principle, using sunflower oil as rejuvenator and Ca–alginate as the encapsulating agent. They showed that asphalt mixtures with 0.5% wt. of capsules reached a healing level up to 53%, being an effective additive for rejuvenating the aged matrix of asphalt materials.

Most encapsulated rejuvenators are synthesised under the one-factor-at-a-time approach. A drawback of this approach is that it does not consider the interactions that could exist between the capsules' design parameters. Such interactions can simultaneously affect the physical and mechanical properties of the capsules to ensure their integrity when incorporated into bituminous materials. Thus, this approach can hardly respond questions such as (1) "Which is the appropriate encapsulation method for the synthesis of homogeneous-sized capsules for healing purposes?"; and (2) "What are the type and concentrations of the capsule components that optimise their physical, thermal, and mechanical performance?" This study proposes using a factorial approach to respond to these questions, providing a better understanding of how the main design parameters and their interactions affect the synthesis and the properties of the capsules. The main objective of this study is to evaluate the main and combined effect of the encapsulation method, alginate biopolymer concentration, and oil type on the morphological, physical, thermal, and mechanical properties of encapsulated rejuvenators for asphalt self-healing purposes. A general factorial design was performed considering two encapsulation methods, two alginate concentrations, and three types of oils to synthesise encapsulated rejuvenators at laboratory scale.

2. Materials and Methods

2.1. Materials

Polynuclear calcium–alginate capsules containing different oils were prepared in this study. Low-viscosity sodium alginate powder (Mannuronic/Guluronic ratio of 0.92, density 1.02 g/cm^3, viscosity $\leq$ 300 mPa·s in a 2% w/w solution), provided by Buchi (Flawil, Switzerland) and calcium–chloride dihydrate ($CaCl_2 \cdot 2H_2O$) at 77% purity, provided by Winkler (Concepción, Chile) were used to form the polymeric matrix of the capsules. On the other hand, virgin cooking oil (VCO) from commercial sunflower oil; waste cooking oil (WCO) from recycled sunflower oil after one-cycle frying at 180 °C; and a commercial synthetic virgin engine oil (VEO)—code SAE 10W-40—were encapsulated as bitumen rejuvenating agents. The main properties of each oil are shown in Table 1.

Table 1. Physical properties of the oil used as bitumen rejuvenators.

Type of Oil	Density (g/cm^3)	Viscosity at 20 °C (cP)	pH at 25 °C
VCO	0.85	70	5.3–5.5
WCO	0.83	89	4.4–4.6
VEO	0.73	274	7.9–8.1

2.2. FTIR-ATR Test of the Encapsulated Oils

Rejuvenating oils, i.e., VCO, WCO and VEO, were chemically analysed by attenuated total reflection Fourier transform infrared spectroscopy (ATR-FTIR). The analyses were performed in a single-reflection Golden Gate^{TM} accessory (Specac, Orpington, UK) equipped with a diamond crystal. The IR spectra of oils were recorded over 32 scans in the range of 4000 cm^{-1} to 600 cm^{-1}, with a resolution of 4 cm^{-1} using a Nicolet iS20 Spectrometer equipped with a DTGS detector and CaF_2 windows (Thermo Fisher Scientific, Waltham, MA, USA). The comparison and analysis of spectra were performed after signal processing, i.e., baseline, ATR correction for Diamond crystal, peak identification, and normalisation (a_{norm}) according to the procedure presented by Hofko et al. [31]. Second-derivative spectra (Savitzky–Golay algorithm) were calculated to resolve the overlapping peaks and remove of the effects of baseline drifts.

2.3. Synthesis and Characterisation of the O/W Emulsions

Synthesis of the oil-based emulsions consisted of the following two steps: (S1) Based on the previous research literature on encapsulated rejuvenators [12], two types of sodium alginate solutions in concentrations of 2% w/w (viscosity of 205.0 cP @ 20 °C) and 3% w/w (viscosity of 602.1 cP @ 20 °C) of deionised water were prepared under continuous agitation using a mechanical stirrer (Scilogex, Model OS40-Pro-LB Pro, Rocky Hill, CO, USA) at 700 rpm for 30 min at room temperature. (S2) After this, VCO, WCO, and VEO were incorporated dropwise into each solution at a fixed biopolymer: oil mass ratio of 1:5, while stirring at 700 rpm for 40 min under mechanical agitation. In total, six O/W emulsions were prepared and labelled as X-Y, with X being the type of oil (VCO, WCO, and VEO) and Y the alginate concentration, i.e., 2% or 3%. Thus, an emulsion synthesised with VEO and 2% alginate concentration was labelled as VEO-2.

The viscosity of each emulsion was measured using a rotational viscometer (Fungilab, Model Smart Series, Barcelona, Spain). The physical stability of the emulsions (i.e., the tendency of the emulsion component to be physically separated over time), was characterised by the creaming index (CI), based on the method proposed by McClements [32] and recently applied by Norambuena-Contreras et al. [27]. CI was determined at: 0 h, 1 h, 2 h, 3 h, 6 h, 12 h, 24 h, 26 h, 28 h, and 30 h. Finally, representative values of viscosity and CI at each time were determined as the average of three measurements.

2.4. Synthesis and Characterisation of the Alginate-Based Capsules

The synthesis of polynuclear capsules was based on the ionic gelation principle of sodium alginate in the presence of calcium ions. For that, two encapsulation methods were compared: the dropping funnel (M1) and the microfluidic pressure pump (M2), based on the parameters established by Norambuena-Contreras et al. [21], see Figure 1a.

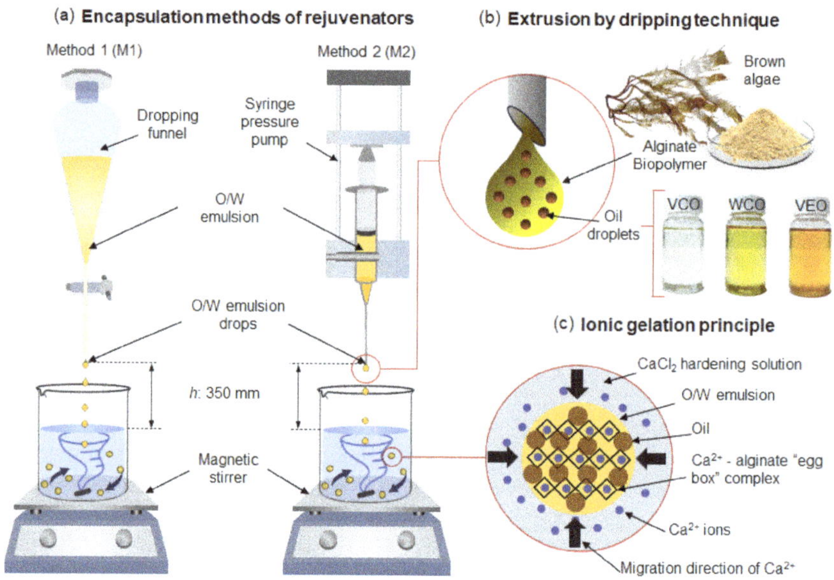

Figure 1. Scheme of the synthesis of the capsules under (**a**) dropping funnel and syringe pump methods. (**b**) Extrusion and releasing of an oil-in-water emulsion droplet. (**c**) Ca^{2+} migration and reaction with the alginate biopolymer by ionic gelation principle, forming the well-known Ca–alginate egg-box complex. As a result, the oil is encapsulated in the biopolymeric internal structure.

First, a hardening solution consisting of 5% w/w of calcium chloride ($CaCl_2$) into deionised water was prepared. Then, each emulsion design was poured into different recipients depending on the method: a 500 mL glass separatory funnel with internal diameter of 5.85 mm for M1, and a 60 mL syringe connected to a hollow steel needle of internal diameter of 1.2 mm for M2. For M1 the dropping mechanism depended on gravity and the viscosity of each emulsion, while for M2, the dropping mechanism depended on the flow rate (2 mL/min) provided by an automatic syringe pump (New Era NE-1010, Farmingdale, NY, USA), see Figure 1b.

For both methods, the emulsions were dropped into the $CaCl_2$ hardening bath at 20 °C in constant agitation at 250 rpm using a magnetic stirrer. At this stage, the Ca^{2+} ions presenting in the calcium chloride solution reacted with the alginate-based emulsion by ionic gelation principle forming the well-known Ca–alginate egg-box complex, see Figure 1c, resulting in the encapsulation of the oil. Once the encapsulation process was finished, the capsules were maintained in the hardening bath for 30 min. Then, the capsules were filtered, rinsed with 300 mL of deionised water, and oven-dried for 24 h at 30 °C.

Finally, the capsules were placed into a glass vial and stored in a freezer at −18°C, preventing the oxidation of the oils. In total, 12 different designs of Ca–alginate capsules were synthesised and identified as X-Y-Z, with X being the type of oil encapsulated (VCO,

WCO, or VEO), Y representing the alginate concentration (2% or 3%), and Z referencing the encapsulation method (M1 or M2). For example, capsules synthesised with VEO, 2% alginate concentration, using the syringe pump method is labelled as VEO-2-M2.

The morphology of the capsules was characterised by their size and shape. The size was statistically evaluated as the diameter of 100 randomly selected capsules using an optical microscope (Leica EZ4, Wetzlar, Germany) with 35× magnification and the image processing software ImageJ®. The shape of each capsule design was also evaluated by the sphericity factor (S_F) as specified by Alpízar-Reyes et al. [19], where S_F values below 0.05 indicate particles tending to a spherical shape, while values closer to 1 indicate particles with an elongated morphology. The encapsulation efficiency ($E.E$), i.e., the proportion between the encapsulated oil and the total oil to synthesise the capsules, was determined in this study based on the method described by Concha et al. [25].

The relative density of the capsules was measured following method B of the ASTM D792-13 [33]. Thermogravimetric analysis (TGA) tests were carried out for each design of capsule and their components. For this, a thermogravimetric analyser (NETZSCH, STA 409 PC, Wittelsbacherstraße, Germany) was used. An amount of ~25 mg was placed in a ceramic cap and heated from ambient temperature to 600 °C at a rate of 10 °C/min. All the tests were performed under a constant N_2 (99.9995%, Airliquide, Santiago, Chile) flow of 50 mL/min, ensuring an inert atmosphere during the test.

The mechanical characterisation of each capsule design was performed by compressive strength and micro-indentation under conditions previously tested by Norambuena-Contreras et al [21]. Compressive tests were performed under ASTM D695-02a [34] using a Universal Testing Machine (Test Resources, Shakopee, MN, USA) equipped with parallel plates and a 1 kN load cell, and operated a at a speed of 0.2 mm/min. The hardness was measured on each capsule design using an HV-1000A Microhardness Tester (Russell Fraser Sales Pty Ltd., Kirrawee, New South Wales, Australia) equipped with a Vickers indenter probe. Representative compression and hardness values were obtained as the average of 10 and 5 measurements, respectively.

2.5. Self-Healing Efficiency of the Oils as Bitumen Rejuvenators

The self-healing efficiency of each oil was quantified in cracked bitumen samples by fluorescence microscopy. For each of the oils, a thin film bitumen sample with dimensions 20 × 20 × 0.5 mm^3 was prepared on a glass petri dish using masking tape and a metallic spatula. A 200 μm-width microcrack was made in the centre of the sample using a blade feeler gauge. Then, using a hollow needle, a drop of ~2 mg of each oil was dropped on the microcrack simulating the release of the rejuvenator from the capsule. The crack closure was recorded periodically every 1 min until its closure using an inverted fluorescence microscope (ICOE IV 5100 FL, Ningbo, China) with phase contrast and a magnification up to 400×. The representative crack width at each time was quantified as the average of 8 measurements along the crack's width using image processing software ImageJ® Fiji distribution, version 1.52 p (MPI-CBG, Dresden, Germany). Finally, the self-healing efficiency of each oil was calculated as the relationship between the average crack width at a specific time measured in μm, and the initial crack width measured in μm.

2.6. Statistical Analysis

The simultaneous effect of the type of oil, the alginate concentration, and the synthesis method on the morphological, physical, and mechanical properties of the capsules was evaluated. A general factorial design 2 × 2 × 3 was proposed for the design of the capsules, consisting of 2 levels of alginate concentrations (2% and 3%), 2 levels associated with the synthesis method (M1-dropping funnel and M2-syringe pump), and 3 levels associated with the type of oil (VCO, WCO, and VEO). Analysis of variance (ANOVA) was used to evaluate the statistically significant effect of the main and combined factors ($\alpha = 0.05$). Tukey HSD Post-Hoc test was conducted to evaluate pairwise mean comparisons with a significance of $\alpha = 0.05$. All the statistical analyses were performed using the software OriginPro 2022, version 9.95-2022b (OriginLab Corporation, Northampton, MA, USA).

3. Results and Discussion

3.1. Effect of the Oil Type and Alginate Concentration on the Physical Properties of the Emulsions

Table 2 shows the average results of the viscosity for the different O/W emulsions. It can be noticed that the viscosity values were significantly affected by the type of oil and the alginate concentration. In particular, the higher the viscosity of the oil (see Table 1) and the alginate concentration, the higher the viscosity of the O/W emulsion was.

Table 2. Average results of viscosity values of the O/W emulsions measured @20 °C, in cP.

VCO-2	VCO-3	WCO-2	WCO-3	VEO-2	VEO-3
245.5 ± 7.2	824.1 ± 9.2	270.0 ± 9.3	940.0 ± 7.7	291.1 ± 11.3	1053.0 ± 8.4

The viscosity of the O/W emulsion is a crucial parameter during the encapsulation process, since it can affect the release frequency of the droplets from each encapsulation device, influencing the morphology of the capsules, e.g., their size and shape. In particular, it is hypothesised that the extrusion of highly viscous emulsions retards the release of the outgoing droplets from each encapsulation device, resulting in bigger and elongated Ca–alginate capsules. This hypothesis is evaluated in Section 3.2.

Another important factor to evaluate in the O/W emulsions is their physical stability during the encapsulation process, i.e., the disperse phase (oil) should be homogeneously distributed into the continuous phase (alginate solution). Thus, the physical stability of the emulsions was estimated by analysing the creaming index (CI) values over time (Figure 2a). Results indicate that CI increased with time, suggesting that after a certain period the emulsion components in all the formulations were separated. It was found that emulsions with the same alginate concentration present similar creaming behaviour, suggesting an increase in the stability with the increase of alginate content in the mixture, regardless the nature of the oil.

For O/W emulsions containing 2% of alginate, the increase in the CI with time can be characterised in two stages: A first stage, from 0 to 3 h, where the CI of the VCO-2, WCO-2, and VEO-2 emulsions sharply increased up to values of 81.14%, 75.2%, and 84.85%, respectively. This indicated an early and fast tendency of the O/W emulsions to be creamed. Then, from 3 to 30 h, there was a quasi-stationary stage with CI values of 84.57%, 83.05%, and 89.14% at 30 h for the VCO-2, WCO-2, and VEO-2 based emulsions, respectively. In contrast, emulsions synthesised with 3% of alginate presented a CI increase defined by four stages: A first stage, from 0 to 6 h was identified, where the CI reached values of 13.71%, 10.8%, and 12.85% for the VCO-3, WCO-3, and VEO-3 emulsions, respectively. Then, for the following three stages (6–12 h; 12–28 h; 28–30 h), the CI was increased with time, but at a lower rate at each stage. At the end, the VCO-3, WCO-3, and VEO-3 emulsions presented CI values of 75.71%, 82.85%, and 76.57%, respectively. Based on the previous analysis, the synthesis of O/W emulsions based on 3% alginate were physically more stable when compared with those based on 2% alginate.

The influence of the alginate concentration on emulsion stability can be explained by several factors. Mainly, the creaming effect involves the coalescence and ascension of larger oil droplets (i.e., disperse phase) to the top of the emulsion over time, as represented in Figure 2b. Since alginate acts as a weighting agent on an O/W emulsion, the higher the alginate concentration in the emulsion, the more difficult the coalescence and ascension of the larger oil droplets to the top of the emulsion. This phenomenon is attributed to the increase in the viscosity of the emulsion, reducing the movement of the oil droplets to coalescence. To prove this point, Figure 2c shows fluorescence microscopy images for the VCO-2 and VCO-3 emulsions taken from the top of the emulsion at 0 h, 3 h, 6 h, 12 h, and 24 h. Figure 2d,e shows the statistical distribution of the droplet size. Since CI was more affected by the alginate concentration than by the type of oil, fluorescence microscopy images of VCO with different alginate concentrations well-represents the creaming phenomenon in all the O/W emulsions. From these images, the increase of the oil droplet size over time can be noticed.

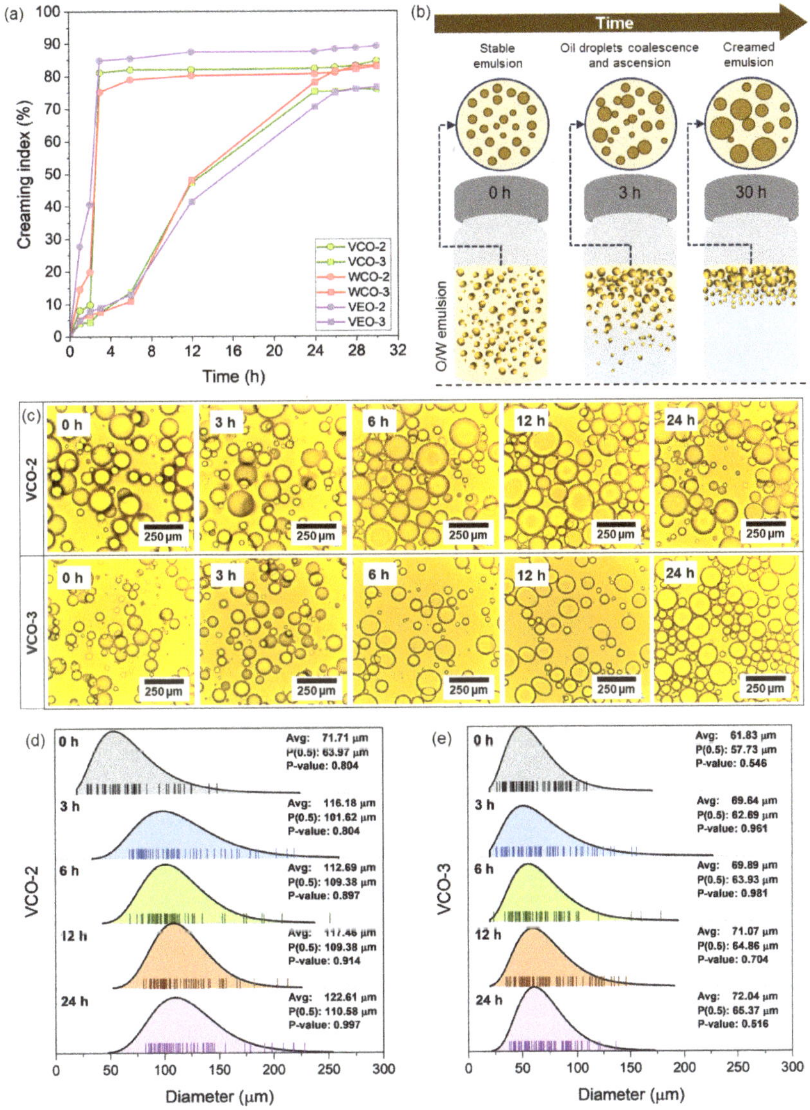

Figure 2. (**a**) Average results of creaming index evaluated at different times for the different O/W emulsions; (**b**) schematical representation of the creaming process occurring in the O/W emulsions over time; (**c**) examples of fluorescence microscopy images for the VCO-2 and VCO-3 emulsions showing the oil droplet growing over time; and statistical distribution of the oil droplet size over time for the case of (**d**) VCO-2 and (**e**) VCO-3 emulsions.

Nonetheless, VCO-3 emulsion presented droplet sizes significantly lower than VCO-2. As example, for the VCO-2 emulsion, the oil droplet size increased from 63.97 μm to 110.58 μm after 24 h, representing an increase of ~73% with respect to the initial size. For the same period of time (24 h), the representative droplet size for the VCO-3 emulsion increased from 57.73 μm to 65.37 μm, representing an increase of ~13% with respect to the initial size. Finally, based on the previous analysis, it can be concluded that the component of the O/W emulsions mainly contributing to their physical stability is the alginate concentration. Additionally, for a successful synthesis of the capsules, the encapsulation should take place once the freshly O/W emulsions are prepared, reducing the occurrence of creaming phenomenon.

3.2. Morphological and Physical Properties of the Biocapsules

Figure 3 shows optical microscopy images and the histograms of the size of each type of capsule fitted to a normal distribution. It can be noticed that the size of the capsules was influenced by the synthesis method, the type of oil, and the concentration of the alginate biopolymer. In particular, (i) capsules synthesised using M1 resulted in bigger sizes than using M2; (ii) capsules with 2% alginate resulted in smaller sizes than those with 3% alginate; and (iii) the VEO capsules presented the biggest sizes followed by the VCO and WCO capsules. As an example, the VEO-3-M1 and the WCO-2-M2 capsules presented the biggest and the smallest sizes with values of 3.95 mm and 1.62 mm, respectively.

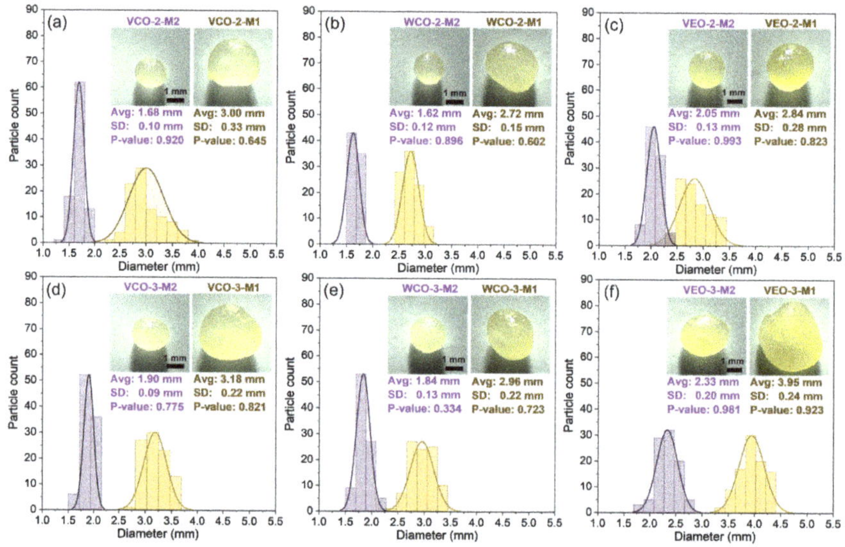

Figure 3. Statistical distribution of the size of each type of capsule fitted to a normal distribution: (**a**) VCO-2-M1 and M2; (**b**) WCO-2-M1 and M2; (**c**) VEO-2-M1 and M2; (**d**) VCO-3-M1 and M2; (**e**) WCO-3-M1 and M2; (**f**) VEO-3-M1 and M2.

In particular, Figure 4a represents the size of the capsules through boxplots, where the effect of the encapsulation method on the size of the capsules can be clearly noticed by identifying two groups: capsules synthesised using M1, and capsules synthesised using M2. Moreover, from this Figure we noticed that M1 capsules presented more dispersion on their size when compared to M2 capsules. This can be attributed to the fact that M2 was an extrusion-controlled process, where a constant pressure was applied during the extrusion of O/W emulsions to form the capsule, while for M1 the extrusion mechanism was gravity-based, meaning low control of the release of the O/W emulsion droplets during the extrusion process. Thus, the encapsulation method M2 results in capsules with a more uniform size than M1, and so, a similar shape is expected for M2-type capsules.

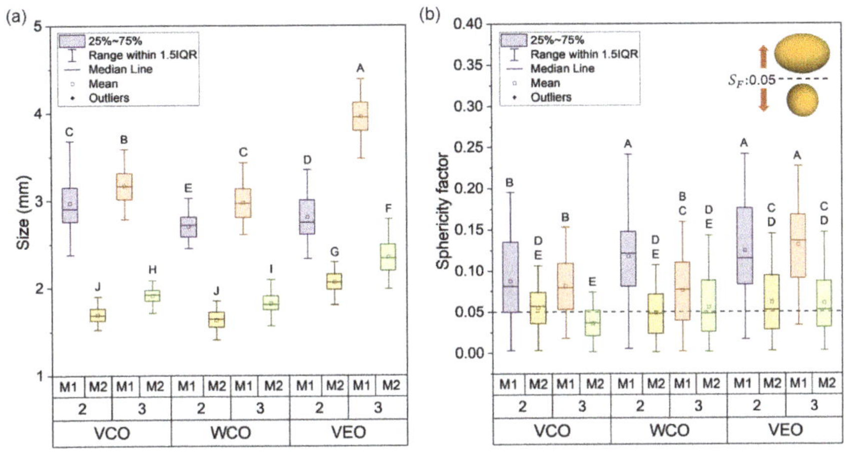

Figure 4. Box plot graphs for (**a**) the size and (**b**) sphericity factor of the different capsule designs.

To prove the previous hypothesis, Figure 4b shows the average S_F results of the capsules. From this Figure, capsules synthesised using M1 tended to S_F values over 0.05, indicating a more elongated morphology, while capsules synthesised under M2 presented lower dispersion of data with S_F values closer to 0.05, indicating that this encapsulation process resulted in capsules with more regular spherical morphology. The alginate concentration and the type of oil also influenced the morphology of the capsules. Thus, capsules synthesised with highly viscous oils and higher concentrations of alginate resulted in more elongated morphologies. Finally, considering that the capsules are added as an additive to asphalt mixtures, bigger sizes with elongated morphology could affect the physical properties of the asphalt mixture more significantly than smaller sizes and regular spherical morphology. Thus, capsules based on VCO or WCO, 2% alginate and method M2 are recommended for addition in asphalt mixtures.

Otherwise, Figure 5a shows the average results of encapsulation efficiency for each capsule design. It is observed that the capsules presented *E.E* values ranging from 85.88% (VEO-2-M1) to 97.03% (VCO-3-M2). Overall, the effect of the type of oil on the efficiency can be noticed, observing that VEO capsules presented the lower *E.E* values when compared to VCO and WCO capsules. This can be attributed to the flow properties, associated with the viscosity of the capsules' components. As characterised in Sections 2.1 and 2.3, the viscosity of the VCO and the WCO was lower than the 2% and 3% alginate solutions, respectively. On the contrary, VEO viscosity was higher than the 2% alginate solution, but lower than the 3% alginate solution. This means that for the extrusion process (using M1 or M2) of an O/W emulsion based on 2% alginate, both VCO and WCO will flow more easily than the alginate solution, increasing their *E.E* values when compared with the VEO. For the extrusion of the O/W emulsion based on 3% alginate, all the oils will flow more easily than the alginate solution, resulting in *E.E* values higher than for the 2% solution.

In contrast, Figure 5b demonstrates that VEO capsules presented lower density values than the VCO and WCO capsules. Moreover, the increase in the alginate concentration leads to an increment in the density of VCO and WCO capsules, while the VEO capsules exhibited the opposite behaviour being their densities reduced with higher alginate concentrations. These differences in density are ascribed to a combined effect of the sizes and volume of capsules, and to the density of confined oils. According to the data in Figure 4a, the capsule VEO-3-M1 has nearly double the volume (32.3 mm^3) of capsule WCO-3-M1 (13.7 mm^3). For the same capsule designs, Figure 5a shows that the VEO-3-M1 and the WCO-3-M1 capsules presented similar *E.E* values, 93.34% (SD: 2.17%) and 92.87% (SD: 0.8%). Nonetheless, as seen in Table 1, density of VEO is lower than that of WCO. In consequence, the density of

VEO-3-M1 capsules is lower than those of WCO-3-M1. From this analysis, it can be noted that the variation in the size of the capsules is higher than that of the *E.E.* This means that differences in the density of the capsules can be mainly explained by significative changes in their volume.

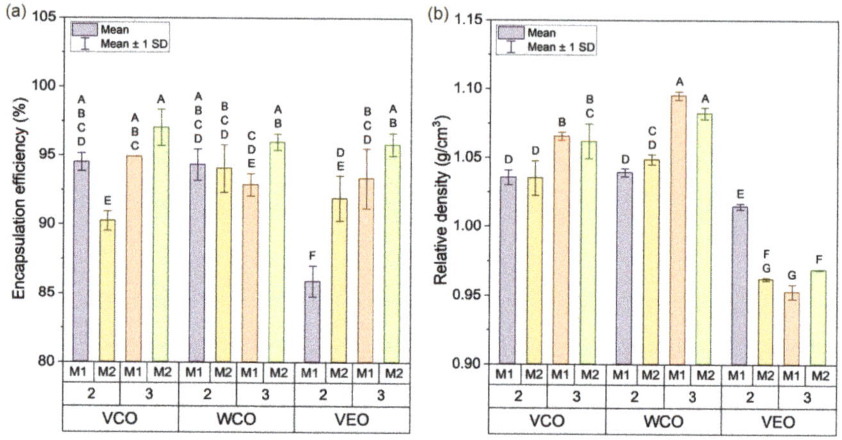

Figure 5. Average results of (**a**) encapsulation efficiency and (**b**) relative density of each capsule.

To evaluate if the differences evidenced in the morphological (size, sphericity) and physical (density and encapsulation efficiency) properties of the capsules are statistically significant, Table 3 resumes the ANOVA p-value results for the main factors (type of oil, alginate concentration, and synthesis method) and their combinations. These results show that significant differences were found on the morphological and physical properties. Particularly, since the type of oil comprises three levels, significant differences detected by ANOVA can be associated with up to three different pairwise means comparisons (VCO-WCO, VCO-VEO, WCO-VEO). Tukey HSD test revealed that the three pairwise means comparisons were significantly different for each of the morphological and physical properties. Therefore, the effect of VCO, WCO, and VEO on the capsule's size can be treated as three independent groups. In a similar way, each level of the alginate concentration (2%, 3%) and the encapsulation method (M1, M2) can be treated as independent groups.

Table 3. ANOVA p-values showing the effect of the main studied factors and their interactions on the morphological and physical properties of the capsules.

Variables	Size	Sphericity	Density	Encapsulation Efficiency
A	<0.001 [1]	<0.001 [1]	<0.001 [1]	<0.001 [1]
B	<0.001	<0.001	<0.001	<0.001
C	<0.001	<0.001	0.0012	<0.001
A * B	<0.001	0.0028	<0.001	<0.001
A * C	<0.001	<0.001	0.0029	<0.001
B * C	<0.001	0.0725	0.0016	0.0160
A * B * C	<0.001	<0.001	<0.001	<0.001

Notes: A: oil type (VCO, WCO, VEO); B: alginate concentration (2%, 3%); C: encapsulation method (M1, M2). [1]: significant differences for VCO-WCO, VCO-VEO, and WCO-VEO by Tukey HSD mean comparisons.

In terms of the interactions between the main factors, Table 3 revealed the presence of significantly different double and triple interactions, excepting the double interaction between the alginate concentration and the encapsulation method for the measure of sphericity. Graphically, Figure 4a,b and Figure 5a,b present the Tukey HSD analysis for pairwise mean comparison represented by letters. Thus, mean values that do not share a letter can be considered significantly different (p-value < 0.05). As an example of the morphological properties of the capsules, Figure 4a shows that the measure of their size was categorised in 10 groups (represented as single letters from A to J), meaning that the size of 10 from the 12 capsule designs can be considered as significantly different. Accordingly, from this Figure, it can be stated that VEO-3-M1 design presented the biggest size, while capsule designs VCO-2-M2 and WCO-2-M2 presented the smaller size values. As an example of the physical properties of the capsules, Figure 5a revealed that their *E.E* values can be categorised in six groups (A–F). Particularly, VEO-2-M1 is individualised as one single group (F), resulting in the lowest *E.E* value.

Based on the previous analysis, it can be concluded that the type of oil, alginate concentration, and encapsulation method, as well as their interactions significantly influenced on the morphology and physical properties of the capsules. In particular, the combined effect of M2 and 2% alginate for each of the oils produced capsules with (i) the smallest and uniform sizes, (ii) regular spherical morphology, and (iii) high encapsulation efficiencies. A similar effect was observed for capsules synthesised using M2 and 3% alginate. Accordingly, M2 could be considered as a potential encapsulation method for the synthesis of homogeneous-sized capsules. To reduce the effect of the capsule addition on the physical properties of the asphalt mixture, capsules based on VCO or WCO, 2% of alginate, and method M2 are recommended for their incorporation in asphalt mixtures. Nonetheless, to select an appropriate alginate concentration and type of oil, thermal–mechanical stability of the capsules and the rejuvenating effect of each oil on cracked bitumen will be analysed within the next sections.

3.3. Thermal–Chemical Characterisation of the Capsules and Their Components

Figure 6 shows the results from the FTIR-ATR characterisation of the rejuvenators used for the encapsulation process. The identification and assignment of spectral bands shown in Figure 6a were represented according to the literature reports [35–37] and considering the nature of the rejuvenating liquids. The spectral information gathered from the bioderived oils (i.e., VCO and WCO) were similar, indicating a negligible effect of the oxidation during one cycle of frying for VCO.

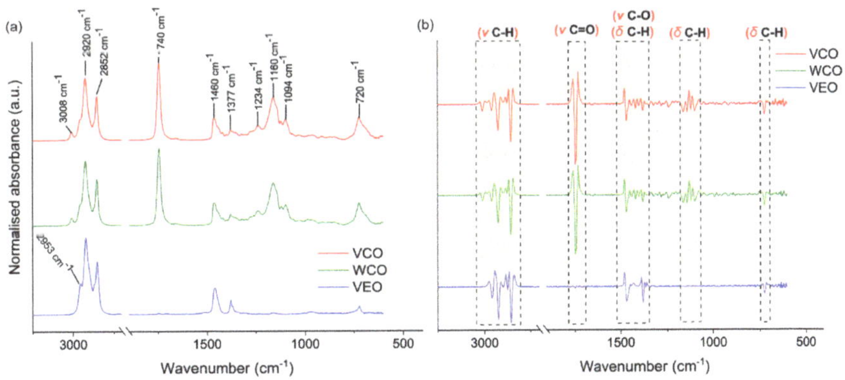

Figure 6. (**a**) FTIR-ATR infrared spectra of VCO, WCO, VEO, and (**b**) second derivative of spectra for VCO, WCO, VEO.

Both VCO and WCO presented multiple signals between 1090 cm^{-1} and 1460 cm^{-1}, ascribed to C–H bending vibration in aliphatic hydrocarbons, and C–O stretching vibrations commonly found in ethers (see Figure 6b). The signal at 720 cm^{-1} was found in the three oils, and it is typical of weak C–H asymmetric bending in alkyl chains (CH_2 and CH_3). In addition, the presence of methylene moieties (CH_2) in the saturated fatty acid backbone in VCO and WCO was confirmed by the C–H in-plane deformation band at 1377 cm^{-1}, and the C–H symmetric and asymmetric stretching bands at 2852 and 2920 cm^{-1}, respectively.

A distinctive sharp signal was found for VCO and WCO at 1743 cm^{-1}. This band is ascribed to the stretching in carbonyl groups (C=O) corresponding to carboxylic acid or triglycerides present in these oils as witnessed before by Goh et al. [37]. In the case of the synthetic VEO, the bands located at 2852 and 2920 cm^{-1} can be associated with the C–H stretching in alkyl chains (CH_2) [35], while the peaks at 1460 and 1377 cm^{-1} correspond to the C–H bending vibrations in methyl groups typical of synthetic engine oils. As previously stated by Norambuena-Contreras et al. [27], the similarities in intensity, position and nature of functional groups identified in VCO and compared with WCO suggest that WCO is thermally stable, which supports its use as a prospective encapsulated rejuvenating agent in asphalt materials usually manufactured at a temperature >150 °C.

Otherwise, Figure 7 shows the TGA-DTG curves for (a) VCO, WCO, VEO, and capsules with, and without, the oils synthesised with (b) 2% and (c) 3% of alginate. As previously recommended by Norambuena-Contreras et al. [30], capsules should be directly added during the asphalt mixing process at a temperature around 150 °C (T_{am}). In consequence, it is expected that the capsules as well as their components are thermally stable at this mixing temperature. As seen in Figure 6a, the three oils proposed here as asphalt rejuvenators presented high thermal stability, showing no thermal degradation at T_{am}. In fact, these oils started to degrade at temperatures higher than T_{am}, with a characteristic single DTG peak identified at temperatures between 200 °C and 450 °C for the WCO and between 320 °C and 500 °C for both VCO and VEO. Although both VCO and WCO come from sunflower, the shifting on the thermal degradation peak of WCO can be attributed to its previous physical–chemical oxidation during cooking. Overall, the single degradation step of VCO and WCO was attributed to the decomposition of volatiles associated to polyunsaturated (linoleic acid) and monounsaturated (oleic acid) fatty acids [38]. From the previous analysis, it is concluded that VCO, WCO, and VEO can be potentially considered as thermally stable rejuvenators for asphalt applications. However, their thermal performance must be proven when encapsulated in the biopolymeric matrix of alginate.

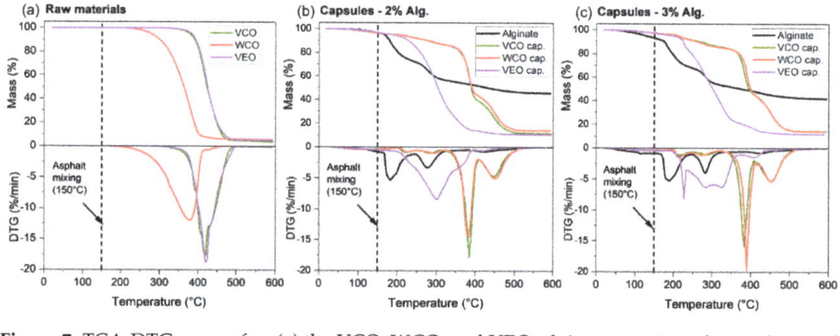

Figure 7. TGA-DTG curves for: (**a**) the VCO, WCO, and VEO; alginate matrix and capsules synthesised with (**b**) 2% and (**c**) 3% of alginate.

In reference to the encapsulation of the oil in the biopolymeric matrix of alginate, Figure 7b,c show that the TGA curves were quite similar regardless the alginate concentration in the capsules. Regarding the alginate-hollow capsule, its TG behaviour is characterised by two noticeable degradation steps: (i) from 150 °C to 250 °C, related to moisture evaporation and the chemical dehydration of the biopolymer; and (ii) from 260 °C to 350 °C, associated with the fracture of glycosidic bonds, decarboxylation and decarbonylation, releasing H_2O, CO_2, and other light compounds [39]. When encapsulating the oils into the alginate biopolymer, Figure 7b,c also show a similar tendency of the capsules with 2% and 3% of alginate when comparing the TGA curves based on the type of oil. Thus, the alginate concentration had no apparent influence on the thermal decomposition of the capsules. Nonetheless, due to the higher thermal stability of the oils, the capsules were significantly degraded at temperatures higher than their respective empty alginate matrices. The particularly earlier thermal degradation for the VEO capsule at ~210 °C can be explained to its low amount of encapsulated oil when compared to VCO and WCO capsules, as demonstrated by the E.E values in Section 2.2. Hence, the enhancement of the thermal stability by effect of the VEO was reduced.

Finally, TGA tests demonstrated that the biopolymeric capsules with VCO, WCO, and VEO can be thermally stable additives for their incorporation to hot mix asphalt (i.e., manufacturing temperatures >150 °C). In particular, capsules based on VCO and 2% alginate could be an optimal design due to: (i) the higher thermal stability of the oil, and (ii) the lower thermal degradation at T_{am}. Moreover, since the alginate concentration had no effect on the thermal stability of the capsules, capsule designs with lower alginate concentration could potentially reduce the associated cost of production. Nonetheless, the decision of the appropriate capsule should also consider additional aspects such as the mechanical performance of the capsules and healing capacity of the oil once released from the capsule.

3.4. Mechanical Stability of the Biocapsules by Compression and Microindentation Tests

In reference to the effect of the main factors studied on the maximum compression load (C_{cap}) resisted by each capsule design (Figure 8a,b) shows: (i) a similar tendency of C_{cap} when comparing the capsules based on the type of encapsulated oil, so the compression curves of the capsules can only be represented depending on the alginate concentration and the encapsulation method as shown in Figure 8b. (ii) Moreover, capsules based on 3% of alginate presented C_{cap} values higher than those with 2% of alginate, and (iii) capsules synthesised using M1 presented C_{cap} values higher than those using M2. In consequence, the combination of the factors M1 and 3% of alginate resulted in the highest C_{cap} values. This result can be attributed to the bigger size reached by M1 capsules, distributing the compressive force on a bigger surface than the M2 capsules, and the potential increase of crosslinking reactions between the alginate and the Ca^{2+} ions, producing a strengthening effect on the multicavity structure of the capsule. Based on the previous analysis and supported by Norambuena-Contreras et al. [21], it can be concluded that C_{cap} mainly depended on the synthesis method and the concentration of alginate used to produce the capsules.

Additionally, since the alginate-based capsules are added during the mixing process of asphalt, they should be strong enough to resist the compaction load of the asphalt mixture referenced in 10 N as stated by Ruiz-Riancho et al. in previous research [23]. Figure 8a shows that all the capsules presented values over this limit indicating their suitability to resist the compaction process and to be later activated during the operation of the road by effect of traffic loads. Nonetheless, it should be considered that C_{cap} values far from the previous compaction load reported could difficult the opportune activation of the capsules, reducing the release of the oil on the cracked-aged bitumen. Accordingly, capsules based on M2 and 2% or 3% of alginate could be more appropriate for resisting the compaction process, while being susceptible to the activation process.

Similarly, Figure 9a indicates that, like in the compressive tests, the synthesis method and the alginate concentration mainly affected the hardness of the capsules, but in a different way. Particularly, capsules synthesised using method M2 presented hardness values higher than those synthesised using M1. Moreover, the addition of 3% of alginate resulted in higher values of hardness than the capsules with 2% of alginate, also reported in [21]. Thus, the combined effect of M2 and 3% alginate resulted in capsules with the highest Vickers hardness, with an average value of 16.58 MPa (SD: 5.1 MPa) while capsule designs based on M2, and 2% alginate presented an average value of 11.88 MPa (SD: 0.75 MPa). As reference, synthetic polymers used for encapsulation applications present Vickers hardness value around 400 MPa.

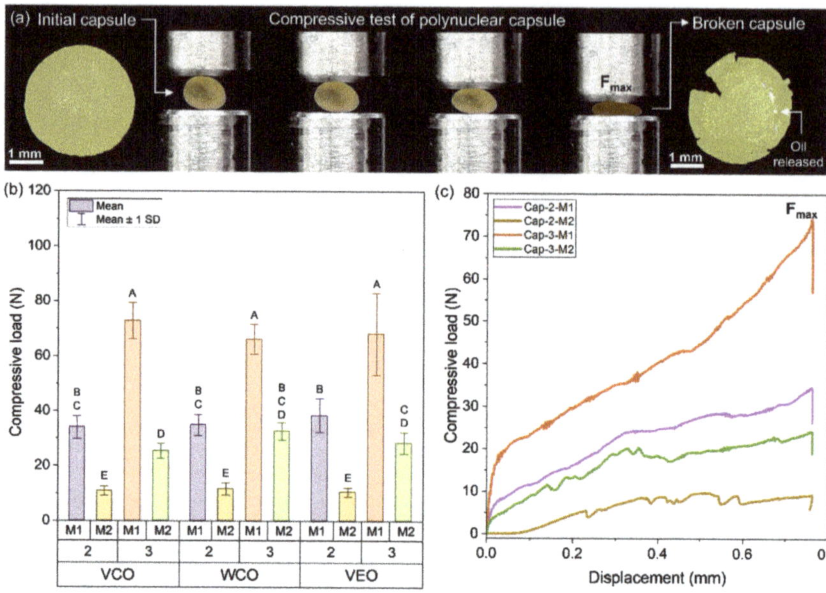

Figure 8. (**a**) Representation of the compressive test carried out on each capsule design; (**b**) average results of the maximum compressive load resisted by each capsule design; (**c**) representative compressive load curves for the capsules depending on the alginate concentration and used method.

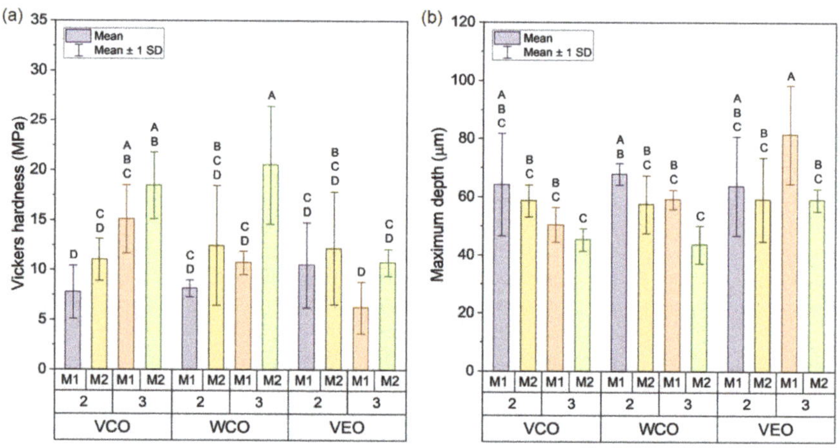

Figure 9. Average results of (**a**) Vickers hardness and (**b**) associated maximum indentation depth.

In contrast, biopolymers used for encapsulations purposes such as polylactic acid, chitin, and chitosan have been stated in 30 MPa, 57 MPa, and 45 MPa, respectively [40]. Therefore, the use of encapsulating soft materials with lower hardness values, such as alginate, could facilitate oil release from the capsule through its breakage or deformation. This behaviour is precisely what is required in a polynuclear capsule with asphalt self-healing purposes, i.e., that is mechanically resistant to the manufacturing processes of asphalt mixtures, but with the ability to be susceptible to the mechanical activation by the effect of the stress concentration trigger on the surface of the capsule.

The overall hardness values of the capsules are in line with their indentation depth as shown in Figure 9b, i.e., the higher the hardness, the lower the indentation depth, so the capsule offered more resistance to be penetrated by the indenter. As an example, the WCO-3-M2 capsule presented the highest average hardness and the lowest indentation depth, being 20.53 MPa (SD: 5.92 MPa) and 43.63 μm (SD: 6.45 μm), respectively. In contrast, the VEO-3-M1 capsule presented the lowest hardness and the highest indentation depth, with values of 6.21 MPa (SD: 2.6 MPa) and 81.50 μm (SD: 6.45 μm), respectively.

Differences in hardness values can be partially attributed to the relative densities of each capsule and to a consequence of their microstructures, see Figure 5b. Thus, the denser the capsule, the lower the indentation depth and so the higher the hardness of the capsule. With this antecedent, it is hypothesised that denser capsules based on 3% of alginate mechanically reinforced their biopolymeric microstructure than those with 2% of alginate, being able to better resist the indentation load. Thus, in relation to the raw materials constituting the capsules, it can be concluded that the alginate was mainly responsible of the compressive strength and hardness of the capsules. Thus, based on the previous recommendation on the use of capsules based on M2 and 2% or 3% of alginate, it can be concluded that capsules synthesised using M2, and 2% of alginate are recommended to activate a capsule for self-healing of crack-aged asphalt mixtures.

Table 4 resumes the ANOVA p-value results for the main factors affecting the mechanical performance of the capsules (type of oil, alginate concentration, and synthesis method) and their combinations. Results indicate that the type of oil had no significant effect on the compression resistance of the capsules, while for the Vickers hardness and the indentation depth, the type of oil led to significant differences. For this test, the Tukey HSD analysis revealed significant differences in the VCO-VEO and WCO-VEO pairwise mean comparisons. For each level of the alginate concentration (2%, 3%) and the encapsulation method (M1, M2) the three mechanical variables presented significant differences, excepting the effect of the alginate concentration for the Vickers depth (p-value > 0.05).

Table 4. ANOVA p-values showing the effect of the main studied factors and their interactions on the mechanical properties of the capsules.

Variables	Compressive Load	Vickers Hardness	Vickers Depth
A	0.8838	0.0141	0.0039
B	<0.001	0.0013	0.0575
C	<0.001	<0.001	<0.001
A * B	0.5128	<0.001	0.0026
A * C	0.0238	0.1800	0.4046
B * C	<0.001	0.1501	0.1767
A * B * C	0.0159	0.5225	0.3859

Notes: A: oil type (VCO, WCO, VEO); B: alginate concentration (2%, 3%); C: encapsulation method (M1, M2).

Additionally, Table 4 shows that the detection of the double and triple interactions between variables was different for the mechanical variables evaluated. In reference to the compressive load, Table 4 revealed that only the interaction between the type of oil and the alginate concentration had no significant effect (A * B interaction). In the case of the Vickers hardness and Vickers depth, significant differences were only detected for the interaction between the type of oil and the alginate concentration (A * B interaction). Figures 7a and 8a,b present the Tukey HSD analysis for pairwise mean comparison represented by letters. Based on the recommendation on the use of capsules based on 2% or 3% alginate and the synthesis method M2, Tukey HSD analysis revealed that the pairwise mean comparison of 2-M2 and 3-M2 double interactions was significantly different for compression and hardness. This result confirms the selection of the capsule design based on 2% alginate and the synthesis method M2 to facilitate the activation of the capsule.

3.5. Healing Capacities of Types of Oils as Asphalt Rejuvenators

Figure 10a–c shows the microscopy fluorescence images of the microcrack closure in bitumen samples by the effect of the rejuvenating oils at 0 min, 20 min, 40 min, and 80 min. From these images, it can be seen that the crack closure was affected by the type of oil used as asphalt rejuvenator, since the crack width over time was not uniform. For a time of 80 min, the µ-crack was completely closed when VCO was used as the rejuvenator. At this time, the bitumen samples with WCO and VEO presented µ-cracks width values of 2.57 µm and 92.62 µm, respectively, suggesting that oil viscosities play a role in this process (Table 1). This behaviour suggests that the more viscous the oil, the lower its diffusion capacity into the cracked zone, resulting in a retardation of its softening effect on the aged bitumen.

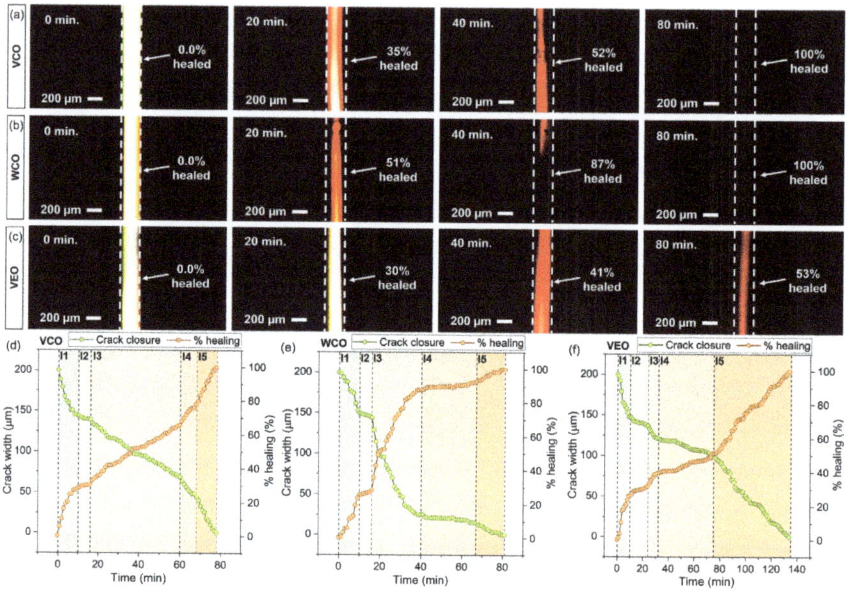

Figure 10. Fluorescence microscopy images for (**a**) VCO, (**b**) WCO, and (**c**) VEO asphalt rejuvenators on asphalt film samples at 0 min, 20 min, 40 min, and 80 min. Results of crack width and healing efficiency of bitumen samples rejuvenated with (**d**) VCO, (**e**) WCO, and (**f**) VEO over time.

To prove the previous hypothesis, Figure 10d–f show the crack width and the healing efficiency curves of bitumen samples with (d) VCO, (e) WCO, and (f) VEO recorded every 1 min until the complete closure of the crack was attained. Overall, shorter times required to close the µ-crack mean a better healing performance of the oil. The complete crack closure of the bitumen samples by effect of VCO and WCO took 78 and 81 min, respectively. Those times were considerably lower than that of the VEO, requiring a time of 135 min for complete crack closure. Based on this analysis, VCO could be more appropriate as an encapsulated rejuvenator for asphalt self-healing purposes.

Nonetheless, Figure 10d–f also show that the µ-crack closure over time was not uniform for the different oils used. The variability of the µ-crack closure over time can be grouped by five different intervals characterised as a linear relationship between the crack width and time. Since each of these intervals took place in different periods of time, a rate of crack closure ($|R_c|$) was calculated for comparison purposes, as depicted in Equation (1):

$$|Rate\ of\ crack-closure\ (R_c)| = \frac{cw_f - cw_i}{tw_f - tw_i} \quad (1)$$

where cw_i and cw_f are the initial and final crack width values for each interval, in μm, respectively, and tw_i and tw_f are the associated initial and final times, in min.

Table 5 summarises the $|R_c|$ values for each bitumen sample. Overall, $|R_c|$ values allow to identify (i) changes in the healing performance of each oil, quantifying them and (ii) the time periods where each of the oils are more efficient to heal the cracked bitumen (e.g., at the beginning or the end of the healing process). In consequence, the higher the $|R_c|$ value, the faster the difussion of the oil and so the more effective the healing process. From Table 5, it was noticed that all the samples presented quite similar crack closure rates during the first 10 min. This means that the oils, during the first interval, presented an initial softening effect on the bitumen in direct contact with them. In this initial stage, the viscosity of the oils was not a determinant factor, since all the oils reduced the initial crack from 200 μm to 150 μm in a period of 10 min (crack closure rate around −5 μm/min).

Table 5. Results of $|R_c|$ values in μm/min for each interval and rejuvenated bitumen.

Sample	Interval 1	Interval 2	Interval 3	Interval 4	Interval 5
Bitumen–VCO	5.62	0.78	1.59	3.12	4.42
Bitumen–WCO	5.09	0.87	4.96	0.37	1.06
Bitumen–VEO	5.25	0.79	1.95	0.54	1.63

During the second interval, the crack closure rate was significantly reduced to a quasi-stationary state. This meant that the oils were diffused into the bitumen surrounding the microcrack to close it. Thus, from the second interval, the viscosity of each oil is a determinant factor for the crack's closure over time. For instance, Table 5 shows that, for each oil, the $|R_c|$ values were quite similar with values varying between 0.78 μm/min and 0.87 μm/min. Nonetheless, this quasi-stationary condition was maintained for B–VCO and B–WCO for 6 min, while during 14 min for the B–VEO. Hence, the higher the viscosity of the oil, the longer the duration of the quasi-stationary state.

From the second interval to the final one, the B–VCO sample increased its crack-closure rate from 0.87 μm/min to a maximum value of 4.42 μm/min following a linear tendency, while for B–WCO and B–VEO samples, the crack closure rate was quite irregular with alternating values. This indicates a more consistent softening effect of VCO, constantly diffusing into the bitumen surrounding the μ-crack over time. Based on these results, it can be concluded that VCO and WCO showed better healing performance than VEO. In particular, the shorter time taken by VCO to close the initial 200 μ-crack and its better performance during the crack-healing process indicate its potential use as an effective rejuvenator to be encapsulated for asphalt self-healing purposes.

4. Conclusions

In this paper, a factorial approach was used to evaluate the effect of the capsules' design parameters on the physical, thermal, and mechanical properties of enhanced encapsulated rejuvenators for asphalt self-healing. The main findings revealed that using an extrusion-controlled method, such as a syringe pump, was the best for producing homogeneously sized capsules. Capsules synthesised with lower alginate concentrations and VCO as rejuvenator meet appropriate thermal and mechanical stability criteria for their incorporation into bituminous materials. The design parameters and their interactions also affected the properties of the biobased capsules and their components, concluding that:

- Variations in the physical stability of an O/W emulsion were mainly explained by changes in the alginate concentration reduced the creaming rate with time. Creaming was not affected when varying the type of oil.
- It was concluded that to avoid any potential phase separation of the emulsion components, it is recommended that the encapsulation process take place once prepared the O/W emulsions.

- The use of the syringe pump method and low alginate concentrations produced capsules with the smallest and uniform sizes (1.60–1.68 mm), regular spherical morphology and high encapsulation efficiencies, up to 94%.
- Encapsulation efficiency of the capsules was affected by the viscosity of the capsule's components. Overall, high encapsulation efficiencies were reached when the viscosity of the oil was lower than the alginate solution (e.g., VCO-based capsules).
- Thermal stability of the capsules mainly depended on the type of oil used as rejuvenator. Capsules based on VCO presented the best thermal stability with mass losses below 5% at the temperature of asphalt mixing (150 °C) as seen by TGA test.
- FTIR-ATR results showed that the spectral information gathered from the bioderived oils (VCO and WCO) was similar, indicating a negligible effect of the oxidation during one cycle of frying for VCO.
- Capsule design based on the syringe pump method (M2) and 2% alginate showed compressive strength and hardness values of 10 N and 12 MPa, respectively, potentially surviving the compaction process of bituminous materials and its later activation under traffic loads.
- Healing tests on cracked bitumen samples revealed the use of VCO reduced the time to close a 200 microcrack with better performance during the crack-healing process, indicating its potential use as an effective rejuvenator for asphalt self-healing.
- Finally, the synthesis of capsules under the syringe pump method, using virgin cooking oil (VCO) and 2% alginate, as an effective encapsulated additive to improve the self-healing properties of bituminous materials is recommended.

Finally, the interactions of the capsules' design parameters discovered in this study open new research on the use of factorial designs to synthesise more capsules for asphalt self-healing focused on: (i) the optimisation of the design parameters of the capsules such as the concentration of $CaCl_2$ and the concentration of the encapsulated oil, (ii) the study on the proper amount of oil to rejuvenate long-term aged bitumen based on rheological tests and its equivalent dosage in capsules, and (iii) the effect of the content of optimised capsules on the physical, mechanical, and healing properties of bituminous materials.

Author Contributions: Conceptualisation, J.N.-C.; methodology, J.L.C., J.N-C., I.G.-T. and L.E.A.-P.; software, J.L.C., J.N.-C. and L.E.A.-P.; validation, J.N.-C., I.G.-T., L.E.A.-P. and Q.L.; formal analysis, J.N.-C., J.L.C. and L.E.A.-P.; investigation, J.N.-C., J.L.C. and I.G.-T.; resources, J.N.-C. and L.E.A.-P.; data curation, J.L.C., J.N.-C. and I.G.-T.; writing—original draft preparation, J.L.C. and J.N.-C.; writing—review and editing, J.L.C., J.N.-C., L.E.A.-P., I.G.-T. and Q.L.; visualisation, J.L.C. and J.N.-C.; supervision, J.N.-C.; project administration, J.N.-C.; funding acquisition, J.N.-C. All authors have read and agreed to the published version of the manuscript.

Funding: This research was funded by the National Research and Development Agency (ANID) from the Government of Chile, through the Research Project FONDECYT Regular 2019 No. 1190027.

Data Availability Statement: Not applicable.

Acknowledgments: The authors express their gratitude to the students Matias Fuentes and Felipe Muñoz from LabMAT at the UBB for their technical support during the development of the testing program. The first author wishes to thank the financial support provided by the University of Bío-Bío for his internal PhD scholarship granted. Finally, the authors thank Jaime Zamorano from Gelymar S.A. for his help in the chemical characterisation of the alginate biopolymer used in this study.

Conflicts of Interest: The authors declare no conflict of interest.

References

1. Hager, M.D. *Self-Healing Materials*; Springer International Publishing AG: Cham, Switzerland, 2016; p. 287.
2. Read, J.; Whiteoak, D. *The Shell Bitumen Handbook*, 6th ed.; Shell Bitumen: London, UK, 2014; p. 6.
3. Kanellopoulos, A.; Norambuena-Contreras, J. *Fundamentals of Self-Healing Construction Materials*; Springer: Cham, Switzerland, 2022; ISBN 9783030868796.
4. Lv, Q.; Huang, W.; Zhu, X.; Xiao, F. On the investigation of self-healing behavior of bitumen and its influencing factors. *Mater. Des.* **2017**, *117*, 7–17. [CrossRef]

5. Gamarra, A.; Ossa, E.A. Thermo-oxidative aging of bitumen. *Int. J. Pavement Eng.* **2018**, *19*, 641–650. [CrossRef]
6. Liang, Y.; Wu, R.; Harvey, J.T.; Jones, D.; Alavi, M.Z. Investigation into the Oxidative Aging of Asphalt Binders. *Transp. Res. Rec.* **2019**, *2673*, 368–378. [CrossRef]
7. Zhang, X.; Wang, Q.; Liu, F.; Zhou, Z.; Wang, G.; Liu, X. Experimental characterization of the oxidative kinetic aging behavior of rejuvenated asphalt binder. *Constr. Build. Mater.* **2022**, *346*, 128488. [CrossRef]
8. Fang, Y.; Zhang, Z.; Yang, J.; Li, X. Comprehensive review on the application of bio-rejuvenator in the regeneration of waste asphalt materials. *Constr. Build. Mater.* **2021**, *295*, 123631. [CrossRef]
9. Behnood, A. Application of rejuvenators to improve the rheological and mechanical properties of asphalt binders and mixtures: A review. *J. Clean. Prod.* **2019**, *231*, 171–182. [CrossRef]
10. Liu, Y.; Su, P.; Li, M.; You, Z.; Zhao, M. Review on Evolution and Evaluation of Asphalt Pavement Structures and Materials. *J. Traffic Transp. Eng. Engl. Ed.* **2020**, *7*, 573–599. [CrossRef]
11. Anupam, B.R.; Sahoo, U.C.; Chandrappa, A.K. A methodological review on self-healing asphalt pavements. *Constr. Build. Mater.* **2022**, *321*, 126395. [CrossRef]
12. Gonzalez-Torre, I.; Norambuena-Contreras, J. Recent advances on self-healing of bituminous materials by the action of encapsulated rejuvenators. *Constr. Build. Mater.* **2020**, *258*, 119568. [CrossRef]
13. Xu, S.; Liu, X.; Tabaković, A.; Schlangen, E. A novel self-healing system: Towards a sustainable porous asphalt. *J. Clean. Prod.* **2020**, *259*, 120815. [CrossRef]
14. Sun, D.; Sun, G.; Zhu, X.; Guarin, A.; Li, B.; Dai, Z.; Ling, J. A comprehensive review on self-healing of asphalt materials: Mechanism, model, characterization and enhancement. *Adv. Colloid Interface Sci.* **2018**, *256*, 65–93. [CrossRef]
15. Li, Y.; Hao, P.; Zhang, M. Fabrication, characterization and assessment of the capsules containing rejuvenator for improving the self-healing performance of asphalt materials: A review. *J. Clean. Prod.* **2021**, *287*, 125079. [CrossRef]
16. Wang, X.; Guo, G.; Zou, F.; Zhao, H.; Li, Y. Enhancing self-healing properties of microcrack on aged asphalt incorporating with microcapsules encapsulating rejuvenator. *Constr. Build. Mater.* **2022**, *344*, 128123. [CrossRef]
17. Shu, B.; Wu, S.; Dong, L.; Norambuena-Contreras, J.; Yang, X.; Li, C.; Liu, Q.; Wang, Q. Microfluidic synthesis of polymeric fibers containing rejuvenating agent for asphalt self-healing. *Constr. Build. Mater.* **2019**, *219*, 176–183. [CrossRef]
18. Casado-Barrasa, R.; López-Montero, T.; Castro-Fresno, D.; Miró, R. Evaluation of the rejuvenation of asphalt by means of oil-saturated porous aggregates. *Constr. Build. Mater.* **2022**, *318*, 125825. [CrossRef]
19. Alpizar-Reyes, E.; Concha, J.L.; Mart, F.J.; Norambuena-Contreras, J. Biobased Spore Microcapsules for Asphalt Self-Healing. *ACS Appl. Mater. Interfaces* **2022**, *14*, 31296–31311. [CrossRef]
20. Yamaç, Ö.E.; Yilmaz, M.; Yalçın, E.; Kök, B.V.; Norambuena-Contreras, J.; Garcia, A. Self-healing of asphalt mastic using capsules containing waste oils. *Constr. Build. Mater.* **2021**, *270*, 121417. [CrossRef]
21. Norambuena-Contreras, J.; Arteaga-Perez, L.E.; Guadarrama-Lezama, A.Y.; Briones, R.; Vivanco, J.F.; Gonzalez-Torre, I. Microencapsulated bio-based rejuvenators for the self-healing of bituminous materials. *Materials* **2020**, *13*, 1446. [CrossRef]
22. Kumar Nayak, A.; Saquib Hasnain, M. *Alginates in Drug Delivery*; Academic Press: Cambridge, MA, USA, 2020.
23. Ruiz-Riancho, N.; Garcia, A.; Grossegger, D.; Saadoon, T.; Hudson-Griffiths, R. Properties of Ca-alginate capsules to maximise asphalt self-healing properties. *Constr. Build. Mater.* **2021**, *284*, 122728. [CrossRef]
24. Xu, S.; Liu, X.; Tabaković, A.; Lin, P.; Zhang, Y.; Nahar, S.; Lommerts, B.J.; Schlangen, E. The role of rejuvenators in embedded damage healing for asphalt pavement. *Mater. Des.* **2021**, *202*, 109564. [CrossRef]
25. Concha, J.L.; Arteaga-Pérez, L.E.; Alpizar-Reyes, E.; Segura, C.; Gonzalez-Torre, I.; Kanellopoulos, A.; Norambuena-Contreras, J. Effect of rejuvenating oil type on the synthesis and properties of alginate-based polynuclear capsules for asphalt self-healing. *Road Mater. Pavement Des.* **2022**, 1–26. [CrossRef]
26. Norambuena-Contreras, J.; Arteaga-Pérez, L.E.; Concha, J.L.; Gonzalez-Torre, I. Pyrolytic oil from waste tyres as a promising encapsulated rejuvenator for the extrinsic self-healing of bituminous materials. *Road Mater. Pavement Des.* **2021**, *22*, S117–S133. [CrossRef]
27. Norambuena-Contreras, J.; Concha, J.L.; Arteaga-Pérez, L.E.; Gonzalez-Torre, I. Synthesis and Characterisation of Alginate-Based Capsules Containing Waste Cooking Oil for Asphalt Self-Healing. *Appl. Sci.* **2022**, *12*, 2739. [CrossRef]
28. Sun, D.; Pang, Q.; Zhu, X.; Tian, Y.; Lu, T.; Yang, Y. Enhanced Self-Healing Process of Sustainable Asphalt Materials Containing Microcapsules. *ACS Sustain. Chem. Eng.* **2017**, *5*, 9881–9893. [CrossRef]
29. Xu, S.; Tabaković, A.; Liu, X.; Palin, D.; Schlangen, E. Optimization of the calcium alginate capsules for self-healing asphalt. *Appl. Sci.* **2019**, *9*, 468. [CrossRef]
30. Norambuena-Contreras, J.; Yalcin, E.; Garcia, A.; Al-Mansoori, T.; Yilmaz, M.; Hudson-Griffiths, R. Effect of mixing and ageing on the mechanical and self-healing properties of asphalt mixtures containing polymeric capsules. *Constr. Build. Mater.* **2018**, *175*, 254–266. [CrossRef]
31. Hofko, B.; Zia Alavi, M.; Grothe, H.; Jones, D.; Harvey, J. Repeatability and Sensitivity of FTIR ATR Spectral Analysis Methods for Bituminous Binders. *Mater. Struct.* **2017**, *50*, 187. [CrossRef]
32. McClements, D.J. Critical review of techniques and methodologies for characterization of emulsion stability. *Crit. Rev. Food Sci. Nutr.* **2007**, *47*, 611–649. [CrossRef]
33. *ASTMD792-13*; Standard Test Methods for Density and Specific Gravity (Relative Density) of Plastics by Displacement. ASTM International: West Conshohocken, PA, USA, 2013.

34. *ASTM D695-02a*; Standard Test Method for Compressive Properties of Rigid Plastics. ASTM International: West Conshohocken, PA, USA, 2002.
35. Chimeno-Trinchet, C.; Murru, C.; Elena Díaz-García, M.; Fernández-González, A.; Badía-Laíño, R. Artificial Intelligence and Fourier-Transform Infrared Spectroscopy for Evaluating Water-Mediated Degradation of Lubricant Oils. *Talanta* **2020**, *219*, 121312. [CrossRef]
36. Stuart, B. *Infrared Spectroscopy: Fundamentals and Applications*, 1st ed.; John Wiley & Sons: West Sussex, UK, 2004; pp. 71–93. [CrossRef]
37. Goh, K.M.; Maulidiani, M.; Rudiyanto, R.; Wong, Y.H.; Ang, M.Y.; Yew, W.M.; Abas, F.; Lai, O.M.; Wang, Y.; Tan, C.P. Rapid Assessment of Total MCPD Esters in Palm-Based Cooking Oil Using ATR-FTIR Application and Chemometric Analysis. *Talanta* **2019**, *198*, 215–223. [CrossRef]
38. Gouveia De Souza, A.; Oliveira Santos, J.C.; Conceição, M.M.; Dantas Silva, M.C.; Prasad, S. A thermoanalytic and kinetic study of sunflower oil. *Braz. J. Chem. Eng.* **2004**, *21*, 265–273. [CrossRef]
39. Zhang, C.J.; Liu, Y.; Cui, L.; Yan, C.; Zhu, P. Bio-based calcium alginate nonwoven fabrics: Flame retardant and thermal degradation properties. *J. Anal. Appl. Pyrolysis* **2016**, *122*, 13–23. [CrossRef]
40. Razali, M.S.; Khimeche, K.; Boudjellal, A.; Ramdani, N. Effect of Newly Developed Sintered Seashell on the Microhardness Properties of Biocomposites. *Mater. Lett.* **2021**, *291*, 129565. [CrossRef]

polymers

MDPI

Article

Preparation and Macro-Micro Properties of SBS/REOB Modified-Rejuvenated Asphalt

Jin Li [1,*], Li Zhu [1], Xiaozhou Yan [1,*], Chongsheng Xin [2], Miaozhang Yu [1] and Degang Cheng [2]

[1] School of Transportation Civil Engineering, Shandong Jiaotong University, Jinan 250357, China
[2] Engineering Department of Jinan Kingyue Highway Engineering Company Limited, Jinan 250013, China
* Correspondence: sdzblijin@163.com (J.L.); 22107017@stu.sdjtu.edu.cn (X.Y.); Tel.: +86-136-7882-4225 (J.L.); +86-158-0672-1768 (X.Y.)

Abstract: To solve the problem that waste oil residues cannot be utilized and to reuse the aged asphalt, suitable modifiers were selected to compound the aged asphalt with waste oil residues to study its performance. SBS/REOB modified-rejuvenated asphalt was prepared by a high-speed shearing mechanism with aged asphalt, Recycled Engine Oil Bottom (REOB), Styrenic Block Copolymers (SBS) modifier, and stabilizer. The effects of SBS content, REOB content, shear time, and shear rate on the conventional physical properties of asphalt were studied by orthogonal grey correlation analysis, and the optimum preparation scheme of SBS/REOB modified-rejuvenated asphalt was determined. The high and low temperature rheological properties of SBS/REOB modified-rejuvenated asphalt were studied using the Multiple Stress Creep Recover (MSCR) test and bending beam rheological (BBR) test. The mechanism of SBS/REOB on the modification and regeneration of aged asphalt was explored through four component tests and Fourier transforms infrared spectroscopy. The results show that the optimum preparation scheme is 4.5% SBS dosage, 9% REOB dosage, 50~60 min shear time, and 4500 r/min shear rate. The addition of SBS improves the elastic recovery performance and high temperature deformation resistance of REOB rejuvenated asphalt. At the same time, the S-value decreases and the m-value increases, which significantly improves the low temperature cracking resistance of REOB rejuvenated asphalt. The addition of REOB achieves component blending and regeneration of aged asphalt by supplementing the light components. After the addition of SBS absorbs the light component and swelling reaction occurs, the whole modification-regeneration process is mainly physical co-mixing and co-compatibility.

Keywords: recycled engine oil bottom; SBS modifier; rheological properties; the modification-rejuvenation mechanism

Citation: Li, J.; Zhu, L.; Yan, X.; Xin, C.; Yu, M.; Cheng, D. Preparation and Macro-Micro Properties of SBS/REOB Modified-Rejuvenated Asphalt. *Polymers* **2022**, *14*, 5071. https://doi.org/10.3390/polym14235071

Academic Editors: Wei Jiang, Quantao Liu, Jose Norambuena-Contreras and Yue Huang

Received: 31 October 2022
Accepted: 18 November 2022
Published: 22 November 2022

1. Introduction

Asphalt regeneration refers to coordinating the components and improving the performance of aged asphalt by using new asphalt or regenerant with appropriate viscosity and cooperating with specific preparation methods. The performance indicators of rejuvenated asphalt at each stage are closely related to the dosage and type of regenerant [1–3]. Incorporating pellet with reclaimed asphalt pavement (RAP) is beneficial in improving the sustainability and cost-efficiency of the asphalt mixtures industry [4]. Some regenerant may induce some differences in one or more of the situations by encouraging the characteristics of asphalt [5]. Based on the principle of creative reuse of waste materials, domestic and foreign experts and scholars began to focus on researching and developing waste oil-based products as regenerating agents [6–8].

As early as 1992, Herrington [9] conducted a study using waste automotive motor oil residues that had undergone reduced pressure distillation as a stabilizing regenerant for road asphalt. Ackbarali D.S et al. [10] comparatively studied the mixture performance of waste motor oil rejuvenated asphalt and normal emulsified asphalt and found that the former has better road performance. Zhang Yan, Chen Meizhu et al. [11] studied the regeneration effect of waste engine oil and waste soybean oil on aged asphalt from macro and micro perspectives comparatively. The results showed that the optimal blending amounts of waste engine oil and waste soybean oil were 5% and 6%, respectively. The regeneration effect of waste engine oil was better than that of waste soybean oil.

Generally, 70–80% of recycled engine oil (REO) can be effectively rejuvenated by the aforementioned processes. However, the remaining residue (accounting for 20–30%) cannot be effectively recycled due to the presence of many impurities; this residue is ultimately called recycled engine oil bottom (REOB). Miaozhang Yu [12] compared the performance and content of each component of three REOB and regulated regenerants to verify the theoretical feasibility of REOB as a regenerant. Li Jin [13–16] et al. used REOB as an asphalt regenerant and investigated the high and low temperature performance of REOB rejuvenated asphalt using asphalt conventional performance test, dynamic shear rheology (DSR) test, and asphalt bending creep stiffness (BBR) test, and found that 7% of REOB could achieve complete regeneration of the three major indexes of aged asphalt, but the low temperature performance recovery was insufficient compared to the original sample base asphalt. It is usually challenging to return aged asphalt regeneration to the level of performance of the new base asphalt, while the use of composite modification can be more targeted to improve and optimize the asphalt performance within a controlled range, improving its road performance [17–19]. The further modification of the rejuvenated asphalt has also been an important research direction in the field of composite modified asphalt in recent years.

In summary, the use of suitable modifiers with REOB regenerator on the aged asphalt composite modification-regeneration, its environmental protection, and asphalt performance optimization has an important double significance—to achieve the "waste" into "treasure". In this paper, an orthogonal test was designed to determine the preparation scheme of SBS/REOB modified-rejuvenated asphalt, and then to investigate its conventional properties, rheological properties, and microstructural composition to evaluate the improvement effect of SBS/REOB on the properties of aged asphalt.

2. Materials and Methods

2.1. Raw Materials

2.1.1. Base Asphalt

The 70# A-grade road petroleum asphalt produced by Santobo Petrochemical Co., Ltd. (Binzhou, China), aged in the laboratory, as the raw asphalt for preparing SBS/REOB modified-rejuvenated asphalt. Based on the requirements of the <<Standard Test Methods of Bitumen and Bituminous Mixtures for Highway Engineering>>, the properties of the original base asphalt were measured as shown in Table 1.

Table 1. Performance index of 70# base asphalt.

Test Project		Test Results	Standard Indicators
Needle penetration 100 g, 5 s, 25 °C/0.1 mm		73.2	60–80
Insertion index PI		−1.1	−1.5~1.0
Softening point/°C		49.7	≥46
60 °C dynamic viscosity/Pa·s		271	≥180
Viscosity (135 °C) /Pa·s		0.266	
10 °C degrees/cm		73.0	≥25
15 °C degrees/cm		>100	≥100
Flash point (COC)/°C		280	≥260
TFOT (Thin Film)	Quality change/%	−0.255	≤±0.8
Oven Test) after	Residual penetration ratio (25 °C)	68.3	≥61
163 °C, 5 h	Residual penetration ratio (25 °C)	10.6	≥6

2.1.2. REOB Regenerant

The REOB selected for this test came from a waste engine oil treatment plant in Zibo, Shandong Province, China, whose primary treatment process was: waste engine oil filtration-thin film distillation-sugar aldehyde refining-white clay process. From the physical and chemical properties in Table 2, the REOB meets the specification of a conventional RA-1 regenerant.

Table 2. Physicochemical properties of REOB.

Inspection Items		R-1	RA-1 Regenerant
Density/(g/cm^3)		0.911	The measured
Flash point/°C		258	≥220
Aromatic content/%		83.1	The measured
Saturation fraction content/%		4.6	≤30
TFOT post	Viscosity ratio	1.29	≤3
	Quality change /%	−1.89	−4~4

2.1.3. SBS Modifier

The modifier was adopted from SBS1401 (YH-792), a thermoplastic styrene-butadiene rubber produced by Yueyang Baling Petrochemical (Yueyang, China), which is linear SBS with an S/B block ratio of 40/60.

2.1.4. Stabilizer

The active ingredient of the asphalt stabilizer used was sulfur, with a content of 99.9% or more, a moisture content of 0.02%, a melting point of 120 °C, and an 80-mesh sieve margin of 0.1%.

2.1.5. Preparation of the Aged Asphalt Binders

Aged asphalt preparation method: we selected 40 cm × 30 cm × 4.8 cm stainless steel plates, 600 g of new asphalt was poured into each plate. The asphalt had a plate thickness of about 0.5 cm, and was put into the 163 °C oven heating insulation for 48 h, and stirred every 2 h to ensure that the material is uniformly heated to prevent the occurrence of surface-crusting phenomenon. The performance test results are shown in Table 3.

Table 3. Comparison of asphalt performance indexes under different aging methods.

Test Project	48 h Oven Aging	PAV	5 Years of Recycling Asphalt	10 Years of Recycling Asphalt
Penetration (25 °C)/mm	22.0	26.6	24.6	17.6
Softening point/°C	67.6	62.3	64.9	76.6
Ductility (10 °C, 5 cm/min)/cm	0.2	1.7	2.1	brittle break
Viscosity (135 °C) /Pa·s	2.247	1.967	2.118	2.763

Table 3 showed that the 48 h oven-aged asphalt was more heavily aged than PAV-aged asphalt, and the performance of "48 h oven" aged asphalt was between "5 years of service" and "10 years of service" compared to the extracted and rejuvenated asphalt. The performance of "48 h oven" aged asphalt was between that of "5 years' service" and "10 years' service" recovered asphalt.

2.2. Design of Orthogonal Test

We set three levels, respectively, designed a L9(3^4) orthogonal test, as shown in Table 4, and prepared 9 groups of SBS/REOB modified-rejuvenated asphalt under different conditions.

Table 4. Orthogonal test design scheme.

Test Group	Horizontal Combinations	Test Factor			
		A	B	C	D
		REOB Dosage /%	SBS Dosage /%	Shearing Time/min	Shear Rate /(r/min)
1	A1B1C1D1	1 (11)	1 (3.5)	1 (40)	1 (3500)
2	A1B2C2D2	1 (11)	2 (4.0)	2 (50)	2 (4500)
3	A1B3C3D3	1 (11)	3 (4.5)	3 (60)	3 (5500)
4	A2B1C3D2	2 (9)	1 (3.5)	2 (50)	3 (5500)
5	A2B2C3D1	2 (9)	2 (4.0)	3 (60)	1 (3500)
6	A2B3C1D2	2 (9)	3 (4.5)	1 (40)	2 (4500)
7	A3B1C3D2	3 (7)	1 (3.5)	3 (60)	2 (4500)
8	A3B2C1D3	3 (7)	2 (4.0)	1 (40)	3 (5500)
9	A3B3C2D1	3 (7)	3 (4.5)	2 (50)	1 (3500)

2.3. *Preparation Process*

(a) REOB was added to the thermally aged asphalt, and a specific mass fraction of REOB (factor A) was dispersed into the asphalt at 150 °C using a mixer at 2500 r/min for 20 min.
(b) Set the shear speed to 3000 r/min and add the required SBS (factor B) within 5 min. Control the temperature to 190 °C.
(c) After shearing at a certain speed (factor D) for a certain time (factor C), add 6% stabilizer of SBS dosage, and then stir at a low speed of 1000 r/min for 10 min to remove air bubbles.
(d) Dissolution development at 160 °C for 1 h.

2.4. *Experimental Method*

2.4.1. Routine Performance Test

Nine groups of modified-rejuvenated asphalt were tested for penetration, softening point, ductility, and elastic recovery rate under the orthogonal test scheme. The results were analyzed to obtain the optimal dosing values of SBS and REOB as well as the optimal shear time and shear rate.

2.4.2. High-Temperature Rheological Property Test

The Multiple Stress Creep Recovery test (MSCR) used the delayed elastic recovery performance of asphalt under applied stress to evaluate the high-temperature performance of the binder, and the cumulative strain of asphalt had a good correlation with the high-temperature performance of asphalt [20–22].

Referring to the American Association of State Highway and Transportation Officials (AASHTO) test protocol, MSCR tests were conducted using a dynamic shear rheometer (DSR) on SBS/REOB modified-rejuvenated asphalt and SBS modified asphalt and REOB-rejuvenated asphalt with the same admixture dose under optimal preparation conditions. The test temperature was set at 64 °C. According to the loading stress, the test was divided into two stages of 0.1 kPa and 3.2 kPa stress loading, and each step was loaded in the form of 10 cycles, 10 s/cycle (loading lasts 1 s, unloading lasts 9 s) to obtain the creep recovery rate R and irrecoverable creep flexibility J_{nr} for each asphalt [23,24].

2.4.3. Low-Temperature Rheological Property Test

Standard size 101.6 mm × 12.7 mm × 6.4 mm beam specimens of PAV aged asphalt was prepared and tested by Bending Beam Rheometer (BBR) at −6 °C, −12 °C, and −18 °C for the bending creep modulus S and slope m of the creep curve.

Among the S and m results of the 8th, 15 s, 30 s, 60 s, 120 s and 240 s automatically tested by the computer, the test result at 60 s was taken as the final value.

2.4.4. Microscopic Test

The four components of asphalt were tested and analyzed using rod-thin-layer chromatography/hydrogen flame ionization detection (TLC-FID) and the functional group changes were analyzed using infrared spectroscopy to reveal the modification-regeneration mechanism of SBS/REOB on aged asphalt [25,26].

3. Results and Discussion

3.1. Determination of the Best Preparation Conditions

The sample preparation and sample performance tests were carried out in groups according to the orthogonal design, and the physical property test results of each group of asphalt samples are shown in Table 5.

Table 5. Results of the orthogonal test.

Test Group	Penetration/mm	Softening Point/°C	Ductility/cm	Elastic Recovery/%
1	84	57.5	23.9	81
2	80.2	61.2	25.2	84
3	63.8	63.9	29.3	83
4	78.3	55.7	26.5	79
5	76	62.3	27.2	88
6	64.5	63.7	33.4	86
7	76.9	58.3	26.8	77
8	79.1	64.1	29.8	89
9	62.7	65.2	30.5	81

Based on the experimental results, the data were processed using grey correlation analysis with the following procedure:

(a) The matrix sequences are listed in Table 5 Orthogonal test results. Here, the mean value of each group of indicators is taken as the reference sequence, i.e., X0 = (62.10, 63.40, 58.75, 60.38, 64.03, 61.15, 59.75, 66.10, 58.35).
(b) Dimensionless treatment of the index series
(c) Solving the difference series with two polar differences
(d) Calculate the correlation coefficient, degree, and proportion of factors.
(e) Results processing: A weighted average score is given to each indicator factor based on obtaining the proportion of each factor. We call this method the grey correlation composite scoring method.

Let the difference between the maximum value and the minimum value of the test result of each test index be h, b_{ij} = the proportion of each factor/h, the score value is Fi = $\sum_{j=1}^{4} b_{ij} \times \textit{test index value}$. Since the smaller the penetration index value is, the better the viscosity is, the value of this index was taken as negative in the scoring calculation here.

The obtained scoring results are processed for polar difference calculation, and the method is called the gray correlation polar difference method. The obtained scores and the results of the polar difference calculation are shown in Table 6.

Table 6. Gray correlation composite score and extreme difference analysis results.

Horizontal Combinations	Test Factor				Rating Fi
	A	B	C	D	
	REOB Dosage /%	SBS Dosage/%	Shearing Time/min	Shear Rate/(r/min)	
A1B1C1D1	1 (11)	1 (3.5)	1 (40)	1 (3500)	313.73
A1B2C2D2	1 (11)	2 (4.0)	2 (50)	2 (4500)	338.18
A1B3C3D3	1 (11)	3 (4.5)	3 (60)	3 (5500)	370.38
A2B1C3D2	2 (9)	1 (3.5)	2 (50)	3 (5500)	408.58
A2B2C3D1	2 (9)	2 (4.0)	3 (60)	1 (3500)	459.61
A2B3C1D2	2 (9)	3 (4.5)	1 (40)	2 (4500)	489.21
A3B1C3D2	3 (7)	1 (3.5)	3 (60)	2 (4500)	409.39
A3B2C1D3	3 (7)	2 (4.0)	1 (40)	3 (5500)	470.65
A3B3C2D1	3 (7)	3 (4.5)	2 (50)	1 (3500)	475.26
$\overline{K_1}$	340.76	377.23	425.53	416.20	
$\overline{K_2}$	452.77	423.81	407.34	412.26	Factor
$\overline{K_3}$	451.77	444.95	413.13	417.54	Priorities
Range R	111.70	67.72	17.19	4.28	A > B > C > D
Optimum	A2	B3	C1	D3	

(f) Analysis of results.

REOB content was the most important factor affecting the performance of asphalt samples, followed by SBS content, which had the greatest influence on the penetration index. The shear time and shear rate mainly affected the elastic recovery performance of materials and had the least influence on the softening point index. The degree of influence of each factor on the product effect is REOB dosage > SBS dosage > shear time > shear rate.

Based on the evaluation principle that the higher the overall score is, the better the product performance, the highest score of 489.21 was obtained for the A2-B3-C1-D2 solution, i.e., "9% REOB, 4.5% SBS dosage, 50 min shear at 4500 r/min" was the preferred preparation condition. The results of the grey correlation ANOVA showed that the preferred solution was A2-B3-C1-D3, i.e., "9% REOB, 4.5% SBS dosage, shear at 4500 r/min for 60 min". The difference between them only lies in the shearing time, so the optimal preparation condition is recommended as "9% mass fraction of REOB + 4.5% SBS, shearing 50~60 min at 4500 r/min".

3.2. High-Temperature Rheological Properties

The SBS/REOB modified-rejuvenated asphalt under two processes was selected as the main object of the study. The MSCR test was carried out in comparison with the as-built base asphalt, aged asphalt, and SBS modified asphalt. Figures 1 and 2 show the MSCR test results for each asphalt at 0.1 kPa and 3.2 kPa stress levels, respectively.

From Figures 1 and 2, it can be seen that the strains generated by 9% REOB + aged asphalt were at the highest level of all stress levels. The peak strains at 0.1 kPa and 3.2 kPa were 12 and 15 times higher than those of SBS modified asphalt, which shows that the SBS modification of REOB rejuvenated asphalt can effectively improve its stability performance under externally applied loads. The difference between the strain values of SBS/REOB modified-rejuvenated asphalt and REOB modified asphalt becomes more prominent as the load becomes stronger, i.e., the resistance to load deformation of the modified-rejuvenated asphalt is more prominent at high stress levels. From the comparison of the strain values generated by the three SBS modified asphalts, 4.5% SBS + 9% REOB + 50 min + aged asphalt produced the smallest strain value and was better than 4.5% SBS + base asphalt, which had the best high temperature resistance to deformation.

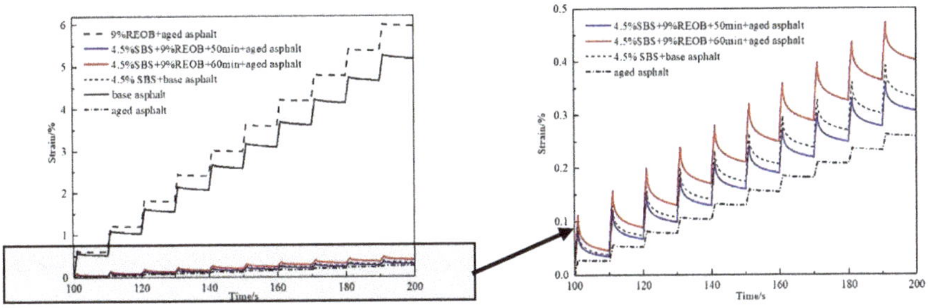

(a) (b)

Figure 1. MSCR test results at 0.1 kPa stress level: (**a**) Test results of each asphalt; (**b**) Test results of aged asphalt, 4.5% SBS + 9% REOB + 50 min + aged asphalt, 4.5% SBS + 9% REOB + 60 min + aged asphalt and 4.5% SBS + base asphalt.

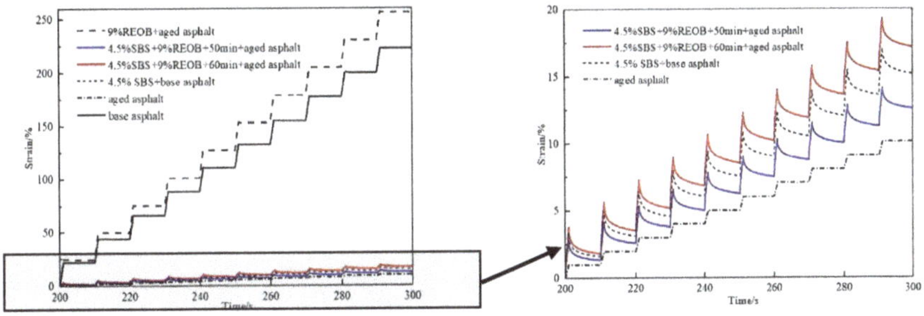

(a) (b)

Figure 2. MSCR test results at 3.2 kPa stress level: (**a**) Test results of each asphalt; (**b**) Test results of aged asphalt, 4.5% SBS + 9% REOB + 50 min + aged asphalt, 4.5% SBS + 9% REOB + 60 min + aged asphalt, and 4.5% SBS + base asphalt.

Additionally, creep recovery rate R, % and irrecoverable creep flexibility J_{nr}, kPa^{-1} are usually used to evaluate the high-temperature creep recovery performance of asphalt to reflect the high temperature anti-rutting performance. Practical pavement applications with a comparative index of stress sensitivity $J_{nr,diff}$ at 3.2 kPa versus 0.1 kPa.

Table 7 gives the creep recovery rates $R_{0.1}$, $R_{3.2}$, and irrecoverable creep flexibility $J_{nr,\,0.1}$, $J_{nr,\,3.2}$ for the above four asphalt bonds at 0.1 kPa and 3.2 kPa stress levels, respectively, according to the results plotted in Figures 3 and 4.

Table 7. Calculated results of MSCR tests for each asphalt at different stress levels.

Number	Asphalt Type	$R_{0.1}$/%	$R_{3.2}$/%	$J_{nr,0.1}$/ kPa^{-1}	$J_{nr,3.2}$/ kPa^{-1}	$J_{nr,diff}$/%
B	Base asphalt	9.26	1.57	5.085	6.811	33.94
A	Aged asphalt	0.93	0.30	0.278	0.310	11.51
9R + A	9% REOB + aged asphalt	3.94	0.57	5.909	7.788	31.80
4.5S + 9R + 50 + A	4.5% SBS + 9% REOB + 50 min + aged asphalt	67.34	59.65	0.375	0.466	24.27
4.5S + 9R + 60 + A	4.5% SBS + 9% REOB + 60 min + aged asphalt	66.22	59.46	0.270	0.342	26.67
4.5S + B	4.5% SBS + base asphalt	71.71	63.92	0.249	0.302	21.28

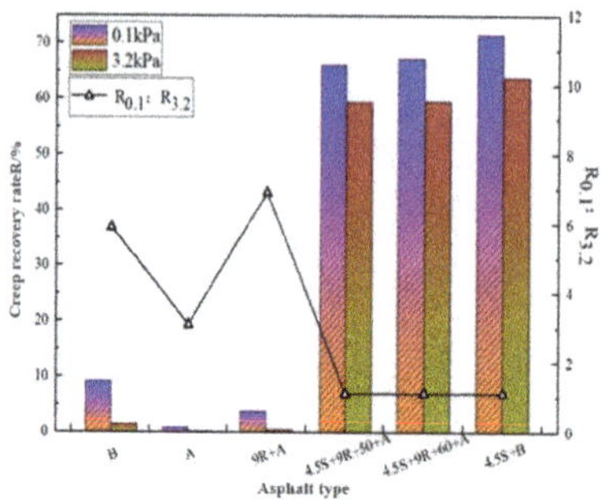

Figure 3. Comparison of R for different asphalt types in the Figure.

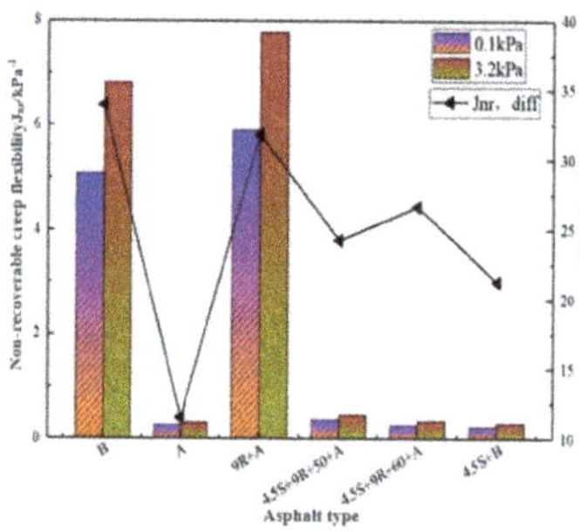

Figure 4. Comparison of J_{nr} for different asphalt types.

Figure 3 combined with Table 7 test results shows that the R of 9% REOB + aged asphalt under all stress levels is more than 1 times higher than that of aged asphalt but cannot recover to the level of base asphalt. Under 0.1 kPa stress level, the modified-aged asphalt R value can be more than 50%, compared to REOB rejuvenated asphalt creep recovery performance improved by more than 16 times. Overall, 3.2kPa under the R reached more than 30%, and 4.5S + 9R + 50 + A can achieve 4.5S + B of its R value of 93.3%. This indicates that SBS makes the REOB rejuvenated asphalt binder in the proportion of elasticity increases, so that the state of the binder is more elastomeric, SBS on rejuvenated asphalt "elasticity" effect is obvious, so the elastic recovery performance is improved, but its "elasticity" ability is limited, and cannot fully reach the level of ordinary SBS-modified asphalt. The sensitivity of different materials to high stress is different, and the base asphalt and 9% REOB rejuvenated asphalt have the highest $R_{0.1}$:$R_{3.2}$ values, 5.9 and 10.1, respectively, while the $R_{0.1}$:$R_{3.2}$ of modified-rejuvenated asphalt is not more than 2, more "high temperature", "heavy load", under the resistance of deformation, with 4.5S + 9R + 50 + A best.

As can be seen from Figure 4, the J_{nr} values of modified-rejuvenated asphalt are significantly lower than those of base asphalt and rejuvenated asphalt. At a high stress level of 3.2 kPa, the J_{nr} values of 4.5S + 9R + 50 + A and 4.5S + 9R + 60 + A decreased by 94.0% and 95.6%, respectively, compared with those of REOB rejuvenated asphalt, which shows that the J_{nr} values of 4.5S + 9R + 60 + A modified-rejuvenated asphalt are the best. This result indicates that the 4.5S + 9R + 60 + A modified-rejuvenated asphalt has the best high temperature deformation resistance. In addition to the aged asphalt, the $J_{nr,diff}$ values of 4.5S + B were the smallest, 4.5S + 9R + 50 + A was the second smallest, less than 25%, and the $J_{nr,diff}$ values of B and 9R + A were more than 30%, indicating that the modified-rejuvenated asphalt materials prepared under this process are less sensitive to stress changes, which is closely related to the good stability of SBS modifier. The $J_{nr,diff}$ value of asphalt binder has excellent stress sensitivity at less than 75%, which can make the asphalt pavement have better rutting resistance, and the materials prepared by the modification-rejuvenated process in this study all meet this requirement.

The differences between R and J_{nr} values of modified-rejuvenated asphalt obtained at different shear times are small, with 4.5S + 9R + 50 + A showing more excellent elastic recovery performance and stress sensitivity, and 4.5S + 9R + 60 + A having more advantages in high temperature deformation resistance.

3.3. Low-Temperature Rheological Properties

The US SHRP specification states that for PAV-aged asphalt, the stiffness modulus value should be less than 300 MPa at 60 s loading and the creep rate m should be not less than 0.30. From the BBR test results of each asphalt at different temperatures in Figure 5, it can be seen that the S and m values of the aged asphalt at −12 °C are far above this specification, so the aging asphalt is not continued to be tested at lower temperatures. Therefore, the test is not continued at lower temperatures.

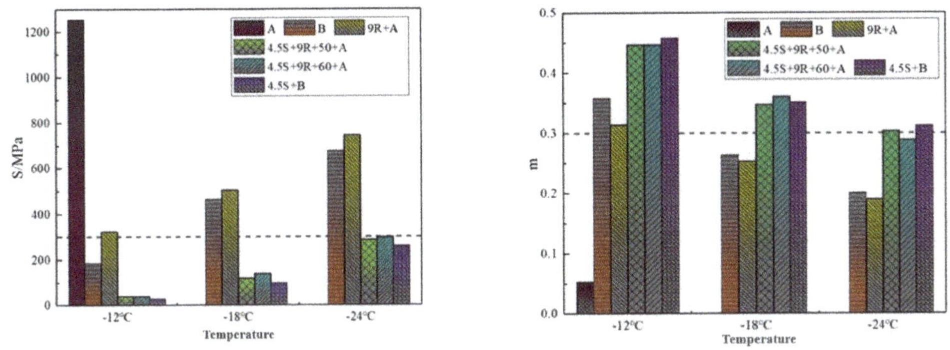

(**a**) (**b**)

Figure 5. BBR test results of each asphalt at different temperatures: (**a**) Modulus of rigidity; (**b**) Creep rate.

As shown in Figure 5, comparing the S-value results of different materials at each temperature, it can be seen that the S-value of 9% REOB rejuvenated asphalt 9R + A is significantly lower and the m-value is significantly higher than that of aged asphalt, indicating that the use of REOB on aged recycled can improve the low-temperature performance of asphalt, but does not reach the low-temperature performance level of base asphalt B. SBS/REOB modified-rejuvenated asphalt S-value and lower than B with 9R + A, m-value increased, indicating that modified-rejuvenated asphalt relative to rejuvenated asphalt low-temperature performance to obtain further improvement, and better than the base asphalt.

At −12 °C and −18 °C test temperatures, the S and m values of the two modified-rejuvenated asphalt 4.5S + 9R + 50 + A and 4.5S + 9R + 60 + A prepared at different shear times can be basically achieved to the level of ordinary SBS modified asphalt 4.5S + B. At −24 °C, the m of 4.5S + 9R + 60 + A cannot meet the SHRP specification, and its low temperature cracking resistance is not as good as 4.5S + 9R + 50 + A. The reason is that the excessive shear time leads to different degrees of aggregation of SBS particles, which destroys the composition of the SBS modified asphalt network structure, or excessive shear makes the SBS polymer chain break, and the asphalt has partially aged. Therefore, from the viewpoint of low temperature performance, 50 min is recommended as the optimal shear time for SBS/REOB modified-rejuvenated asphalt preparation.

3.4. *Micro-Mechanism Analysis*

3.4.1. Four Components

Aged asphalt, 70# base asphalt, 9% REOB rejuvenated asphalt, 4.5% SBS modified asphalt, and 4.5% SBS + 9% REOB modified-rejuvenated asphalt were selected for four-component analysis, and the results are shown in Figure 6.

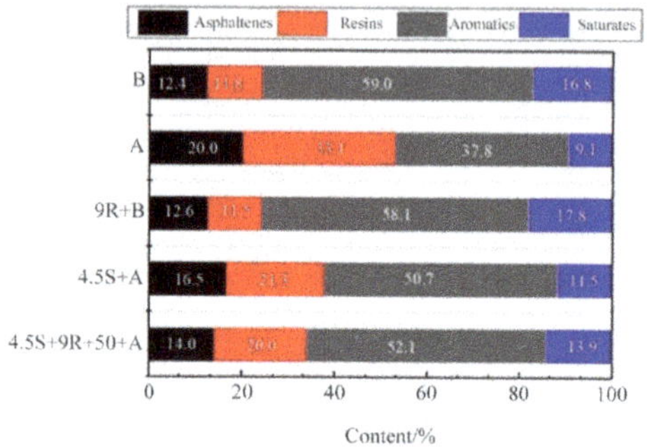

Figure 6. Percentage of four components of different types of asphalt.

(1) As shown in Figure 6, compared with the base asphalt, the change amounts of asphaltenes resins, aromatics, and saturates in the aged asphalt are 7.6, 21.3, −22.2, and −7.7, respectively, which means that the saturates and aromatics of light components are reduced, and the asphaltene and resins of a recombinant fraction are significantly increased, and component migration occurs. The change of four groups of rejuvenated asphalt compared with aging asphalt is −7.4, −22.6, 20.3, 8.7, the light component increases, and aging asphalt light component loss, which indicates that the high proportion of aromatics contained in REOB can supplement the asphalt in the aging process of the lack of aromatics part, so that the light component content increased, the recombination fraction decreased, to achieve asphalt regeneration, the regeneration has restored the composition of the asphalt components that had changed during the aging process, which resulted in macroscopic performance recovery.

Based on the blending component theory, assuming that REOB regenerates the aged asphalt and its mechanism is pure component blending, the content of each component of the rejuvenated asphalt should satisfy the calculation result of Equation (1).

$$P = m_a P_a + m_r P_r, \quad (1)$$

where P is the proportion of a component in REOB rejuvenated asphalt (%); P_a is the proportion of a component in aged asphalt (%); P_r is the proportion of a component in REOB (%); m_a is the blending ratio of aged asphalt in rejuvenated asphalt; m_r is the blending ratio of REOB in rejuvenated asphalt.

The results of the measured and calculated values of the four components of 9% REOB rejuvenated asphalt are shown in Table 8.

Table 8. Calculated results of the four components of 9% REOB.

Object		Asphaltenes/%	Resins/%	Aromatics/%	Saturates/%
Aged asphalt		20.0	33.1	37.8	9.1
REOB		3.6	8.7	83.1	4.6
9% REOB	Measured value	12.6	11.5	58.1	17.8
	Calculated values	18.6	31.1	41.5	8.7

Comparing the measured values with the theoretical calculated values, it is found that there are apparent differences between them. Therefore, the regeneration process is not a pure component reconciliation theory, and there are physical or chemical reactions and migration among the four components that cannot be ignored.

(2) The SBS modification of base asphalt and REOB asphalt also caused the redistribution of the components in asphalt. The proportion of each component is reduced in the light component and increased in the recombination component, in which the light component is mostly aromatics and a small part of saturates, which corresponds to the macroscopic properties of ductility, softening point increase, and penetration decrease, further verifying from the microscopic chemical components that the modified-renewed asphalt is improved by SBS absorption of light component swelling reaction. The same amount of SBS is used in the modified-rejuvenated asphalt, which is less migratory than the directly modified asphalt, indicating that it is more difficult to modify the rejuvenated asphalt, so its performance improvement is limited.

3.4.2. Infrared Spectrum

The transmittance-wave number infrared absorption spectra (FTIR) plots of the regenerant REOB with modified asphalt, rejuvenated asphalt, and modified-rejuvenated asphalt with base asphalt are given in Figures 7 and 8, respectively.

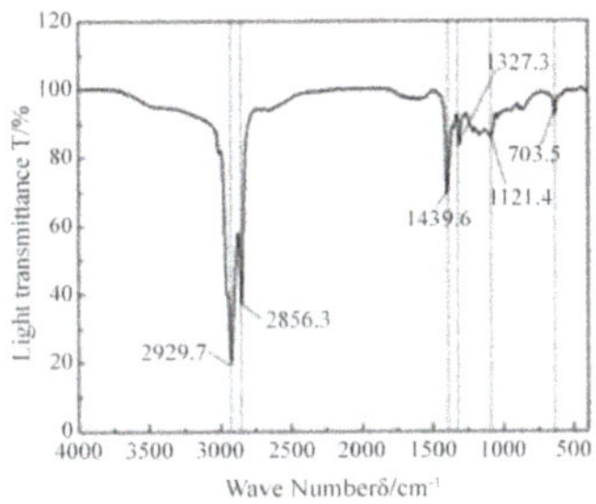

Figure 7. IR spectrum of REOB.

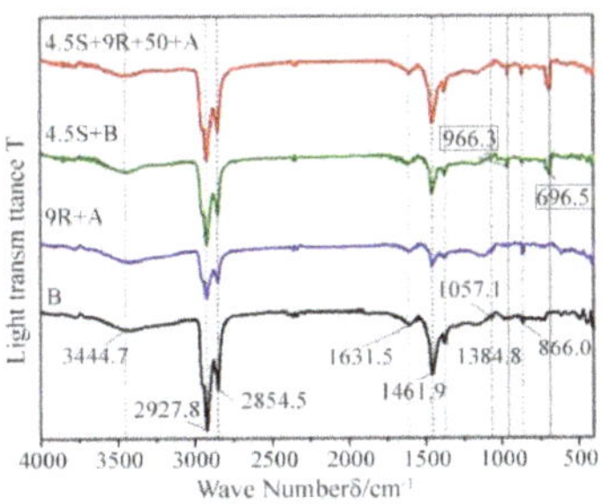

Figure 8. Infrared spectra of different types of asphalt.

From Figure 7, it can be seen that the REOB showed sharp strong absorption peaks at 2929.7 cm^{-1} and 2856.3 cm^{-1}, which were generated by CH_2 asymmetric vibration and symmetric vibration, respectively, indicating that the regenerant contains non-polar methylene; the asymmetric bending vibration peak of CH_3 appeared at 1439.6 cm^{-1}, and the generation of these characteristic peaks indicates that the REOB contains a higher proportion of saturated hydrocarbons. A strong absorption peak appears at 703 cm^{-1} in the fingerprint area, resulting from the vibration of the outer C=H bending surface of the aromatic ring, indicating that the REOB contains light components of aromatic hydrocarbons. In summary, it can be inferred that REOB mainly comprises alkanes and saturated hydrocarbons such as cycloalkanes and aromatic compounds.

Figure 8 shows that the positions of the peaks appearing in different asphalt are generally consistent, and there are differences in the intensity of the absorption peaks. The four asphalts have broad and weak absorption peaks at 3444.7 cm^{-1}, which are considered to be generated by N-H bonding vibrations or O-H bonding vibrations in alcohols and phenols. The absorption peaks at 2927.8 cm^{-1} and 2854.5 cm^{-1} are generated by methylene CH_2 vibrations, while 1461.9 cm^{-1} and 1384.8 cm^{-1} show characteristic absorption peaks generated by methyl CH_3 bending vibrations. 4.5S + 9R + 50 + A modified-rejuvenated asphalt has a characteristic peak at 866.0 cm^{-1} caused by =C-H-bonds on the benzene ring, which indicates the presence of benzene ring aromatic hydrocarbons in the tested material.

Both 4.5S + B and 4.5S + 9R + 50 + A show distinct SBS characteristic peaks at 966.3 cm^{-1} and 696.5 cm^{-1}, generated by C=C bond distortion and C-H bond vibration in the benzene ring, respectively, where the C=C bond is present in the polybutadiene of SBS and the C-H bond is present in the polystyrene segment. This result indicates that the modified-rejuvenated asphalt obtained is consistent with the conventional modified asphalt in terms of composition.

9R + A and 4.5S + 9R + 50 + A have more substantial aromatic absorption peaks at 866.0 cm^{-1} than B and 4.5S + B. In comparison, B and 4.5S + B have higher intensity alkyl absorption peaks, which means that 9% REOB can replenish the lost saturates, aromatics, and other components in the aging asphalt, i.e., the mutual migration and transformation of functional groups, which is eventually reflected in the recovery of asphalt properties. However, there is a limit to the regeneration ability.

In summary, the composition of the chemical components of asphalt mainly includes alkanes, cycloalkanes, aromatic compounds and heteroatomic derivatives, etc. The composition of REOB regenerant is similar to that of REOB, thus proving the regeneration effect of REOB on aging asphalt from the perspective of the composition. However, the results of the four-component analysis do not exclude a weak chemical reaction. The IR spectrum of the asphalt modified with SBS for REOB, i.e., 4.5% SBS + 9% REOB, shows only the new characteristic peaks of SBS, which is a simple superposition of SBS modifier and REOB rejuvenated asphalt, and the process involves physical co-mixing and co-compatibility.

4. Conclusions

(1) The degree of influence of SBS/REOB modified-rejuvenated asphalt performance using grey correlation analysis is ranked as: REOB dosage > SBS dosage > shear time > shear rate. The recommended vital preparation parameters are "9% REOB + 4.5% SBS dosage, shear at 4500 r/min for 50 min~60 min".

(2) The addition of SBS modifier helps to improve the creep recovery rate R and reduce the irrecoverable creep flexibility J_{nr} and $J_{nr, diff}$ of REOB rejuvenated asphalt, so that the obtained SBS/REOB modified-rejuvenated asphalt has higher elastic recovery performance, higher temperature deformation resistance, and lower stress sensitivity. SBS/REOB modified-rejuvenated asphalt has a lower S value and higher m value than REOB rejuvenated asphalt, which has better low-temperature ductility and flexibility.

(3) The four-component and FTIR tests show that the regeneration of aged asphalt by REOB is a non-complete component reconciliation, and the presence and location of the characteristic peaks of modified-rejuvenated asphalt is a simple superposition of SBS modifier and REOB rejuvenated asphalt. The modification-regeneration mechanism of SBS/REOB on aged asphalt is physically dominated, accompanied by weak chemical reactions.

Author Contributions: Conceptualization, J.L. and L.Z.; methodology, C.X.; software, X.Y.; validation, M.Y., X.Y. and D.C.; formal analysis, L.Z.; investigation, L.Z.; resources, J.L.; data curation, J.L.; writing—original draft preparation, L.Z.; writing—review and editing, X.Y.; visualization, M.Y.; supervision, J.L.; project administration, L.Z.; funding acquisition, J.L. All authors have read and agreed to the published version of the manuscript.

Funding: This research was funded by Shandong Provincial Transportation Science and Technology Program, grant number 2021B11.

Institutional Review Board Statement: Not applicable.

Data Availability Statement: The experimental data in this paper are from the pavement material laboratory of Shandong Jiaotong University, which is the provincial key laboratory.

Conflicts of Interest: The authors declare no conflict of interest.

References

1. Zaumanis, M.; Mallick, R.B.; Frank, R. 100% recycled hot mix asphalt: A review and analysis. *Resour. Conserv. Recycl.* **2014**, *92*, 230–245. [CrossRef]
2. Yuan, D.; Jiang, W.; Hou, Y.; Xiao, J.; Ling, X.; Xing, C. Fractional derivative viscoelastic response of high-viscosity modified asphalt. *Constr. Build. Mater.* **2022**, *350*, 128915. [CrossRef]
3. Yuan, D.; Jiang, W.; Xiao, J.; Tong, Z.; Jia, M.; Shan, J.; Ogbon, A.W. Assessment of the Aging Process of Finished Product–Modified Asphalt Binder and Its Aging Mechanism. *J. Mater. Civ. Eng.* **2022**, *34*, 04022174. [CrossRef]
4. Al-Saffar, Z.H.; Yaacob, H.; Mohd Satar, M.K.I. The Influence of PelletRAPTM Rejuvenator on the Mechanical Properties of Reclaimed Asphalt Pavement. *IOP Conf. Ser. Earth Environ. Sci.* **2021**, *682*, 012071. [CrossRef]
5. Chelovian, A.; Shafabakhsh, G. Laboratory evaluation of Nano Al_2O_3 effect on the dynamic performance of stone mastic asphalt. *Int. J. Pavement Res. Technol.* **2017**, *10*, 131–138. [CrossRef]
6. Zhang, X.Y.; Xu, C.J.; Kong, Q.M. Research review of waste rubber powder modified asphalt(1). *Pet. Asph.* **2004**, *4*, 1–5.
7. Feng, Z.G.; Zhao, P.X.; Yao, D.D.; Li, X.J. Research progress on application of re-refined engine oil bottom in asphalt mixture. *China Sci.* **2021**, *16*, 121–127+143.
8. Jia, M.; Sha, A.; Jiang, W.; Wang, W. Laboratory evaluation of poly(ethylene glycol) for cooling of asphalt pavements. *Constr. Build. Mater.* **2020**, *273*, 121774. [CrossRef]
9. Herrington, P.R. Use of re-refined oil distillation bottoms as extenders for roading bitumens. *J. Mater. Sci.* **1992**, *27*, 6615–6626. [CrossRef]
10. Ackbarali, D.S.; Maharaj, R. The Viscoelastic Properties of Trinidad Lake Asphalt-Used Engine Oil Blends. *Int. J. Appl. Chem.* **2011**, *7*, 1–8.
11. Zhang, Y.; Chen, M.Z.; Wu, S.P.; Liu, S.Q.; Liu, J.X. Different physical properties and structures of waste grease recycled asphalt. *J. Wuhan Univ. Technol. (Transp. Sci. Eng.)* **2017**, *41*, 104–108.
12. Yu, M.Z. Design and Pavement Performance Test of Recycled Engine Oil Bottom Rejuvenated Asphalt Mixture. Master's Thesis, Shandong Jiaotong University, Jinan, China, 2019.
13. Li, J.; Yu, M.; Cui, X.; Li, Y.; Cheng, Y. Anti-aging Performance of Recycled Engine Oil Bottom Rejuvenated Asphalt. *J. Build. Mater.* **2021**, *24*, 224–230.
14. Yu, M.; Li, J.; Wang, W.T.; Guo, D. Analysis of physical rheological properties and components of asphalt recycled from waste motor oil residues. *Highway* **2020**, *65*, 6.
15. Yu, M.; Li, J.; Cui, X.; Guo, D.; Li, X. Antiaging Performance Evaluation of Recycled Engine Oil Bottom Used in Asphalt Rejuvenation. *Adv. Mater. Sci. Eng.* **2019**, *2019*, 2947170. [CrossRef]
16. Li, J.; Yu, M.; Cui, X.; Wang, W. Properties and Components of Recycled Engine Oil Bottom Rejuvenated Asphalt and Its Grey Relationship Analysis. *Adv. Mater. Sci. Eng.* **2019**, *2019*, 2462487. [CrossRef]
17. Zhou, Z.G.; Chen, G.H.; Zhang, H.B.; Ling, Y.Y. Study on the Preparation and Properties of Modified Asphalt by Rubber Powder/SBS and High Viscosity Modifier. *Compos. Mater. Rep.* **2021**, *35*, 6093–6099.
18. Lu, Y.Y.; Yu, L.M.; Yang, J.K.; Zeng, W.S.; Lu, J.Y. Preparation of hydrotalcite/waste rubber powder composite modified asphalt and its performance study. *Pet. Process. Petrochem.* **2020**, *51*, 81–86.
19. Dong, F.; Yu, X.; Liu, S.; Wei, J. Rheological behaviors and microstructure of SBS/CR composite modified hard asphalt. *Constr. Build. Mater.* **2016**, *115*, 285–293. [CrossRef]
20. Angelo, J.D. *Guidance on the Use of the MSCR Test with the M320 Specification*; Asphalt Institute Technical Advisory Committee: Lexington, KY, USA, 2010.
21. Guo, Y.M.; Xu, L.; Wu, L.; Shen, X. High-Temperature Performance Evaluation of Modified Asphalts Based on Multiple Stress Creep Recovery Test. *J. Build. Mater.* **2018**, *21*, 5.
22. Jiang, W.; Yuan, D.; Shan, J.; Ye, W.; Lu, H.; Sha, A. Experimental study of the performance of porous ultra-thin asphalt overlay. *Int. J. Pavement Eng.* **2022**, *23*, 2049–2061. [CrossRef]

23. Lei, J.A.; Zheng, N.X.; Xu, X.Q.; Wu, C.; Lü, D. High Temperature Rheological Properties of Warm Mixed Asphalt. *J. Build. Mater.* **2020**, *23*, 904–911.
24. Li, X.; Jian, C.; Que, G. Microstructure and performance of crumb rubber modified asphalt. *Constr. Build. Mater.* **2009**, *23*, 3586–3590.
25. Sha, A.; Jiang, W.; Wang, W.; Lou, B.; Jia, M.; Cao, Y. Design and prospect of new pavement materials for smart road. *Chin. Sci. Bull.* **2020**, *65*, 3259–3269. [CrossRef]
26. Li, L.; Zhang, H.; Chen, Z.; Wei, C. Physical and Theological evaluation of aging behaviors of SBS modified asphalt with thermochromic powders. *Constr. Build. Mater.* **2018**, *193*, 135–141. [CrossRef]

polymers

Article

Study on Low-Temperature and Fatigue Performance of High RAP Content Hot Recycled Asphalt Mixture Based on the Degree of Blending (DOB)

Jianming Wu [1], Hao Sun [1], Lixia Wan [1], Jiangang Yang [2,*] and Shuyi Wang [2]

[1] Jiangxi Ganyue Expressway Engineering Co., Ltd., Nanchang 330069, China
[2] School of Civil Engineering and Architecture, East China Jiaotong University, Nanchang 330013, China
* Correspondence: 2851@ecjtu.edu.cn

Abstract: This paper selected three kinds of AC-20 hot-mix recycled asphalt mixtures with high RAP content (30%, 40%, and 50%). It obtains a mixture of different degrees of miscibility by changing RAP preheating temperatures and mixing temperatures. The calculation formula of the degree of blending (DOB) of RAP asphalt interface recycling is proposed. The DSR test quantitatively characterized the DOB mixture's low temperature, and fatigue properties were tested by beam bending test and four-point bending fatigue test. The prediction models of the recycled mixture's low temperature and fatigue properties were proposed. The RAP preheating temperature is the most critical factor that dominates both transfers of RAP asphalt to the surface of new aggregate and the effective blending of old and new asphalt. DOB has a significant great influence on low-temperature performance and fatigue performance. The DOB of recycled asphalt can be improved by adjusting and optimizing the process parameters of plant hot recycled mixture to effectively improve the recycled mixture's low-temperature crack resistance and fatigue lifetime. The optimal RAP dosage and mixing process of required performance can be obtained based on the prediction models to save experimental time and cost.

Keywords: hot recycled asphalt mixture; RAP; degree of blending (DOB); low-temperature performance; fatigue lifetime; prediction model

Citation: Wu, J.; Sun, H.; Wan, L.; Yang, J.; Wang, S. Study on Low-Temperature and Fatigue Performance of High RAP Content Hot Recycled Asphalt Mixture Based on the Degree of Blending (DOB). *Polymers* **2022**, *14*, 4520. https://doi.org/10.3390/polym14214520

Academic Editors: Wei Jiang and Yue Huang

Received: 15 July 2022
Accepted: 18 October 2022
Published: 25 October 2022

1. Introduction

Hot central plant recycling refers to a maintenance technology, which treats the waste materials of maintenance pavement by milling, recycling, crushing, filing, and so on. It then adds them to the new asphalt mixture in proportion to remix, pave, and compact into the asphalt pavement [1,2]. The hot central plant recycling technology has advantages, such as controllable quality, mature technology, etc., which are widely applied in road engineering [3,4]. It is a maintenance technology with a relatively wide application range and good performance of recycled pavement in all recycling methods [5].

In the hot mix recycled asphalt mixture, the old asphalt wrapped outside the RAP material will be miscible with the added new asphalt in the mixing process, and the degree of miscibility between the two determines the content and properties of effective asphalt in the mixture, thus affecting the pavement performance of the mixture [6]. In the National Cooperative Highway Research Program [7] report, the degree of miscibility between old and new asphalts is defined as completely insoluble (black stone state), partially soluble, and completely soluble. The most acceptable and innovative theory about the miscibility of old and new asphalt is the "partially miscible" theory [8,9]. The core idea of this theory is that in all the old asphalt wrapped on the surface of RAP old aggregate during the process of mixing, transportation, paving, and rolling of recycled asphalt mixture, only the outer part of the old asphalt which is far away from the surface of the old aggregate is blended with the new asphalt. In contrast, the inner layer of the old asphalt which is close to the surface of the old aggregate is still a part of the old aggregate. The inner layer of the old asphalt plays a role, together with the old aggregate, as the skeleton of the recycled mixture, without participating in the mixing of the old and new asphalt, that is, the partial miscible state [10]. Many studies have shown that when RAP content is too high, the two will be in a partially mixed state [7]. Therefore, quantitatively characterizing the mixing state of old and new asphalts in recycled mixture under high RAP blending ratio is the primary key issue to proposing the design method of hot recycled asphalt mixture with high RAP content [11].

Currently, the evaluation indexes of the interfacial blending degree are mostly the conventional performance indexes of the asphalt obtained by extraction test and rotary evaporation test [12,13], such as penetration, softening point, low-temperature flexibility, rheological properties of Brookfield, viscosity, and so on. The test results collected by testing these conventional performance indexes are often relatively discrete, which further indirectly results in the calculated value of the degree of blending (DOB) seriously deviating from the actual value so as not to facilitate quantitative characterization. The creep stiffness modulus S and the curvature of the m value measured by the low-temperature bending beam rheometer (BBR) are not very sensitive under low-temperature test conditions, but the changing range of test results is extremely limited [14], which thereby results in the accuracy of the final test results being difficult to be effectively guaranteed. Due to the small amount of asphalt required for the dynamic shear rheological test (DSR) and the relatively high accuracy of the data collected from the test [9,15], it can be considered for the calculation of the degree of blending (DOB) of interface recycling.

In highway maintenance engineering, the proportion of used materials in the plant-mixed hot recycled mixture is generally not more than 30% [16,17]. The high content of recycled asphalt pavement will cause a serious decline in some road performance, especially the low-temperature crack resistance and durability of the mixture. The low-temperature crack disease is more likely to occur at the early stage of pavement, and then with the repeated fatigue of load, cracks gradually expand to form larger cracks [18], which overweighs the economic benefits obtained from saving materials [19]. Therefore, in order to improve the low-temperature cracking resistance of recycled asphalt mixture, reduce the occurrence of low-temperature cracks in the recycled pavement, improve the using quality of hot recycled asphalt pavement, and prolong the service life of the road, it is necessary to research the low-temperature cracking resistance of hot recycled asphalt pavement [20,21]. In addition, in the preparation of recycled mixture with high RAP content, the change of mixing process will largely affect the mixing state of old and new asphalt in the recycled mixture, thus affecting the road performance of the recycled mixture. At present, there is little research on the influence of interface regeneration and fusion degree of old and new asphalt on road performance of hot recycled asphalt mixture, especially the influence of interface regeneration and fusion degree on low-temperature performance and fatigue life of recycled mixture under high content. Therefore, it is another crucial problem to put forward the design standard of high RAP content recycled asphalt mixture road performance to evaluate the influence of different mixing states of new and old asphalt on road performance.

Given the above two fundamental problems, this paper proposes a mathematical calculation method to characterize the interface degree of blending (DOB) by designing a three-factor and three-level test scheme and calculating the DOB under different RAP content, RAP preheating temperature, and mixing temperatures. The influence of various factors on the interface DOB is analyzed. The low-temperature and fatigue performance of recycled mixture under different mixing processes are studied, and the prediction formulas are proposed. This provides a scientific basis for improving the design method of high RAP content hot recycled asphalt mixture and the promotion and application of plant mixing hot recycling technology on the highway.

2. Materials and Methods

2.1. Materials

2.1.1. Asphalt

The new asphalt used in this paper during the experimental research is Sinopec Donghai brand SBS-modified asphalt, of which the brand number is I-D. The selected RAP is SBS modified asphalt mixture milled from the municipal roads in Nanchang City. As stipulated in "Technical Specifications for Construction of Highway Asphalt Pavement" (JTG F40-2004), conventional properties of SBS-modified asphalt are tested. Due to the severe aging of old asphalt, the viscosity test at 60 °C cannot be carried out, so the viscosity at 135 °C on it is tested instead. The quality test results and technical indexes of the conventional properties of SBS-modified asphalt are shown in Table 1.

Table 1. Test results of conventional indexes of new and old asphalt.

Categories of Materials	25 °C Penetration (0.1 mm)	5 °C Ductility (cm)	Softening Point (°C)	135 °C Viscosity (Pa·s)
Old asphalt	26.9	brittleness	87.2	16.8
New asphalt	58.6	27	73.5	2.9
SBS modified asphalt (I-D) Standard requirements	40~60	≥20	≥60	<3

2.1.2. Aggregate and Filler

Limestone is selected as the new aggregate, and the properties of the new and old aggregates are tested according to "Test Methods of Aggregate for Highway Engineering" (JTG E42-2005). The technical indexes are shown in Table 2.

Table 2. Conventional technical indexes of new and old aggregates.

Categories of Materials	Crushed Value (%)	Apparent Density ($g·cm^{-3}$)	Water Absorption (%)	Water Washing Treatment <0.075 mm (%)	Flat and Elongated Particle Content (%)
Old aggregate	16.7	2.715	1.2	0.8	5.5
New aggregate	16	2.713	0.474	0.71	9.7
Standard requirements	≤26	≥2.6	≤2	<5	≤15

2.1.3. Regenerant

In this paper, combined with practical engineering experience, the RA series thermal regeneration agent is selected, and the dosage of the regeneration agent is 8% of the weight of the total binder required for the mixture. According to the "Technical specification for highway asphalt pavement regeneration" (JTG F41-2008), the performance of the regeneration agent is tested, and the results are shown in Table 3.

Table 3. Property indexes of regenerant.

Categories of Materials	60 °C Viscosity (mPa·s)	Flashpoint (°C)	Saturates Content (%)	Aromatics Content (%)	Viscosity Contrast before and after Thin Film Oven Test (%)	Quality Changes before and after Thin Film Oven Test (%)	Density ($g·cm^{-3}$)
Regenerant	372.4	263	22.8	42.98	2.48	−3.17	1.013
Standard requirements	176~900	≥220	≤30	measured	≤3	≤4, ≥−4	measured

2.1.4. Mineral Gradation

In this paper, the recycled mixture with each RAP content (30%, 40%, and 50%) is graded with the same mineral aggregate. On the one hand, the influence caused by different gradations is excluded. In addition, the content of old asphalt in RAP materials is guaranteed to be a single variable, which is convenient for the comparative analysis of various properties of the recycled mixture. Relying on the actual needs of the project, according to the mineral grading range of the targeted AC-20 asphalt mixture, and combined with practical engineering experience, the synthetic gradation of minerals is determined, and gradations are as shown in Figure 1.

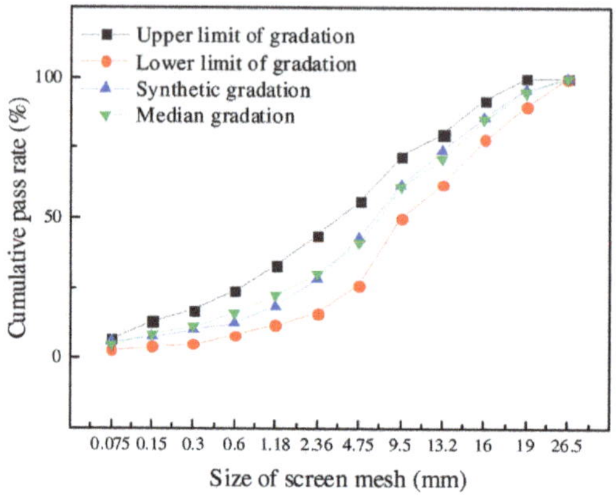

Figure 1. Mineral gradation curves.

2.2. Methods

2.2.1. DOB Theory

In this paper, based on the "partial miscible" theory, in consideration of the change of complex shear modulus before and after the blending of old and new asphalts, and aiming at the problem that the degree of interaction and blending on the interface among old RAP asphalt, new asphalt, and regenerant cannot be quantitatively characterized in the recycling process, the evaluation indexes of interfacial recycling and blending of RAP asphalt, i.e., Degree of Blending (DOB), is adopted. The specific formula is shown in Equation (1).

$$\text{DOB} = \frac{G_{new} - G_{old}}{G_{old} - G_{design}} \tag{1}$$

In the Equation, G_{new} refers to the complex shear modulus of asphalt wrapped on the surface of new aggregate; G_{old} refers to the complex shear modulus of asphalt wrapped on the surface of old aggregate; G_{design} refers to the complex shear modulus of asphalt in the recycled mixture.

2.2.2. DOB Test

To avoid the unfavorable situation that the residual cohesion of RAP fines is too large to form a large-size particle cluster, which cannot effectively distinguish RAP and new aggregates, 4.75 mm is used as the dividing sieve. The RAP fines with particle size below 4.75 mm are selected for the old aggregate, and the new coarse aggregate with a particle size above 9.5 mm is selected for the new aggregate. This prevents the residual cohesive force of RAP fine particles from being too large to form false aggregates with large particle sizes, thus the unfavorable conditions that RAP fine particles and new aggregates cannot be effectively separated are avoided. The DOB test scheme is shown in Figure 2. RAP heating time is 2 h, new asphalt heating temperature is 165 °C, the new aggregate heating temperature is 180 °C, and stirring time is 3 min.

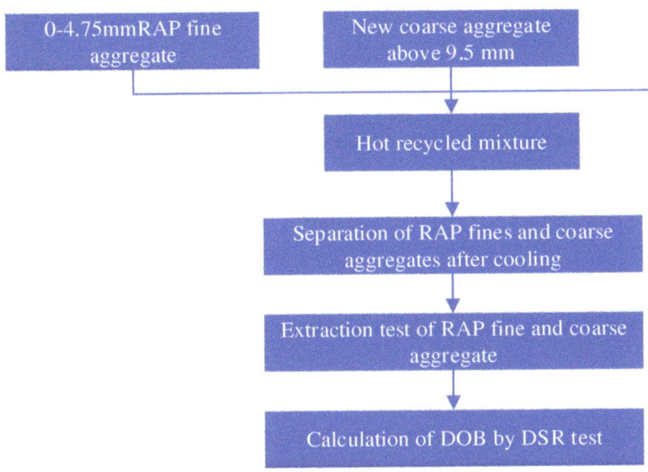

Figure 2. Diagram of DOB test scheme.

Trichloroethylene solution is used to carry out the extraction and rotary evaporation tests of recycled asphalt wrapped on the surface of new and old aggregates, respectively. Then, the DSR test is carried out on the extracted asphalt and the asphalt in the recycled mixture to obtain the complex shear moduli G_{new}, G_{old}, and G_{design} of three kinds of asphalt.

The RT-DSR 200 series dynamic shear rheometer is used in the test, and the strain control mode is adopted. The test temperature is 64 °C, and the loading frequency is 10 Hz. The asphalt sample is applied by sinusoidal dynamic shear strain with a strain of 12%. The thickness of the asphalt sample is 2 mm, and the main shaft is 8 mm.

The 100% DOB recycled asphalt mixture is prepared to completely extract the old asphalt from the RAP material to be mixed, so that the old asphalt in RAP is wholly separated from the old aggregate, and then the recycled agent, new asphalt, and new aggregate are added to remix. The preheating temperature of RAP is 120 °C, the mixing temperature is 165 °C, and other mixing parameters are unchanged. At this point, the new and old asphalt in the mixture are thoroughly mixed.

2.2.3. Beam Bending Test

According to "Standard Test Methods of Bitumen and Bituminous Mixtures for Highway Engineering" (JTJ052-2000), the rut plate of 300 mm × 300 mm × 50 mm is formed by using the wheel rolling method, and then cut into the prismatic beam specimens of 250 mm × 30 mm × 35 mm. The flexural–tensile failure strain and the beam flexural–tensile strength are tested is tested by using UTM-25 equipment at the test temperature of −10 °C and the loading rate of 50 mm/min.

2.2.4. Four-Point Bending Fatigue Test

According to the "Standard Test Methods of Bitumen and Bituminous Mixtures for Highway Engineering" (JTJ052-2000), the specimens were first formed by a shear compaction instrument, and then cut into 380 mm × 63.5 mm × 50 mm middle beam specimens. Then, three strain levels of 200 με, 400 με, and 600 με were used. The test temperature was 15 °C, the loading frequency was 10 Hz, and the continuous sinusoidal loading mode was applied. Finally, the four-point bending fatigue test was carried out by UTM-25.

3. Results and Discussion

3.1. DOB

Recycled asphalt mixture contains a certain proportion of RAP, and hot recycling conditions mainly include RAP content, RAP preheating temperature, and RAP mixing temperature. The difference in the performance of recycled mixture under different recycling conditions is mainly caused by the difference in the blending degree of old and new asphalt. The RAP preheating temperature, the mixing temperature, and the content will affect the blending of old and new asphalt in recycled asphalt mixture. Appropriate recycling conditions can improve the blending degree of old and new asphalt to improve the performance of the recycled asphalt mixture.

In order to quickly select the construction parameters with high degree of fusion between new and old asphalt from various horizontal combinations, this paper uses three factors and three levels of orthogonal tests to test RAP asphalt interface recycling mixture. The table of factors and levels is shown in Table 4.

Table 4. Factors and levels of orthogonal test.

Levels \ Factors	RAP Content (%)	RAP Preheating Temperature (°C)	Mixing Temperature (°C)
Level 1	30	100	150
Level 2	40	120	165
Level 3	50	140	180

The DOB is calculated and analyzed according to the DSR test results, and the table of DOB test results is then obtained (see Table 5). The DOB results are statistically analyzed, and the average values and the range table of DOB under various hot recycling conditions are then obtained (see Table 6). The trend diagram of DOB results, as shown in Figure 2, is obtained according to factors and levels in Table 6.

Table 5. DOB test results.

RAP Content (%)	RAP Preheating Temperature (°C)	Mixing Temperature (°C)	DOB (%)
30	100	150	68.7
30	120	165	85.1
30	140	180	91.9
40	100	165	62.4
40	120	180	76.7
40	140	150	86.7
50	100	180	36.1
50	120	150	54.3
50	140	165	74.8

Table 6. Range analysis results.

Analysis Value	Factors		
	RAP Content (%)	RAP Preheating Temperature (°C)	Mixing Temperature (°C)
$\overline{K1}$	81.9	56.6	69.1
$\overline{K2}$	75.3	72.0	74.2
$\overline{K3}$	55.1	83.6	68.9
R	26.8	27	5.3

From Table 6, it can be determined that among the three factors affecting the degree of blending of RAP asphalt interface recycling, the range value under the factor of the RAP content is 26.8, the range value under the factor of the RAP preheating temperature is 27, and the range value under the factor of the RAP mixing temperature is 5.3, which indicates that the RAP preheating temperature has the most significant influence on the degree of blending of RAP asphalt interface recycling. The influence of various factors on the degree of blending of RAP asphalt interface recycling from the highest to the least is RAP preheating temperature > RAP content > mixing temperature, fully showing that the RAP preheating temperature is the most essential factor leading to the effective blending of old and new asphalt.

It can be seen from Figure 3 that DOB continues to decrease with the increase of RAP content, indicating that if the RAP content is higher, then the older asphalt in the mixture cannot be thoroughly mixed with new asphalt and regenerant, which reduces the cementation between asphalt and aggregate. The reasons lie in the fact that with the increase of RAP content in the recycled mixture, during the whole mixing stage of hot central plant recycling, the old asphalt wrapped on the surface of RAP old mineral material is forced to absorb more material energy required in freeing itself from the bondage of the residual cohesion of the original old mineral material. Therefore, under the same mixing process for the recycled asphalt mixture with different RAP contents, the proportion that effectively activates the old asphalt wrapped on the RAP mineral materials will inevitably vary to varying degrees, further directly leading to the difference in the DOB calculation results. Therefore, the effective recycling rate of the old asphalt decreases instead. An appropriate increase in the mixing temperature can increase the DOB, but if it continues to increase, it will decrease the DOB. The greater the RAP preheating temperature is, the greater the DOB is. However, the excessively high RAP preheating temperature and mixing temperature will aggravate the secondary aging of old asphalt and also lead to the loss of volatilization by heating regenerant. The excessively high heating temperature of new aggregate may also deteriorate the mechanical properties of the new aggregate itself.

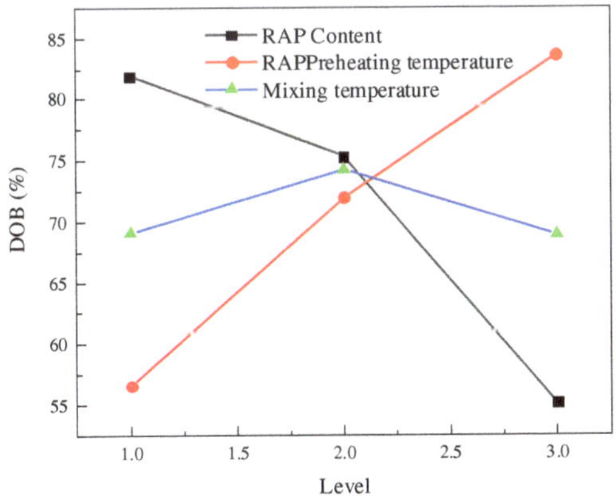

Figure 3. Levels and DOB results changing trend.

In the laboratory, during the mixing and actual production process of the hot recycled mixture, it is necessary to add a certain amount of new asphalt. Then, the old and new asphalt and regenerant (if necessary) infiltrate, diffuse, invade, interact, and blend to form blending recycled asphalt, and finally achieve the purpose of effectively recycling aging asphalt. In order to activate the old asphalt in RAP old material to the maximum extent as much as possible and effectively improve the degree of blending of RAP asphalt interface recycling, the preheating temperature of RAP should be appropriately increased, and the mixing temperature should be reasonably increased under the appropriate RAP content.

3.2. Low-Temperature Performance of RAP Mixture

The beam bending test of recycled asphalt mixture produced under preset mixing process parameters is carried out using the universal testing machine. The test results are shown in Table 7.

Table 7. Test results of the low-temperature beam bending test.

RAP Content (%)	RAP Preheating Temperature (°C)	Mixing Temperature (°C)	DOB (%)	Maximum Flexural–Tensile Strain (με)	Flexural–Tensile Strength (MPa)
30	100	150	68.7	2915	8.65
30	120	165	85.1	3207	8.71
30	140	180	91.9	3359	8.66
30	120	165	100	3235	8.64
40	100	165	62.4	2568	9.49
40	120	180	76.7	2880	9.15
40	140	150	86.7	2798	9.07
40	120	165	100	3269	8.75
50	100	180	36.1	2086	9.71
50	120	150	54.3	1673	9.42
50	140	165	74.8	2288	9.59
50	120	165	100	3098	8.35

According to the test results in Table 7, the variation of maximum flexural-tensile strain and flexural-tensile strength of recycled mixture with the variation of DOB under different RAP contents are plotted, as shown in Figures 4 and 5.

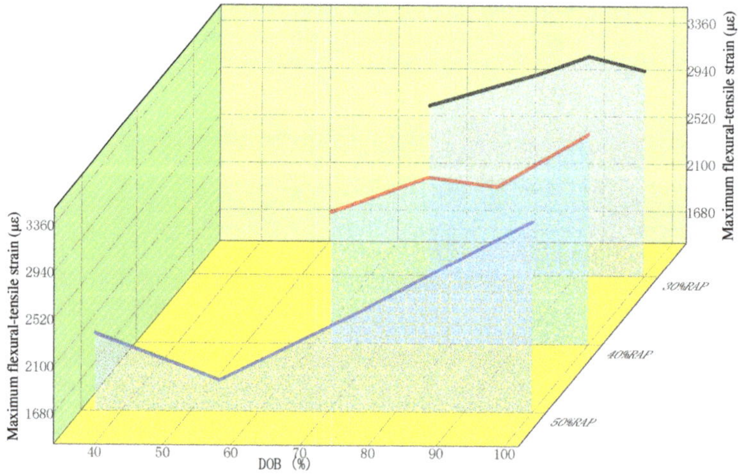

Figure 4. Variation of maximum flexural–tensile strain with the variation of DOB under different RAP contents.

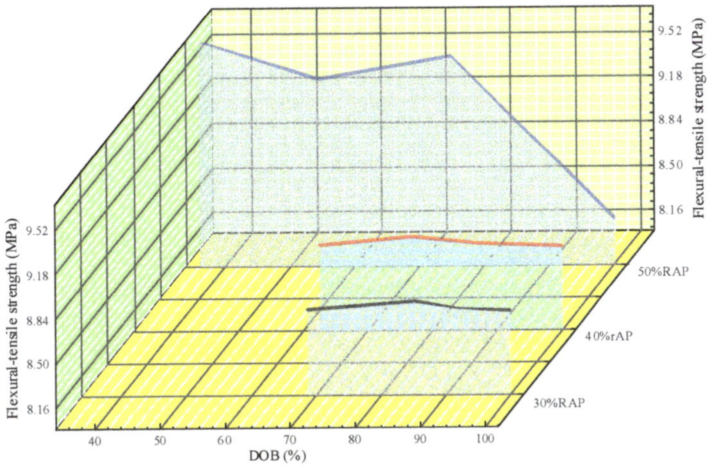

Figure 5. Variation of flexural–tensile strength with the variation of DOB under different RAP contents.

From Table 7 and Figure 4, it can be determined that in the low-temperature beam test of the recycled mixture under three RAP contents, i.e., 30%, 40%, and 50%, the overall trend of the maximum flexural–tensile strain of the mixture increases by varying degrees with the increase of DOB. The higher the RAP content, the greater the maximum flexural–tensile strain's growth rate. The reasons lie in the fact that when the miscibility state of old and new asphalt increases, the old asphalt in the recycled mixture is continuously stripped off from the surface of the old aggregate and then transformed into free asphalt. After being miscible with the new asphalt and regenerant, the old asphalt can restore to the new asphalt's performance standard to enhance the mixture's overall tenacity, so it shows the phenomenon that the maximum flexural–tensile strain constantly increases. The test results indicate that the maximum flexural–tensile strains of the recycled mixture with 30% and 40% RAP contents are greater than the lower limit of the maximum flexural–tensile strain (2500 $\mu\varepsilon$) specified in the current asphalt pavement regeneration technical specification for the winter cool zone, that is, all the low-temperature cracking resistance performance of the recycled mixture is qualified. It indicates that compared with the ordinary hot mix asphalt mixture, the hot central plant recycled mixture with high RAP content (the content of the old material is not more than 40%) can also serve the actual engineering projects. However, the maximum flexural–tensile strains do not meet the specification requirements for three kinds of the recycled mixture under the partial blending states of 50% RAP content. The reasons lie in the fact that under the long-term effect of load and environment, asphalt of RAP material becomes brittle and hard with reduced plasticity and is likely to fracture under low-temperature conditions.

It can be seen from Figure 5 that the flexural–tensile strength of recycled mixture shows a decreasing trend in varying degrees with the increase of the degree of blending (DOB) of RAP asphalt interface recycling. Specifically, with the gradual increase of RAP content, the decreasing amplitude of the flexural–tensile strength is more prominent. The reason is that with the continuous increase of free asphalt, the plasticity of the overall asphalt in the recycled mixture increases, which leads to a decrease in the overall strength of the mixture. Therefore, the flexural–tensile strength thereof decreases with the increase of the miscibility state.

By observing the curve ends in Figures 4 and 5, it is not difficult to find that when the degree of blending (DOB) of RAP asphalt interface recycling reaches the situation of complete blending, the maximum flexural–tensile strain and flexural–tensile strength of recycled mixture under different RAP contents tend to be basically consistent. The reason is that the degree of blending (DOB) of RAP asphalt interface recycling in the recycled mixture gradually increases under the combined effect of more excellent mixing process parameters and other factors, which results in that a larger proportion of aging asphalt in the RAP is heated and softened. Then, it gradually freed itself from the bondage of RAP mineral under the effect of mechanical stirring, transformed into active asphalt, and then further produced a greater degree of blending effect with new asphalt and regenerant (if necessary), which promotes the engineering performance of aging asphalt to be closer to the level of original asphalt and further strengthens the overall tenacity of the recycled mixture. Therefore, the maximum flexural–tensile strain of recycled mixture shows a continuously increasing trend. On the contrary, the proportion of active asphalt in the recycled asphalt mixture gradually increases, resulting in the plastic physical characteristics of its structural asphalt film being further strengthened, so that decay of the strength characteristics of the recycled mixture deteriorated. Finally, the flexural–tensile strain decreases with the increased degree of blending (DOB) of RAP asphalt interface recycling.

In summary, with the increase of DOB, the low-temperature performance of the mixture is improved. According to the requirements of the current specifications, any RAP content below 40% can meet the low-temperature performance requirements of the mixture. Therefore, given the situation that the RAP content is greater than 40%, it can be considered to take specific technical measures to adjust and optimize the mixing process parameters of hot central plant recycling, to promote the degree of blending (DOB) of RAP asphalt interface recycling to a certain extent. At this time, a slight increase in the degree of blending (DOB) of the old and new asphalt interface recycling can also play a significant role in recovering and improving the performance of the low-temperature cracking resistance of the recycled mixture.

3.3. Fatigue Performance of RAP Mixture

For the recycled asphalt mixture produced under preset mixing process parameters, the four-point bending fatigue test was carried out by using the UTM-25 servo-hydraulic multifunctional material test system. The test results are shown in Table 8. The strain modes are 200, 400, and 600 με, respectively. According to the test results in Table 8, the changes in fatigue life of recycled mixture with DOB under different RAP contents are plotted, as shown in Figure 6.

Table 8. Test results of the four-point bending fatigue test.

RAP Content (%)	RAP Preheating Temperature (°C)	Mixing Temperature (°C)	DOB (%)	200 με Fatigue Lifetime (Times)	400 με Fatigue Lifetime (Times)	600 με Fatigue Lifetime (Times)
30	100	150	68.7	354,216	90,349	8573
30	120	165	85.1	363,048	94,817	9894
30	140	180	91.9	366,499	98,006	10,917
30	120	165	100	380,437	99,854	11,425
40	100	165	62.4	232,458	61,063	4919
40	120	180	76.7	236,661	66,205	6094
40	140	150	86.7	238,978	67,946	6144
40	120	165	100	345,267	81,766	6627
50	100	180	36.1	96,527	29,559	926
50	120	150	54.3	91,389	29,737	1414
50	140	165	74.8	96,944	31,749	3482
50	120	165	100	235,818	57,510	3695

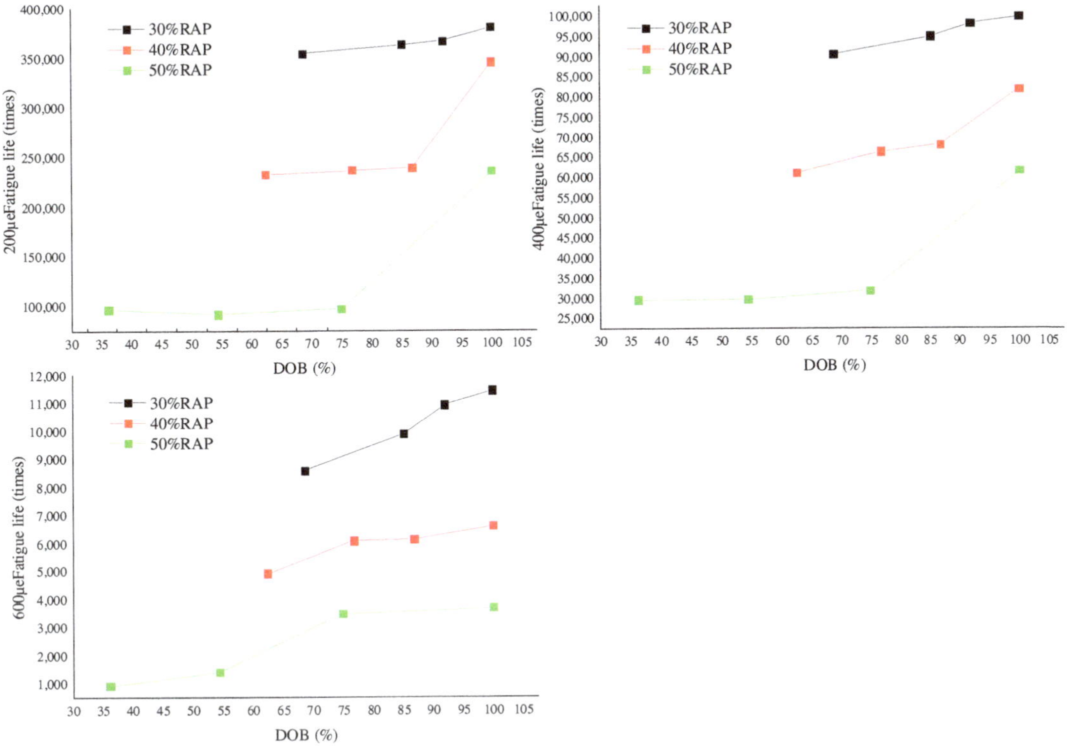

Figure 6. Variation of fatigue lifetimes with the variation of DOB under different RAP contents.

It can be seen from Table 8 that under the same strain conditions, the fatigue times of each recycled mixture gradually increase with the increase of the mixing state of old and new asphalts. With the increase of RAP preheating temperature and mixing temperature, the fatigue times of the mixture also increase. Especially when the RAP content reaches 50%, the fatigue times of the mixture increase obviously, indicating that increasing the RAP preheating temperature and mixing temperature can improve the fatigue lifetimes of the recycled mixture. This is because the increase in the mixing state increases the content of free asphalt in the mixture and improves the adhesion of asphalt to aggregate, indicating that the increase in the mixing state of old and new asphalt has improved the fatigue lifetimes of the mixture.

It can also be seen from Figure 6 that when the RAP content is higher, the fatigue number of the mixture decreases more obviously, indicating that the effect of RAP content on the fatigue lifetimes of the recycled mixture is more obvious. The fatigue number of the 50% RAP content mixture at 200 με is significantly lower than that of 30% RAP content, and the greater the strain, the more obvious the decrease. This is because the increase in RAP content leads to an increase in the content of old asphalt mixed with new asphalt, which reduces the adhesion of asphalt mortar and reduces the toughness of the recycled mixture, thus affecting the fatigue performance of the mixture significantly.

In summary, for the recycled mixture with high RAP content, the mixing state of old and new asphalts in the mixture can be improved as much as possible by changing the mixing process to improve the fatigue performance of the recycled mixture.

3.4. Prediction of Low-Temperature and Fatigue Performance of RAP Mixture

3.4.1. Flexural–Tensile Strength

It can be determined from Table 7 that RAP preheating temperature, mixing temperature, and content influence the low-temperature performance of recycled mixture (excluding three groups of data with 100% DOB). This section takes RAP content, the RAP preheating temperature, and mixing temperature as independent variables to analyze the change of flexural–tensile strength of recycled mixture under three factors. From the experimental data, RAP preheating temperature, mixing temperature, and RAP content have no clear linear law on flexural strength, but the interaction of three factors has an impact on flexural strength. Therefore, nonlinear fitting is considered to predict the bending strength. After many attempts, the bending strength is fitted by the Levenberg Marquardt method through Matlab software. The model after multiple fittings is shown in Formula (2).

$$Y_1 = 19.707 - 71.273{x_1}^{-1.02} - \frac{10.007{x_2}^{0.061}}{{x_3}^{0.082}} \tag{2}$$

In the formula, Y_1 refers to flexural–tensile strength, MPa; x_1 refers to RAP content, %; x_2 refers to RAP preheating temperature, °C; x_3 refers to mixing temperature, °C.

The model correlation coefficient R^2 = 0.951, indicating that the prediction model can well reflect the changing trend of the flexural–tensile strength of recycled mixture under different RAP contents and different construction technologies. It can be determined from Formula (2) that the higher the RAP content and the mixing temperature are, the greater the flexural–tensile strength of the recycled mixture is. The higher the RAP preheating temperature, the smaller the compressive strength, but the overall effect is relatively weak.

3.4.2. Maximum Flexural–Tensile Strain

This section takes RAP content, RAP preheating temperature, and mixing temperature as independent variables to analyze the change of maximum flexural–tensile strain of the recycled mixture under the influence of three factors. Since the three factors have no clear linear influence rules on the maximum flexural–tensile strain, the nonlinear model is also used for fitting. The model after multiple fittings is shown in Formula (3).

$$Y_2 = 1831.5 - 0.0034{x_1}^{3.306} + 3.2833 \times 10^{-9}{x_2}^{5.144} + 2.52{x_3}^{1.236} \tag{3}$$

In the formula, Y_2 refers to flexural-tensile strength, με; x_1 refers to RAP content, %; x_2 refers to RAP preheating temperature, °C; x_3 refers to mixing temperature, °C.

The correlation coefficient of the model R^2 = 0.973 indicates that the prediction model can well reflect the changing trend of the maximum flexural–tensile strain of recycled mixture under different RAP contents and different construction technologies. It can be determined from Formula (3) that the higher the RAP preheating temperature and the mixing temperature are, the larger the maximum flexural–tensile strain of the recycled mixture is.

3.4.3. Fatigue Lifetime

In this section, the change of fatigue life of the recycled mixture under 200 με, 400 με, and 600 με strains was analyzed under the influence of three factors, including RAP content, RAP preheating temperature, and mixing temperature. The nonlinear model is used for fitting, and the model after multiple fitting is shown in Formulas (4)–(6). Formulas (4)–(6) are fatigue life prediction models under the strain of 200 με, 400 με, and 600 με, respectively.

$$L_{200} = 52490 - 1558.4{x_1}^{1.477} + 3087.605{x_2}^{0.259}{x_3}^{0.376} \tag{4}$$

$$L_{400} = 119540 - 153.46{x_1}^{1.683} + 161.97{x_2}^{0.974} + 0.033{x_3}^{2.32} \tag{5}$$

$$L_{600} = 15220 - 835.53{x_1}^{0.841} + 7.179{x_2}^{1.348} + 115.96{x_3}^{0.721} \quad (6)$$

In these formulas, L refers to fatigue lifetime, times; x_1 refers to RAP content, %; x_2 refers to RAP preheating temperature, °C; x_3 refers to mixing temperature, °C.

The correlation coefficients R^2 of models (4)–(6) are 0.999, 0.999, and 0.994, respectively, indicating that these three fatigue prediction models can well reflect the variation trend of fatigue life of the recycled mixture under different RAP contents and different construction technologies. These prediction models will be different due to different materials such as asphalt, RAP, and regenerant. However, the test and prediction methods are similar, and factors such as mixing time and mixing temperature can also be included in the model in the follow-up study.

According to these prediction models, in practical engineering applications, if the RAP content is determined, the bending tensile strength, bending tensile strain, and fatigue lifetime change diagrams can be plotted at the RAP preheating temperature and mixing temperature. Based on this diagram, the most suitable RAP preheating and mixing temperature can be obtained by comprehensively considering the economic cost and actual performance, thus saving much test time and cost.

4. Conclusions

In this paper, an experimental scheme of three factors and three levels is designed, and a mathematical calculation method to characterize the interface regeneration and fusion degree is proposed. The interface regeneration and fusion degree were analyzed under different RAP content, preheating, and mixing temperatures. Through the beam bending and four-point bending tests, the low-temperature and fatigue performance of recycled asphalt mixture under different conditions is evaluated, and its low-temperature and fatigue performance are predicted. The main conclusions are as follows.

(1) The influence on the DOB of RAP asphalt interface recycling from the highest to the least is in the order of RAP preheating temperature > RAP content > mixing temperature. The RAP preheating temperature is the most critical factor that dominates the effective blending of old and new asphalt. In order to activate the old asphalt in RAP old materials to the maximum extent possible and effectively improve the degree of blending of RAP asphalt interface recycling, the RAP preheating temperature should be appropriately increased, and the mixing temperature should be reasonably increased under the appropriate RAP content.

(2) The overall trend of the maximum flexural–tensile strain of recycled mixture is that it increases to varying degrees with the increase of DOB. In contrast, the flexural-tensile strength decreases with increased RAP asphalt interface DOB. When the RAP asphalt interface DOB reaches the complete fusion situation, the recycled mixture's maximum flexural strain and flexural strength under different RAP content are basically consistent. Under the same strain condition, the fatigue lifetime of each recycled mixture increases gradually with the increase of the mixing state of old and new asphalt. The influence of RAP content on the fatigue lifetime of the recycled mixture is relatively apparent, and the greater the strain is, the more pronounced the reduction is. With the increase of RAP preheating temperature and mixing temperature, the fatigue lifetime of the mixture increases.

(3) For the recycled mixture with high RAP content, it can be considered to take specific technical measures to adjust and optimize the process parameters of plant mixing the hot recycled mixture, and improve the DOB of RAP asphalt, to improve the low-temperature performance and fatigue performance of recycled mixture. According to the low-temperature performance and fatigue performance prediction model, in practical engineering applications, if the RAP content is determined, the bending tensile strength, bending tensile strain, and fatigue life change diagram can be drawn by RAP preheating temperature and mixing temperature. According to the diagram, the economic cost and actual performance can be comprehensively considered to obtain the most suitable RAP preheating temperature and mixing temperature to save much test time and cost.

Author Contributions: J.W. is responsible for data curation, formal analysis, writing—original draft, and writing—review and editing. H.S. is responsible for methodology and writing—review and editing. L.W. is responsible for methodology. J.Y. is responsible for conceptualization. S.W. is responsible for data curation. All authors discussed and contributed to the manuscript. All authors have read and agreed to the published version of the manuscript.

Funding: This research received no external funding.

Institutional Review Board Statement: This research did not require ethical approval.

Data Availability Statement: Data is contained within this article.

Conflicts of Interest: The authors declare no conflict of interest.

References

1. Kim, S.; Sholar, G.A.; Byron, T.; Kim, J. Performance of Polymer-Modified Asphalt Mixture with Reclaimed Asphalt Pavement. *Transp. Res. Rec. J. Transp. Res. Board* **2009**, *2126*, 109–114. [CrossRef]
2. Yang, L.Y.; Tan, Y.Q.; Liu, H.; Li, E.G. Research and Application of Warm Recycled Asphalt Mixture. *Adv. Mater. Res.* **2012**, *446–449*, 2412–2417. [CrossRef]
3. Jamshidi, A.; White, G. Evaluation of Performance and Challenges of Use of Waste Materials in Pavement Construction: A Critical Review. *Appl. Sci.* **2019**, *10*, 226. [CrossRef]
4. Martišius, M. Reclaimed Asphalt Usage: Handling, Processing, Management and Future Trends in Lithuania. In Proceedings of the 5th International Symposium on Asphalt Pavements & Environment (APE), Padua, Italy, 11–13 September 2019; pp. 294–302.
5. Montañez, J.; Caro, S.; Carrizosa, D.; Calvo, A.; Sánchez, X. Variability of the mechanical properties of Reclaimed Asphalt Pavement (RAP) obtained from different sources. *Constr. Build. Mater.* **2020**, *230*, 116968. [CrossRef]
6. Yan, Y.; Roque, R.; Hernando, D.; Chun, S. Cracking performance characterisation of asphalt mixtures containing reclaimed asphalt pavement with hybrid binder. *Road Mater. Pavement Des.* **2017**, *20*, 347–366. [CrossRef]
7. Al-Qadi, I.L.; Elseifi, M.; Carpenter, S.H. *Reclaimed Asphalt Pavement—A Literature Review*; Research Report FHWA-ICT-07-001; Illinois Center for Transportation: Champaign, IL, USA, 2007; Available online: http://hdl.handle.net/2142/46007 (accessed on 14 July 2022).
8. Jeong, K.-D.; Lee, S.-J.; Kim, K.W. Laboratory evaluation of flexible pavement materials containing waste polyethylene (WPE) film. *Constr. Build. Mater.* **2011**, *25*, 1890–1894. [CrossRef]
9. Zhang, Q.; Goh, S.W.; You, Z.P. Study on Dynamic Modulus of Waste Plastic Modified Asphalt Mixture Using Waste Plastic Bag Chips. *Adv. Mater. Res.* **2011**, *261–263*, 824–828. [CrossRef]
10. Bressi, S.; Cavalli, M.C.; Partl, M.N.; Tebaldi, G.; Dumont, A.G.; Poulikakos, L.D. Particle clustering phenomena in hot asphalt mixtures with high content of reclaimed asphalt pavements. *Constr. Build. Mater.* **2015**, *100*, 207–217. [CrossRef]
11. Orešković, M.; Pires, G.M.; Bressi, S.; Vasconcelos, K.; Presti, D.L. Quantitative assessment of the parameters linked to the blending between reclaimed asphalt binder and recycling agent: A literature review. *Constr. Build. Mater.* **2020**, *234*, 117323. [CrossRef]
12. Abdalfattah, I.A.; Mogawer, W.S.; Stuart, K. Quantification of the degree of blending in hot-mix asphalt (HMA) with reclaimed asphalt pavement (RAP) using Energy Dispersive X-Ray Spectroscopy (EDX) analysis. *J. Clean. Prod.* **2021**, *294*, 126261. [CrossRef]
13. Zhou, Z.; Gu, X.; Dong, Q.; Ni, F.; Jiang, Y. Rutting and fatigue cracking performance of SBS-RAP blended binders with a rejuvenator. *Constr. Build. Mater.* **2019**, *203*, 294–303. [CrossRef]
14. Moghaddam, T.B.; Soltani, M.; Karim, M.R. Evaluation of permanent deformation characteristics of unmodified and Polyethylene Terephthalate modified asphalt mixtures using dynamic creep test. *Mater. Des.* **2014**, *53*, 317–324. [CrossRef]
15. Ravichandran, M.; Castorena, C. Critical Evaluation of Dynamic Shear Rheometer (DSR)-Based Methods to Measure Asphalt Binder Diffusion. *J. Test. Eval.* **2021**, *49*, 4014–4031. [CrossRef]
16. Abed, A.; Thom, N.; Lo Presti, D. Design considerations of high RAP-content asphalt produced at reduced temperatures. *Mater. Struct.* **2018**, *51*, 91. [CrossRef]
17. Yang, J.; Tao, W.; Gao, J.; Yu, D.; Zhou, J.; He, L.; Yao, Y. Measurement of particle agglomeration and aggregate breakdown of reclaimed asphalt pavement. *Constr. Build. Mater.* **2021**, *296*, 123681. [CrossRef]
18. Quresh, N.A.; Khurshid, B. Investigation of Hot In-Place Recycling Effects on Hot Mix Asphalt Pavement. *Int. J. Pavement Eng. Asph. Technol.* **2015**, *16*, 7–20. [CrossRef]
19. Mansourkhaki, A.; Ameri, M.; Habibpour, M.; Underwood, B.S. Relations between colloidal indices and low-temperature properties of reclaimed binder modified with softer binder, oil-rejuvenator and polybutadiene rubber. *Constr. Build. Mater.* **2020**, *239*, 117800. [CrossRef]
20. Singh, D.; Girimath, S. Influence of RAP sources and proportions on fracture and low temperature cracking performance of polymer modified binder. *Constr. Build. Mater.* **2016**, *120*, 10–18. [CrossRef]
21. Xiao, F.; Li, R.; Zhang, H.; Amirkhanian, S. Low Temperature Performance Characteristics of Reclaimed Asphalt Pavement (RAP) Mortars with Virgin and Aged Soft Binders. *Appl. Sci.* **2017**, *7*, 304. [CrossRef]

polymers

MDPI

Article

Study on the Influence of Nano-OvPOSS on the Compatibility, Molecular Structure, and Properties of SBS Modified Asphalt by Molecular Dynamics Simulation

Lei Feng, Peng Zhao *, Tongdan Chen and Minghai Jing

School of Materials Science and Engineering, Chang'an University, Xi'an 710061, China
* Correspondence: zyzhaop@chd.edu.cn

Abstract: The present research is carried out to inspect the influence of nano-OvPOSS (octavinyl oligomeric silsesquioxane) with different particle sizes on styrene-butadiene-styrene (SBS) modified asphalt through the method of molecular dynamics simulation. This nanomaterial is investigated for the first time to be used in asphalt modification. With the construction of modified asphalt simulation models and the analysis of their mixing energy, radius of gyration (R_g), radial distribution function (RDF), ratio of free volume (RFV), heat capacity, bulk modulus, and shear modulus, this study elucidates the influence of nano-OvPOSS on the compatibility between SBS and asphalt, on the structure of SBS as well as that of asphalt molecules and on the temperature stability and mechanical properties of SBS modified asphalt. The results show that nano-OvPOSS not only is compatible with SBS as well as with asphalt, but also is able to improve the compatibility between SBS and asphalt. Nano-OvPOSS is able to reinforce the tractility of branched chains of SBS and make SBS easier to wrap the surrounding asphalt molecules. The free movement space of molecules in the SBS modified asphalt system also shrinks. Moreover, the addition of nano-OvPOSS into SBS modified asphalt results in higher heat capacity, bulk modulus, and shear modulus of modified asphalt. All of these effects contribute to a more stable colloidal structure as well as more desirable temperature stability and deformation resistance of the modified asphalt system. The overall results of the study show that nano-OvPOSS can be used as a viable modifier to better the performance of conventional SBS modified asphalt.

Keywords: nano-OvPOSS; SBS; modified asphalt; compatibility; structure; mechanical properties

Citation: Feng, L.; Zhao, P.; Chen, T.; Jing, M. Study on the Influence of Nano-OvPOSS on the Compatibility, Molecular Structure, and Properties of SBS Modified Asphalt by Molecular Dynamics Simulation. *Polymers* **2022**, *14*, 4121. https://doi.org/10.3390/polym14194121

Academic Editor: Khalil A. Khalil

Received: 29 August 2022
Accepted: 28 September 2022
Published: 1 October 2022

1. Introduction

Polymer modifiers are widely used to improve the properties of asphalt to obtain low-temperature crack resistance, high-temperature rutting resistance, and fatigue resistance [1–4]. Styrene-butadiene-styrene (SBS), a kind of thermoplastic elastomer, is universally employed as an asphalt modifier. SBS modified asphalt is characterized as neither being viscous when heated nor being fragile when cooled, which is of great plasticity and property of aging resistance [5–7]. It has been found that the properties of the modified asphalt are closely related to the amount of SBS, considering that it has a close relation with the microstructure of the SBS modified asphalt. With a small amount, SBS disperses evenly in the asphalt in the form of dispersion medium. As the amount increases, cross-linked SBS networks are synthesized, forming the ideal microstructure of polymer-modified asphalt—the interlock network structure—thereby improving the mechanical and rheological properties of asphalt. The amount of SBS added into the asphalt is normally lower than 10% [8–10].

Generally, SBS modified asphalt still needs some improvements. Due to the relatively large difference in solubility and density between SBS and asphalt, this type of system is thermodynamically unstable and a phase separation (SBS-rich phase and asphalt-rich phase) is prone to occur under the influence of gravitational fields [11,12]. Meanwhile, its susceptibility to degradation and aging under the influence of sunlight, oxygen, and heat significantly also minimizes its positive effect on asphalt [13–15]. To meet all-round requirements of asphalt modifier, the composite modification method is carried out in which nanomaterial is added with polymer to modify asphalt, which not only makes good use of the skeleton structure of nanomaterial, but also gives full play to polymer's advantages [16].

Various studies conducted previously have provided positive proof that the addition of nanomaterials to SBS modified asphalt is able to better the performance of SBS modified asphalt. Han et al. [17–20] researched nano-montmorillonite/SBS composites modified asphalt and discovered that the addition of nano-montmorillonite improves the properties of SBS in high or low temperature and effectively prevents the dissociation situation and at the same time refines the ageing resistance of asphalt. Yet, these studies have only investigated macroscopic properties without the exploring modification mechanism from the molecular level. In 2005, Ouyang et al. [21] studied SBS/Kaolin clay composites modified asphalt and found that Kaolin clay can efficaciously adjust the density difference between SBS and asphalt so that the gathering trend of the SBS particles is slowed. However, as is mentioned in their study, Kaolin clay plays an insignificant role in improving the mechanical properties of asphalt. In 2016, Rezaei et al. [22] discovered that the dynamic shear modulus and rutting resistance factor of compound modified asphalt become prominently higher because of the addition of nano-SiO_2/SBS. However, this research was limited to only studying the rheological properties of the influence of nano-SiO_2/SBS on asphalt. In 2020, Su et al. [23] used the molecular dynamics method to study nano-ZnO/SBS compound modified asphalt and found that the addition of nano-ZnO ameliorates the property of SBS modified asphalt in high temperature as well as the shear-resistant property of SBS modified asphalt. What is worth mentioning is that these studies all adopted the kind of nanomaterial that is either metal oxide material or non-metal oxide material, which in fact is unfavorably compatible with asphalt and easy to aggregate in the asphalt system. Thus, despite the fact that the performance of modified asphalt is improved, some negative situations such as the less desired compatibility of nanomaterial with asphalt and the aggregation of nanomaterial in asphalt systems are very likely to occur at the same time.

To offset such deficiency, POSS (polyhedral oligomeric silsesquioxane), a kind of organic-inorganic nanohybrid material possessing both the properties of organic and inorganic materials, was explored as an asphalt modifier in this study in an attempt to avoid the occurrence of the above-mentioned unfavorable situations and at the same time better the performance of modified asphalt on the whole. This kind of nanomaterial was inspected for the first time in the field of asphalt modification.

POSS was first synthesized in 1946 by Scott [24]. Afterwards, many kinds of POSS were prepared and utilized, such as OvPOSS, CpPOSS, and CyPOSS [25–30]. Composed of silicon and oxygen and with a cage-like skeleton structure whose surface is covered by organic groups, POSS is easy to functionalize by changing the organic groups, which bolsters up its miscibility with asphalt and allows for extra convenience of its incorporation into polymer through blending. Moreover, after the hybrid reaction between the polymer and POSS, the properties of the oxidation resistance, flame retardancy, and aging resistance of the polymer are consolidated [31–34].

Nano-OvPOSS (octavinyl oligomeric silsesquioxane), containing a vinyl group that is also a common organic group in SBS, was selected as the study object in this research owing to its similarity in structure with SBS so that the reciprocal effect between nano-OvPOSS and SBS in the modified asphalt system can be smoothed [9,28,31]. The effect of nano-OvPOSS on the compatibility, structure, and properties of SBS modified asphalt was investigated using the molecular dynamics (MD) simulation method presenting corresponding molecular models of asphalt materials on the basis of experimental data. The MD method is based on Newton's law and predicts the macroscopic performance of materials by calculating the intra- and intermolecular interactions. Properties derived from the simulation algorithm are similar to those of real asphalt. In this study, the structure-property relationship in the organic-inorganic asphalt blends was analyzed by MD simulation which is intended to bridge the microscopic mechanism and macroscopic properties of materials in the field of asphalt modifier research.

In a nutshell, through MD simulation method, a series of modified asphalt models were constructed and the mixing energy, radius of gyration (R_g), radial distribution function (RDF), ratio of free volume (RFV), the heat capacity, the bulk modulus, and shear modulus were utilized in this research to analyze the compatibility, molecular structure, temperature stability, and mechanical properties of modified asphalt system, with the intention of gaining a brief glimpse of the application of nano-OvPOSS in modified asphalt system.

2. Simulation Models

2.1. Molecular Model of Matrix Asphalt

This study underwent a separation test to obtain the proportion of the four components of the asphalt simulation model. The PG 64-16 matrix asphalt, which is common in the construction of asphalt pavement in China, was used for the separation test. The physical properties of matrix asphalt are shown in Table 1. According to ASTM D4124 standard, the separation test was carried out, and the four components of asphalt were defined as asphaltene, resin, saturate, and aromatic. The separation test results are listed in Table 2.

Table 1. Physical properties of PG64-16 matrix asphalt.

Properties	Test Results	Test Standard
Penetration (0.1 mm) (25 °C, 100 g, 5 s)	73.3	ASTM D5-06
Softening point (°C)	47.5	ASTM D36-06
Ductility (cm) (5 cm/min, 10 °C)	67.2	ASTM D113-07

Table 2. Separation test results of PG64-16 matrix asphalt.

Parameter	Asphaltene	Resin	Saturate	Aromatic
Weight (g)	0.074	0.278	0.222	0.386
Percent (%)	7.71	28.96	23.13	40.20

The four components of asphalt model were constructed with reference to the results conducted by Hansen et al. [35]. From the separation test results of matrix asphalt, the ratio of each molecule in asphalt simulation system was calculated, as is listed in Table 3. The four components of matrix asphalt and the constructed matrix asphalt simulation model are shown in Figure 1.

Table 3. The information about molecular models of four asphalt components.

Name	Chemical Formula	Number of Molecules	Number of Atoms	Content (%)
Asphaltene	$C_{64}H_{52}S_2$	2	236	7.34
Resin	$C_{41}H_{54}S$	12	1152	28.81
Saturate	$C_{22}H_{46}$	18	1224	23.17
Aromatic	$C_{24}H_{28}$	31	1612	40.68

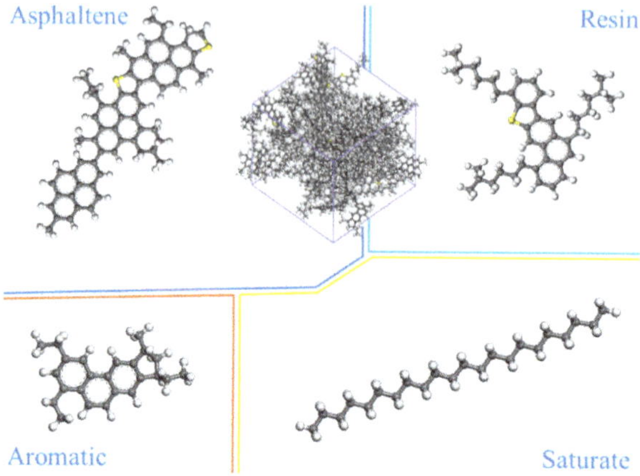

Figure 1. Molecular structure models of four asphalt components and matrix asphalt (carbon atoms are gray, sulfur atoms are yellow, and hydrogen atoms are white).

2.2. Molecular Model of Nano-OvPOSS

Using the cluster building tool, nanoclusters with three different sphere diameters were built, that are 5 Å, 7 Å, and 9 Å. The sphere diameters of nanoclusters are referred to as the size of nanoclusters in this research. The parameters of OvPOSS are listed in Table 4. The constructed nano-OvPOSS nanoclusters with three kinds of diameters are shown in Figure 2.

Table 4. The parameters of OvPOSS.

Parameter	Results
Molecular formula	$C_{16}H_{24}O_{12}Si_8$
Molecular weight	633
Density (g/cm^3)	1.22
Melting point (°C)	>350
Flash point (°C)	148.4
Toxicity	Non-toxic

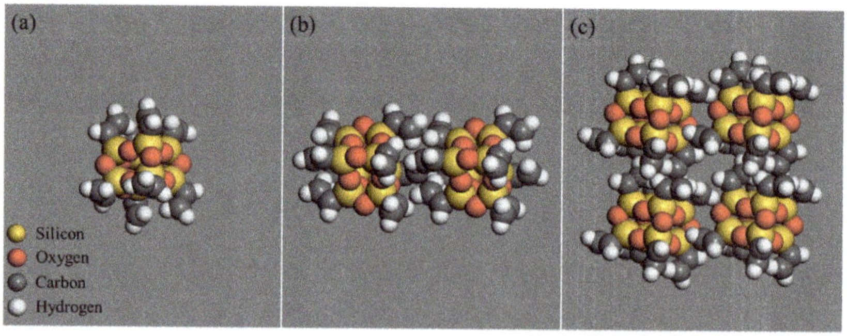

Figure 2. OvPOSS nanoclusters with different sizes: (**a**) 5 Å; (**b**) 7 Å; (**c**) 9 Å (silicon atoms are orange and oxygen atoms are red).

2.3. Molecular Model of Linear SBS

The linear SBS, derived from styrene and 1,3-butadienem (Figure 3), was used in this research, whose molecular formula and molecular model are shown in Figures 4 and 5.

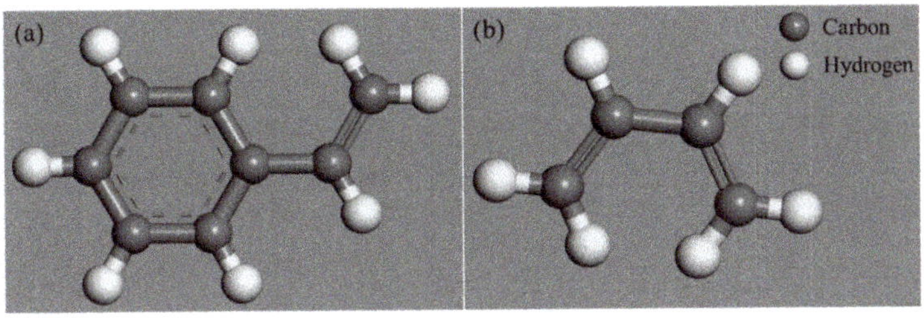

Figure 3. Monomer molecular models: (**a**) styrene; (**b**) 1,3-butadiene.

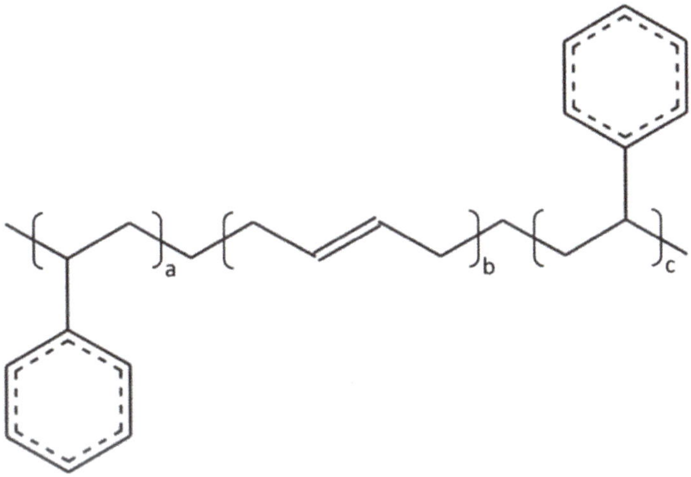

Figure 4. Molecular formula of linear SBS.

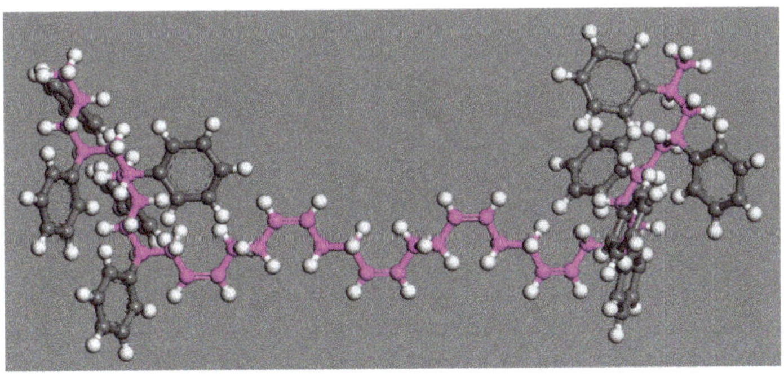

Figure 5. Molecular model of linear SBS.

2.4. Molecular Model of Nano-OvPOSS/SBS Modified Asphalt

The nano-OvPOSS/SBS modified asphalt models with different nano-OvPOSS diameters were constructed by Amorphous cell module, as is shown in Figure 6. In most of SBS modified asphalt experiments, the dosage of SBS was often chosen to be around 5% [4–6,36]. In this research, the dosage of SBS was chosen to be 4.699%, which was able to ensure the successful building of the OvPOSS/SBS modified asphalt model with one SBS molecule. Through several trials, it was found that if two SBS molecules were added to build the OvPOSS/SBS modified asphalt model, the content of SBS turned out to be 8.975%, which is much more than the commonly used dosage (5%). Meanwhile, the dosage of nano-OvPOSS was chosen to be 9.054% to meet the least amount requirement for the model construction of nano-OvPOSS/SBS modified asphalt. Detailed information is shown in Table 5.

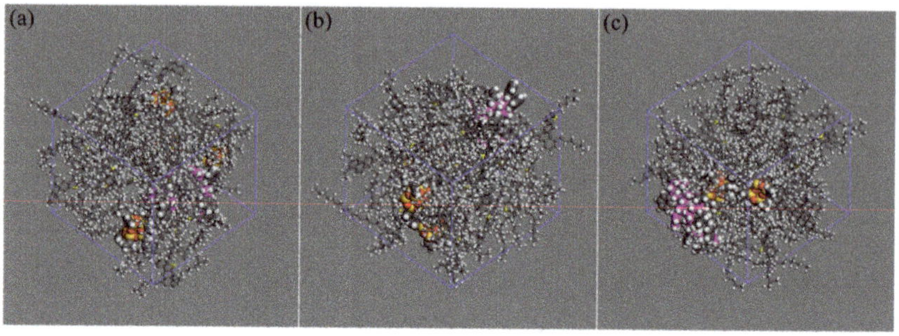

Figure 6. Nano-OvPOSS/SBS modified asphalt models with different nano-OvPOSS diameters: (**a**) 5 Å nano-OvPOSS; (**b**) 7 Å nano-OvPOSS; (**c**) 9 Å nano-OvPOSS.

Table 5. Detailed information of modified nano-OvPOSS/SBS asphalt system.

Name	Chemical Formula	Content (%)
Asphaltene	$C_{64}H_{52}S_2$	6.331
Resin	$C_{41}H_{54}S$	24.842
Saturate	$C_{22}H_{46}$	19.992
Aromatic	$C_{24}H_{28}$	35.082
OvPOSS	$C_{16}H_{24}O_{12}Si_8$	9.054
SBS	$C_{100}H_{112}$	4.699

3. MD Simulation Theory and Method

3.1. Simulation Task

3.1.1. Mixing Energy

According to the Flory–Huggins model, the expression of free energy of mixing of a binary system can be expressed as Equations (1) and (2):

$$\frac{\Delta G}{RT} = \frac{\phi_b}{n_b} ln\phi_b + \frac{\phi_s}{n_s} ln\phi_s + x\phi_b\phi_s \quad (1)$$

$$x = \frac{E_{mix}}{k_B T} \quad (2)$$

where ΔG is the free energy of mixing; R is the gas constant; T is the temperature; ϕ_i is the volume fraction of i; n_i is the degree of polymerization of i; E_{mix} is the mixing energy; x is the interaction parameter; and k_B is the Boltzmann constant.

In the traditional Flory–Huggins model, each component occupies a lattice site. For a lattice with coordination number Z, the mixing energy is calculated by Equation (3):

$$E_{mix} = \frac{1}{2} Z(E_{bs} + E_{sb} - E_{bb} - E_{ss}) \quad (3)$$

where E_{ij} is the binding energy between a unit of component i and a unit of component j.

3.1.2. R_g

R_g is used to reflect the changes of the shape of polymer molecular. The expression of R_g is shown in Equation (4).

$$R_g = \left(\frac{\sum r^2 m}{\sum m}\right)^{\frac{1}{2}} \quad (4)$$

where r is the distance; m is the molecular mass.

3.1.3. RDF

RDF can be calculated by the ratio of regional density to the average density of the system, the expression of RDF is shown in Equation (5).

$$g(r) = \frac{dN}{\rho 4\pi r^2 dr} \tag{5}$$

where g(r) is the RDF; N is the total number of atoms; ρ is the density of system; r is the distance.

3.1.4. RFV

Free volume is the unoccupied space between molecules. RFV is the parameter to express the percentage of the volume not occupied by molecules, which can be calculated by Equation (6).

$$\text{RFV} = \frac{V_f}{V_f + V_o} \times 100\% \tag{6}$$

where V_f is the free volume; V_o is the occupied volume.

3.2. Simulation Method

The asphalt models with three dimensional periodic conditions were built according to related parameters by the Amorphous cell tool. The geometry optimization was employed to optimize the structure of model, and then the dynamic process for 2000 ps was employed to ensure a stable asphalt molecular configuration. To obtain the true global potential energy minimum configuration, the anneal process was used to overcome the migration energy barrier in the modified asphalt system. In this process, the prepared asphalt models were initially heated up to 500 K and then cooled to 298 K, and this anneal process was carried out four times for a total of 2000 ps. After these relaxation procedures, the geometry optimization was conducted for a second time, then the molecular dynamic process with the NPT ensemble for 1000 ps at 298 K was followed. This molecular dynamic process generated various equilibrium models to be used for subsequent studies. The time step is 1.0 fs. The COMPASS force field [37,38] was adopted to describe the atomistic interactions. The temperature and pressure of the molecular dynamic system were controlled by Nose [39] thermostat and Berendsen [40] barostat separately, and the decay constant of temperature control method and pressure control method were both 0.1 ps.

3.3. Model Validation

The accuracy of simulation models and methods can be verified through two important indicators—density and energy. Thus, these two indicators of SBS modified asphalt system and nano-OvPOSS/SBS modified asphalt systems were studied. Figure 7 shows the density and the energy of the four asphalt systems as a function of simulation time. The density and the energy of the four systems can be seen in this study to have reached the equilibrium state within 1000 ps. The dynamic results of the last 600 ps (400–1000 ps) were chosen as the data reference for subsequent calculation, during which time the density profile and the energy profile were able to reach a stable state. It can be seen that the separate addition of the three kinds of nano-OvPOSS increased the density of the SBS modified asphalt system and decreased the total energy of the SBS modified asphalt system.

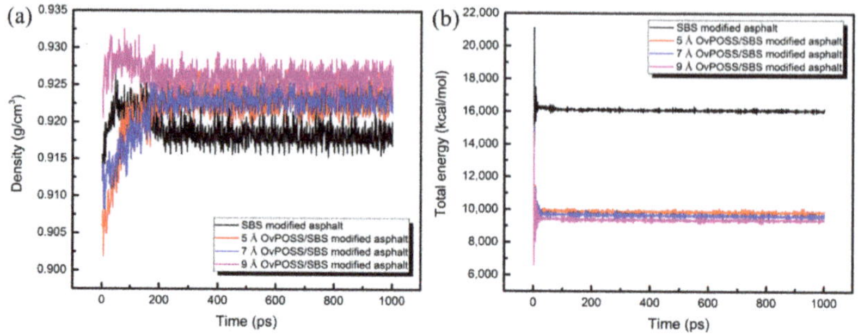

Figure 7. Model validation parameters: (**a**) density; (**b**) energy.

4. Results and Discussion

4.1. Compatibility Analysis

The eight vinyl groups of nano-OvPOSS endow this nanomaterial with distinctive potential to be compatible with polymer. This can be supported by the molecular dynamics method that simulates the probability distribution of the mixing energy of SBS with nano-OvPOSS of three different particle sizes, the mixing energy of four asphalt components with nano-OvPOSS of three different particle sizes, and the effect of nano-OvPOSS on the mixing energy of SBS with asphalt. The compatibility analysis is based on the Flory–Huggins model [41–45]. Details are as the following.

4.1.1. The Compatibility of Nano-OvPOSS with SBS

Nano-OvPOSS was used as the screen and SBS as the base. The probability distributions of mixing energy of SBS with nano-OvPOSS (5 Å, 7 Å, and 9 Å) were calculated by using Blends tool, as is shown in Figure 8. For further clarification, study of the mixing energy of SBS with traditional nanomaterials of SiO_2, ZnO, and TiO_2 (4.4 Å) was cited in this part to facilitate the comparison, which is shown in Figure 9.

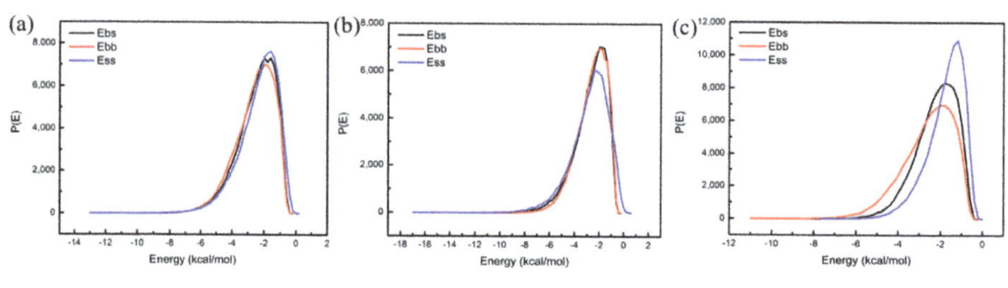

Figure 8. The probability distribution of mixing energy of nano-OvPOSS with SBS: (**a**) 5 Å nano-OvPOSS; (**b**) 7 Å nano-OvPOSS; (**c**) 9 Å nano-OvPOSS.

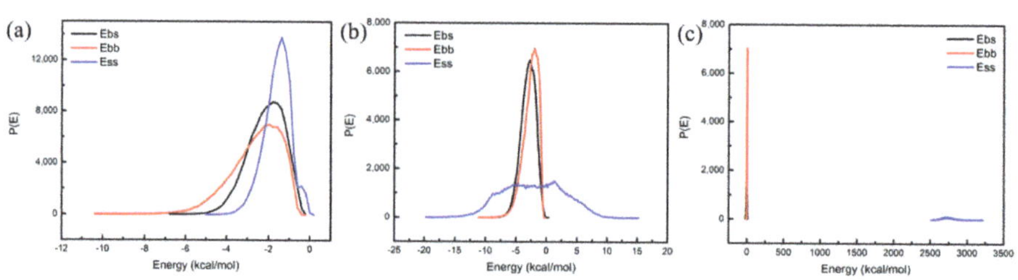

Figure 9. The probability distribution of mixing energy of traditional nanomaterials with SBS: (**a**) SiO_2; (**b**) ZnO; (**c**) TiO_2.

It can be seen that the curves of E_{bb} (base–base), E_{bs} (base–screen), and E_{ss} (screen–screen) of SBS with nano-OvPOSS (5 Å, 7 Å, and 9 Å) in Figure 8 are very similar to each other. In contrast, the curves of that of SBS with SiO_2, ZnO, and TiO_2 (4.4 Å) in the three respective pictures in Figure 9 are obviously different from each other. Generally, the similarity of the curves of E_{bb}, E_{bs}, and E_{ss} reflects the degree of the compatibility of two materials. It is usually accepted that the more similar the probability distributions of E_{bb}, E_{bs}, and E_{ss} of two materials are, the better the compatibility of the two materials is. Therefore, the above simulation results show that nano-OvPOSS has desirable compatibility with SBS while SiO_2, ZnO, and TiO_2 do not. Obviously, nano-OvPOSS, which is organic-inorganic, outperforms traditional nanomaterials SiO_2, ZnO, and TiO_2, which are metal oxide or non-metal oxide, in its compatibility with SBS. This demonstration of poor compatibility of SBS with SiO_2, ZnO, and TiO_2 is ascribed to their total difference from polymer (SBS) in the structure and composition, while nano-OvPOSS is endowed with desired performance of compatibility thanks to its similar structure and composition with polymer (SBS) in accordance with the similar dissolve mutually theory.

4.1.2. The Compatibility of Nano-OvPOSS with Four Asphalt Components

Nano-OvPOSS was used as the screen, and four asphalt components as the base. The probability distributions of mixing energy of nano-OvPOSS (5 Å, 7 Å, and 9 Å) with the four asphalt components were calculated as is shown in Figure 10. Figure 10a represents the probability distribution of mixing energy of asphaltene with 5 Å nano-OvPOSS, 7 Å nano-OvPOSS, and 9 Å nano-OvPOSS; subsequently, Figure 10b–d show the probability distribution of mixing energy of resin with 5 Å nano-OvPOSS, 7 Å nano-OvPOSS, and 9 Å nano-OvPOSS, the probability distribution of mixing energy of saturate with 5 Å nano-OvPOSS, 7 Å nano-OvPOSS, and 9 Å nano-OvPOSS, and the probability distribution of mixing energy of aromatic with 5 Å nano-OvPOSS, 7 Å nano-OvPOSS, and 9 Å nano-OvPOSS, respectively.

In comparison with Figure 9, each of the twelve pictures of Figure 10 shows obvious similarity of E_{bb}, E_{bs}, and E_{ss} with each other, which means 5 Å nano-OvPOSS, 7 Å nano-OvPOSS, and 9 Å nano-OvPOSS are favorably compatible with the four components of asphalt.

As is shown in the four pictures of 5 Å nano-OvPOSS with asphaltene, resin, saturate, and aromatic, the three curves of E_{bb}, E_{bs}, and E_{ss} in each picture express the most favorable similarity with each other, which are compared with the respective four pictures of 7 Å nano-OvPOSS with asphaltene, resin, saturate, and aromatic as well as with the respective four pictures of 9 Å nano-OvPOSS with asphaltene, resin, saturate, and aromatic. It can be concluded that 5 Å nano-OvPOSS has the most favorable compatibility with asphaltene, resin, saturate, and aromatic when compared with 7 Å nano-OvPOSS and 9 Å nano-OvPOSS.

Meanwhile, Figure 10a,d express clear variation from similarity to difference of the three curves of E_{bb}, E_{bs}, and E_{ss} with the particle size of nano-OvPOSS increasing from 5 Å to 9 Å. Conclusions can be drawn that the compatibility between nano-OvPOSS and asphaltene as well as the compatibility between nano-OvPOSS and aromatic are influenced to a large extent by the particle size of nano-OvPOSS. The smaller the particle size of nano-OvPOSS is, the better the compatibility with asphaltene and aromatic is.

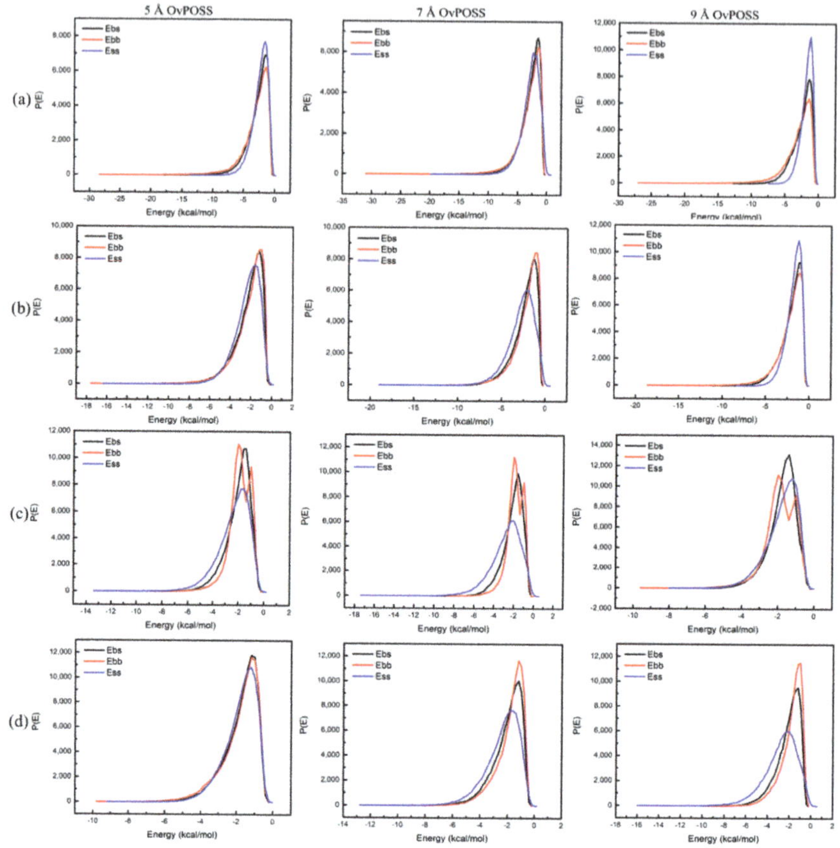

Figure 10. The probability distribution of mixing energy of nano-OvPOSS (5 Å OvPOSS, 7 Å OvPOSS, and 9 Å OvPOSS) with four asphalt components: (**a**) asphaltene; (**b**) resin; (**c**) saturate; (**d**) aromatic.

4.1.3. The Effect of Nano-OvPOSS on the Compatibility of SBS Modified Asphalt

The asphalt blends were used as the screen and SBS as the base. Figure 11a shows the probability distribution of mixing energy of matrix asphalt/SBS without nano-OvPOSS; Figure 11b–d show the probability distribution of mixing energy of 5 Å nano-OvPOSS/asphalt/SBS, 7 Å nano-OvPOSS/asphalt/SBS, and 9 Å nano-OvPOSS/asphalt/SBS, respectively. As can be seen from Figure 11a, the probability distribution curves of E_{bb}, E_{bs}, and E_{ss} of matrix asphalt/SBS are very similar to each other, which signifies that SBS has good compatibility with asphalt. The similarity of the curves of E_{bb}, E_{bs}, and E_{ss} of 5 Å nano-OvPOSS/asphalt/SBS can also be found in Figure 11b. Such similarity indicates the compatibility between SBS and asphalt are not changed by the addition of nano-OvPOSS and 5 Å nano-OvPOSS/asphalt/SBS are compatible with each other. The same conclusion can also be drawn from Figure 11c that 7 Å nano-OvPOSS/asphalt/SBS are compatible with each other and from Figure 11d that 9 Å nano-OvPOSS/asphalt/SBS are compatible with each other.

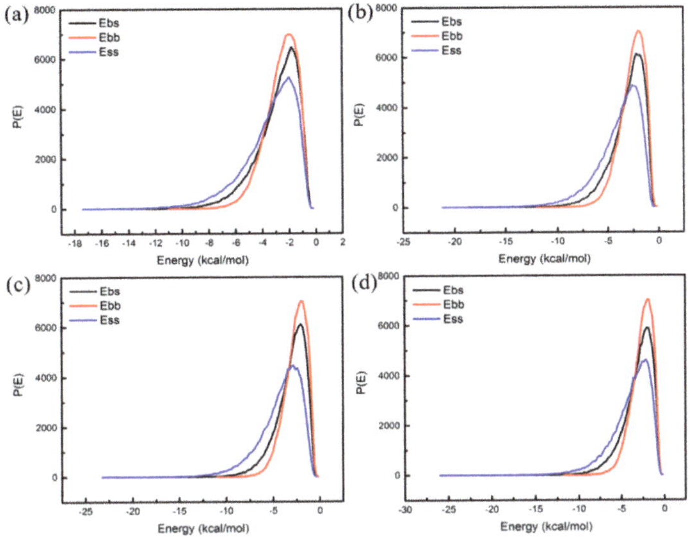

Figure 11. The probability distribution of mixing energy of SBS with asphalt blends: (**a**) matrix asphalt/SBS; (**b**) 5 Å nano-OvPOSS/asphalt/SBS; (**c**) 7 Å nano-OvPOSS/asphalt/SBS; (**d**) 9 Å nano-OvPOSS/asphalt/SBS.

The quantitative analysis of the effect of nano-OvPOSS on the compatibility of SBS modified asphalt can be taken from Figure 12 which expresses the compatibility indicator E_{mix} of modified asphalt. If the absolute value of E_{mix} is the smallest, the compatibility of two materials is the best. It can be seen from Figure 12 that the absolute value of E_{mix} of 5 Å nano-OvPOSS/asphalt/SBS is the lowest and that of SBS/asphalt is the highest. Moreover, as the particle size of nano-OvPOSS becomes larger, the absolute values of E_{mix} of (5 Å, 7 Å, and 9 Å) nano-OvPOSS/SBS/asphalt increase sequentially. The results show that nano-OvPOSS is able to improve the compatibility between SBS and asphalt, thus easing off the phase separation and guaranteeing SBS modified asphalt a more stable state. Moreover, 5 Å nano-OvPOSS displays the best performance in improving the compatibility between SBS and asphalt.

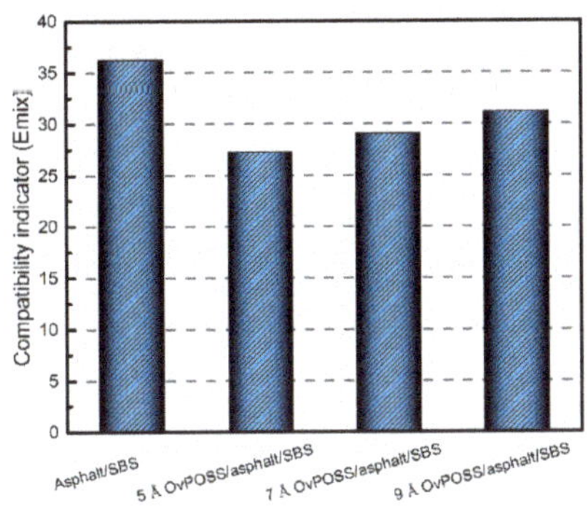

Figure 12. Compatibility indicator E_{mix} of modified asphalt.

The eight ethylene organic branches of nano-OvPOSS are capable of interacting with the active groups of polymers, which is why nano-OvPOSS is able to improve the compatibility between SBS and asphalt. It is well accepted that the outstanding performance of nanomaterial, such as their large specific surface area and high strain resistance, attributes to their unique small size [46–48]. It is also accepted that the smaller the size of nanomaterial, the larger their specific surface area will be. Therefore, the smallest particle size of nano-OvPOSS in this study, 5 Å nano-OvPOSS, has the largest contact range with polymer molecules, which gives it the best compatibility between SBS and asphalt compared with 7 Å nano-OvPOSS and 9 Å nano-OvPOSS.

The validation of the favorable compatibility between the organically processed nanomaterials and asphalt has actually been confirmed previously by several related experiments. In 2020, Fanourakis et al. [49] found that the solubility of the organically treated oxide MoO_3 in organic matter was improved, and its aggregation behavior was significantly reduced. In 2020, Zhang et al. [50,51] used a high-temperature storage stability test to estimate the compatibility between nanomaterial and asphalt. They found that after organic modification, the ΔS were further reduced to 0.5 °C in ODBA-REC (octadecyl dimethyl benzyl ammonium chloride-rectorite) modified asphalt since the benzyl chain in ODBA increases the adsorption effect of asphalt molecules on individual REC layers, thereby contributing to the desired compatibility between the ODBA-REC and asphalt. In this study, that nano-OvPOSS is viable in improving the compatibility with asphalt not only can be confirmed by the MD studies stated and analyzed above, but also can be drawn from its structural advantages. Basically, the structural equivalence guarantees the compatibility of different materials. Nano-OvPOSS is capable of displaying desired compatibility with SBS as well as with asphalt because it possesses both the properties of organic material and those of inorganic material as an organic-inorganic nano-hybrid material. Furthermore, it can improve the compatibility between SBS and asphalt in the meantime. Consequently, nano-OvPOSS outperforms the traditionally used nanomaterials such as SiO_2, ZnO, and TiO_2 as an ideal asphalt modifier when taking the promotion of the compatibility with polymer into account.

4.2. *Influence of Nano-OvPOSS on the Structure of SBS Modified Asphalt*

4.2.1. Influence of Nano-OvPOSS on R_g of SBS

R_g is often used to characterize the dynamic trajectory of flexible systems for molecular systems. A small R_g indicates the polymer is relatively compact, meaning throughout its trajectory the polymer spends most of its time as a folded structure. Figures 13 and 14 present respectively the changes of R_g and structures of SBS before and after the addition of nano-OvPOSS with different particle sizes into SBS modified asphalt. As is shown, without nano-OvPOSS, the R_g peak position of SBS is 7.28 Å and the width of R_g peak is 0.33 Å. With the separate addition of nano-OvPOSS with three different particle sizes, the R_g peak position of SBS shifts to 7.15 Å, 6.91 Å, and 7.20 Å. It reveals that the addition of nano-OvPOSS leads to the collapse of SBS molecule and increases the compactness of SBS. It is also notable that the 7 Å nano-OvPOSS has a more discernible effect on the compactness of SBS.

On the one hand, the SBS chain collapses in a crowded environment after the addition of nano-OvPOSS; on the other hand, the different shapes of the crowded particles bring out the different structures of SBS after the collapse [52], as is shown in Figure 14. Notably, the R_g peak width of SBS is narrowed to 0.31 Å, 0.26 Å, and 0.29 Å, respectively. The decrease of the peak width indicates that SBS branched chain reinforces its tractility and becomes easier to wrap the surrounding asphalt molecules after the addition of nano-OvPOSS, which promotes its attraction of the surrounding asphalt molecular [23].

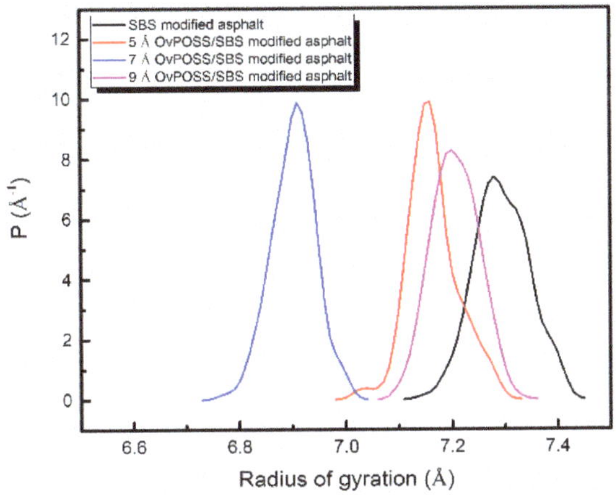

Figure 13. R_g of SBS in different modified asphalt systems.

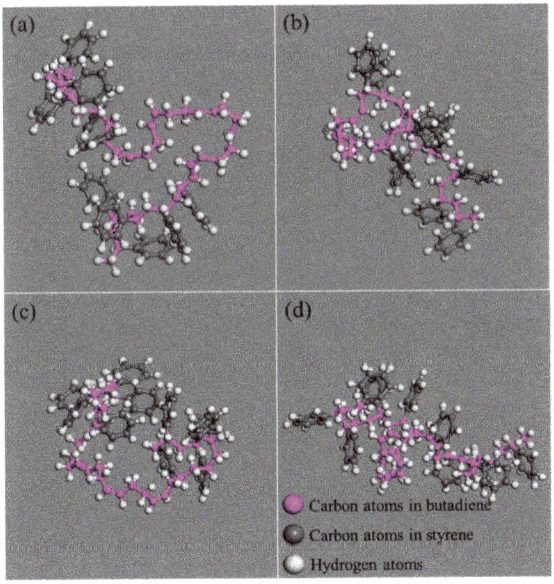

Figure 14. Structure of SBS in different modified asphalt systems: (**a**) SBS modified asphalt; (**b**) 5 Å nano-OvPOSS/SBS modified asphalt; (**c**) 7 Å nano-OvPOSS/SBS modified asphalt; (**d**) 9 Å nano-OvPOSS/SBS modified asphalt.

4.2.2. Influence of Nano-OvPOSS on RDF of SBS Modified Asphalt

The RDF is a measure of the relative probability of finding a particle at a distance r from a reference particle, and it mainly characterizes the packing state of atoms and the distance between atoms. To study the influence of nano-OvPOSS on RDF of atoms in SBS modified asphalt system, a carbon atom and hydrogen atom were selected as reference and selection for research. Figure 15 shows the RDF of atoms in a SBS modified asphalt system before and after the addition of nano-OvPOSS (5 Å, 7 Å, and 9 Å).

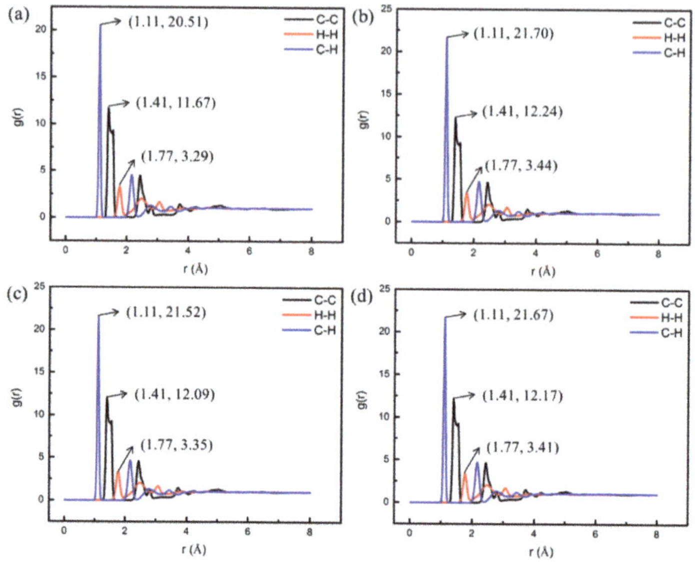

Figure 15. RDF of atoms in four modified asphalt systems: (**a**) SBS modified asphalt; (**b**) 5 Å OvPOSS/SBS modified asphalt; (**c**) 7 Å OvPOSS/SBS modified asphalt; (**d**) 9 Å OvPOSS/SBS modified asphalt.

According to Figure 15, before the addition of nano-OvPOSS, the curve of RDF has the first peak at 1.11 Å, representing the distance between carbon atom and hydrogen atom in the system; the second peak of the curve of RDF is at 1.41 Å, representing the distance between carbon atoms in the system; the third peak of the curve of RDF is at 1.77 Å, representing the distance between hydrogen atoms in the system. After adding 5 Å, 7 Å, and 9 Å nano-OvPOSS separately, the positions of the first peak, second peak, and third peak of the curve of RDF remain unchanged, which means the addition of nano-OvPOSS has no influence on the chemical structure of SBS modified asphalt system. On the other hand, before the addition of nano-OvPOSS, the intensity of the first peak, second peak, and third peak of SBS modified asphalt system is 20.51%, 11.67%, and 3.29%, respectively. With the addition of 5 Å nano-OvPOSS, the intensity increases to 21.70%, 12.24%, and 3.44% respectively; with the addition of 7 Å nano-OvPOSS, the intensity increases to 21.52%, 12.09%, and 3.35%; with the addition of 9 Å nano-OvPOSS, the intensity increases to 21.67%, 12.17%, and 3.41%. These increments suggest that the atomic packing density of C-H, C-C, and H-H is enhanced because of the addition of nano-OvPOSS. Among them, 5 Å nano-OvPOSS results in the most prominent effect in enhancing the atomic packing density. Such strengthened interaction between atoms in asphalt molecules is basically derived from the interaction of the eight vinyl groups of nano-OvPOSS with asphalt molecules.

4.2.3. Influence of Nano-OvPOSS on RFV of SBS Modified Asphalt

Free volume theory has been applied to predict the variation of viscosity and diffusivity of polymeric materials [53,54]. If the free volume fraction of asphalt reduces, its fluidity declines and thus its viscosity becomes higher, which in turn fortifies the deformation resistance of asphalt. Figure 16 describes the RFV of modified asphalt system in which the value of nano-OvPOSS/SBS modified asphalt of different nano-OvPOSS particle sizes (5 Å, 7 Å, and 9 Å) decreases by 3.39%, 5.85%, and 3.13% after the addition of nano-OvPOSS. This indicates that the free movement space of molecules in the modified asphalt system shrinks and the fluidity is hindered, so the addition of nano-OvPOSS increases the viscosity and enhances the deformation resistance of asphalt. This beneficial influence of nano-OvPOSS also derives from its unique structure of eight vinyl groups as an organic-inorganic nanohybrid material, because of which nano-OvPOSS reinforces its combination with molecules in asphalt system, boosts the non-bonding interactions between molecules, and then constructs a stable structural asphalt system. Notably, the 7 Å nano-OvPOSS performs better in reducing the RFV than 5 Å nano-OvPOSS and 9 Å nano-OvPOSS.

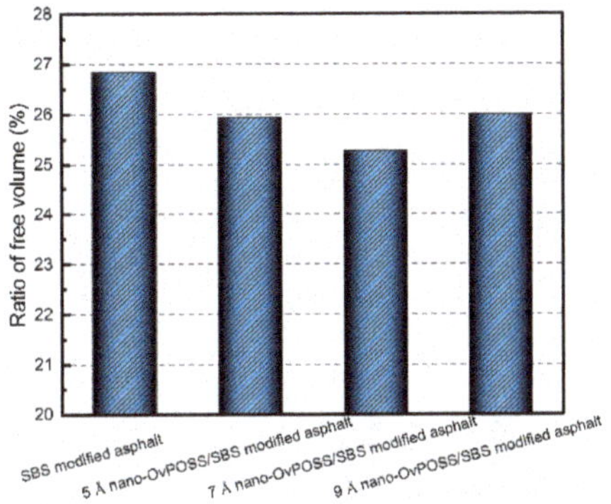

Figure 16. RFV of modified asphalt systems.

4.3. The Effect of Nano-OvPOSS on the Temperature Stability of SBS Modified Asphalt

Heat capacity is defined as the amount of heat energy required to raise the temperature of a given quantity of matter by one degree Celsius. The larger the heat capacity, the more heat it absorbs. Therefore, the laws of thermodynamics are used to design new processes for reactions that would have high efficiency. In this study, the isochoric specific heat capacity (C_v) of modified asphalt systems was simulated at room temperature. The simulation results of C_v are shown in Figure 17. It can be seen that the results of the C_v of the three nano-OvPOSS/SBS modified asphalt systems are basically the same (because of the same addition content of nano-OvPOSS), and all are obviously higher than that of SBS modified asphalt. This simulation result shows that the nano-OvPOSS improves the C_v of SBS modified asphalt, which suggests nano-OvPOSS/SBS modified asphalt system has better temperature stability. That is to say, the addition of nano-OvPOSS is able to alleviate the thermodynamic instability of SBS modified asphalt, which is mainly because the inorganic body of nano-OvPOSS is capable of absorbing more heat.

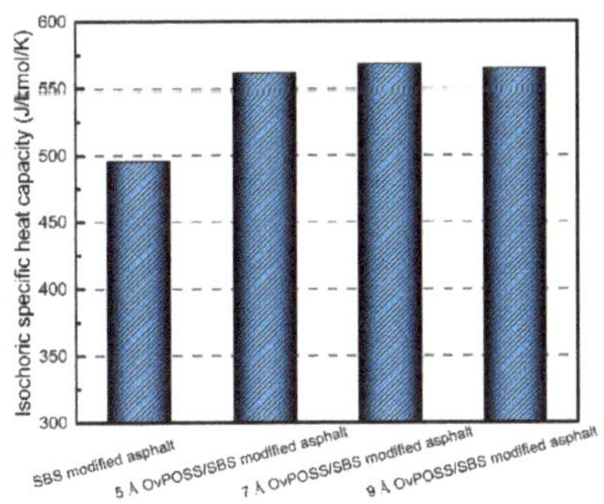

Figure 17. The isochoric specific heat capacity of modified asphalt.

4.4. The Effect of Nano-OvPOSS on the Mechanical Properties of SBS Modified Asphalt

The bulk modulus and shear modulus were simulated to investigate the effect of nano-OvPOSS on the mechanical properties of SBS modified asphalt. The results are shown in Figure 18. As can be seen, compared with SBS modified asphalt, the bulk modulus of nano-OvPOSS/SBS modified asphalt with different nano-OvPOSS particle size (5 Å, 7 Å, and 9 Å) increases by 30.15%, 47.11%, and 50.57%, respectively, and the shear modulus of these asphalt blends increases by 32.58%, 49.15%, and 57.11%, respectively. Thus, nano-OvPOSS has a positive effect on the two moduli of SBS modified asphalt. Such increment helps to equip asphalt with better mechanical properties and exceptional ability to resist deformation, thereby bringing out the desired performance in asphalt pavement. These simulation results are in agreement with laboratory test results [36]. In addition, the maximum values of bulk modulus and shear modulus of nano-OvPOSS/SBS modified asphalt are both obtained when the particle size of nano-OvPOSS is 9 Å. Therefore, the combination of 9 Å nano-OvPOSS and SBS in modifying asphalt is able to acquire the best deformation resistance ability of asphalt.

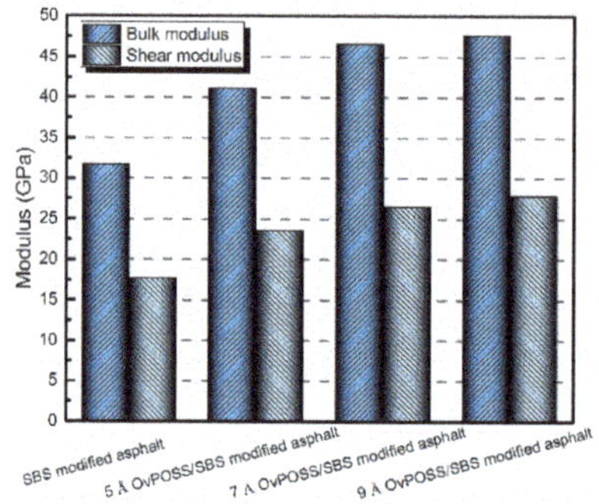

Figure 18. Physical modulus of different modified asphalt systems.

5. Conclusions

In this research, the influence of nano-OvPOSS with different particle sizes on SBS modified asphalt was inspected through the MD method. Conclusions are the following.

- Nano-OvPOSS is compatible with SBS as well as with four asphalt components, and is also able to improve the compatibility between SBS and asphalt. This is because the eight ethylene organic branches of nano-OvPOSS are capable of interacting with the active groups of polymers. Furthermore, 5 Å nano-OvPOSS has better ability to improve the compatibility between SBS and asphalt than 7 Å nano-OvPOSS and 9 Å nano-OvPOSS since it has the larger contact range with polymer molecules.
- The addition of nano-OvPOSS enhances the inter atomic bonding strength of asphalt component molecules. Meanwhile, it can reinforce the tractility of branched chains of SBS and make SBS easier to wrap and adsorb the surrounding asphalt molecules. Moreover, the free movement space of molecules in the SBS modified asphalt system declines. All of these effects result in a more stable colloidal structure of asphalt.
- The improvement of the C_V of SBS modified asphalt suggests that nano-OvPOSS/SBS modified asphalt system has better temperature stability. The reason why the addition of nano-OvPOSS is able to alleviate the thermodynamic instability of SBS modified asphalt is that the inorganic body of nano-OvPOSS is capable of absorbing more heat.
- The nano-OvPOSS modifier has a positive effect on bulk modulus and shear modulus of SBS modified asphalt, which renders better mechanical properties and deformation resistance of SBS modified asphalt in asphalt pavement.
- The overall results of the study prove the feasibility of nano-OvPOSS as an ideal asphalt modifier to attain a well-rounded performance of conventional SBS modified asphalt. As a functional material, POSS is capable of changing its eight substituents to realize more functions, one of which is nano-OvPOSS. Hence, the complexity and diversity of the structures of POSS allow for quite a few possibilities of its application in the field of modified asphalt. It is earnestly anticipated that more and more modifiers

taking the advantage of POSS can be explicitly explored by many a researcher in the near future.

Author Contributions: Conceptualization, L.F. and P.Z.; methodology, P.Z.; software, L.F.; validation, T.C., M.J. and P.Z.; formal analysis, L.F.; investigation, T.C. and M.J.; resources, M.J.; data curation, T.C.; writing—original draft preparation, L.F.; writing—review and editing, L.F.; visualization, T.C.; supervision, P.Z.; project administration, L.F.; funding acquisition, L.F. All authors have read and agreed to the published version of the manuscript.

Funding: This research was funded by the Fundamental Research Funds for the Central Universities, CHD (300102319707).

Conflicts of Interest: The authors declare no conflict of interest.

References

1. Yildirim, Y. Polymer modified asphalt binders. *Constr. Build. Mater.* **2007**, *21*, 66–72. [CrossRef]
2. Zhu, J.; Birgisson, B.; Kringos, N. Polymer modification of bitumen: Advances and challenges. *Eur. Polym. J.* **2014**, *54*, 18–38. [CrossRef]
3. Yut, I.; Zofka, A. Correlation between rheology and chemical composition of aged polymer-modified asphalts. *Constr. Build. Mater.* **2014**, *62*, 109–117. [CrossRef]
4. Airey, G.D.; Singleton, T.M.; Collop, A.C. Properties of polymer modified bitumen after rubber-bitumen interaction. *J. Mater. Civ. Eng.* **2002**, *14*, 344–354. [CrossRef]
5. Sengoz, B.; Isikyakar, G. Analysis of styrene-butadiene-styrene polymer modified bitumen using fluorescent microscopy and conventional test methods. *J. Hazard. Mater.* **2008**, *150*, 424–432. [CrossRef]
6. Ye, F.; Yin, W.; Lu, H. A model for the quantitative relationship between temperature and microstructure of styrene-butadiene-styrene modified asphalt. *Constr. Build. Mater.* **2015**, *79*, 397–401. [CrossRef]
7. Liang, M.; Liang, P.; Fan, W.; Qian, C.; Xin, X.; Shi, J.; Nan, G. Thermo-rheological behavior and compatibility of modified asphalt with various styrene-butadiene structures in SBS copolymers. *Mater. Des.* **2015**, *88*, 177–185. [CrossRef]
8. Cong, P.; Xu, P.; Chen, S. Effects of carbon black on the anti aging, rheological and conductive properties of SBS/asphalt/carbon black composites. *Constr. Build. Mater.* **2014**, *52*, 306–313. [CrossRef]
9. Chen, J.; Liao, M.; Shiah, M.S. Asphalt modified by styrene-butadiene-styrene triblock copolymer: Morphology and model. *J. Mater. Civ. Eng.* **2002**, *14*, 224–229. [CrossRef]
10. Qian, C.; Fan, W.; Ming, L.; Nan, G.; Luo, H. Influence of compatibilizer composition on performance of SBS modified asphalt. *Mater. Sci. Energy Technol. Power Eng. II* **2018**, *1971*, 050011.
11. Ouyang, C.; Gao, Q.; Shi, Y.; Shan, X. Compatibilizer in waste tire powder and low-density polyethylene blends and the blends modified asphalt. *J. Appl. Polym. Sci.* **2011**, *123*, 485–492. [CrossRef]
12. Lu, X.; Isacsson, U.; Ekblad, J. Phase separation of SBS polymer modified bitumens. *J. Mater. Civ. Eng.* **1999**, *11*, 51–57. [CrossRef]
13. Wang, S.; Yuan, C.; Deng, J. Crumb tire rubber and polyethylene mutually stabilized in asphalt by screw extrusion. *J. Appl. Polym. Sci.* **2014**, *131*, 81–86. [CrossRef]
14. Sugano, M.; Kajita, J.; Ochiai, M.; Takagi, N.; Iwai, S.; Hirano, K. Mechanisms for chemical reactivity of two kinds of polymer modified asphalts during thermal degradation. *Chem. Eng. J.* **2011**, *176*, 231–236. [CrossRef]
15. Li, J.; Jia, W.; Yuan, W. Effect of polyethylene grafted with maleic anhydride on asphalt properties. *J. Perform. Constr. Facil.* **2014**, *28*, 04014012. [CrossRef]
16. Fang, C.; Yu, R.; Liu, S.; Li, Y. Nanomaterials applied in asphalt modification: A review. *J. Mater. Sci. Technol.* **2013**, *29*, 589–594. [CrossRef]
17. Lu, H.; Ye, F.; Yuan, J.; Yin, W. Properties comparison and mechanism analysis of naphthenic oil/SBS and nano-MMT/SBS modified asphalt. *Constr. Build. Mater.* **2018**, *187*, 1147–1157. [CrossRef]
18. Galooyak, S.S.; Dabir, B.; Nazarbeygi, A.E.; Moeini, A. Rheological properties and storage stability of bitumen/SBS/montmorillonite composites. *Constr. Build. Mater.* **2010**, *24*, 300–307. [CrossRef]
19. Yu, J.; Wang, L.; Zeng, X.; Wu, S.; Li, B. Effect of montmorillonite on properties of styrene-butadiene-styrene copolymer modified bitumen. *Polym. Eng. Sci.* **2007**, *47*, 1289–1295. [CrossRef]
20. Zhang, H.; Yu, J.; Wu, S. Effect of montmorillonite organic modification on ultraviolet aging properties of SBS modified bitumen. *Constr. Build. Mater.* **2012**, *27*, 553–559. [CrossRef]
21. Ouyang, C.; Wang, S.; Zhang, Y.; Zhang, Y. Preparation and properties of styrene-butadiene-styrene copolymer/Kaolinite clay compound and asphalt modified with the compound. *Polym. Degrad. Stab.* **2005**, *87*, 309–317. [CrossRef]
22. Rezaei, S.; Ziari, H.; Nowbakht, S. High-temperature functional analysis of bitumen modified with composite of nano-SiO_2 and styrene butadiene styrene polymer. *Pet. Sci. Technol.* **2016**, *34*, 1195–1203. [CrossRef]
23. Su, M.; Si, C.; Zhang, Z.; Zhang, H. Molecular dynamics study on influence of nano-ZnO/SBS on physical properties and molecular structure of asphalt binder. *Fuel* **2020**, *263*, 116777.1–116777.13. [CrossRef]

24. Scott, D.W. Thermal rearrangement of branched-chain methylpolysiloxanes. *J. Am. Chem. Soc.* **1946**, *68*, 356–358. [CrossRef]
25. Zhou, Z.; Cui, L.; Zhang, Y.; Zhang, Y.; Yin, N. Isothermal crystallization kinetics of polypropylene/POSS composites. *J. Polym. Sci. Part B Polym. Phys.* **2008**, *46*, 1762–1772. [CrossRef]
26. Zhou, Z.; Zhang, Y.; Zeng, Z.; Zhang, Y. Properties of POSS-filled polypropylene: Comparison of physical blending and reactive blending. *J. Appl. Polym. Sci.* **2008**, *110*, 3745–3751. [CrossRef]
27. Zhou, Z.; Cui, L.; Zhang, Y.; Zhang, Y.; Yin, N. Preparation and properties of POSS grafted polypropylene by reactive blending. *Eur. Polym. J.* **2008**, *44*, 3057–3066. [CrossRef]
28. Zhou, Z.; Zhang, Y.; Zhang, Y.; Yin, N. Rheological behavior of polypropylene/octavinyl polyhedral oligomeric silsesquioxane composites. *J. Polym. Sci. Part B Polym. Phys.* **2008**, *46*, 526–533. [CrossRef]
29. Zheng, L.; Hong, S.; Cardoen, G. Polymer nanocomposites through controlled self-assembly of cubic silsesquioxane scaffolds. *Macromolecules* **2004**, *37*, 8606–8611. [CrossRef]
30. Wu, J.; Haddad, T.S.; Mather, P.T. Vertex group effects in entangled polystyrene-polyhedral oligosilsesquioxane (POSS) copolymers. *Macromolecules* **2009**, *42*, 1142–1152. [CrossRef]
31. Li, G.; Wang, L.; Ni, H.; Pittman, C.U. Polyhedral oligomeric silsesquioxane (POSS) polymers and copolymers: A review. *J. Inorg. Organomet. Polym.* **2001**, *11*, 123–154. [CrossRef]
32. Phillips, S.H.; Haddad, T.S.; Tomczak, S.J. Developments in nanoscience: Polyhedral oligomeric silsesquioxane (POSS)-polymers. *Curr. Opin. Solid State Mater. Sci.* **2004**, *8*, 21–29. [CrossRef]
33. Kuo, S.W.; Chang, F.C. POSS related polymer nanocomposites. *Prog. Polym. Sci.* **2011**, *36*, 1649–1696. [CrossRef]
34. Li, G.; Pittman, C.U. Polyhedral oligomeric silsesquioxane (POSS) polymers, copolymers, and resin nanocomposites. In *Macromolecules Containing Metal and Metal-like Elements: Group IVA Polymers*; John Wiley & Sons: Hoboken, NJ, USA, 2005; pp. 79–131.
35. Hansen, J.S.; Lemarchand, C.A.; Nielsen, E.; Dyre, J.C.; Schroder, T. Four-component united-atom model of bitumen. *J. Chem. Phys.* **2013**, *138*, 94508. [CrossRef] [PubMed]
36. Zhang, H.L.; Su, M.M.; Zhao, S.F.; Zhang, Y.P.; Zhang, Z.P. High and low temperature properties of nano-particles/polymer modified asphalt. *Constr. Build. Mater.* **2016**, *114*, 323–332. [CrossRef]
37. Sun, H.; Ren, P.; Fried, J.R. The COMPASS force field: Parameterization and validation for phosphazenes. *Comput. Theor. Polym. Sci.* **1998**, *8*, 229–246. [CrossRef]
38. Rigby, D.; Sun, H.; Eichinger, B.E. Computer simulations of poly (ethylene oxide): Force field, pvt diagram and cyclization behaviour. *Polym. Int.* **1997**, *44*, 311–330. [CrossRef]
39. Nose, S. A unified formulation of the constant temperature molecular dynamics methods. *J. Chem. Phys.* **1984**, *81*, 511–519. [CrossRef]
40. Berendsen, H.J.C.P.; Postma, J.; Gunsteren, W.; Dinola, A.D.; Haak, J.R. Molecular-dynamics with coupling to an external bath. *J. Chem. Phys.* **1984**, *81*, 3684. [CrossRef]
41. Huggins, M.L. Principles of polymer chemistry. *J. Am. Chem. Soc.* **1954**, *76*, 2854. [CrossRef]
42. Bawendi, M.G.; Freed, K.F. A lattice model for self- and mutually avoiding semiflexible polymer chains. *J. Chem. Phys.* **1987**, *86*, 3720–3730. [CrossRef]
43. Fan, C.F.; Olafson, B.D.; Blanco, M.; Hsu, S.L. Application of molecular simulation to derive phase diagrams of binary mixtures. *Macromolecules* **1992**, *25*, 3667–3676. [CrossRef]
44. Freed, K.F.; Bawendi, M.G. Lattice theories of polymeric fluids. *J. Phys. Chem.* **1989**, *93*, 2194–2203. [CrossRef]
45. Pesci, A.I.; Freed, K.F. Lattice models of polymer fluids: Monomers occupying several lattice sites. II. Interaction energies. *J. Chem. Phys.* **1989**, *90*, 2003–2016. [CrossRef]
46. Choi, W.S.; Lee, H. Nanostructured materials for water purification: Adsorption of heavy metal ions and organic dyes. *Polymers* **2022**, *14*, 2183. [CrossRef]
47. Vu, D.; Ahn, K. Triboelectric enhancement of polyvinylidene fluoride membrane using magnetic nanoparticle for water-based energy harvesting. *Polymers* **2022**, *14*, 1547. [CrossRef]
48. Zhang, Z.; Cao, H.; Quan, Y.; Ma, R.; Pentzer, E.B.; Green, M.J.; Wang, Q. Thermal stability and flammability studies of mxene-organic hybrid polystyrene nanocomposites. *Polymers* **2022**, *14*, 1213. [CrossRef] [PubMed]
49. Fanourakis, S.K.; Pea-Bahamonde, J.; Rodrigues, D.F. Inorganic salts and organic matter effects on nanorod, nanowire, and nanoplate MoO_3 aggregation, dissolution, and photocatalysis. *Environ. Sci. Nano* **2020**, *7*, 3794–3804. [CrossRef]
50. Zhang, H.; Zhu, C.; Wei, C.; Duan, H.; Yu, J. Application of functionalized nanomaterials in asphalt road construction materials. In *Handbook of Functionalized Nanomaterials for Industrial Applications*; Elsevier: Amsterdam, The Netherlands, 2020; pp. 865–907.
51. Zhang, H.; Zhu, C.; Yan, K.; Yu, J. Effect of rectorite and its organic modification on properties of bitumen. *J. Mater. Civ. Eng.* **2015**, *27*, C4014002.1–C4014002.5. [CrossRef]
52. Chen, A.; Zhao, N. Comparative study of the crowding-induced collapse effect in hard-sphere, flexible polymer and rod-like polymer systems. *Phys. Chem. Chem. Phys.* **2019**, *21*, 12335–12345. [CrossRef]
53. Vrentas, J.S.; Duda, J.L.; Ling, H.C. Free-volume theories for self-diffusion in polymer-solvent systems. I. Conceptual differences in theories. *J. Polym. Sci. Part B Polym. Phys. Ed.* **1985**, *23*, 275–288. [CrossRef]
54. Rad, F.Y.; Sefidmazgi, N.R.; Bahia, H. Application of diffusion mechanism. *Transp. Res. Rec. J. Transp. Res. Board* **2014**, *2444*, 71–77. [CrossRef]

polymers

MDPI

Article

Study of the Microscopic Mechanism of Natural Rubber (Cis-1, 4-Polyisoprene, NR)/Polyethylene (PE) Modified Asphalt from the Perspective of Simulation

Yujing Chen [1], Kui Hu [2], Caihua Yu [3,*], Dongdong Yuan [1] and Xiaoyi Ban [1]

1 School of Highway, Chang'an University, Xi'an 710064, China
2 College of Civil Engineering, Henan University of Technology, Zhengzhou 450001, China
3 Department of Structural Engineering, College of Civil Engineering, Tongji University, Shanghai 200092, China
* Correspondence: ychce0919@tongji.edu.cn

Abstract: This paper aims to study the interaction mechanism of waste tire/plastic modified asphalt from the microscopic perspective of molecules. Based on BIOVIA Materials Studio, a classic four-component asphalt model consisting of asphaltene ($C_{149}H_{177}N_3O_2S_2$), resin ($C_{59}H_{85}NOS$), aromatic (C46H50S), and saturate ($C_{22}H_{46}$) was constructed. Waste tires are represented by natural rubber (NR), which uses cis-1, 4-polyisoprene as a repeating unit. In contrast, waste plastics are characterized by polyethylene (PE), whose optimum degree of polymerization is determined by the difference in solubility parameters. Then, the above molecular models are changed to a stable equilibrium state through the molecular dynamics process. Finally, the interaction process is analyzed and inferred using the indexes of radial distribution function, diffusion coefficient, and concentration distribution; further, the interaction mechanism is revealed. The results show that the optimal degree of polymerization of PE is 12, so the solubility parameter between PE and NR-modified asphalt is the lowest at 0.14 $(J/cm^3)^{1/2}$. These models are in agreement with the characteristics of amorphous materials with the structures ordered in the short-range and long-range disordered. For NR-modified asphalt, the saturate moves fastest, and its diffusion coefficient reaches 0.0201, followed by that of the aromatic (0.0039). However, the molecule of NR ranks the slowest in the NR-modified asphalt. After the addition of PE, the diffusion coefficient of resin increased most significantly from 0.0020 to 0.0127. NR, PE, and asphaltene have a particular attraction with the lightweight components, thus changing to a more stable spatial structure. Therefore, using NR and PE-modified asphalt can change the interaction between asphalt molecules to form a more stable system. This method not only reduces the large waste disposal task but also provides a reference for the application of polymer materials in modified asphalt.

Keywords: asphalt pavements; modified asphalt; functional polymer; cis-1, 4-polyisoprene; polyethylene; molecular dynamics; microscopic mechanism

Citation: Chen, Y.; Hu, K.; Yu, C.; Yuan, D.; Ban, X. Study of the Microscopic Mechanism of Natural Rubber (Cis-1, 4-Polyisoprene, NR)/Polyethylene (PE) Modified Asphalt from the Perspective of Simulation. *Polymers* **2022**, *14*, 4087. https://doi.org/10.3390/polym14194087

Academic Editor: Jens-Uwe Sommer

Received: 28 August 2022
Accepted: 21 September 2022
Published: 29 September 2022

1. Introduction

Petroleum asphalt is an essential raw material in road engineering, especially in asphalt pavements, which are used widely. At present, petroleum is gradually becoming exhausted, so asphalt, a by-product of petroleum, will also face the problem of exhaustion. Against this background, it is indispensable and crucial to lower the consumption of asphalt in road engineering and decrease dependence on petroleum. In addition, the deteriorating environment and traffic conditions put higher requirements on the performance of asphalt pavements. Therefore, the modification of asphalt to reduce cost and improve performance is one of the leading research directions for researchers, which is very significant. The use of modified asphalt has recently increased, and its application scope and applications are becoming broader [1–5].

There are many kinds of materials that can be used as modifiers, and the current research mainly includes the following aspects. Some researchers use biomaterials as modifiers to prepare bio-asphalt [6–8], such as beans [9], waste cooking oil [10,11], biochar [12,13], and so on. Forestry resources and animal excreta are liquefied and separated to prepare modifiers. In addition, in the process of biomass raw materials processing to produce bio-based materials (furfural, furfuryl alcohol, dimethylfuran, biodiesel, etc.), the remaining polymer carbohydrates can also be used to create bio-asphalt. Some researchers use soil as a modifier, such as nano clay [14], diatomite [15,16], organically modified bentonite [17,18] and so on. Take bentonite as an example. Its electron microscope diagram is shown in Figure 1, where (a) represents the original bentonite and (b) represents the nano-bentonite, which shows a clear layered structure and good continuity. Bentonite itself belongs to the octahedral structure. If added to the asphalt, it can fill the pores between the micelle and micelle and further increase the anti-skid resistance of asphalt. Fibers are also adopted as modifiers, such as polyester [19,20], lignin [21,22], basalt fibers [23,24] and so on [25]. This kind of modifier can enhance the cohesion between aggregate particles so that the asphalt pavement can adapt to high wear requirements, increased skid resistance, and other sections. Guo [26,27], Zheng [28], and some other researchers [29,30] used waste tires as raw materials to prepare rubber-modified asphalt and conducted research. Some researchers use plastics as modifiers, such as polyethylene, polyurethane, polypropylene, etc. [4,31,32]. Some nano-modifiers are carbon nanotubes, graphene, and other nanomaterials, but the cost is high.

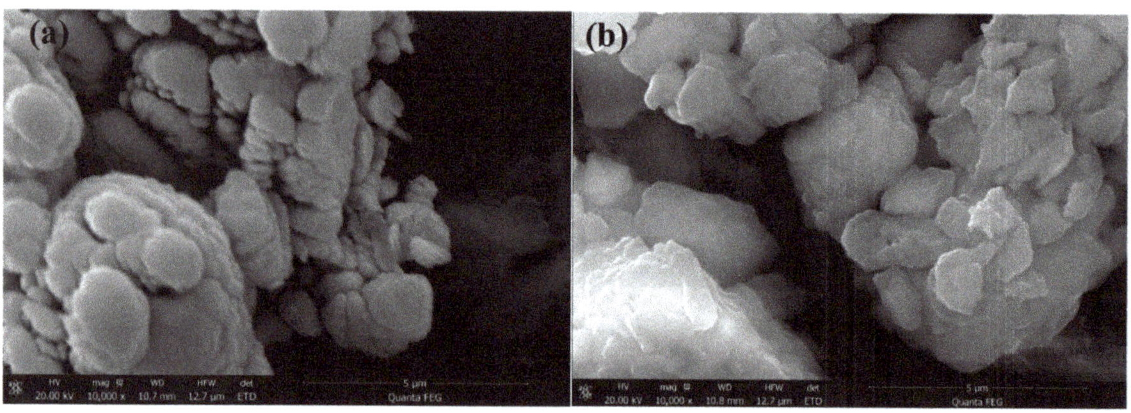

Figure 1. The SEM images of (**a**) BT and (**b**) OBT.

It should be noted that as asphalt modifiers, waste tires and plastics have a significant negative influence on the ecological environment [33]. Waste tires and plastics are non-biodegradable substances, which brings difficulties to waste management [27]. If incinerated, a lot of smoke and other harmful substances will be produced, which will significantly impact the air, soil, rivers, and other environments. However, waste plastics and rubber have good reuse value as asphalt modifiers and can improve asphalt performance [28]. Generally speaking, plastic polymers can significantly enhance asphalt performance at high temperatures. In contrast, rubber polymers can upgrade the deformation resistance of asphalt at low temperatures, such as styrene-butadiene rubber (SBR). Therefore, the use of the two as asphalt modifiers simultaneously becomes one of the best choices for the recycling and reuse of waste material. At present, only one kind of modified agent is doped separately. Still, there are few studies on adding two types of waste materials simultaneously. Hence, it is not easy to give full play to the synergistic effect of the two and improve the high- and low-temperature performance of asphalt simultaneously.

It is a much more cost-effective and scientific to perform research using molecular dynamics. The traditional test has a long test cycle and a high cost. In addition, many conventional experiments only describe and compare the experimental results, which fails to explain some empirical phenomena in depth. Molecular dynamics, on the other hand, refers to the study of microscopic phenomena from the perspective of molecules or atoms. Currently, many researchers are studying the molecular dynamics of asphalt, such as the modification mechanism of modifiers [34]. For example, Hu et al. [35] studied the interaction mechanism between styrene-butadiene-styrene (SBS) and asphalt to explain the formation and separation mechanism of an SBS-modified asphalt two-phase structure from a microscopic perspective.

The purpose of this paper is to study the molecular interaction mechanism after two modifiers simultaneously using molecular dynamics. In this paper, the original asphalt and rubber-modified asphalt models were built from the perspective of molecular dynamics. Then, plastic was added to create a plastic/rubber asphalt model, and the influence of plastic molecules on rubber asphalt was analyzed. Among the models, the classical four-component molecular model was taken to represent asphalt, and rubber is expressed by natural rubber molecules composed of polyisoprene. Additionally, plastics are represented by polyethylene molecules, the optimal polymerization degree of which is measured by solubility parameters. After the molecular dynamics process, all the above models reach a state of equilibrium and stability. The radial distribution function (RDF) was used to verify the molecular model, and the mean square displacement (MSD) and diffusion coefficient were used to study the diffusion of molecules in modified asphalt. Finally, concentration distribution was taken to study the spatial distribution of different molecules.

2. Simulation

2.1. Model of Asphalt

Bitumen is a highly complex mixture of hydrocarbons of different molecular weights and their derivatives of non-metallic components (oxygen, nitrogen, sulfur, etc.). Petroleum bitumen, a vital type of bitumen, is widely used in road engineering. From a scientific research perspective, the parts of petroleum asphalt with some common characteristics, such as similar chemical composition and physical properties, are divided into the same group, called components. The American Society for Testing and Materials (ASTM) D4124-09 has proposed and recommended a four-component analysis method. This method divides petroleum asphalt into asphaltene, aromatic, resin, and saturate.

The simplified model adopted in this study consists of four representative molecules, and this method has been widely used because of its simplicity, convenience, and conformity with the actual situation [36]. The four representative molecules are asphaltene ($C_{149}H_{177}N_3O_2S_2$), resin (C59H85NOS), aromatic (C46H50S), and saturate ($C_{22}H_{46}$), as shown in Figure 2. The proportion of the four representative molecules in this model is calculated according to the mass ratio and molecular weight of each component in asphalt, in line with previous studies [36,37], as shown in Table 1.

Table 1. Specific information about the asphalt model.

Components	Molecular Formula	Number Ratio	Relative Molecular Mass	Mass Fraction (%)	NumAtoms
Asphaltene	$C_{149}H_{177}N_3O_2S_2$	1	2106.19	19.734	333
Resin	$C_{59}H_{85}NOS$	3	856.395	24.071	147
Saturate	$C_{22}H_{46}$	5	310.610	14.551	68
Aromatic	$C_{46}H_{50}S$	7	634.966	41.645	97

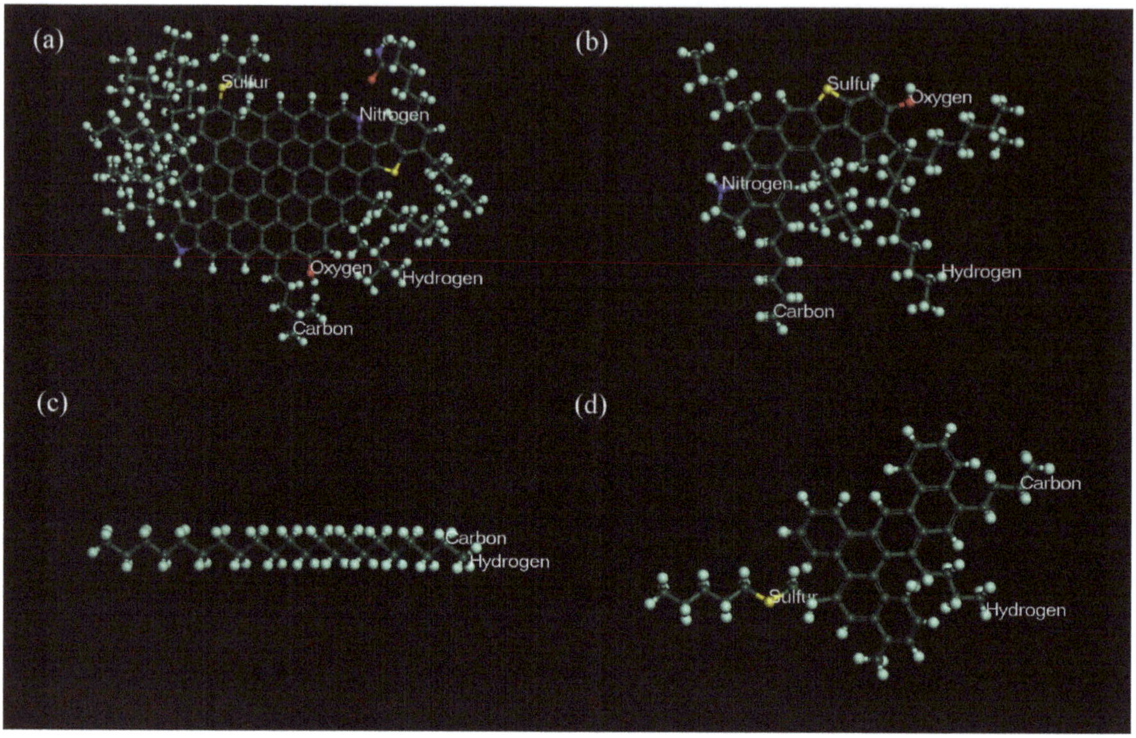

Figure 2. Models of representative molecules in bitumen. (**a**) asphaltene; (**b**) resin; (**c**) saturate; (**d**) aromatic.

2.2. Models of Rubber

Rubber powder mainly comes from waste tires, and its composition is complicated. The three most common types of rubber are NR (natural rubber), BR (cis-polybutadiene rubber), and SBR (styrene-butadiene rubber). Since natural rubber is the most widely used, it was selected as the research object in this study.

Natural rubber mainly comprises cis-1, 4-polyisoprene, accounting for more than 97% and 2–3% of the 3- and 4-bonded structure. Therefore, when building the molecular model of rubber, NR was regarded as a homopolymer, which was polymerized from the isoprene mentioned above. The relative molecular weight range of NRs is relatively wide, generally between 30,000 and 30 million, because of the different degrees of polymerization. The degree of polymerization affects the results of the experiments and leads to different simulation costs, so a proper selection is needed. Referring to previous studies [29,38], the degree of polymerization selected in this paper is 16, and the number of repeating units is set at 16. The molecular chain for NR is shown in Figure 3.

2.3. Models of PE

Polyethylene, short for PE, is a thermoplastic resin prepared by the polymerization of ethylene, including ethylene and a small amount of α-olefin copolymer. The degree of polymerization, a key factor, affects the molecular weight and the amount of simulation, and it is also closely related to the solubility parameters. The independent variables of the polymerization degree for polyethylene used in this article are 6, 12, 18, 24, 30, 36, 42, 48, 54, 60, 66, 72, 78, 84, 90, 96, 102, 108, 114, 120, and there are 20 different degrees of polymerization in total. Firstly, the degree of polymerization is closely related to the amount of simulation calculations because the degree of polymerization is a direct factor affecting the molecular weight; the greater the degree of polymerization, and the greater the molecular weight, the larger the amount of simulation calculations. Secondly, the solubility parameters of polyethylene with different degrees of polymerization are different. If the difference in the solubility parameters is less than 1.3~2.1 $(J/cm^3)^{1/2}$, the two materials are compatible. The smaller the difference is, the more easily the two are compatible. Therefore, in this paper, the solubility parameter difference is selected to evaluate the solubilization effect of PE on rubber-modified asphalt.

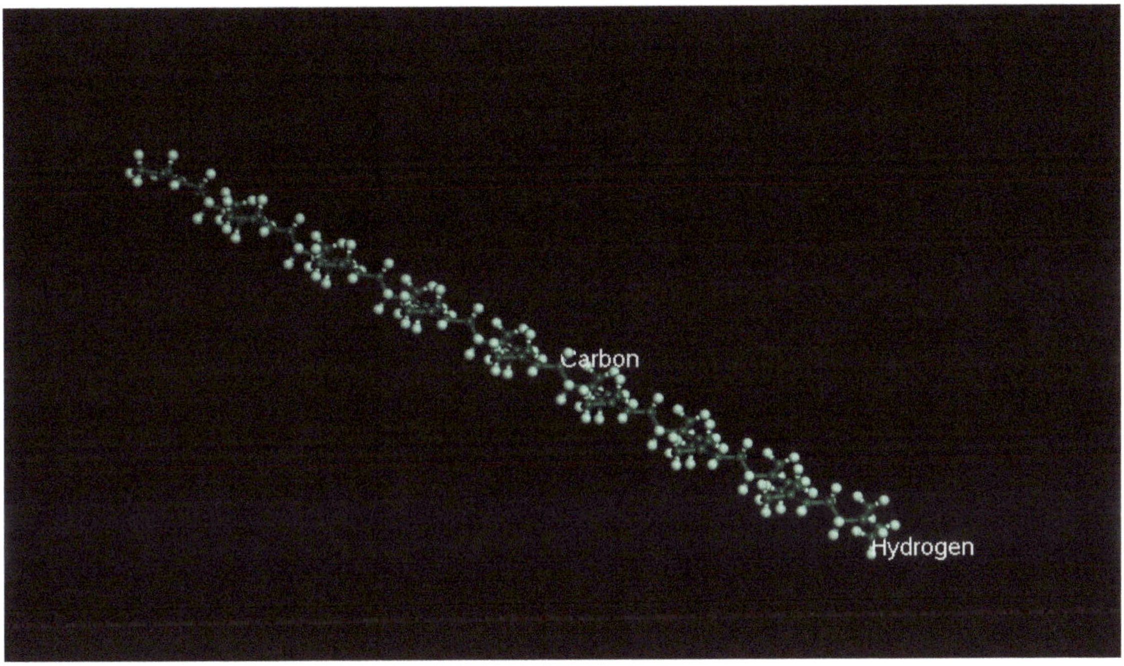

Figure 3. The molecular chain for NR.

2.4. Models of the Mixture

The simulation software used in this study is the BIOVIA Materials Studio, an integrated computer simulation platform involving quantum mechanics, molecular dynamics, and mesoscopic dynamics. Its function is to reveal the interaction mechanism of substances or predict the properties of materials from the microscopic perspective of molecules or atoms; therefore, it is widely used in various research fields.

The basic principles of molecular dynamics simulation were introduced. A system with N particles and the description of classical mechanics depends on 3N degrees of freedom $(q,p) \equiv \{q_i, p_i | i = 1\ (1)\ N\}$. The position $q \equiv \{q_i | i = 1\ (1)\ N\}$, momentum $P \equiv \{p_i | i = 1(1)N\}$. q_i is the position of the particle $No.\ i$ and p_i is the linear momentum. (q,p) determines the system's state, and each microscopic state is a phase point in phase space.

In molecular dynamics studies, the laws of Newtonian mechanics are usually adopted, as shown in the related literature [39].

$$\frac{dv_i}{dt} = \frac{F_i\ (t)}{m_i} = -\frac{1}{m_i}\nabla_i U \quad \forall i = 1(1)N \tag{1}$$

$$\frac{dr_i(t)}{dt} = v_i(t) \quad \forall i = 1(1)N \tag{2}$$

where, r_i and v_i are the position and velocity of particle $No.\ i$, respectively. The bond action, similarly to the quantum effect in the above equation, is reflected in the potential energy U. The potential energy U, which depends on the position of all of the nuclei, can be calculated precisely by quantum chemistry or approximated by force-field methods.

Formula (1) and (2) are a set of $3N$ differential equations, which should be solved by the numerical integration method of differential equations. Therefore, the position $\{r_i(t)\}$ and velocity $\{v_i(t)\}$ of the system at each moment can be solved by the integration algorithm. In this case, the trajectory of all the nuclei of the system is given, that is, the time evolution of the system. Then, the change of all system properties with time can be obtained.

COMPASS is a powerful force field that supports the atomic level simulation of condensed matter materials, short for "Condensed-Phase Optimized Molecular Potential for Atomistic Simulation Study." It is the first ab initio force field parameterized and verified by condensed matter properties and various ab initio and empirical data of isolated molecules. Therefore, the structure, conformation, vibration, and thermophysical properties of different molecules in isolated or condensed systems could be accurately predicted for a wide range of temperatures and pressures by this field [40–44].

The specific steps of building a stable model are as follows.

1. The establishment of initial polymer model.
2. Geometric optimization.
3. The simulation of annealing.
4. In this procedure, the pressure (1.01×10^{-4} Gpa) and temperature of the system remain constant [45].
5. NVT+NPT dynamics simulation.
6. At ambient temperature and standard atmospheric pressure, 50 ps of MD simulation was performed to bring the model closer to its natural state [46,47]. Taking the original asphalt model as an example, after the above steps were performed, parameters such as the energy, density, and others for the system reached a stable state, as shown in Figure 4. For example, the density of the original asphalt model increased and stabilized at about 1.0, which coincided with the actual situation.
7. The calculation and analysis of parameters.
8. After the molecular dynamic simulation, parameters related to this study were analyzed and calculated, such as the solubility parameters and radial distribution function, etc., according to the stable model generated at the end of the simulation process.

The specific composition of the two modified asphalt models, including NR-modified asphalt and PE-modified asphalt, is shown in Tables 2 and 3. Both the original asphalt model and the modified asphalt model obey the above steps. In the NR-modified asphalt, one NR molecule was included, and its content was 9.28%. The original asphalt model and NR-modified asphalt model in an equilibrium state are shown in Figure 5.

In PE/NR-modified asphalt, the degree of polymerization of the PE molecule should first be determined. In this study, different molecular models were constructed for PE molecules with varying degrees of polymerization, and the kinetic simulation process was performed according to the steps mentioned above. The equilibrium model was finally obtained, as shown in Figure 5.

Table 2. NR-modified asphalt model.

Serial Number	Representative Molecule	Chemical Composition	Molecule Number	Relative Molecular Mass	Mass Fraction
1	Asphaltene	$C_{149}H_{177}N_3O_2S_2$	1	2106.190	17.90%
2	Resin	$C_{59}H_{85}NOS$	3	856.395	21.84%
3	Saturate	$C_{22}H_{46}$	5	310.610	13.20%
4	Aromatic	$C_{46}H_{50}S$	7	634.966	37.78%
5	NR	$C_{80}H_{130}$	1	1091.920	9.28%

Table 3. PE/NR-modified asphalt model.

Serial Number	Representative Molecule	Chemical Composition	Molecule Number	Relative Molecular Mass	Mass Fraction
1	Asphaltene	$C_{149}H_{177}N_3O_2S_2$	1	2106.190	17.40%
2	Resin	$C_{59}H_{85}NOS$	3	856.395	21.23%
3	Saturate	$C_{22}H_{46}$	5	310.610	12.83%
4	Aromatic	$C_{46}H_{50}S$	7	634.966	36.72%
5	NR	$C_{80}H_{130}$	1	1091.920	9.02%
6	PE	$C_{24}H_{50}$	1	338.664	2.80%

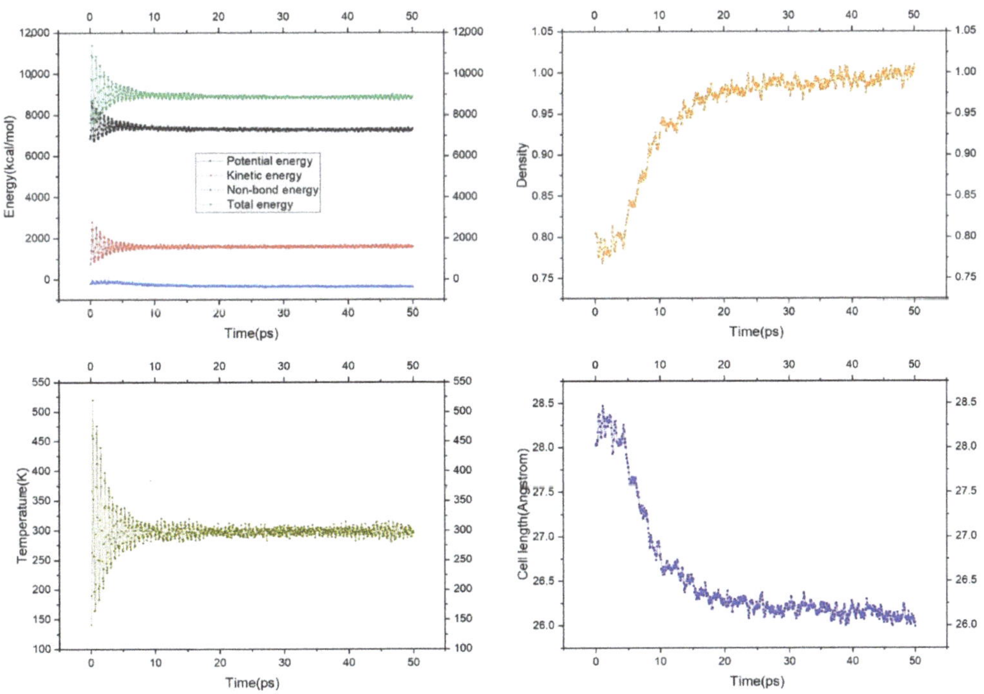

Figure 4. Dynamic equilibrium process.

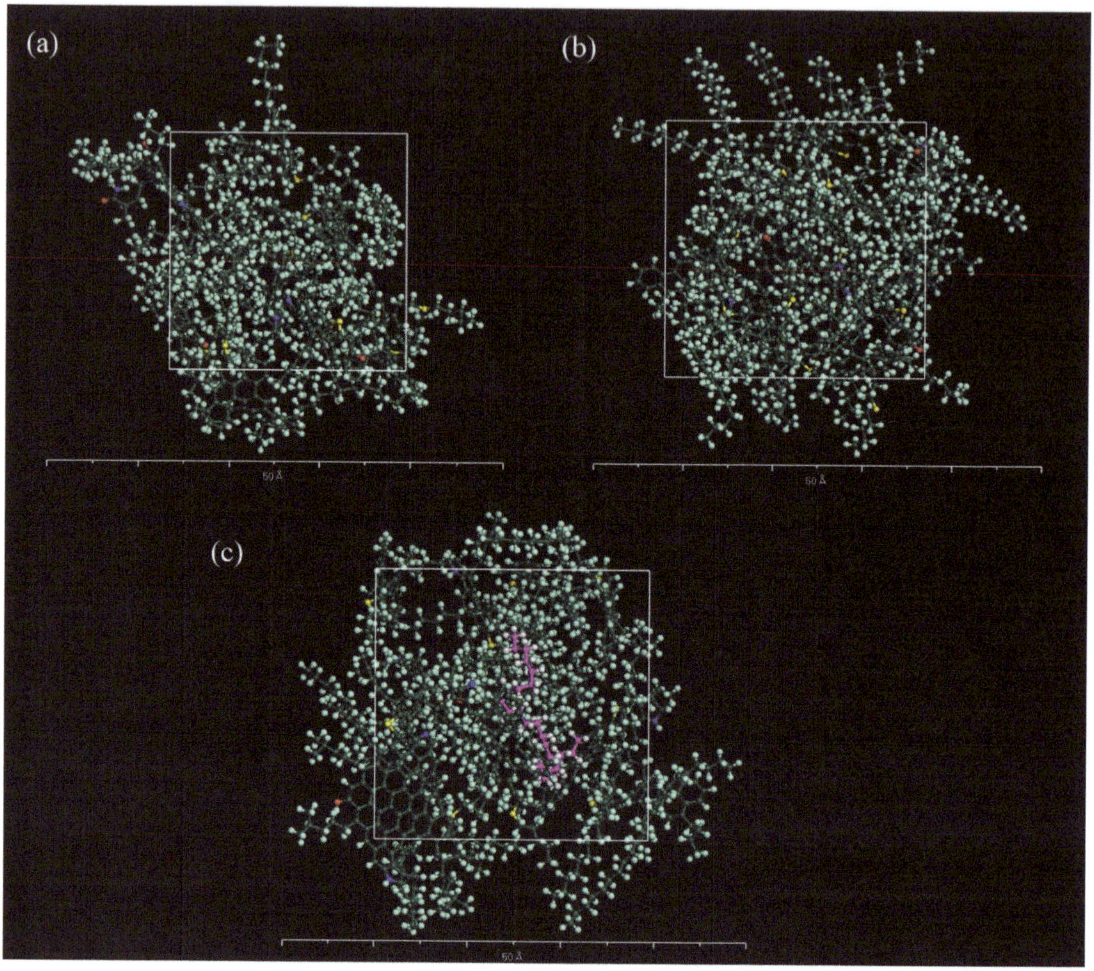

Figure 5. Models (**a**) original asphalt; (**b**) NR-modified asphalt; (**c**) PE/NR-modified asphalt.

3. Results and Discussion

3.1. Solubility Parameters

Molecular models of PE with different degrees of polymerization are shown in Figure 6. The solubility parameters were used to evaluate the compatibility between polyethylene with different degrees of polymerization and NR-modified asphalt. If the solubility parameters for two polymer materials become closer, the blending effect is better [42]. In this study, solubility parameters were adopted to evaluate the compatibility between different polymers, primarily to determine the degree of polymerization of PE molecules. Its value is the square root of the cohesive energy density, as shown in the following Formula (3):

$$\delta = \sqrt{CED} = \sqrt{\frac{\Delta E}{V}} \quad (3)$$

where ΔE represents cohesive energy and V represents volume.

If the action of the hydrogen bond is ignored, the intermolecular interaction force mainly consists of a van der Waals force and electrostatic force; the CED is composed of the van der Waals cohesive energy density (CED_v) and electrostatic cohesive energy density (CED_e).

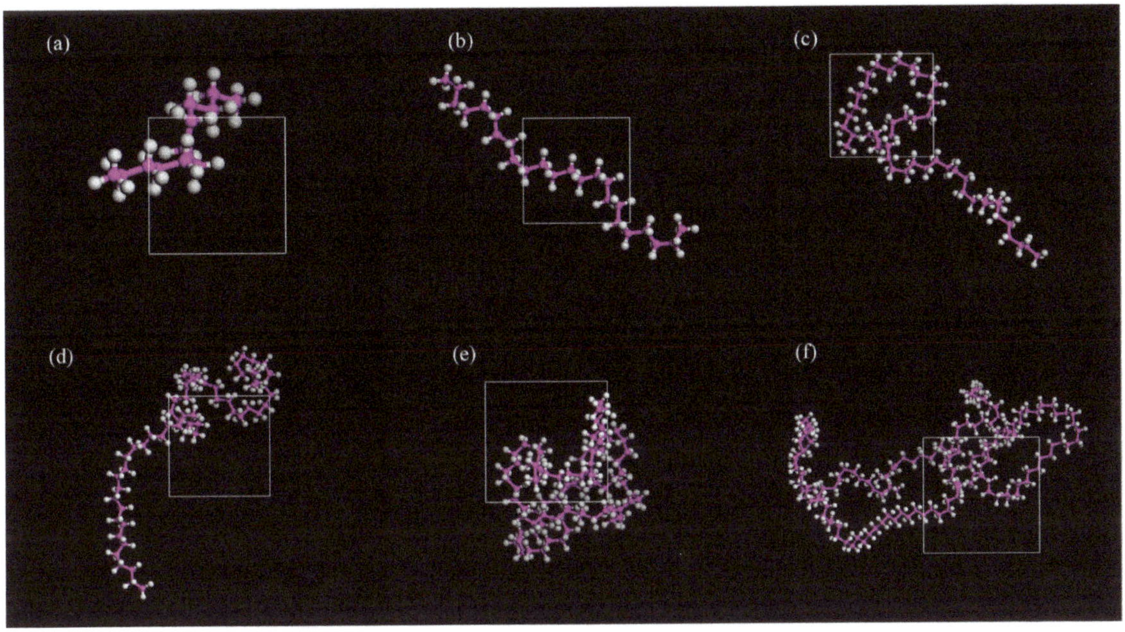

Figure 6. PE molecular model with different degrees of polymerization (**a**) 6 (**b**) 12 (**c**) 24 (**d**) 30 (**e**) 42 (**f**) 78.

Then, the related cohesive energy index is obtained and used to estimate the solubility parameter. The solubility parameters of the PE additives with individual degrees of polymerization are shown in Table 4, while that of the NR-modified asphalt is 17.23. The difference between the two is shown in Figure 7. According to the principle of similarity-compatibility, when the degree of polymerization is 12, the difference between the two is the smallest, which is 0.14 < 1.3~2.1 $(J/cm^3)^{1/2}$. Therefore, the optimum polymerization degree of the PE molecule in this paper is 12.

Table 4. Solubility parameters of different PE molecules.

Number	1	2	3	4	5	6	7	8	9	10	11	12	13	14
Polymerization degree	6	12	18	24	30	36	42	48	54	60	66	72	78	84
Solubility parameter	17.85	17.09	16.66	15.66	14.33	15.56	14.99	15.71	15.47	16.08	14.70	15.92	15.79	15.02

A PE molecular chain with a polymerization degree of 12 was adopted to establish a modified asphalt model. After the steps of dynamics simulation mentioned above were performed, the balanced model of PE/NR modified asphalt was finally obtained for subsequent analysis and calculation.

3.2. Radial Distribution Function

RDF, an essential indicator in molecular dynamics, can be employed to verify the correctness of molecular models.

The radial distribution function (RDF) can be applied to analyze the interaction and microscopic distribution of particles [39,43,44]. The radial distribution function is the spatial probability distribution of particle B approaching the center of particle A, which is in the r to $r + d_r$ shell. It can also be understood as the ratio of the number density of particle B to the mean density of particle B in this region.

$$g(r) = \frac{dN}{\rho V} \tag{4}$$

dN is the number of particles B in the shell r to $r + d_r$; ρ is the numerical density of particle B.

V is the volume of the spherical shell, centered on atom A, with the radius R and the thickness d_r. The formula is as follows:

$$V = \frac{4}{3}\pi\,(r + dr)^{\,3} - \frac{4}{3}\pi r^3 = 4\pi r^2 dr + 4\pi r dr^2 + \frac{4}{3}\pi dr^3 \approx 4\pi r^2 dr \tag{5}$$

For the RDF of amorphous crystals, there are a few peaks with different heights and sharpness at close distances, but the peak height decreases rapidly with an increase in space. At long distances, the radial distribution function tends to be evenly distributed, $g\,(r) = 1$.

The RDF of the crystal still maintains a sharp peak shape at a long distance, indicating that the crystal has a long-range ordered structure, whereas the amorphous crystal has a short-range ordered structure and a long-range disorder structure.

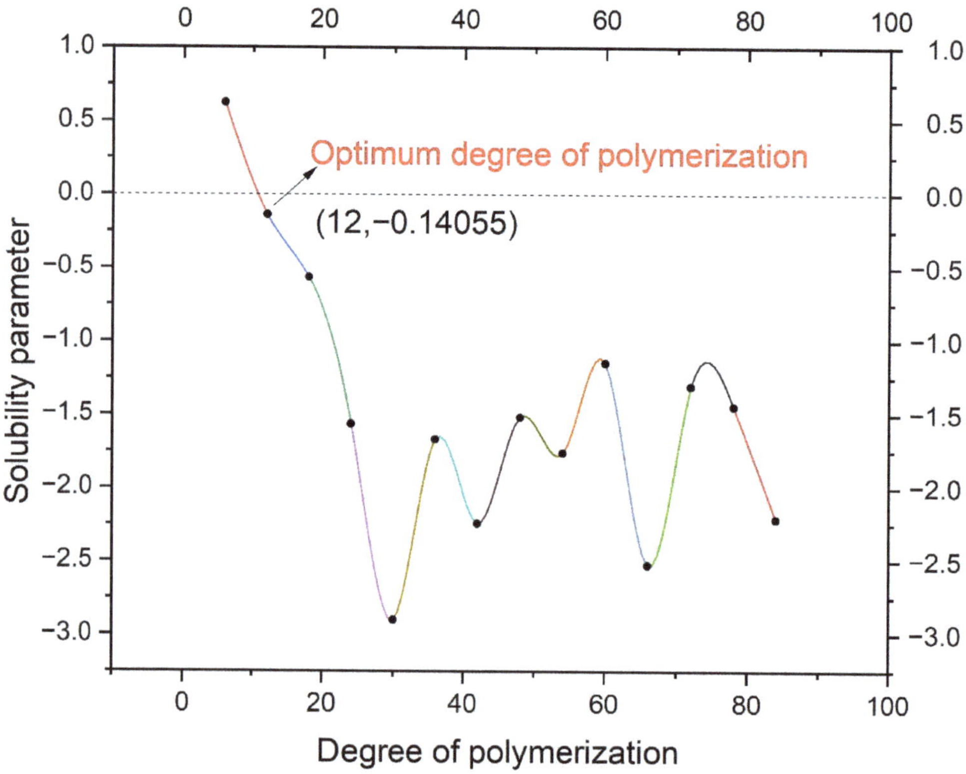

Figure 7. The difference in the solubility parameter.

3.3. Diffusion Coefficient

The diffusion coefficient can represent the molecular movement rate and is employed to study the changes of diverse molecular movements in NR-modified asphalt systems with and without PE. According to Einstein's law [48], if the simulation time is long, the slope of the curve of mean square displacement (*MSD*) to time is six times the diffusion coefficient; thus, the diffusion coefficient of particles can be gained, as shown in formula (6). In molecular dynamics calculation, molecules move continuously from the initial position, and the position of each moment is different. If $r\ (t)$ represents the position of particle i at time t, MSD represents the average square of displacement. The formula of the diffusion coefficient is shown in formula (6).

$$D = \lim_{t \to \infty} \frac{MSD\ (t)}{6t} = \lim_{t \to \infty} \frac{|r(t) - r(0)|^2}{6t} \tag{6}$$

To analyze the movement of various molecules in NR-modified asphalt, the MSD results of different molecules were obtained through molecular dynamics calculation, as shown in Figure 8. According to the functional relationship between the diffusion coefficient and MSD, the diffusion coefficient is obtained, as shown in Figure 9. The analysis shows that the saturate has the fastest movement speed, which is much higher than other types of molecules, and its diffusion coefficient reaches 0.0201. This phenomenon is most likely related to its negligible molecular weight. Aromatic ranks second, at about 0.0039.

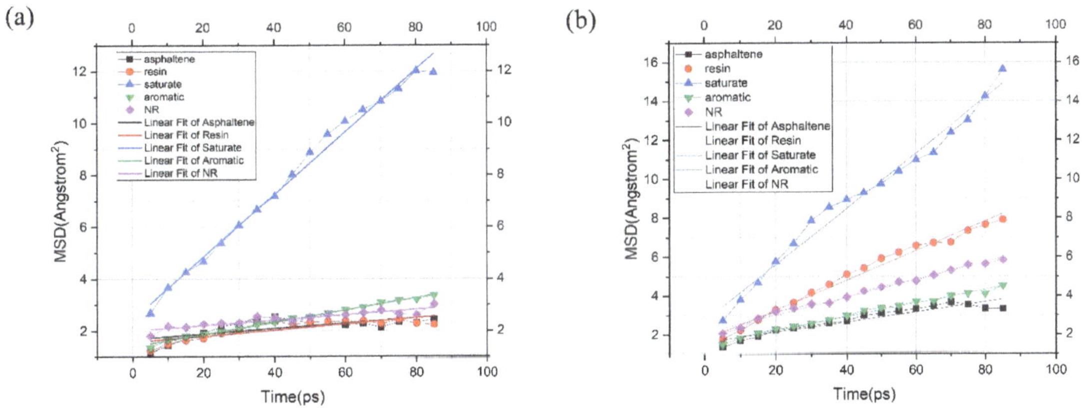

Figure 8. *MSD* for modified asphalt (**a**) NR-modified asphalt; (**b**) PE/NR-modified asphalt.

In addition, the molecular movement velocity of asphaltene, resin, and rubber are roughly similar, showing that their diffusion coefficients are about 0.0018, likely connected with the higher molecular weight and unique shapes to some extent.

For the NR-modified asphalt system, the diffusion coefficient of NR is the lowest. More specifically, the long-chain shape of NR and its poor compatibility with the polar components of asphalt may be the main reasons restricting the movement of NR. Therefore, NR is the slowest molecule in the NR-modified asphalt system, with a diffusion coefficient of 0.00171. However, NR attracts non-polar components in asphalt (aromatic and saturate), which results in a more aromatic and saturated distribution around NR molecules. Some light components are attracted by NR molecules and transferred to the vicinity of NR molecules, so it could be noted that the light components have the highest diffusion coefficients, which are 0.0201 and 0.0039, respectively, in NR-modified asphalt.

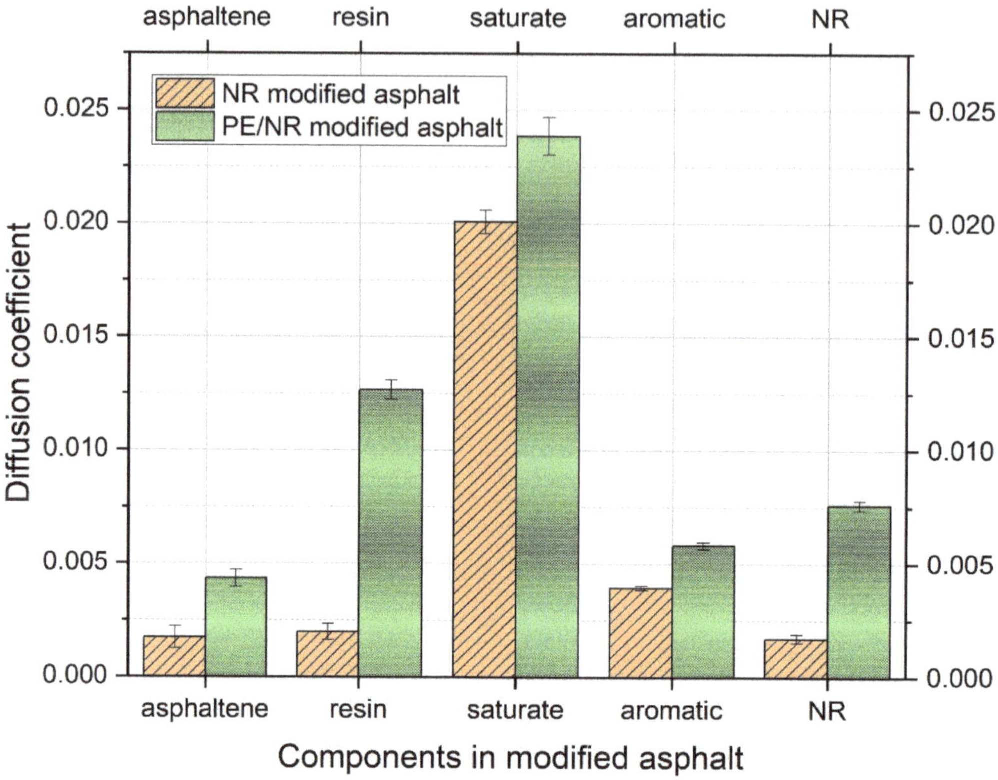

Figure 9. Diffusion coefficient of each component in NR modified asphalt and PE/NR modified asphalt.

As shown in Figure 9, the movement of molecules changed significantly due to the addition of PE. The diffusion coefficient of resin molecules increased dramatically from 0.0020 to 0.0127. This may be because there is a specific force between the PE and resin, thus accelerating the movement of resin molecules. According to the colloid theory for asphalt, resins attach molecules in asphalt to form micelles. In this study, the transfer of resin molecules caused the asphaltene molecules to lose part of the constraint, resulting in a slight increase in the diffusion coefficient of asphaltene from 0.00173 to 0.0043. Because of the rearrangement and combination of the dispersed phase, it is inevitable to promote the corresponding transfer of light components, including saturate and aromatic, that is, the corresponding increase of diffusion coefficient; for example, the diffusion coefficient of saturate increases from 0.0201 to 0.0239, and that of the aromatic component increases from 0.0039 to 0.0058.

In the statistical data analysis, the Pearson correlation coefficient is adopted to reflect the linear correlation of two random variables. For NR-modified asphalt and PE/NR-modified asphalt, *MSD* increased with time. Linear fitting was performed, and the Pearson correlation coefficient was analyzed. For NR-modified asphalt, the Pearson correlation coefficient of asphaltene and resin was 0.67805 and 0.82105, respectively, as shown in Table 5. Meanwhile, the PE/ NR modified asphalt coefficient is above 90%, 0.94514, and 0.99194, respectively, as shown in Table 6.

Table 5. Data analysis of MSD for NR-modified asphalt.

Equation			$y=a+b*x$		
Plot	asphaltene	resin	saturate	aromatic	NR
Weight			No Weighting		
Intercept	1.68015 ± 0.14844	1.55818 ± 0.1093	2.44376 ± 0.15638	1.38694 ± 0.02799	2.007 ± 0.06071
Slope	0.01035 ± 0.0029	0.01188 ± 0.00213	0.12047 ± 0.00305	0.02361 ± 5.46305E-4	0.01028 ± 0.00118
Residual Sum of Squares	1.28424	0.69631	1.42537	0.04566	0.21482
Pearson's r	0.67805	0.82105	0.99522	0.99601	0.91308
R-Square (COD)	0.45975	0.67412	0.99046	0.99204	0.83371
Adj. R-Square	0.42373	0.65239	0.98983	0.9915	0.82263

Table 6. MSD data analysis for PE/NR modified asphalt.

Equation			$y=a+b*x$		
Plot	asphaltene	resin	saturate	aromatic	NR
Weight			No Weighting		
Intercept	1.61155 ± 0.11877	1.75603 ± 0.12832	2.74225 ± 0.25882	1.53794 ± 0.05243	2.10403 ± 0.06517
Slope	0.02598 ± 0.00232	0.07596 ± 0.0025	0.14313 ± 0.00505	0.03484 ± 0.00102	0.0455 ± 0.00127
Residual Sum of Squares	0.82214	0.95978	3.90458	0.1602	0.24752
Pearson's r	0.94514	0.99194	0.99079	0.99359	0.99419
R-Square (COD)	0.89329	0.98395	0.98166	0.98722	0.98842
Adj. R-Square	0.88618	0.98288	0.98044	0.98637	0.98764

3.4. Concentration Distribution

Concentration distribution, an important indicator, can show how concentration varies with position in a region at a given time. This parameter can be acquired through the Forcite Analysis module in Materials Studio. The spatial distribution of each molecule in the NR modified asphalt in three different directions (100,010,001) and the spatial distribution of each molecule in the PE/NR modified asphalt were evaluated, and the interaction between each molecule was further analyzed and predicted. In the concentration distribution curve of a molecule, the peak represents the location at which the molecule is most abundant.

As shown in Figure 10a–c, the four components of asphalt basically follow the colloid structure theory to form a blend with an ordered arrangement. After careful analysis, the following specific rules can be found. Firstly, it might be suggested that there is adsorption between the asphaltene, NR molecules, and the light components. Specifically, the concentration curve peak of asphaltene is surrounded by the concentration curve peak of saturate because the saturate is adsorbed around by asphaltene, and the mutual attraction between the two is considerable. Again, it is surrounded by a peak of aromatic. Both indicate adsorption between the asphaltene and light components. At the same time, the concentration curve peak of NR was also surrounded by the concentration curve peak of the light component, indicating that the light component was also adsorbed around the NR molecule. Second, the peak of the NR concentration curve is far away from that of resin, as shown in Figure 10b, which may be closely related to the incompatibility between the NR and resin, thus preventing them from coexisting in the same position. Similarly, the same rule can be found in the relationship between NR and asphaltene, which is more pronounced. As shown in Figure 10a, the concentration curves of the two have apparent peaks, and the distance between the two peaks is greater, indicating that there is probably a more vigorous competition for light components and incompatibility between NR and asphaltene. This is consistent with a previous study [49]. NR has the highest binding strength to aromatic and saturate, followed by resins, and has the lowest binding capacity to asphaltenes.

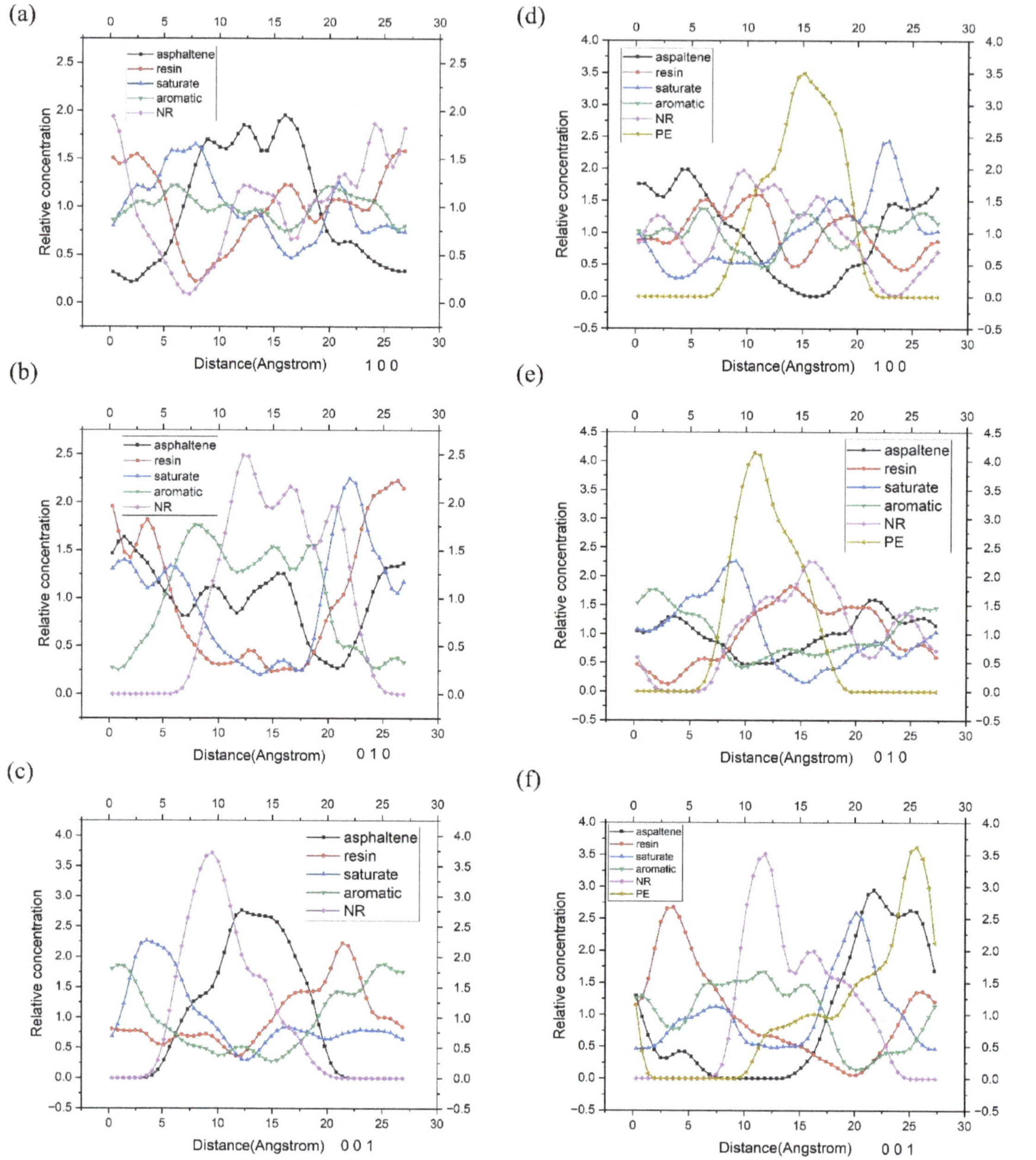

Figure 10. Concentration distribution of each component for NR-modified asphalt and PE/NR-modified asphalt. (The relative concentration of NR-modified asphalt in different directions, including (**a**) 1 0 0, (**b**) 0 1 0, and (**c**) 0 0 1; The relative concentration of PE/NR-modified asphalt in different directions includes (**d**) 1 0 0, (**e**) 0 1 0, and (**f**) 0 0 1).

Figure 10d–f shows the concentration distribution curve of PE/ NR-modified asphalt. It could be predicted from this group diagram that the addition of PE changes the colloid structure composition of asphalt. As shown in Figure 10d, both the PE concentration curve and that of asphaltene have evident peak values, and the peaks of them are far away. It could be the case that the PE molecule has a specific repulsive force with asphaltene. It can also be seen from Figure 10d that the concentration curve of the PE molecule overlaps with that of the saturate to a certain extent, and the saturation fraction is found near the periphery of the peak, which might be related to the intense attraction between the two. It is likely that the adsorption of the saturate could make the PE molecule move and form a more stable system. Therefore, PE molecules tend to further promote the formation of a new balanced and stable system by changing the interaction between the original system.

4. Conclusions

In this paper, two types of polymer materials, PE (polyethylene) and NR (cis-1, 4-polyisoprene), were used as modifiers to study the interaction mechanism in the two modified asphalts. Specifically, the interaction of molecules in the original asphalt and NR-modified asphalt was analyzed before and after the addition of PE. The change of the interaction can help the modified asphalt to become a more stable in structure and thus have better storage performance, which has a reference value for changing the molecular agglomeration phenomenon. On the other hand, the reuse of PE and NR, two typical wastes, is of great significance for the protection of the ecological environment.

The specific conclusions are as follows.

1. The optimal degree of polymerization of PE in this study model is 12. In this case, the solubility parameter between PE and NR-modified asphalt is the smallest at $0.14 < 1.3\text{~}2.1\ (\mathrm{J/cm^3})^{1/2}$.
2. The three models in the paper are typical amorphous substances with structures of short-range order and long-range disorder. According to the RDF diagram, when $r = 1.11$, the $g\ (r)$ functions for the three models all have sharp peaks, which are 11.76, 11.74, and 11.8, respectively.
3. In NR-modified asphalt, the diffusion coefficient of saturate is the largest, at far higher than that of other types of molecules, and its value reaches 0.0201. In addition, the molecular velocity of asphaltenes, resins, and rubbers are about the same, showing that their diffusion coefficients are about 0.0018. NR is the slowest molecule.
4. In PE/NR-modified asphalt, the movement of molecules changed significantly. The diffusion coefficient of resin molecules increased considerably from 0.0020 to 0.0127. The most noticeable feature is a specific adsorption phenomenon between the PE molecule and light components.

Author Contributions: Conceptualization, Y.C. and K.H.; methodology, K.H.; software, Y.C.; validation, Y.C.; formal analysis, X.B.; investigation, Y.C.; resources, Y.C.; writing—original draft preparation, Y.C.; writing—review and editing, D.Y.; supervision, C.Y.; All authors have read and agreed to the published version of the manuscript.

Funding: This research was funded by the National Key R&D Program of China (2018YFB1600200), Fundamental Research Fund of Natural Science Foundation of China (52208434), China Postdoctoral Science Foundation (2022M711079), Science and Technology Project of Henan Province (192102310229), Natural Science Foundation of Henan Province (222300420142).

Data Availability Statement: Not applicable.

Acknowledgments: We would like to thank all the editors and reviewers for correcting the manuscript very carefully.

Conflicts of Interest: The authors declare no conflict of interest.

References

1. Porto, M.; Caputo, P.; Loise, V.; Eskandarsefat, S.; Teltayev, B.; Rossi, C.O. Bitumen and bitumen modification: A review on latest advances. *Appl. Sci.* **2019**, *9*, 742. [CrossRef]
2. Presti, D.L. Recycled tyre rubber modified bitumens for road asphalt mixtures: A literature review. *Constr. Build. Mater.* **2013**, *49*, 863–881. [CrossRef]

3. Jiang, W.; Yuan, D.D.; Shan, J.H.; Ye, W.L.; Lu, H.H.; Sha, A.M. Experimental study of the performance of porous ultra-thin asphalt overlay. *Int. J. Pavement Eng.* **2022**, *23*, 2049–2061. [CrossRef]
4. Wu, S.; Montalvo, L. Repurposing waste plastics into cleaner asphalt pavement materials: A critical literature review. *J. Clean. Prod.* **2021**, *280*, 124355. [CrossRef]
5. Jiang, W.; Xiao, J.J.; Yuan, D.D.; Lu, H.H.; Xu, S.D.; Huang, Y. Design and experiment of thermoelectric asphalt pavements with power-generation and temperature-reduction functions. *Energy Build.* **2018**, *169*, 39–47. [CrossRef]
6. Fini, E.H.; Al-Qadi, I.L.; You, Z.P.; Zada, B.; Mills-Beale, J. Partial replacement of asphalt binder with bio-binder: Characterisation and modification. *Int. J. Pavement Eng.* **2012**, *13*, 515–522. [CrossRef]
7. Sihombing, A.V.R.; Subagio, B.S.; Hariyadi, E.S.; Yamin, A. Mechanical Properties of Bio-Asphalt on Recycled Asphalt Pavement Binder. In Proceedings of the 9th International Conference on Maintenance and Rehabilitation of Pavements—Mairepav9, Zurich, Switzerland, 1–3 July 2020; Springer: Cham, Switzerland, 2020; pp. 529–538.
8. Ren, J.L.; Zang, G.Y.; Wang, S.Y.; Shi, J.; Wang, Y.Y. Investigating the pavement performance and aging resistance of modified bio-asphalt with nano-particles. *PLoS ONE* **2020**, *15*, e0238817.
9. Yan, K.Z.; Zhang, M.; You, L.Y.; Wu, S.H.; Ji, H.Y. Performance and optimization of castor beans-based bio-asphalt and European rock-asphalt modified asphalt binder. *Constr. Build. Mater.* **2020**, *240*, 117951. [CrossRef]
10. Elahi, Z.; Mohd Jakarni, F.; Muniandy, R.; Hassim, S.; Ab Razak, M.S.; Ansari, A.H.; Ben Zair, M.M. Waste Cooking Oil as a Sustainable Bio Modifier for Asphalt Modification: A Review. *Sustainability* **2021**, *13*, 11506. [CrossRef]
11. Jain, S.; Chandrappa, A.K. Rheological and chemical investigation on asphalt binder incorporating high recycled asphalt with waste cooking oil as rejuvenator. *Innov. Infrastruct. Solut.* **2022**, *7*, 1–19. [CrossRef]
12. Zhou, X.X.; Zhao, G.Y.; Wu, S.P.; Tighe, S.; Pickel, D.; Chen, M.Z.; Adhikari, S.; Gao, Y.M. Effects of biochar on the chemical changes and phase separation of bio-asphalt under different aging conditions. *J. Clean. Prod.* **2020**, *263*, 121532. [CrossRef]
13. Zhou, X.X.; Zhao, G.Y.; Miljković, M.; Tighe, S.; Chen, M.Z.; Wu, S.P. Crystallization kinetics and morphology of biochar modified bio-asphalt binder. *J. Clean. Prod.* **2022**, *349*, 131495. [CrossRef]
14. Joohari, M.I.; Aziz, N.A.; Daud, N.M.; Mansor, S.; Halim, M.A. Performance of porous asphalt pavement using clay brick dust as mineral filler[C]//Journal of Physics: Conference Series. *IOP Publ.* **2019**, *1349*, 012098.
15. Abdelsalam, M.; Yue, Y.C.; Khater, A.; Luo, D.; Musanyufu, J.; Qin, X.L. Laboratory study on the performance of asphalt mixes modified with a novel composite of diatomite powder and lignin fiber. *Appl. Sci.* **2020**, *10*, 5517. [CrossRef]
16. Du, T.; Song, P.; Liu, L. Experimental Study on Activated Diatomite Modified Asphalt Pavement in Deep Loess Area. *Processes* **2022**, *10*, 1227. [CrossRef]
17. Chen, X.L.; Sun, Y.S.; Han, Y.X.; Tian, Z.F. Microstructure and DSC of organic bentonite modified bitumen. *J. Northeast. Univer.* **2012**, *33*, 743.
18. Chen, S.; Zhang, B.; He, X.Y.; Su, Y.; Liu, Q.; Xu, H. Research on mechanical-activated nanoscale bentonite and surface aging behavior of its modified asphalt. *Constr. Build. Mater.* **2022**, *321*, 126356. [CrossRef]
19. Du, D.C.; Li, X.L.; Zheng, G.Y.; Zhang, L.Q. Study on the Performance of Polyester Fiber Modified Asphalt Mastic. *Plastics* **2012**, *41*, 42–45.
20. Wu, S.; Ye, Q.; Li, N. Investigation of rheological and fatigue properties of asphalt mixtures containing polyester fibers. *Constr. Build. Mater.* **2008**, *22*, 2111–2115. [CrossRef]
21. Cheng, C.; Sun, W.W.; Hu, B.; Tao, G.X.; Peng, C.; Tian, Y.J.; Wu, S.J. Analysis of the mechanism and effectiveness of lignin in improving the high-temperature thermal stability of asphalt. *J. Renew. Mater.* **2020**, *8*, 1243–1255. [CrossRef]
22. Zhang, Y.; Wang, X.C.; Ji, G.Y.; Fan, Z.Y.; Guo, Y.C.; Gao, W.Z.; Xin, L. Mechanical performance characterization of lignin-modified asphalt mixture. *Appl. Sci.* **2020**, *10*, 3324. [CrossRef]
23. Zheng, Y.X.; Cai, Y.C.; Zhang, G.H.; Fang, H.Y. Fatigue property of basalt fiber-modified asphalt mixture under complicated environment. *J. Wuhan Univ. Technol. Mater. Sci. Ed.* **2014**, *29*, 996–1004. [CrossRef]
24. Wang, W.; Cheng, Y.; Ma, G.; Tan, G.; Sun, X.; Yang, S. Further investigation on damage model of eco-friendly basalt fiber modified asphalt mixture under freeze-thaw cycles. *Appl. Sci.* **2018**, *9*, 60. [CrossRef]
25. Arabani, M.; Shabani, A. Evaluation of the ceramic fiber modified asphalt binder. *Constr. Build. Mater.* **2019**, *205*, 377–386. [CrossRef]
26. Guo, F.C.; Zhang, J.P.; Pei, J.Z.; Zhou, B.C.; Falchetto, A.C.; Hu, Z. Investigating the interaction behavior between asphalt binder and rubber in rubber asphalt by molecular dynamics simulation. *Constr. Build. Mater.* **2020**, *252*, 118956. [CrossRef]
27. Guo, F.C.; Zhang, J.P.; Pei, J.Z.; Ma, W.S.; Hu, Z.; Guan, Y.S. Evaluation of the compatibility between rubber and asphalt based on molecular dynamics simulation. *Front. Struct. Civ. Eng.* **2020**, *14*, 435–445. [CrossRef]
28. Zheng, W.H.; Wang, H.N.; Chen, Y.; Ji, J.; You, Z.P.; Zhang, Y.Q. A review on compatibility between crumb rubber and asphalt binder. *Constr. Build. Mater.* **2021**, *297*, 123820. [CrossRef]
29. Wang, W.S.; Cheng, Y.C.; Chen, H.P.; Tan, G.J.; Lv, Z.H.; Bai, Y.S. Study on the performances of waste crumb rubber modified asphalt mixture with eco-friendly diatomite and basalt fiber. *Sustainability* **2019**, *11*, 5282. [CrossRef]
30. Li, J.; Chen, Z.X.; Xiao, F.P.; Amirkhanian, S.N. Surface activation of scrap tire crumb rubber to improve compatibility of rubberized asphalt. *Resour. Conserv. Recycl.* **2021**, *169*, 105518. [CrossRef]
31. Adiyanto, O.; Mohamad, E.; Abd Razak, J. Systematic Review of Plastic Waste as Eco-Friendly Aggregate for Sustainable Construction. *Int. J. Sustain. Constr. Eng. Technol.* **2022**, *13*, 243–257.

32. Ben Zair, M.M.; Jakarni, F.M.; Muniandy, R.; Hassim, S. A brief review: Application of recycled polyethylene terephthalate in asphalt pavement reinforcement. *Sustainability* **2021**, *13*, 1303. [CrossRef]
33. Leng, Z.; Padhan, R.K.; Sreeram, A. Production of a sustainable paving material through chemical recycling of waste PET into crumb rubber modified asphalt. *J. Clean. Prod.* **2018**, *180*, 682–688. [CrossRef]
34. Qu, X.; Wang, D.W.; Wang, L.B.; Huang, Y.C.; Hou, Y.; Oeser, M. The state-of-the-art review on molecular dynamics simulation of asphalt binder. *Adv. Civ. Eng.* **2018**, *2018*, 4546191. [CrossRef]
35. Yu, C.H.; Hu, K.; Chen, Y.J.; Li, Q.; Zhang, T.L.; Chen, G.X. The Influence of Styrene/butadiene Ratio on the Self-assembly Behavior of SBS Modified Asphalt[C]//IOP Conference Series: Earth and Environmental Science. *IOP Publ.* **2021**, *668*, 012060.
36. Chu, L.; Luo, L.; Fwa, T.F. Effects of aggregate mineral surface anisotropy on asphalt-aggregate interfacial bonding using molecular dynamics (MD) simulation. *Constr. Build. Mater.* **2019**, *225*, 1–12. [CrossRef]
37. Hu, B.; Huang, W.K.; Yu, J.L.; Xiao, Z.C.; Wu, K.H. Study on the adhesion performance of asphalt-calcium silicate hydrate gel interface in semi-flexible pavement materials based on molecular dynamics. *Materials* **2021**, *14*, 4406. [CrossRef]
38. Hu, D.L.; Pei, J.Z.; Li, R.; Zhang, J.P.; Jia, Y.S.; Fan, Z.P. Using thermodynamic parameters to study self-healing and interface properties of crumb rubber modified asphalt based on molecular dynamics simulation. *Front. Struct. Civ. Eng.* **2020**, *14*, 109–122. [CrossRef]
39. Yan, L.M.; Zhu, S.H. *Theory and Practice of Molecular Dynamics Simulation*; Science Press: Beijing, China, 2013.
40. Sun, H.; Rigby, D. Polysiloxanes: Ab initio force field and structural, conformational and thermophysical properties. *Spectrochim. Acta Part A Mol. Biomol. Spectrosc.* **1997**, *53*, 1301–1323. [CrossRef]
41. Sun, H. An ab initio force-field optimized for condensed-phase applications overview with details on alkane and benzene compounds. *J. Phys. Chem.* **1998**, *102*, 7338–7364. [CrossRef]
42. Sun, H.; Ren, P.; Fried, J.R. The COMPASS force field: Parameterization and validation for phosphazenes. *Comput. Theor. Polym. Sci.* **1998**, *8*, 229–246. [CrossRef]
43. Bunte, S.W.; Sun, H. Molecular modeling of energetic materials: The parameterization and validation of nitrate esters in the COMPASS force field. *J. Phys. Chem. B* **2000**, *104*, 2477–2489. [CrossRef]
44. Sun, H.; Jin, Z.; Yang, C.W.; Akkermans, R.L.C.; Robertson, S.H.; Spenley, N.A.; Miller, S.; Todd, S.M. COMPASS II: Extended coverage for polymer and drug-like molecule databases. *J. Mol. Model.* **2016**, *22*, 1–10. [CrossRef] [PubMed]
45. Su, M.M.; Si, C.D.; Zhang, Z.P.; Zhang, H.L. Molecular dynamics study on influence of Nano-ZnO/SBS on physical properties and molecular structure of asphalt binder. *Fuel* **2020**, *263*, 116777. [CrossRef]
46. Zhang, Y. *Fundamentals of Computational Materials Science*; Beijing University of Aeronautics and Astronautics Press: Beijing, China, 2007.
47. Lan, Y.H.; Li, D.H.; Yang, R.J.; Liang, W.S.; Zhou, L.X.; Chen, Z.W. Computer simulation study on the compatibility of cyclotriphosphazene containing aminopropylsilicone functional group in flame retarded polypropylene/ammonium polyphosphate composites. *Compos. Sci. Technol.* **2013**, *88*, 9–15. [CrossRef]
48. D'hooge, D.R.; Van Steenberge, P.H.M.; Reyniers, M.F.; Marin, G.B. The strength of multi-scale modeling to unveil the complexity of radical polymerization. *Prog. Polym. Sci.* **2016**, *58*, 59–89. [CrossRef]
49. Hu, K.; Yu, C.H.; Chen, Y.J.; Li, W.; Wang, D.D.; Zhang, W.G. Multiscale mechanisms of asphalt performance enhancement by crumbed waste tire rubber: Insight from molecular dynamics simulation. *J. Mol. Model.* **2021**, *27*, 1–14. [CrossRef]

polymers

MDPI

Article

Comprehensive Study on the Performance of Waste HDPE and LDPE Modified Asphalt Binders for Construction of Asphalt Pavements Application

Usman Ghani [1], Bakht Zamin [2,*], Muhammad Tariq Bashir [2], Mahmood Ahmad [3,*], Mohanad Muayad Sabri Sabri [4] and Suraparb Keawsawasvong [5]

1 Department of Engineering, Ed.8, University of Palermo, Viale Delle Scienze, 90128 Palermo, Italy
2 Department of Civil Engineering, CECOS University of IT & Emerging Sciences, Peshawar 25000, Pakistan
3 Department of Civil Engineering, University of Engineering and Technology Peshawar (Bannu Campus), Bannu 28100, Pakistan
4 Peter the Great St. Petersburg Polytechnic University, 195251 St. Petersburg, Russia
5 Department of Civil Engineering, Thammasat School of Engineering, Thammasat University, Pathumthani 12120, Thailand
* Correspondence: bakht@cecos.edu.pk (B.Z.); ahmadm@uetpeshawar.edu.pk (M.A.)

Abstract: This research is aimed at investigating the mechanical behavior of the bitumen by the addition of high-density polyethylene (HDPE) and low-density polyethylene (LDPE) obtained from waste plastic bottles and bags. Polymers (HDPE and LDPE) with percentages of 0%, 2%, 4%, and 6% in shredded form by weight of bitumen were used to evaluate the spectroscopic, structural, morphological, and rheological properties of polymer-modified binders. The rheological properties for different factors; viscosity ($\dot{\eta}$) from Rotational Viscometer (RV), rutting factor $G^*/\text{Sin}(\delta)$, fatigue characteristics $G^*.\ \text{Sin}(\delta)$, for the modified binder from dynamic shear rheometer (DSR), Short and long-term aging from rolling thin film oven (RTFO), and pressure aging vessel (PAV) was determined. The thermal characteristics, grain size, and texture of polymers for both LDPE and HDPE were found using bending beam rheometer (BBR) and X-ray diffraction (XRD), respectively. Fourier transform infrared (FTIR) analysis revealed the presence of polymer contents in the modified binder. Scanning electron microscopy (SEM) images revealed the presence of HDPE and LDPE particles on the surface of the binder. Creep Rate (m) and Stiffness (S) analysis in relationship with temperature showed a deduction in stress rate relaxation. Results have revealed the best rutting resistance for 6% HDPE. It also showed an improvement of 95.27% in $G^*/\text{Sin}(\delta)$ which increased the performance of the bituminous mix. Similarly, the addition of 4% LDPE resulted in maximum dynamic viscosity irrespective of the temperatures. Moreover, fatigue resistance has shown a significant change with the HDPE and LDPE. The festinating features of waste plastic modified binder make it important to be used in the new construction of roads to address the high viscosity and mixing problems produced by plastic waste and to improve the performance of flexible pavements all over the world.

Keywords: waste polyethylene; spectroscopic analysis; morphological analysis; XRD; creep analysis; SEM

Citation: Ghani, U.; Zamin, B.; Tariq Bashir, M.; Ahmad, M.; Sabri, M.M.S.; Keawsawasvong, S. Comprehensive Study on the Performance of Waste HDPE and LDPE Modified Asphalt Binders for Construction of Asphalt Pavements Application. *Polymers* **2022**, *14*, 3673. https://doi.org/10.3390/polym14173673

Academic Editors: Quantao Liu, Jose Norambuena-Contreras, Yue Huang and Wei Jiang

Received: 6 July 2022
Accepted: 31 August 2022
Published: 4 September 2022

1. Introduction

The heavy traffic and loading over time adversely affect the rheological performance of the flexible pavements. The upper part of the flexible pavements is mainly composed of bitumen, aggregates, and filler materials to bear the stresses imposed by adverse traffic conditions. Bitumen as a binding material is made up of hydrocarbons and has a strong influence on the performance of asphalt pavements [1].

The rheological properties of bitumen can enhance the structural and functional performance of asphalt pavements [2]. Material characterization is one of the major factors affecting pavement design. Supposing the materials are unable to provide better resistance to fatigue and permanent deformation; in this case, distress may likely occur in asphalt pavements, which may affect the structural and functional properties of asphalt pavements. When the performance of the pavement reaches a level where the desired function of the pavement is no longer available, or the pavement is not optimally providing, the desired service is termed as failure. In flexible pavements, three major types of distress cause failure in the structure, i.e., rutting, fatigue cracking, and to an extent, thermal cracking [3]. For the last 40 years, researchers have been putting their efforts to modify bitumen by employing the addition of polymers to enhance its physical and rheological properties [4]. For developing countries such as Pakistan, the addition of waste plastic in bituminous materials is one of the easiest available resources to strengthen the bitumen and reduce the environmental hazards produced by these wastes. According to the Environmental Protection Agency (EPA), Pakistan has produced 3.9 million tonnes of plastic waste during the year 2020, in which 70 percent (2.6 million tonnes) was mismanaged. According to the EPA, this amount of plastic waste may be tripled by 2040 [5].

The modified bitumen obtained from waste plastic is used as either elastomers or plastomers [6,7]. The rheological properties and viscosity function can be highly influenced by the addition of elastomers and plastomers from 2% to 6% by the weight of bitumen [8]. The process of adding bitumen with different polymers is termed a wet process [9]. Depending on the nature, size, the type of equipment used for mixing, and the shape of polymers [10], several properties such as resistance to permanent deformation, resistance to moisture-related distresses, fatigue life, and the achievement of high stiffness at a high temperature, can be improved by adding waste plastics [11–14]. By substituting the HDPE and LDPE, the thinner pavement cross-sections can be developed more efficiently due to their mechanical strength and chemical compatibility with the original binder [15].

The addition of nano-silica from 1% to 6% by weight of bitumen in the polymer-modified binder has improved its viscoelastic properties and the rutting factor $G^*/\text{Sin}\ (\delta)$ and fatigue characteristics $G^*.\text{Sin}\ (\delta)$ [16]. Styrene–butadiene–styrene (SBS) polymer modified asphalt mixtures and analysis of strain distribution showed that SBS polymer modified bitumen is improving the stress levels within the asphalt mastic [17–20]. The low percentage of SBS revealed the dispersion of polymers particles in a continuous phase of bitumen, as the original binder has swollen the small polymer's globules. As a result, compatible fractions are spread in the homogenous form in a continuous phase of bitumen [17]. In another investigation of adding 2% to 6% nano-silica in polymer-modified bitumen, results showed a delay in the aging of bitumen. The rutting and fatigue parameters were improved along with the viscoelastic properties of a polymer-modified binder [21,22]. Aging and regeneration can change the microstructural performance and morphological performance of bitumen [23–25]. The microstructures investigations of SBS polymer modified binder in addition to Montmorillonite (MMT) using X-ray diffraction (XRD) and FTIR indicated that the softening point and aging index have decreased due to the addition of Na+. The introduction of Na+ in SBS polymer-modified binder has created a phase-separated structure [26]. The surface properties of polymer-modified bitumen were determined using scanning electron microscopy and AT-FTIR. The contact angle showed a decrease from 107.7° to 4.7° while the oxygen atomic percentage was increased from 7.12% to 13.15%, respectively. This exhibits the chemical interaction of the polymer-modified binders. Chemical capabilities in terms of oxidation in aged bitumen can be seen without the evolution of interactions [27].

Although the addition of nano-silica and SBS has its advantages in polymer modification there is still a lack of research on the affinity of polymers with bitumen. At a low strain level, there is no slippage of polymer chains or change in morphological properties but at a higher strain and higher cyclic loads, the research question still remains uncertain [28].

Another study for investigating the physical and rheological properties of asphalt binders was done by introducing the nanoparticles of aluminum oxide (Al_2O_3) by weight of 3%, 5%, and 7% bitumen. The addition of 5% aluminum oxide by weight of bitumen, reduced the phase angle (δ), and the complex shear modulus (G*) was increased, respectively [29]. However, the bond between nanoparticles and bitumen still needs further investigation in terms of morphological analysis.

Although extensive research has been conducted so far on polymer-modified bitumen, there is a need for improvement in the polymer in terms of its rheology, morphology, and creep assessment [30]. Recently, researchers have reported some critical issues with using polymers as modifiers utilizing the wet process. These issues include the mixing problems related to high viscosity at the higher temperature, the affinity of polymers with bitumen, and the high cost of its modification [9]. Due to the ample amount of plastic waste available in Pakistan, this research uses two waste plastics, i.e., HDPE obtained from waste bottles and LDPE obtained from waste plastic bags in shredded form, intending to investigate the above critical issues.

The morphological properties of neat binders have been compared with HDPE and LDPE modified binders in order to predict the bond linkage between neat binders and polymers. The performance of a neat and modified binder has been investigated in the Section 2.4.2 rheological investigation section. Dynamic Viscosity measurements have discussed the high viscosity and mixing issues of adding polymers during high temperatures. The comparison of the stiffness for neat and modified binders has been discussed in creep analysis. The schematic representation of the study has been shown in Figure 1.

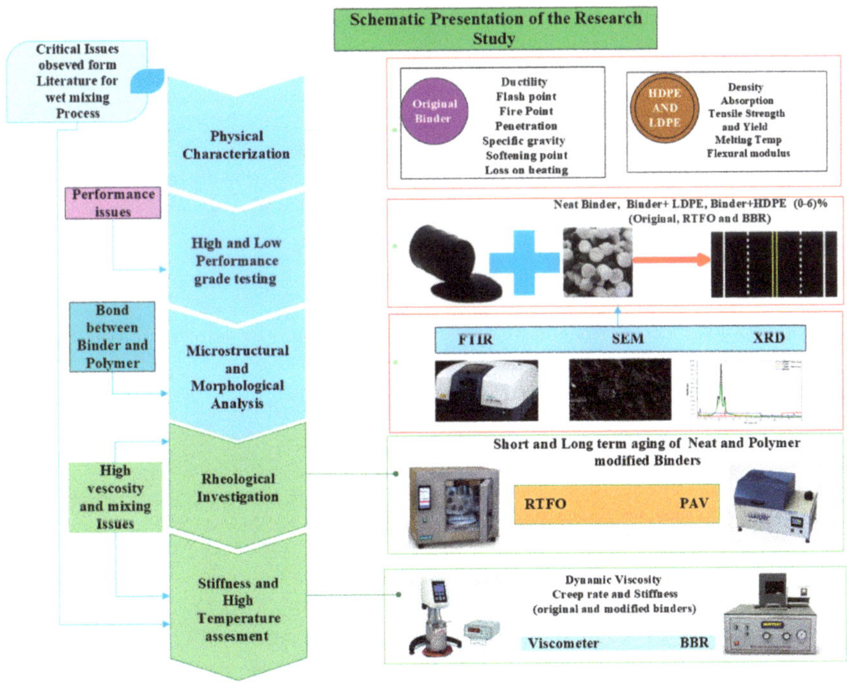

Figure 1. A schematic presentation of the critical issues observed in the current literature.

2. Materials and Methods

2.1. Materials

The fundamental materials used in this research work were bitumen, high-density polyethylene (HDPE), and low-density polyethylene (LDPE) obtained from a local depository in Peshawar Pakistan. The waste plastic bottles and bags were first washed, dried, and shredded to pass through sieve No.4 [30]. The performance and penetration grade of bitumen was PG 58-22 and grade 60/70, respectively, obtained from Attock Oil Refinery

Limited (ARL), Pakistan. Table 1 represents the physical properties of the asphalt binder, while the physical properties of HDPE and LDPE used in this investigation are presented in Table 2.

Table 1. The physical properties of bitumen PG 58-22.

Test Type	Standard	Result
Ductility @ 25 °C,	ASTM D 113	75
Flash point Test (°C)	AASHTO T 48-96	250
Fire Point Test (°C)	AASHTO T 48-96	271
Penetration @ 25 °C	ASTM D 5-97	67
Specific gravity (g/cm^3)	AASHTO T 228	1.017
Softening point (°C)	ASTM D 36-95	43
Loss on heating (%)	AASHTO T 47	0.011

Table 2. The physical properties of polymers.

Parameters	Specifications	HDPE	LDPE
Density (g/cm^3)	ASTM D 792	0.037	0.033
Absorption (%)	ASTM D 570	0	<0.01
Tensile Strength (psi)	ASTM D 638	4400	1900
Yield (%)	ASTM D 638	900	710
Melting Temp. (°C)	ASTM D 3418	125	110
Flexural modulus (psi)	ASTM D 790	20,000	-

2.2. *Performance Grade Testing*

Performance grade (PG) of conventional and HDPE and LDPE modified bitumen was determined by following AASHTO M 320 specifications, as is presented in Table 3.

Table 3. Low- and high-performance grade testing of HDPE and LDPE modified bitumen.

Bitumen + LDPE	Bitumen + HDPE	High Performance		Low Performance	PG
		Original	RTFO *	BBR	
ARL + 0% LDPE	ARL + 0% HDPE	58	58	−12	58–22
ARL + 2% LPDE	ARL + 2% HPDE	64	64	−12	64–22
ARL + 4% LDPE	ARL + 4% HDPE	70	70	−12	70–22
ARL + 6% LDPE	ARL + 6% HDPE	76	70	−12	70–22

* Rolling Thin Film Oven (RTFO), Bending Beam Rheometer (BBR), Attock Refinery Limited (ARL), High-Density Polyethylene (HDPE). Low-Density Polyethylene (LDPE).

2.3. *Preparation of PE Modified Bitumen*

The HDPE and LDPE were mixed separately with bitumen in an agitator at 3000 rpm after applying the heat treatment (163 °C for 1 h in an oven). The HDPE and LDPE were mixed separately with different percentages of 2%, 4%, and 6% by weight of bitumen. The mixture was reheated at 163 °C after the addition of the modifier for 10 min.

2.4. *Testing Procedures*

2.4.1. SEM, FTIR, and XRD Analysis

As the chemistry of the polymers is quite different from that of bitumen, the interaction of HDPE and LDPE with bitumen was determined using different characterization tools. The FTIR spectroscopy was done to identify the functional group in neat and polymer-modified binders. The specifications followed for high- and low-density polyethylene were AASHTO T 302–15. The chemical bonding of HDPE and LDPE with bitumen was investigated by FTIR (Perkin Elmer L1600). This test was performed in the wavenumber range (4000–400) cm^{-1} by using the KBr technique. For performing the FTIR, the testing temperature was 25 degrees Celsius. The bond concentration or the identification of the functional group can be identified by the intensity of the peak spectra in FTIR [31]. The polymer's interaction with the neat bitumen enhances the morphological properties of the HDPE and LDPE modified binder and needs to be understood using scanning electron microscopy (SEM-JEOL JSM 5910).

The XRD analysis of neat bitumen and modified binder has been studied to examine their structural features. The crystalline features of the materials were studied by using a dedicated software, Bruker AXS, D8 advance. X-ray diffraction (XRD) was used to investigate the diffraction patterns, grain size, and texture of the modified bitumen. The X-Ray Diffraction, abbreviated commonly as XRD, is a non-destructive research tool used for the study of crystalline material structure. It studies the structure of the crystal used to classify the crystalline phases contained in a substance and thus discloses information about the chemical composition of the material [32].

2.4.2. Rheological Investigation

The rheological assessments were carried out to determine the viscosities, rutting resistance along with short- and long-term aging of HDPE and LDPE modified asphalt binders using the dynamic shear rheometer (DSR AASHTO T 315). All tests were performed at different temperatures (58 °C, 64 °C, 70 °C, and 76 °C) following the specifications of (AASHTO T 240-09) for short-term aging RTFO-aged and (ASTM D 6521) for long-term aging PAV-aged, the rutting factor $G^*/\text{Sin}(\delta)$ and fatigue characteristics $G^*\text{Sin}(\delta)$ of original and polymers modified binders were determined. The frequency of testing samples was 10 rad/s according to the standard specifications. According to the (AASHTO T-315), the temperature was maintained at 58 °C, 64 °C, 70 °C, and 76 °C for rutting resistance determination. The original samples for $G^*/\text{Sin}(\delta)$ were tested from a minimum value of 1.0 KPa and RTFO aged binder samples at a minimum value of 2.2 KPa at maximum temperature. However, the minimum value of G^*. Sin (δ) was 5000 KPa. Schematics of rheological investigations of neat and modified bitumen are depicted in Figure 2 [33–35].

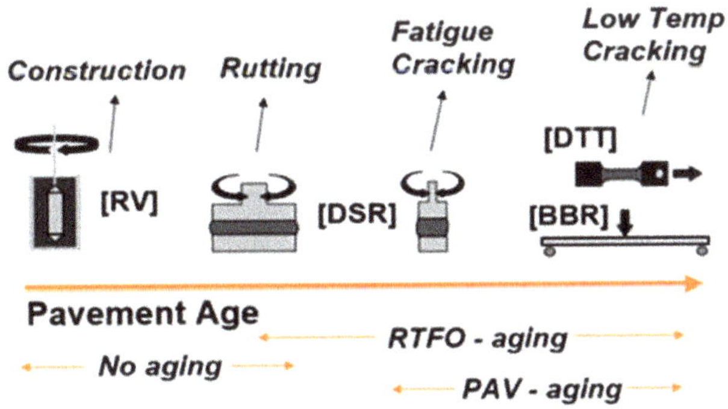

Figure 2. The schematics of rheological investigations of neat and modified bitumen.

2.4.3. Short-Term and Long-Term Aging of a Binder

The Rolling Thin Film Oven Test (RTFO) was performed to discover the short-term aging of bitumen according to the standard specifications of AASHTO T-240. The total duration of the test was 85 min and the temperature was 163 °C in the RTFO bottle [34]. The volatile particles involved in bitumen cause short-term aging. For long-term aging, the sample was put into the plates and Pressure Aging Vessel (PAV) for 20 h. The PAV can predict the binder up to 10 years of its service life. Further physical testing was performed by storing the binder in cans.

2.4.4. Dynamic Viscosity Test

The rotational viscometer (RVDV-111) was used to find out the dynamic viscosity of original and modified samples. The percentages of polymer-modified bitumen were from 2% to 6% by weight of bitumen. According to the specifications of AASHTO T-316, the high temperature for the viscosity test was from 135 °C to 165 °C. The changing interval of temperature was 10 °C, respectively. The rotational speed of the cylindrical spindle was 20 rpm [36].

2.4.5. Bending Beam Rheometer Tests for Creep

The low-temperature cracking or thermal cracking was measured by BBR. According to the specifications of AASHTO T-313 at three different temperatures of 0 °C, −6 °C, and −12 °C, the BBR test was performed. The length, thickness, and width of the beams of bitumen were 127 mm, 6.4 mm, and 12.7 mm, respectively. The temperature was constantly maintained for 60 min. After the preloading conditions, a 100 g load was applied to the rectangular beam at a constant rate for measuring the deflection at the center [37]. The loading time was from 8 to 240 s to discover the creep stiffness (S) and creep rate (m). Schematics for creep rate (m) and creep stiffness (S) are shown in Figure 3.

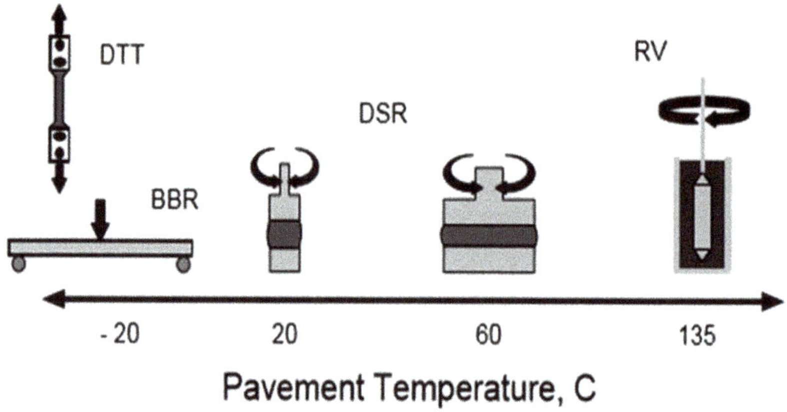

Figure 3. The schematics of creep stiffness (S) and creep rate (m) analysis.

3. Results and Discussions

3.1. Scanning Electron Microscopy

The scanning electron microscopy (SEM) of controlled and HDPE and LDPE modified bitumen is shown in Figure 4. The micro-cracks that appeared on the surface of a neat binder can be seen in Figure 4a. This crack growth may be increased under the cyclic loading imposed by the heavy traffic and will cause severe microstructural disorder in flexible pavements [25]. As the molecular chains of the 6%, HDPE interacts with the neat binder as shown in Figure 4b; the surface of the modified bitumen became smooth and showed no appearance of micro-cracks on the surface. The continuous matrix of polymers, with the addition of 6% by weight of bitumen, is clearly visible in Figure 4b. The packed surface of HDPE-modified bitumen without micro-cracks may provide better resistance against rutting and fatigue characteristics caused by the heavy loading in flexible pavements. As depicted in Figure 4c, some of the micro-cracks still appeared with the same amount of LDPE (6%) added by the weight of bitumen. With the addition of 6%, LDPE the discontinuous matrix of LDPE is observed. Similarly, phase dispersion of bitumen can be seen when the percentage of LDPE is greater than 6%.

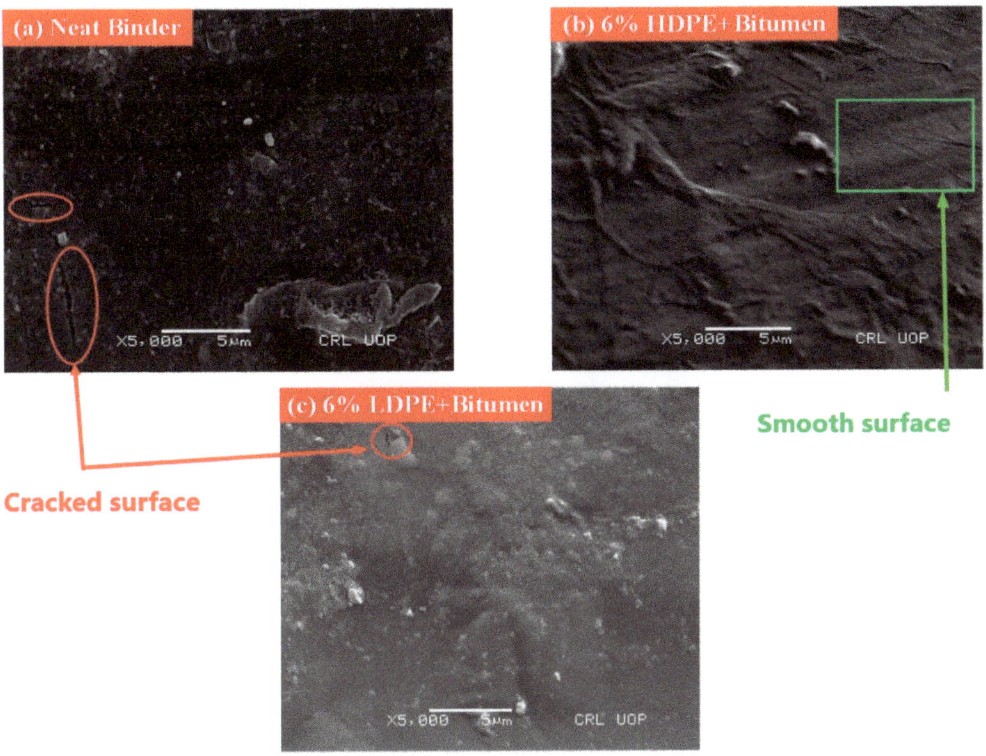

Figure 4. SEM images: (**a**) Neat Binder; (**b**) HDPE modified bitumen; (**c**) LDPE modified bitumen.

3.2. Fourier Transform Infrared (FTIR) Analysis

The FTIR analysis for neat bitumen, HDPE, and LDPE modified bitumen is presented in Figure 5 below. The HDPE and LDPE modified bitumen showed the emergence of peaks and the presence of OH groups shown in Figure 5. Other peaks developed were due to the absorption of HDPE and LDPE in bitumen. The overtones around 2000–1800 cm^{-1} can be clearly seen in Figure 5, which confirms the presence of aromatic carbons. The C-H stretching of organic compounds represented in the range of the peaks in Figure 5 is between 2850–3000 cm^{-1} [32–38]. Values for the other peaks in Figure 5 are between 3000 cm^{-1} and 1500 cm^{-1}, proving the presence of polar groups. The polar group is relatively responsible for creating a stronger bond between a neat binder and dispersed polymer [39]. The bending of the C-H group can be seen at 1570 cm^{-1}. It can also be noted in the figure that the neat HDPE act as apolar or production of alkene or olefin with a highly crystalline structure. An alkene or olefin is a type of unsaturated molecule that contains one carbon to carbon double bond. As compared to the HDPE, the LDPE is less apolar and has a less crystalline structure. Due to the high crystallinity found inside the polymers, there will be a sufficient amount of improvement in the blended mix of polymers with neat bitumen [40].

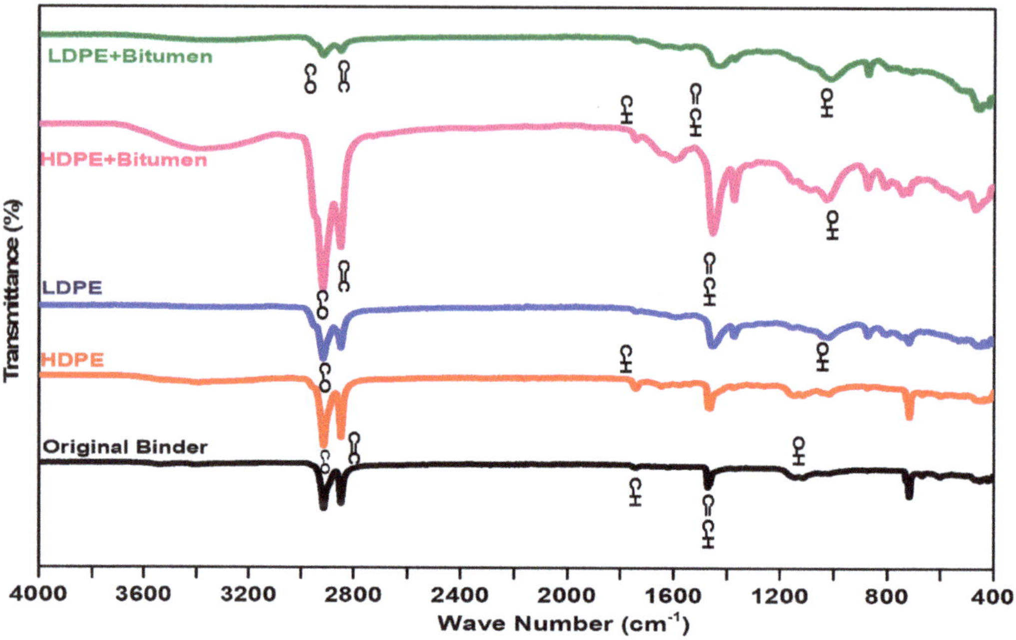

Figure 5. An FTIR analysis of neat, HDPE, and LDPE modified bitumen.

3.3. X-Rays Diffraction (XRD) Analysis

As can be seen in Figure 6, the black line represents the neat bitumen that is not crystallized. The peak 2θ degree for HDPE in addition to bitumen was 26.71° on a scale of 0 to 60 with a scan speed of 2 deg/min with an intensity of 5000 (a.u). The LDPE in addition to bitumen 25.20° with an intensity of 4189 (a.u). As can be seen from Figure 6, at the lower angle the modified bitumen shows a semi-crystalline phase. As the percentage of absorption increases the crystalline phase also increases. Previous research has reported the crystalline behavior of LDPE at 4% by weight of bitumen showed a significant decrease in permanent deformation, fatigue, and thermal crack resistance at high, intermediate, and low temperatures [41]. Another research confirmed that the highly crystalline bitumen-modified mixture can improve the thermal stability of the mixture which has a direct effect on the rheological properties of the asphalt mixture [42]. The degree of crystallinity for HDPE is more as compared to the LDPE but at a higher temperature in the wet mixing process, the HDPE has a high viscosity and mixing issues with the neat binder. From the conducted XRD on neat and modified bitumen, it can be concluded that the crystalline structure of the polymers and their modification enhance a key characterization of the chemical properties of the modified bitumen. Through the addition of polymers, there is an improvement in the elastic properties of a modified binder. As a result, it will provide better resistance to permanent deformation.

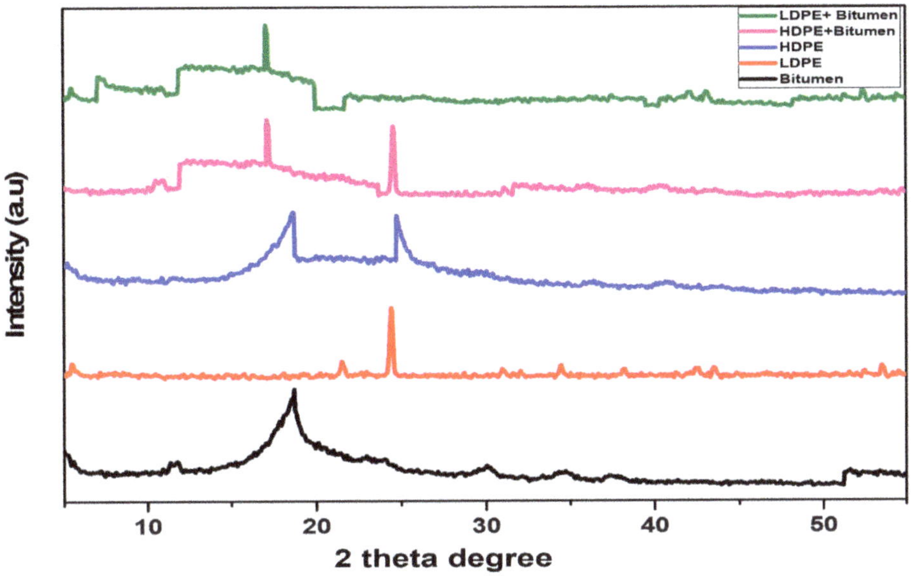

Figure 6. An XRD analysis of neat, HDPE, and LDPE modified bitumen.

4. Rheological Performance Analysis

From 2% to 6% addition of HDPE and LDPE, the rheological performance of neat binder, i.e., PG 58–22 has been compared. Figure 7a,b depicted the rheological performance of the unaged binder. The temperatures 58, 64, 70, and 76 °C were used for finding the rutting factor i.e., G*/Sin (δ). The addition of HDPE and LDPE depicted the increase in G*/Sin (δ). As a result, the rutting factor was increased with the addition of 2%, 4%, and 6% HDPE and LDPE. The phase angle (θ) for the temperature of 58, 64, 70, and 76 °C was determined using DSR.

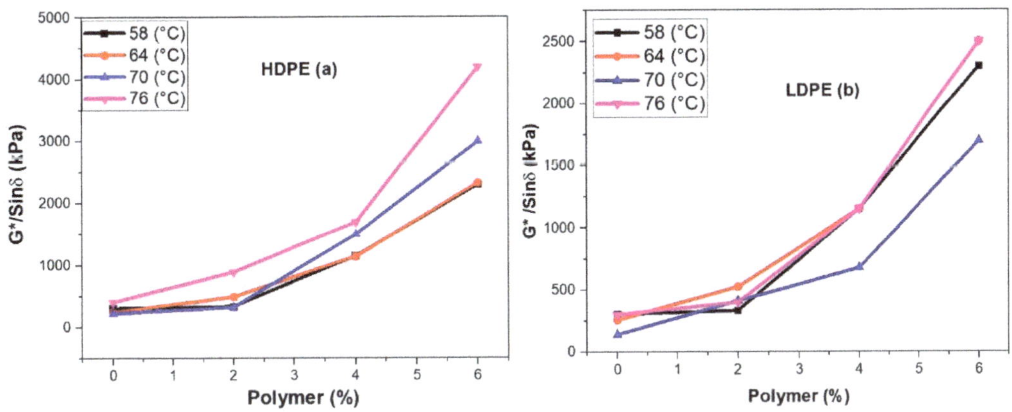

(a) (b)

Figure 7. (**a**) A comparison between HDPE modified bitumen and G*/Sin (δ) in unaged bi tumen, (**b**) comparison between LDPE modified bitumen and G*/Sin (δ) in unaged bitumen.

The rutting resistance as compared to the neat binder was increased with the addition of HDPE and LDPE. High values of rutting parameters, i.e., G*/Sin (δ) may decrease the chances of rutting in asphalt, also summarized by previous research [43,44]. As compared to the original binder after the addition of polymers in Figure 7a, the values of 480, 940, and 1970 Pa showed 13.8%, 26.8%, and 56.2% improvement in rutting resistance. The temperature was maintained at 64 °C for adding HDPE from 2 to 6% in the bitumen. Figure 7b depicts the results of LDPE, and it can be noted that the values 500, 1100, and 2250 Pa showed 19.23%, 42.3%, and 86.5% improvement in rutting resistance. After conducting the short-term aging test, Table 4 shows the values obtained from RTFO residues reused in DSR for different phase angles. As the percentage of polymers increases the values of phase angle decrease. Moreover, after the conduction of the Long-term aging test, Table 5 showed the values obtained from PAV residues reused in DSR for different phase angles at 25 °C. Figure 8a,b depicted the results for long-term aging after the addition of HDPE and LDPE. It can also be noted that according to Table 5 the percentage of polymers increases the phase angle decreases, which shows an increase in elastic properties of the bitumen.

Table 4. RTFO of HDPE and LDPE modified bitumen (PG 58-22).

Bitumen + Polymer	Temperature (°C)	Phase Angle (θ)	G*/Sin δ (kPa)
PG + 0%	25	55.32	4000
PG + 2% HDPE	25	54.20	5900
PG + 4% HDPE	25	51.12	7503
PG +6% HDPE	25	49.59	11,200
PG + 2% LDPE	25	53.88	4700
PG + 4% LDPE	25	52.47	5620
PG +6% LDPE	25	54.97	9700

Table 5. RTFO of HDPE and LDPE modified bitumen (PG 58-22).

Bitumen + Polymer	Temp (°C)	Phase Angle (θ)	G*/Sinδ (kPa)
PG + 0%	58	86.28	2000
PG + 2% HDPE	58	85.23	6203
PG + 4% HDPE	58	83.28	7500
PG + 6% HDPE	58	82.21	3200
PG + 2% LDPE	58	84.70	2400
PG + 4% LDPE	58	85.78	1550
PG + 6% LDPE	58	85.90	1950

The phase angle of pure polymer content is less than the neat binder, irrespective of the temperature conditions. Irrespective of the temperature change, the reduction of phase angle is due to the rich phase of the polymer that behaves like an elastic filler in a neat binder matrix [45]. This shows that the polymer's addition improves the high-temperature performance of asphalt mortar. A decrease in phase angle (θ) was observed with the usage of the polymers, as the phase angle is acting as a lag between the applied shear stress and resulting in the shear strain. The lesser the phase angle, the more asphalt mortar can resist permanent deformation, as shown in Tables 4 and 5, respectively.

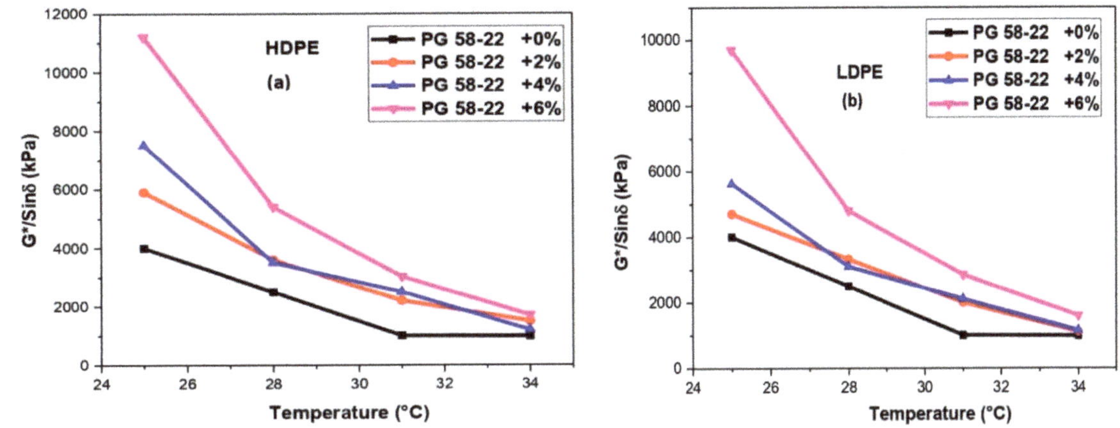

(a) (b)

Figure 8. (**a**) An analysis of G*/Sin (δ) and temperature with HDPE modified bitumen for long-term aging, (**b**) an analysis of G*/Sin (δ) and temperature with LDPE modified bitumen for long-term aging.

5. Dynamic Viscosity Analysis

The results of viscosity for HDPE and LDPE are shown in Figure 9a,b, respectively. The increase in the viscosity values showed a marked difference with the addition of polymers. The maximum dynamic viscosity is obtained by adding 2% of HDPE at all the temperatures, as shown in Figure 9a. The percentage of Dynamic viscosity increases at a temperature of 135, 145 155, and 165 °C was 9.67%, 15.8%, 25%, and 18.1%, respectively. The maximum dynamic viscosity is obtained by adding 4% of LDPE at all the temperatures, as shown in Figure 9b. The percentage of Dynamic viscosity increases at temperatures of 135.0, 145.0, 155.0 and 165.0 °C was 5.90%, 21.80%, 5.0% and 18.10%, respectively [36].

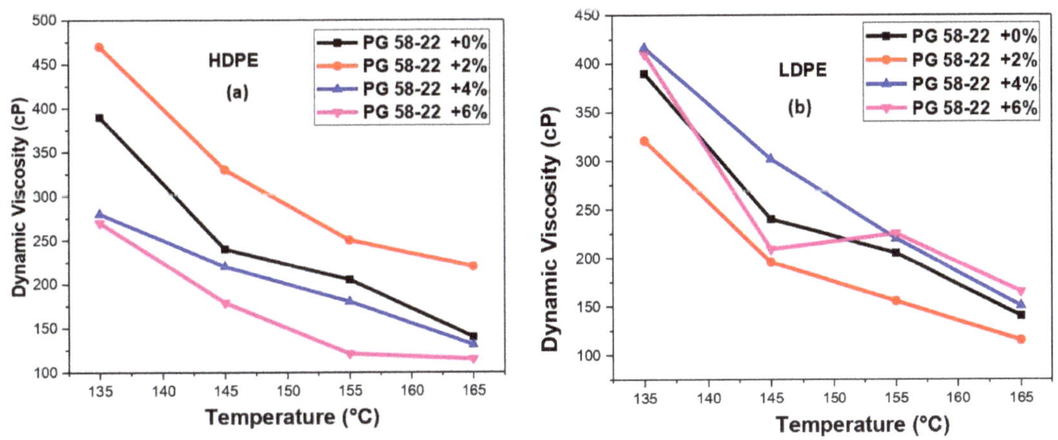

(a) (b)

Figure 9. (**a**) The relation of dynamic viscosity with temperature for HDPE modified bitumen, (**b**) relation of dynamic viscosity with temperature for LDPE modified bitumen.

6. Creep Rate and Creep Stiffness Analysis

By applying the loading time of 60 s for HDPE and LDPE modified bitumen, creep rate (m) and creep stiffness (S) were found using the software as depicted in Figures 10 and 11. The range for (m) should be greater than 0.300 while the (S) should below 300 MPa [46]. The relation between temperature and creep rate (m) for HDPE and LDPE can be seen in Figure 10a,b. The trend of the graph was decreasing at −6 °C for LDPE, but the HDPE was according to the desired criterion.

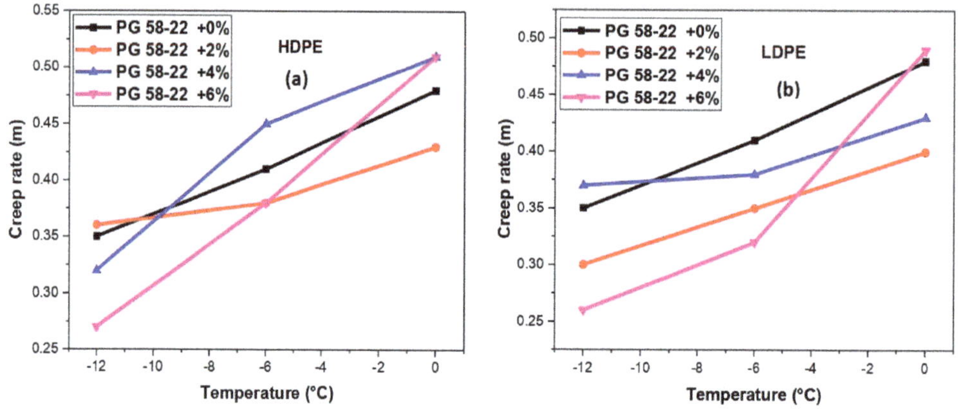

(a) (b)

Figure 10. (**a**) An analysis of creep rate (m) and temperature with HDPE-modified bitumen, (**b**) analysis of creep rate (m) and temperature with LDPE modified bitumen.

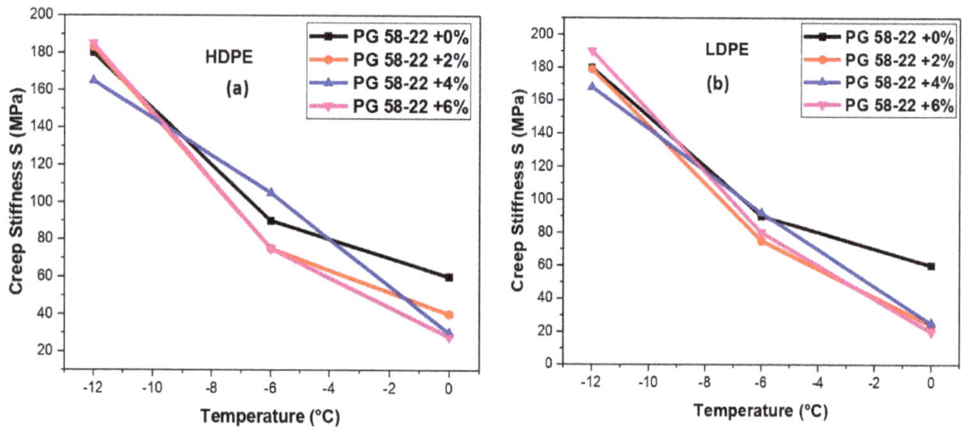

(a) (b)

Figure 11. (**a**) An analysis of creep stiffness (S) and temperature with HDPE-modified bitumen, (**b**) analysis of creep stiffness (S) and temperature with LDPE-modified bitumen.

Figure 11a,b represented the relationship between S and temperature for HDPE and LDPE, respectively. The polymer-modified bitumen becomes stiffer at low temperatures. At low temperatures, adding both HDPE and LDPE at 6% has shown higher creep stiffness values than the original binder, i.e., 185 MPa [47]. The values of S have a relationship with thermal stress, which is developed in pavements that undergo shrinkage. Hence, there could be thermal cracking in the pavement structure. The stress rate has a relationship with the (m) values; whenever it decreases, the relaxation in the stress rate also decreases. So, it can be concluded that for long-term pavement performance Low (S) values and Higher (m) values are required.

7. Summarized Conclusions

This research was conducted to assess the morphological, rheological and dynamic viscosity, and creep characteristics of waste HDPE and LDPE modified asphalt binders. The major conclusions drawn from this investigation are summarized as follows:

- This study concluded that the induction of both the HDPE and LDPE decreased the phase angle (θ) which shows that the asphalt mortar is more resistant to permanent deformation. The high-temperature viscosity of the asphalt binder increased, and a marginal increase was noticed when 2.0% HDPE, and 4.0% LDPE were used in the mix.
- The decreased viscosity facilitates the bituminous mix during mixing and compaction. This could be due to a decrease in sphere particles, which gives a micro bearing effect due to the reduction of sphere particles. On the other hand, the effect of spherical particles on the rutting parameter showed that the greater the rutting parameter would be the rutting susceptibility. The 4.0% LDPE with greater sphere particles and less viscosity than HDPE can be utilized for road construction.
- The addition of 2% HDPE and 4% LDPE by weight of bitumen showed improvement in rutting resistance of 13.8% and 42.3%, respectively. These percentages also reported the maximum dynamic viscosity measurements, i.e., 21.80% at 145 °C. However, increasing the percentage of polymers can result in high viscosity and may have a mixing issue on plant-level modified bitumen preparation.
- The trend of the graph for creep rate (m) was decreasing at −6 °C for LDPE but the trend of HDPE was according to the desired criterion. For the modified bitumen, the Creep stress (S) and its relationship with the temperature showed a decreasing trend, which shows relaxation in stress rate, also decreases.
- The FTIR peaks showed the addition of HDPE and LDPE. The polymers modified bitumen by HDPE and LDPE showed the emergence of peaks. As compared to the HDPE, the LDPE is less apolar and has a less crystalline structure. Due to the high crystallinity found inside the polymers, there will be a sufficient amount of improvement in the blended mix of polymers with neat bitumen.
- By the addition of 6% LDPE, the discontinuous matrix of LDPE is observed. However, it can be concluded that by increasing the percentage of LDPE from 6% the morphological improvement, along with phase dispersion of bitumen can be observed.
- By the X-Rays diffraction, the degree of crystallinity for HDPE is more as compared to the LDPE but at a higher temperature in the wet mixing process the HDPE has a high viscosity and the mixing issues with the neat binder.
- The above study recommended the use of waste plastics obtained from waste plastic bottles and bags for the under-construction asphalt pavements in order to improve the bearing capacity of the wearing surface, especially the under-construction projects all over the world, like the China–Pakistan Economic Corridor (CPEC).
- For the high viscosity issues and mixing problem using a wet mixing process, this research concluded that at a high temperature the waste LDPE proved to be the better modifier as compared to the HDPE, but there is a still a need for further studies on microstructural, morphological and dynamic viscosity of HDPE and LDPE modified binders.

Author Contributions: Conceptualization B.Z., U.G. and M.A.; Data curation, U.G. and B.Z.; Formal analysis, M.T.B., M.A., M.M.S.S. and U.G.; Funding acquisition, M.M.S.S.; Investigation, B.Z., U.G. and M.A.; Methodology, U.G., B.Z., M.M.S.S. and S.K.; Project administration, B.Z. and S.K.; Resources, M.M.S.S. and M.A.; Writing—original draft, U.G., B.Z. and M.A.; Writing—review and editing, U.G., B.Z., M.A. and M.M.S.S. All authors have read and agreed to the published version of the manuscript.

Funding: The research is partially funded by the Ministry of Science and Higher Education of the Russian Federation under the strategic academic leadership program "Priority 2030" (Agreement 075-15-2021-1333 dated 30 September 2021).

Institutional Review Board Statement: Not Applicable.

Informed Consent Statement: Not Applicable.

Data Availability Statement: All the necessary data required for supporting this research work are included in this paper.

Conflicts of Interest: All the authors have seen the final version of the paper and declared no potential conflict of interest.

References

1. Read, J.; Whiteoak, D. *The Shell Bitumen Handbook*; Thomas Telford: London, UK, 2003.
2. Yang, Z.; Zhang, X.; Zhang, Z.; Zou, B.; Zhu, Z.; Lu, G.; Xu, W.; Yu, J.; Yu, H. Effect of aging on chemical and rheological properties of bitumen. *Polymers* **2018**, *10*, 1345.
3. Peshkin, D.G.; Hoerner, T.E.; Zimmerman, K.A. *Optimal Timing of Pavement Preventive Maintenance Treatment Applications*; Transportation Research Board: Washington, DC, USA, 2004; Volume 523.
4. Zhu, J.; Birgisson, B.; Kringos, N. Polymer modification of bitumen: Advances and challenges. *Eur. Polym. J.* **2014**, *54*, 18–38.
5. Ahmed, S.; Mahmood, Q.; Elahi, N.; Nawab, B. Current practices and futuristic options in plastic waste management in Pakistan. *Cent. Asian J. Environ. Sci. Technol. Innov.* **2020**, *1*, 237–244.
6. Bala, N.; Kamaruddin, I. Physical and storage stability properties of linear low density polyethylene at optimum content. In Proceedings of the Engineering Challenges for Sustainable Future: Proceedings of the 3rd International Conference on Civil, Offshore and Environmental Engineering (ICCOEE 2016), Seri Iskandar, Perak, Malaysia, 15–17 August 2016; CRC Press: Boca Raton, FL, USA, 2016; p. 395.
7. Munera, J.; Ossa, E. Polymer modified bitumen: Optimization and selection. *Mater. Des. (1980–2015)* **2014**, *62*, 91–97.
8. Cardone, F.; Ferrotti, G.; Frigio, F.; Canestrari, F. Influence of polymer modification on asphalt binder dynamic and steady flow viscosities. *Constr. Build. Mater.* **2014**, *71*, 435–443.
9. Viscione, N.; Presti, D.L.; Veropalumbo, R.; Oreto, C.; Biancardo, S.A.; Russo, F. Performance-based characterization of recycled polymer modified asphalt mixture. *Constr. Build. Mater.* **2021**, *310*, 125243.
10. Duarte, G.M.; Faxina, A.L. Asphalt concrete mixtures modified with polymeric waste by the wet and dry processes: A literature review. *Constr. Build. Mater.* **2021**, *312*, 125408.
11. Airey, G.D. Rheological properties of styrene butadiene styrene polymer modified road bitumens. *Fuel* **2003**, *82*, 1709–1719.
12. Moghadas Nejad, F.; Azarhoosh, A.; Hamedi, G.H. Effect of high density polyethylene on the fatigue and rutting performance of hot mix asphalt–a laboratory study. *Road Mater. Pavement Des.* **2014**, *15*, 746–756.
13. Bamigboye, G.O.; Bassey, D.E.; Olukanni, D.O.; Ngene, B.U.; Adegoke, D.; Odetoyan, A.O.; Kareem, M.A.; Enabulele, D.O.; Nworgu, A.T. Waste materials in highway applications: An overview on generation and utilization implications on sustainability. *J. Clean. Prod.* **2021**, *283*, 124581.
14. Hanif Khan, M.; Zhu, H.; Ali Sikandar, M.; Zamin, B.; Ahmad, M.; Muayad Sabri Sabri, M. Effects of Various Mineral Admixtures and Fibrillated Polypropylene Fibers on the Properties of Engineered Cementitious Composite (ECC) Based Mortars. *Materials* **2022**, *15*, 2880. [PubMed]
15. Varamini, S. Utilization of Recycled Plastics as Binder Modifiers for Use in Hot-mix Asphalt Pavement. Doctoral Dissertation, Dalhousie University, Halifax, NS, Canada, 2013.
16. Bala, N.; Napiah, M.; Kamaruddin, I.; Danlami, N. Rheological properties investigation of bitumen modified with nanosilica and polyethylene polymer. *Int. J. Adv. Appl. Sci.* **2017**, *4*, 165–174.
17. Sengoz, B.; Isikyakar, G. Evaluation of the properties and microstructure of SBS and EVA polymer modified bitumen. *Constr. Build. Mater.* **2008**, *22*, 1897–1905.
18. Jeong, K.-D.; Lee, S.-J.; Kim, K.W. Laboratory evaluation of flexible pavement materials containing waste polyethylene (WPE) film. *Constr. Build. Mater.* **2011**, *25*, 1890–1894.
19. Ouyang, C.; Wang, S.; Zhang, Y.; Zhang, Y. Improving the aging resistance of styrene–butadiene–styrene tri-block copolymer modified asphalt by addition of antioxidants. *Polym. Degrad. Stab.* **2006**, *91*, 795–804.
20. Jasso, M.; Bakos, D.; MacLeod, D.; Zanzotto, L. Preparation and properties of conventional asphalt modified by physical mixtures of linear SBS and montmorillonite clay. *Constr. Build. Mater.* **2013**, *38*, 759–765.

21. Yusoff, N.I.M.; Alhamali, D.I.; Ibrahim, A.N.H.; Rosyidi, S.A.P.; Hassan, N.A. Engineering characteristics of nanosilica/polymer-modified bitumen and predicting their rheological properties using multilayer perceptron neural network model. *Constr. Build. Mater.* **2019**, *204*, 781–799.
22. Martínez-Anzures, J.D.; Zapién-Castillo, S.; Salazar-Cruz, B.A.; Rivera-Armenta, J.L.; Antonio-Cruz, R.d.C.; Hernández-Zamora, G.; Méndez-Hernández, M.L. Preparation and properties of modified asphalt using branch SBS/nanoclay nanocomposite as a modifier. *Road Mater. Pavement Des.* **2019**, *20*, 1275–1290.
23. Zhou, T.; Cao, L.; Fini, E.H.; Li, L.; Liu, Z.; Dong, Z. Behaviors of asphalt under certain aging levels and effects of rejuvenation. *Constr. Build. Mater.* **2020**, *249*, 118748.
24. Behjat, Y.; Cheng, J.J.; Polak, M.A.; Penlidis, A. Effect of Molecular Structure on the Short-Term and Long-Term Mechanical Behavior of High-Density Polyethylene. *J. Mater. Civ. Eng.* **2014**, *26*, 795–802.
25. Sengoz, B.; Topal, A.; Isikyakar, G. Morphology and image analysis of polymer modified bitumens. *Constr. Build. Mater.* **2009**, *23*, 1986–1992.
26. Zhang, H.; Yu, J.; Wu, S. Effect of montmorillonite organic modification on ultraviolet aging properties of SBS modified bitumen. *Constr. Build. Mater.* **2012**, *27*, 553–559.
27. Kebritchi, A.; Jalali-Arani, A.; Roghanizad, A.A. Rheological behavior and properties of bitumen modified with polymeric coated precipitated calcium carbonate. *Constr. Build. Mater.* **2011**, *25*, 2875–2882.
28. Cuciniello, G.; Leandri, P.; Filippi, S.; Lo Presti, D.; Losa, M.; Airey, G. Effect of ageing on the morphology and creep and recovery of polymer-modified bitumens. *Mater. Struct.* **2018**, *51*, 1–12.
29. Ali, S.I.A.; Ismail, A.; Karim, M.R.; Yusoff, N.I.M.; Al-Mansob, R.A.; Aburkaba, E. Performance evaluation of Al2O3 nanoparticle-modified asphalt binder. *Road Mater. Pavement Des.* **2017**, *18*, 1251–1268.
30. Suksiripattanapong, C.; Uraikhot, K.; Tiyasangthong, S.; Wonglakorn, N.; Tabyang, W.; Jomnonkwao, S.; Phetchuay, C. Performance of asphalt concrete pavement reinforced with high-density polyethylene plastic waste. *Infrastructures* **2022**, *7*, 72.
31. Wang, T.; Xu, G.; Shi, C.; Xu, X.; Yu, Y.; Gong, M.; Yang, J. Rheological properties of aged bitumen rejuvenated by polymer modified bio-derived rejuvenator. *Constr. Build. Mater.* **2021**, *298*, 123249.
32. Mubaraki, M. The Effect of Modified Asphalt Binders by Fourier Transform Infrared Spectroscopy, X-Ray Diffraction, and Scanning Electron Microscopy. *J. Mater. Eng. Struct. «JMES»* **2019**, *6*, 5–14.
33. Padhan, R.K.; Sreeram, A.; Gupta, A. Evaluation of trans-polyoctenamer and cross-linking agents on the performance of waste polystyrene modified asphalt. *Road Mater. Pavement Des.* **2020**, *21*, 1170–1182.
34. Hossain, R.; Wasiuddin, N.M. Evaluation of degradation of SBS modified asphalt binder because of RTFO, PAV, and UV aging using a novel extensional deformation test. *Transp. Res. Rec.* **2019**, *2673*, 447–457.
35. Tian, Y.; Li, H.; Zhang, H.; Yang, B.; Zuo, X.; Wang, H. Comparative investigation on three laboratory testing methods for short-term aging of asphalt binder. *Constr. Build. Mater.* **2021**, *266*, 121204. [CrossRef]
36. Zhang, F.; Yu, J.; Han, J. Effects of thermal oxidative ageing on dynamic viscosity, TG/DTG, DTA and FTIR of SBS-and SBS/sulfur-modified asphalts. *Constr. Build. Mater.* **2011**, *25*, 129–137. [CrossRef]
37. Kommidi, S.R.; Kim, Y.-R. Dynamic shear rheometer testing and mechanistic conversion to predict bending beam rheometer low temperature behavior of bituminous binder. *Constr. Build. Mater.* **2021**, *267*, 120563. [CrossRef]
38. Yan, C.; Huang, W.; Ma, J.; Xu, J.; Lv, Q.; Lin, P. Characterizing the SBS polymer degradation within high content polymer modified asphalt using ATR-FTIR. *Constr. Build. Mater.* **2020**, *233*, 117708. [CrossRef]
39. Polacco, G.; Berlincioni, S.; Biondi, D.; Stastna, J.; Zanzotto, L. Asphalt modification with different polyethylene-based polymers. *Eur. Polym. J.* **2005**, *41*, 2831–2844. [CrossRef]
40. Saroufim, E.; Celauro, C.; Mistretta, M.C. A simple interpretation of the effect of the polymer type on the properties of PMBs for road paving applications. *Constr. Build. Mater.* **2018**, *158*, 114–123. [CrossRef]
41. Geçkil, T.; Ince, C.B.; Tanyıldızı, M.M. Physical, rheological and microstructural properties of waste LDPE and TEOA modified bitumens. *J. Chin. Inst. Eng.* **2022**, *45*, 477–487. [CrossRef]
42. Xia, T.; Zhou, L.; Xu, J.; Qin, Y.; Chen, W.; Dai, J. Rheology and thermal stability of polymer modified bitumen with coexistence of amorphous phase and crystalline phase. *Constr. Build. Mater.* **2018**, *178*, 272–279. [CrossRef]
43. Pérez-Lepe, A.; Martınez-Boza, F.; Gallegos, C.; González, O.; Muñoz, M.; Santamarıa, A. Influence of the processing conditions on the rheological behaviour of polymer-modified bitumen. *Fuel* **2003**, *82*, 1339–1348. [CrossRef]
44. Mazumder, M.; Kim, H.; Lee, S.-J. Performance properties of polymer modified asphalt binders containing wax additives. *Int. J. Pavement Res. Technol.* **2016**, *9*, 128–139. [CrossRef]
45. Wieser, M.; Schaur, A.; Unterberger, S.H. Polymer-Bitumen Interaction: A Correlation Study with Six Different Bitumens to Investigate the Influence of SARA Fractions on the Phase Stability, Swelling, and Thermo-Rheological Properties of SBS-PmB. *Materials* **2021**, *14*, 1273. [CrossRef]
46. Santagata, E.; Baglieri, O.; Dalmazzo, D.; Tsantilis, L. Evaluation of the anti-rutting potential of polymer-modified binders by means of creep-recovery shear tests. *Mater. Struct.* **2013**, *46*, 1673–1682. [CrossRef]
47. Ali, A.W. Multiple Stress Creep Recovery (MSCR) characterization of polymer modified asphalt binder containing wax additives. Master's Thesis, Texas State University, San Marcos, TX, USA, 2018.

polymers

Article

Rheological Properties of Styrene-Butadiene-Styrene Asphalt Mastic Containing High Elastic Polymer and Snow Melting Salt

Yangsen Cao [1], Jiarong Li [1], Zhuangzhuang Liu [1,2,*], Xinzhou Li [1], Fan Zhang [1,*] and Baozeng Shan [1]

1 School of Highway, Chang'an University, Xi'an 710064, China
2 Key Laboratory for Special Area Highway Engineering of Ministry of Education, Chang'an University, South 2nd Ring Road Middle Section, Xi'an 710064, China
* Correspondence: zzliu@chd.edu.cn (Z.L.); fanzhang@chd.edu.cn (F.Z.)

Abstract: Sprinkled snow melting salt (SMS) exerts a snow melting effect and also has a negative impact on the asphalt pavement and the environment. Salt storage pavement technology can alleviate these two problems. However, non-alkaline SMSs may have the risk of affecting asphalt mastic properties and further affecting the mechanical properties of asphalt pavements. Therefore, the general properties and rheological properties of two styrene-butadiene-styrene-modified asphalts with and without high elastic polymer were studied after adding SMS. The asphalt mastic without a high elastic agent is defined as the SBS group, and the other group is the HEA group. Our results show that the HEA group shows a lower penetration and a higher softening point, ductility, and viscosity than the SBS group. The more the SMS, the more the reduction effect of the general performance. The elastic recovery of asphalt mastic decreases with the content of SMS. SMS has no obvious effect on the ratio of the viscous and elastic composition of asphalt mastic. The creep of asphalt mastic increases with the content of SMS. The high elastic polymer can significantly reduce the creep, and even the strain of HEA100 is smaller than that of SBS00. SMS increases the creep stiffness and reduces the creep rate at low temperature. Although SMS increases the potential of asphalt pavement to melt ice and snow, it also reduces the high-temperature rutting resistance and low-temperature crack resistance of asphalt mastic. Salt storage pavement materials can be used in combination with high elastic polymers to reduce the negative effects brought by SMSs.

Keywords: road engineering; salt storage pavement; salt storage asphalt mastic; high elastic polymer; rheological properties

Citation: Cao, Y.; Li, J.; Liu, Z.; Li, X.; Zhang, F.; Shan, B. Rheological Properties of Styrene-Butadiene-Styrene Asphalt Mastic Containing High Elastic Polymer and Snow Melting Salt. *Polymers* **2022**, *14*, 3651. https://doi.org/10.3390/polym14173651

Academic Editors: Dino Leporini and Hiroshi Yoshihara

Received: 13 July 2022
Accepted: 31 August 2022
Published: 2 September 2022

1. Introduction

Economic development is inseparable from the support of transportation. As one of the two most widely used types of pavements, asphalt pavement has the characteristics of driving comfort and low noise. In winter, however, snowfall or standing water can easily lead to icy roads. The icing on the road reduces the friction between the pavement and the tires, further causing brake failure. The icing of the asphalt road brings a huge safety hazard to traffic. Ensuring the safety and smoothness of traffic in ice and snow weather has always attracted the attention of road practitioners.

To keep traffic safe in winter, researchers have tried a variety of strategies to clear snow and ice from roads. In general, these strategies can be roughly divided into two types, one is the external snow removal method, and the other is the internal snow removal method [1]. The external ice and snow removal method adopts the traditional ice and snow removal technology, which mainly involves manual ice and snow removal, mechanical ice and snow removal, and spreading snow melting agents. Internal ice and snow removal methods mainly include the thermal snow melting method [2], the elastic deicing method [3], the suppress freeze method [4], and the conductive concrete method [5]. These three methods of the traditional ice and snow melting methods are generally used in synergy to improve the efficiency of ice and snow removal. However, the large-scale distribution of SMS is labor-intensive, especially in less mechanized areas. In addition, the amount of SMS is difficult to control, which leads to the waste of SMS on the one hand, and the surge of maintenance costs on the other hand [6]. Regardless of the cost, the high-concentration salt solution formed by the ice and snow melting water combined with the SMS will not only have an irreversible impact on pavement materials but also harm the soil, causing soil salinization and endangering the ecological environment. Furthermore, is that this top-down ice-melting strategy has difficulty solving the freezing between the pavement and the ice. The distribution of SMS needs to be combined with the weather forecast; otherwise, traffic jams will still occur even if snow removal machinery is used [7]. To reduce the impact of traditional ice and snow melting technology on the pavement and the environment, researchers have gradually turned their attention to the more promising internal ice and snow removal method, that is, active ice and snow removal technology [8].

The thermal ice melting and conductive concrete technology in the active ice and snow removal technology belong to the energy conversion ice and snow melting technology. Thermal self-melting technology is widely used in heating cable ice melting technology [9] and heat pipe ice melting technology [10]. The conductive concrete technology mainly adds conductive substances such as graphite [11], carbon fiber [12], and steel fiber [13] to the pavement material, and then increases the temperature of the pavement through resistance heat dissipation to melt ice and snow. These two types of energy-converting ice and snow melting technologies have remarkable effects, but they have not been widely used due to the disadvantages of the required large investment and high energy consumption. The elastic de-icing and snow pavement is to add elastic material to the mixture to break the icing of the pavement through the elastic deformation of the pavement. Elastic de-icing is theoretically feasible, but the effect is limited in practical application [14]. The freeze-inhibiting pavement to remove ice and snow is to achieve the purpose of removing ice and snow by inhibiting the formation of ice on the pavement surface. Specific practices include the use of superhydrophobic surface coatings to reduce ice adhesion and accumulation [15], and the use of rough surface structures to inhibit ice formation [16]. The method of inhibiting the formation of ice layers reduces the possibility of pavement surface freezing from the root, but it has the characteristics of high cost and poor wear resistance.

The traditional method of de-icing and snow removal is effective, but there are problems such as untimely de-icing, environmental pollution, and large labor consumption. In active ice and snow removal technology, the energy conversion type of ice and snow melting technology has high energy consumption and high cost. Resilient pavement de-icing increases pavement flexibility and may increase vehicle fuel consumption. The pavement technology to inhibit freezing is still immature and insufficiently popularized. In view of the various problems existing in the above-mentioned ice and snow melting technology, salt storage pavement ice and snow removal technology came into being [17–19]. This technology adds slow-release salt storage materials to pavement materials in the form of fillers. The formation of ice is delayed or inhibited for a period of time by the gradual migration of salts to the pavement surface through the compression of vehicle tires, vibration, and the action of water. Salt storage pavement has great potential in the field of ice and snow melting due to its timely removal of ice and snow, controlled amount of SMS, reasonable construction cost, and the same construction method as traditional pavement [20].

In the research on salt storage pavement, Tan [21] studied the performance of a salt storage asphalt mixture to remove ice and snow and pointed out that the adhesion between the salt storage asphalt mixture and ice decreased. The adhesive force between the open-graded asphalt mixture and the ice is greater than that of the densely-graded asphalt mixture, and the adhesive force increases with the nominal maximum particle size. The salt storage asphalt mixture has good road performance, and its high-temperature performance and low-temperature performance are improved compared with ordinary asphalt mixture. Although moisture resistance is slightly decreased, it still meets the specification requirements. In the following year, Tan [22] developed a four-component slow-release complex salt filler. The carrier of the filler was optimized and the production process was determined according to the slow-release degree of the filler. The test of melting ice and snow shows that the pavement material mixed with slow-release complex salt filler has good de-icing and snow performance at the local temperature of −10 to 0 °C, and can reduce the adhesion of snow to the pavement at −20 to −10 °C. Yu [23] prepared a high-elasticity salt storage asphalt mixture using high-elasticity modified asphalt, and its high-temperature stability and low-temperature crack resistance were better than the SBS-modified asphalt mixture and salt storage asphalt mixture. The ice-breaking rate of high elastic salt storage asphalt mixture is 42% higher than that of ordinary SBS asphalt mixture, and it has the potential to remove an ice layer with 12 mm thickness. Guo [24] observed the microscopic morphology of the snow melting agent and studied the effect of the agent on the moisture resistance of the asphalt mixture. The porous structure on the surface of the de-icer is helpful for the release of NaCl, the main anti-condensation component. SMS can lower the freezing point to −2 °C. The salt solution in which the de-icing agent is dissolved is more likely to wet the aggregate, which reduces the moisture resistance of the asphalt mixture under the dual action of the dynamic water pressure and the salt solution. Wu [25] compared the storage and slow-release performance of six kinds of salt storage carriers for SMS and pointed out that the salt-storage carrier had little effect on the slow-release effect of SMS. The smaller the mass ratio of the salt storage carrier to the SMS, the better the snow melting effect of the pavement, but the worse the moisture resistance of the pavement at the same time. Considering the snow melting performance and mechanical properties of the asphalt mixture comprehensively, it is recommended that the mass ratio of SMS and carrier be 3.5:1.

At present, most of the research on salt storage pavement focuses on the performance of the asphalt mixture or melting ice and snow, which promotes the development of salt storage pavement technology to a certain extent. In salt storage pavement materials, SMS is mainly dispersed in asphalt, which makes SMS and asphalt generally exist in the form of asphalt mastic. Asphalt mastic is an important component of the asphalt mixture, and its performance directly affects the service of pavement [26–29]. However, there are few reports on the effect of SMS on the performance of asphalt mastic. Therefore, this paper uses two different modified asphalts, namely, SBS modified asphalt and HEA, to explore the effect of asphalt and SMS on the performance of asphalt mastic through traditional performance tests, temperature sweep tests, multiple stress creep and recovery tests, and bending beam rheological tests. This work can provide a reference for the research and application of salt storage asphalt pavement materials.

2. Materials and Methods

2.1. Materials

2.1.1. Asphalt

The asphalt used in this study is SBS I-D, and its basic properties are shown in Table 1.

Table 1. Basic technical indexes of SBS I-D asphalt.

Indexes	Unit	Test Result	Standard	Test Method
Penetration (100 g, 5 s, 25 °C)	0.1 mm	54	40–60	T0604-2011
Ductility (5 °C, 5 cm/min)	cm	31	≥20	T0605-2011
Softening Point	°C	80	≥60	T0606-2011
Dynamic viscosity (135 °C)	Pa·s	1.773	≤3	T0620-2011
Elastic recovery (25 °C)	%	90	≥75	T0662-2000
Residues after TFOT				
Mass change	%	−0.213	±1.0	T0609-2011
Penetration ratio (25 °C)	%	70.5	≥65	T0604-2011
Ductility (5 °C, 5 cm/min)	cm	16	≥15	T0605-2011

2.1.2. Modifier

The mechanical properties of salt storage asphalt mixtures are often lost to varying degrees due to the addition of slow-release salt-storage materials, especially the loss of moisture resistance. Therefore, high elastic asphalt (HEA) was selected to improve the performance of the salt storage asphalt mixture. HEA is a modified asphalt made by adding plasticizers and cross-linking agents to SBS-modified asphalt. Compared with SBS modified asphalt, HEA has a higher elastic recovery rate, ductility, and softening point due to the addition of the high elastic agent, so HEA asphalt mixture also has better high-temperature performance, low-temperature performance, and fatigue performance.

The high elastic agent used in this paper is TAFPACK-Super (TPS), and the appearance is orange-yellow transparent particles as shown in Figure 1. The main component of TPS is thermoplastic rubber, which is made with auxiliary component adhesive resin, and then mixed with other stabilizers. The basic properties of TPS are shown in Table 2.

Figure 1. TPS modifier, Icebane, and limestone filler.

Table 2. Basic properties of TPS.

Particle Size/mm	Color	Relative Density	Water Absorption Rate/%	Melting Point/°C
2–4	Yellow	0.96	<1	170

2.1.3. Filler

There are two kinds of fillers used in this paper, one is limestone filler and the other is SMS filler. The basic properties of limestone fillers are shown in Table 3. The SMS filler is Icebane, new ecological SMS, whose main chemical components include silicon dioxide, sodium chloride, calcium oxide, magnesium oxide, etc. Icebane is more than half of the effective snowmelt composition, and the chloride salts are adsorbed in the porous structure. In the rain and snow environment, the active ingredients can be gradually released to achieve the snow melting effect. The basic properties of Icebane are shown in Table 4. The comparison between Icebane and ordinary limestone filler is shown in Figure 1. The white powder particles on the left are Icebane, and the powder on the right is limestone filler.

Table 3. Basic properties of limestone fillers.

Indexes	Test Result	Standard	Test Method
Moisture content (%)	0.2	≤1.0	T0103-1993
Apparent relative density	2.762	≥2.5	T0352-2000
Gross bulk relative density	2.667	/	T0352-2000
Hydrophilic coefficient	0.73	<1	T0353-2000
Plasticity index (%)	2.7	<4	T0354-2000
Appearance	Qualified	No agglomeration	T0355-2000

Table 4. Basic properties of Icebane.

Indexes	Test Result	Standard	Test Method
Appearance	White powder	No agglomeration	T0355-2000
Moisture content (%)	0.2	≤0.5	T0103-1993
Apparent relative density	2.170	/	T0352-2000
Gross bulk relative density	2.136	/	T0352-2000
Salt content (%)	56	50 ± 10	/
pH	8.3	8–8.5	/

2.2. *Preparation of Modified Asphalt and Asphalt Mastic*

2.2.1. Preparation of HEA

A small amount of TPS high-elasticity modifier was added to the SBS modified asphalt at 175 °C several times. To better control the temperature and avoid asphalt aging, oil bath heating equipment is used to control the temperature throughout the process. Agitation was performed using an asphalt shearing machine at a shear rate of 2000 rpm. The addition ratio of high elastic agent and asphalt is 8:92. After the TPS was added to the SBS asphalt, it was kept at 2000 rpm for 10 min to make the modifier completely dissolve into the SBS modified asphalt. After that, the asphalt temperature was raised to 180 °C, and the asphalt was sheared for 35 min under the shearing speed of 3000 rpm to ensure a sufficient reaction between TPS and asphalt. The basic indicators of the HEA are shown in Table 5.

Table 5. Basic indicators of HEA.

Indexes	Unit	Test Result	Standard	Test Method
Penetration (100 g, 5 s, 25 °C)	0.1 mm	49	40–60	T0604-2011
Ductility (5 °C, 5 cm/min)	cm	55	≥20	T0605-2011
Softening Point	°C	94	≥60	T0606-2011
Dynamic viscosity (135 °C)	Pa·s	1.861	≤3	T0620-2011
Elastic recovery (25 °C)	%	98	≥75	T0662-2000
	Residues after TFOT			
Mass change	%	−0.13	±1.0	T0609-2011
Penetration ratio (25 °C)	%	76.6	≥65	T0604-2011
Ductility (5 °C, 5 cm/min)	cm	31	≥15	T0605-2011

2.2.2. Preparation of Salt Storage Asphalt Mastic

The amount of filler in the mixture is small. To maintain long-term snowmelt performance, a higher filler to asphalt ratio is required. However, an excessively large filler-to-asphalt ratio will lead to a reduction in free asphalt in the mastic, and the surface of some fillers cannot be covered by free asphalt. Some studies have pointed out that when the filler and asphalt ratio is 1.2:1, the mechanical properties of asphalt mastic can be maintained at a good level. Therefore, asphalt mastic was prepared with a filler to asphalt ratio of 1.2:1 [30]. In addition, the density of limestone filler is slightly higher than that of Icebane. If the SMS is added in an equal mass, the volume of the SMS should be greater than that of the limestone filler of equal mass. Therefore, Icebane is added to the asphalt matrix in the form of equal volume replacement to prepare salt storage asphalt mastic.

Due to the difference in density between asphalt and filler, the two are prone to segregation, which affects the performance of asphalt mastic. To this end, the glass rod is used for continuous stirring to ensure the uniformity of the asphalt mastic. The preparation process is detailed as follows:

(1) Both the limestone filler and the salt storage material were passed through a 0.075 mm sieve, and the bottom part of the sieve was taken for testing. The two fillers were placed in a dry and ventilated place, and the fillers were dried and ground before the test to avoid agglomeration affecting the accuracy of the test.
(2) The pitch was heated to 175 °C to ensure that the pitch was picked out by the glass rod drips in the form of droplets. The asphalt, preheated to the same temperature as the asphalt in portions, was added. The glass rod was kept agitated to prevent the filler from sticking to the container walls. The heating temperature and stirring time were strictly controlled to avoid asphalt aging.
(3) The different types of asphalt mastics were named SBS00, SBS50, HEA50, SBS75, HEA75, SBS100, and HEA100 according to the proportion of Icebane replacing limestone filler. For example, SBS00 means that the replacement rate of limestone filler is 0%, and the base asphalt is SBS-modified asphalt.

2.3. Testing Methods

2.3.1. Testing of Three Major Indicators: Penetration, Softening Point, and Ductility

To compare the effects of asphalt and SMS on the basic properties of asphalt mastic, the penetration, softening point, and ductility of salt storage asphalt mastic were tested with regard to the "Standard Test Methods of asphalt and Bituminous Mixtures for Highway Engineering (JTG E20-2011)" [31]. Penetration reflects the consistency and hardness of asphalt. Generally speaking, the smaller the penetration, the better the high-temperature rheological resistance of asphalt. The softening point also reflects the high-temperature resistance of asphalt. The higher the softening point, the stronger the thermal storage capacity of the asphalt and the better the thermal stability. Ductility reflects the low-temperature deformation properties of asphalt; that is, the plastic properties of asphalt. The higher the ductility, the better the plasticity of the asphalt and the better the low-temperature crack resistance. Sample information is shown in Table 6.

Table 6. Samples information.

Tests	Specimens	Replication
Penetration test	SBS00, SBS50, SBS100, HEA50, HEA100	4
Softening point test	SBS00, SBS50, SBS100, HEA50, HEA100	4
Ductility test	SBS00, SBS50, SBS100, HEA50, HEA100	4
Brookfield viscosity test	SBS00, SBS50, SBS100, HEA50, HEA100	3
Temperature sweep test	SBS00, SBS50, SBS100, HEA50, HEA100	3
Multiple stress creep and recovery test	SBS00, SBS50, SBS100, HEA50, HEA100	3
Bending beam rheological test	SBS00, SBS50, SBS100, HEA50, HEA100	3

2.3.2. Brookfield Viscosity Test

The viscosity-temperature characteristics of asphalt mastic not only affect the mixing and compaction process of the mixture but are also closely related to the high and low-temperature performance and durability of the mixture. For the salt storage asphalt mixture, the selection of the engineering mixing temperature and the rolling temperature should refer to the viscosity of the asphalt mastic. If the viscosity of the asphalt mastic formed by Icebane and asphalt is not fully studied, it will increase the difficulty of mixing and compacting the asphalt mixture. For example, the poor adhesion between Icebane and asphalt will increase the fluidity of the mixture, cause leakage during the transportation and paving stages of the mixture, and endanger the flatness after compaction. Therefore, to explore the advantages and disadvantages of salt storage asphalt mixture and ordinary mixture, it is necessary to study the viscosity–temperature characteristics of asphalt mastic. Referring to the specification [31], a 28# rotor was selected to test the Brookfield viscosity of asphalt mastic with different substitution rates of SMS.

2.3.3. Temperature Sweep Test

By studying the viscoelastic properties of salt storage asphalt mastic, the effect of SMS on the performance of salt storage asphalt mixture can be further understood. The SHRP specification uses dynamic shear rheological testing as a measure of asphalt performance. The mechanical properties of the salt storage asphalt mastic are analyzed by the complex modulus, phase angle, and other indicators, to make a more scientific and reasonable evaluation of the characteristics of the mixture. In this paper, the DHR-1 dynamic shear rheometer produced by TA company in the United States is used for testing. During the test, it should be noted that the distance between the upper and lower parallel plate fixtures is selected to be 1 mm, which is suitable for the test above 30 °C, and the size of the matching parallel plate is selected to be 25 mm [32,33].

When the temperature sweep test was performed on the salt storage asphalt mastic, the temperature sweep range of was set to 40–80 °C. The temperature interval was 5 °C, the strain was set to 10%, and the angular frequency parameter was set to 10 rad/s. The storage modulus G', complex modulus G^*, phase angle δ, rutting factor $G^*/\sin\delta$, and Z are used as indicators to evaluate the high-temperature performance of salt storage asphalt mastic.

2.3.4. Multiple Stress Creep and Recovery Test

For modified asphalt, simply using the temperature sweep test to evaluate the high-temperature performance is insufficient [29]. AASHTO TP70 proposes to use multiple stress creep and recovery tests to evaluate the high-temperature performance of asphalt. Therefore, according to the repeated creep test procedure, repeated loading tests were carried out on the asphalt mastic under two stress levels of 0.1 and 3.2 kPa at 58, 64, and 70 °C, to simulate the repeated creep effect of light and heavy traffic loads on the asphalt pavement. In each loading cycle, the loading time is 1 s and the recovery time is 9 s. The repetitions for each test are 10 times and the total time is 200 s.

As two indicators for evaluating the repeated creep characteristics of asphalt, the creep recovery rate R and the non-recoverable creep compliance J_{nr} are calculated according to Equation (1) and Equation (2), respectively. The higher the creep recovery rate, the smaller the non-recoverable creep compliance, indicating that the asphalt has sufficient deformation recovery ability under repeated stress.

$$R = \frac{\gamma_p - \gamma_{nr}}{\gamma_p - \gamma_0} \times 100\% \tag{1}$$

$$J_{nr} = \frac{\gamma_{nr} - \gamma_0}{\tau} \tag{2}$$

where γ_0 is the initial strain; γ_p is the peak strain; γ_{nr} is the irrecoverable strain; and τ is the creep stress.

2.3.5. Bending Beam Rheological Test

In the SHRP program of the United States, it is proposed to use the bending beam rheological test to evaluate the low-temperature cracking resistance of asphalt. Two parameters, the modulus of creep stiffness S and the creep slope m, can be obtained through the bending beam rheological test. The smaller the stiffness modulus and the larger the creep rate, the better flexibility and low-temperature crack resistance of the asphalt. This test method is more accurate and is also widely used for the performance grading of asphalt. Since the asphalt pavement will be affected by the coupling of traffic load and temperature during the service period, the stiffness modulus of the asphalt will increase after aging, and the asphalt pavement will be more prone to cracks. Test pieces with standard dimensions of $102 \times 12.5 \times 6.25$ mm^3 were made by pouring asphalt mastic into a special mold. The specimens were then placed in anhydrous ethanol at the specified temperature for incubation. The test temperature is selected as −24, −18, and −12 °C. During the test, the mid-span load of the trabecular was 980 mN and the duration was 240 s. The creep stiffness $S(t)$ can be calculated by Equation (3).

$$S(t) = \frac{PL^3}{4bh^3\delta(t)} \quad (3)$$

where L is the distance between the fulcrums, m; P is the applied load, kN; h is the thickness of the specimen, m; b is the width of the specimen, m; $\delta(t)$ is the maximum deflection, m; and $S(t)$ is the Creep stiffness at time t, MPa.

3. Results and Discussion

3.1. Penetration, Softening Point, and Ductility

Penetration, softening point, and ductility are used to characterize the basic properties of asphalt. Generally speaking, the greater the penetration, the smaller the asphalt consistency, the weaker its rutting resistance, and the better the low-temperature toughness. The higher the softening point, the more resistance of the asphalt to high temperatures. The higher the ductility, the less likely the asphalt will break at low temperatures. With the help of these three indicators, the performance of salt storage asphalt mastic is preliminarily evaluated. Figure 2 shows the penetration, softening point, and ductility of the asphalt mastics of the HEA group and the SBS group. Overall, the HEA asphalt mastic has smaller penetration, and a higher softening point and ductility than the SBS asphalt mastic, indicating that the HEA group has excellent low-temperature and high-temperature performance. This may be because the high elastic modifier TPS forms an inter-crosslinked network structure in the asphalt. This network structure binds the thermal motion of asphalt molecules at room temperature and high temperature and increases the toughness of asphalt at low temperature [34–36]. Therefore, the asphalt mastic of the HEA group has better high-temperature and low-temperature performance than that of the SBS group [14,37].

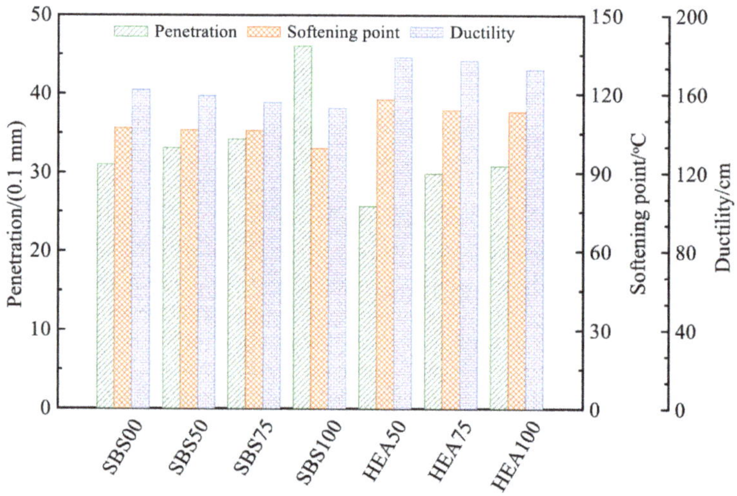

Figure 2. Basic properties of asphalt mastic.

Figure 2 shows the effect of SMS on penetration, softening point, and ductility of asphalt mastic. In terms of penetration, in both the SBS group and the HEA group of salt storage asphalt mastic, the penetration increased with the replacement amount of SMS. This indicates that the addition of SMS softened the asphalt mastic. The SMS replaces the limestone filler with the equal volume replacement method. The particle size of the SMS is generally larger than that of the limestone filler, so the number of SMS particles may be less than that of the limestone filler. SMS filler is coarser, the proportion of structural asphalt in asphalt mastic decreases, and the proportion of free asphalt increases [20,38,39]. Smaller salt particles can enhance asphalt mastic performance [37]. Therefore, as the content of SMS increases, the penetration also gradually increases. Differences in the properties of the two fillers may also be responsible for the increased penetration. The main component of SMS is chloride salt, which has weaker interaction with asphalt than limestone fillers [14,40,41]. In addition, the poor adhesion of SMS to asphalt changes the continuous state of asphalt, and the reduced viscosity of asphalt mastic may also be the reason for the increased penetration [42,43]. In terms of softening point, the replacement amount of SMS increased, and the softening point of the two groups of salt storage asphalt mastics decreased. The decrease in softening point indicates that the heat required for asphalt mastic to obtain flow deformation ability decreases; that is, the energy required for asphalt molecules to overcome intermolecular forces decreases [36,44]. The reduction in the interaction between filler and asphalt due to the reduction in structural asphalt can also be explained as the lowering of the softening point. In terms of ductility, the ductility of the salt-storing asphalt mastic in the SBS group and the HEA group has a decreasing trend with the increase in the SMS content, but the decrease is not obvious. When the content of SMS increases, the content of free asphalt increases, and the low-temperature ductility of asphalt mastic should increase. However, the ductility is reduced here, probably because the reduction in structural asphalt leads to a reduction in the asphalt bond strength. The reduction in the ductility of asphalt by SMS was also demonstrated in the literature [45]. The salt-storing asphalt mastic is in a discontinuous state, and the shrinkage coefficients of SMS and asphalt are different. The higher the content of SMS, the worse the interface continuity of the asphalt mastic, the worse the activity of the mastic, and the lower the ductility [45,46].

From the three basic index results of penetration, softening point, and ductility, it can be preliminarily concluded that HEA has better high-temperature and low-temperature performance than SBS-modified asphalt. SMS has adverse effects on the high-temperature and low-temperature properties of asphalt mastic. The basic performance of HEA100 is slightly better than that of SBS00 without SMS, so adding TPS can make up for the negative impact of SMS on asphalt mastic to a certain extent. Nevertheless, based on these three indicators, the performance of asphalt mastic can only be preliminarily evaluated. The influence of asphalt type and SMS on the performance of asphalt mastic requires further research on viscosity and rheological properties.

3.2. Brookfield Viscosity

In Figure 3, the viscosity of asphalt mastic shows a decreasing trend with temperature. For example, at 135 °C, the viscosity of HEA100 is 3.15, 7.28, and 11.94 times that of 155, 175, and 195 °C, respectively. This is because the increase in temperature accelerates the thermal motion of the asphalt molecules in the asphalt mastic, making the asphalt flow more easily [36,47]. Nonetheless, the higher the temperature, the smaller the viscosity difference between the individual asphalt mastics. The asphalt mastic of the HEA group showed higher viscosity than that of the SBS group at all four temperatures. This may be because the interaction of TPS and SBS-modified asphalt increases the content of macromolecules, making it difficult for HEA mastic to flow [47]. Furthermore, the network structure formed after TPS swelling hinders the thermal motion of asphalt molecules [34–36]. Whether it is HEA group or SBS group salt storage asphalt mastic, the higher the replacement rate of Icebane, the lower the viscosity of the asphalt mastic. Taking the HEA group as an example, when the content of Icebane in HEA is 75% and 100%, their viscosity at 135 °C is 4.47% and 11.32% lower than that of HEA50, respectively. The decrease in viscosity with the content of SMS may be because the bond between SMS and asphalt is weaker than that of limestone fillers [14,40,41]. Furthermore, the reduction in structural asphalt may also be one of the factors leading to the reduction in asphalt viscosity [19,20,39]. In a word, there are two rules for the viscosity of salt storage asphalt mastic, one is that the viscosity of the HEA group is higher than that of the SBS group, and the other is that limestone filler contributes more to asphalt viscosity than Icebane.

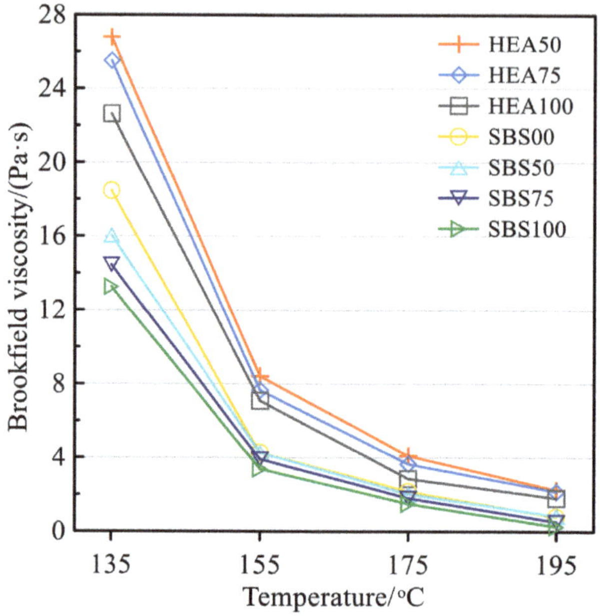

Figure 3. The viscosity of asphalt mastic at different temperatures.

3.3. Analysis of Temperature Sweep Test Results

3.3.1. Storage Modulus, Phase Angle, Complex Modulus, and Rutting Factor

As shown in Figure 4, the storage modulus of asphalt mastic shows a decreasing trend with increasing temperature. This is because the plasticity of the asphalt increases and the elastic properties decrease as the temperature increases [44,47]. The storage modulus of asphalt mastic becomes larger after aging because the light components are gradually transformed into resins and asphaltenes after asphalt aging [36,44]. The heavy component has a larger molecular weight, small molecular distance, strong interaction force, and good elastic recovery performance [36,44,47,48]. Overall, low temperature and aging will increase the elastic components in asphalt, which is conducive to the elastic recovery of mastic. In addition, the content of Icebane also affects the elastic component in the asphalt. The higher the replacement rate of SMS, the worse the elastic recovery of asphalt mastic, which was confirmed in both the SBS group and the HEA group. This may be because the increase in SMS reduces the structural asphalt in asphalt mastic, while structural asphalt is beneficial to improving the elastic recovery performance of asphalt mastic [20,39]. The storage modulus of the asphalt mastic in the HEA group was higher than that of the SBS group. This is because TPS directly increases the elastic component in SBS-modified asphalt as a high elastic agent. In addition, the storage modulus of HEA100 with the highest content of SMS in the HEA group was even higher than that of SBS00 without SMS. This indicates that the performance of the asphalt could compensate for the unfavorable effect of the SMS filler on the elastic deformation capacity of the mastic.

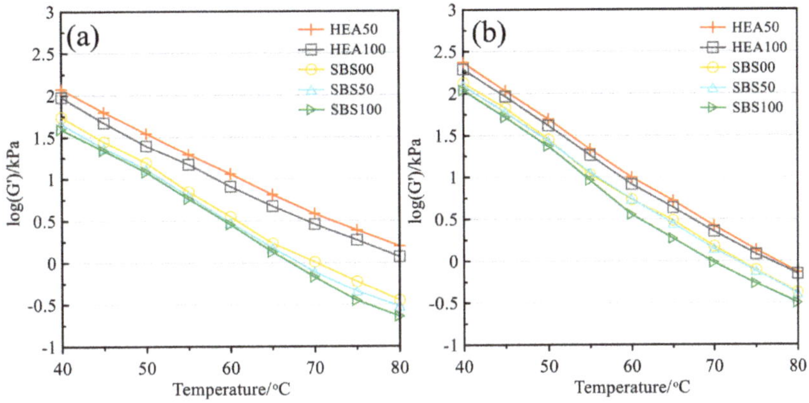

Figure 4. Storage modulus: (**a**) Before TFOT; (**b**) After TFOT.

Figure 5 shows the phase angle of the asphalt mastic before and after aging as a function of temperature. As an index of asphalt viscoelasticity, the phase angle can reflect the proportion of viscoelastic components in asphalt [36]. As the temperature increases, the phase angle of the asphalt mastic increases gradually, because as the temperature increases, the viscous component in the asphalt increases, while the elastic component decreases [49]. Since TPS increases the elastic component in the SBS-modified asphalt, the phase angle of the HEA asphalt mastic is much smaller than that of the SBS group [50]. The phase angle decreases slightly after the asphalt is aged, mainly because the asphalt becomes brittle and hard after aging, the elastic component increases, and the viscous component decreases. As for the influence of the SMS on the phase angle, since the phase angle curves of different amounts of SMS in the two groups of asphalt mastic intersect with each other, the influence of the content of Icebane on the viscoelastic component proportion of asphalt mastic is relatively complex [49,51,52].

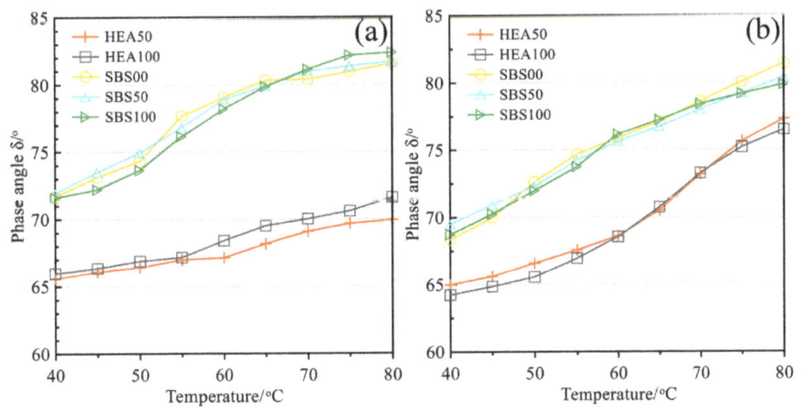

Figure 5. Phase angle: (**a**) Before TFOT; (**b**) After TFOT.

Figures 6 and 7 are the changes in the complex modulus and rutting factor with temperature before and after the aging of asphalt mastic, respectively. These two indicators can be used to measure the rheology of asphalt mastic at high temperatures [36,53]. The complex modulus and rutting factor have similar trends, so the high-temperature performance of salt storage asphalt mastic is analyzed by taking the rutting factor as an example. It can be seen from Figure 7 that with the increase in temperature, the rutting factor shows a decreasing trend, which indicates the ability of asphalt to resist high-temperature rutting decreases. This is because the intermolecular force weakens at high temperatures and the molecular motion of asphalt accelerates [36,49]. After aging, the rutting factor increases, and aging contributes to improved high-temperature performance [44,49,54]. Regardless of the test temperature, the order of rutting factors from high to low is HEA50, HEA100, SBS00, SBS50, and SBS100. Therefore, the high-temperature performance of salt storage asphalt mastic before and after aging will be affected by the asphalt type and the Icebane content. The inclusion of TPS in HEA can increase the rutting resistance of the asphalt mastic, while the higher the SMS content, the smaller the deformation resistance. The high-temperature rheological properties of salt-storing asphalt mastic are consistent with Sections 3.1 and 3.2.

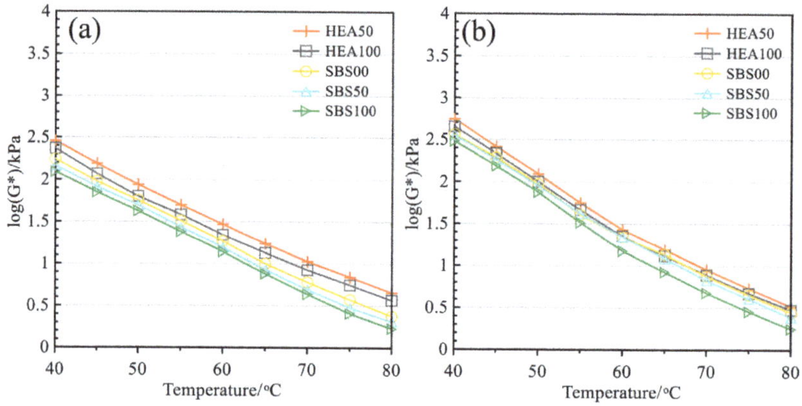

Figure 6. Complex modulus: (**a**) Before TFOT; (**b**) After TFOT.

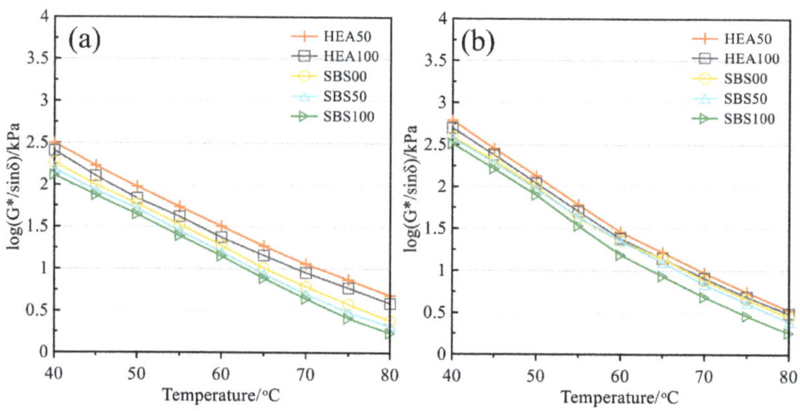

Figure 7. Rutting factor: (**a**) Before TFOT; (**b**) After TFOT.

3.3.2. Viscoelastic Index Z

Similar to the rutting factor, the viscoelastic index Z [$Z=\sin\delta\cdot(1-\cos\delta)/\log(G^*)$] can also reflect the rutting resistance of asphalt mastic [36,55]. The smaller the Z value, the stronger the high-temperature performance of asphalt mastic. Figure 8 shows the viscoelastic index Z of the salt storage asphalt mastic before and after aging. The index Z of the asphalt mastic in the HEA group and the SBS group increases with the temperature, the viscosity of the mastic decreases, and the mastic gradually changes to a viscous-fluid state [47]. Under the same amount of SMS, the Z value of the HEA group was always lower than that of the

SBS group. This shows that the high elastic agent increases the rutting resistance of the asphalt mastic. The Z value increases with the content of SMS, which reflects that SMS has a hindering effect on the high-temperature performance of salt-storing asphalt mastic. Comparing the Z values of SBS00 and HEA100, it can be concluded that the improvement effect of the high elastic agent on the high-temperature performance of asphalt mastic is much higher than the negative effect of SMS on the high-temperature performance. This analysis is consistent with the previous analysis. The Z value of the aged asphalt mastic becomes smaller because the asphalt becomes hard and brittle after aging [36,56].

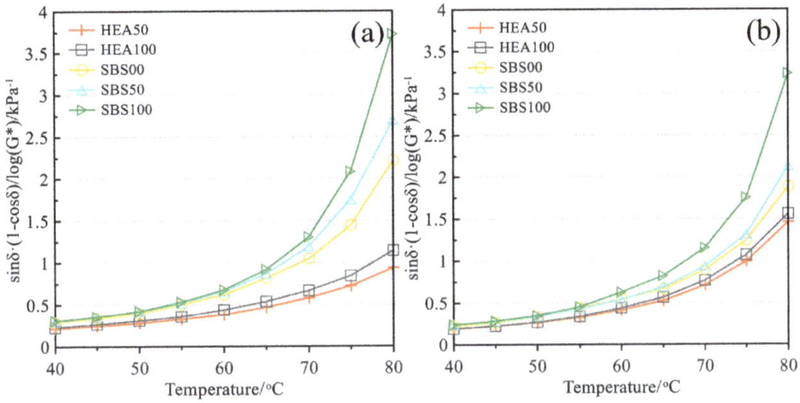

Figure 8. Viscoelastic index Z: (**a**) Before TFOT; (**b**) After TFOT.

3.4. Analysis of MSCR Test Results

3.4.1. Accumulated Strain

Rutting is the accumulation of pavement deformation under repeated traffic loads. The cumulative strain of asphalt mastic can also characterize the service performance of the pavement. The higher the value, the worse the rutting resistance of the pavement [57]. The strain of asphalt mastic under cyclic loading is shown in Figure 9. The higher the temperature, the higher the strain of the asphalt mastic. The elevated temperature enhances the thermal motion of the asphalt molecules, while the elastic component in the asphalt mastic gradually transforms into the viscous component [36]. Short-term aging will make the asphalt mastic hard and brittle, thereby reducing the strain of the mastic. In addition, with the increase in the amount of SMS, the strain of asphalt mastic under load also increases gradually. The interaction between SMS and asphalt is weak, and the addition of SMS may increase the free asphalt content in asphalt mastic, so SMS promotes the deformation of asphalt mastic compared to limestone filler [20,40,41]. Under each stress condition, the order of strain from high to low is SBS100 > SBS50 > SBS00 > HEA100 > HEA50. The cumulative strain of the HEA group was smaller than that of the SBS group. The high elastic properties of HEA can well reduce the strain of mastic under creep conditions, which may be due to the network structure formed after the swelling of the high elastic modifier in the asphalt, which limits the deformation of the asphalt mastic [36]. Except for Figure 9d, the cumulative deformation of HEA100 is always lower than that of SBS00. Even in Figure 8b, the cumulative strain of HEA100 is not higher than that of SBS00. Therefore, the use of SMS in combination with high elastic agents not only provides snow melting potential to the pavement but also improves the rutting resistance.

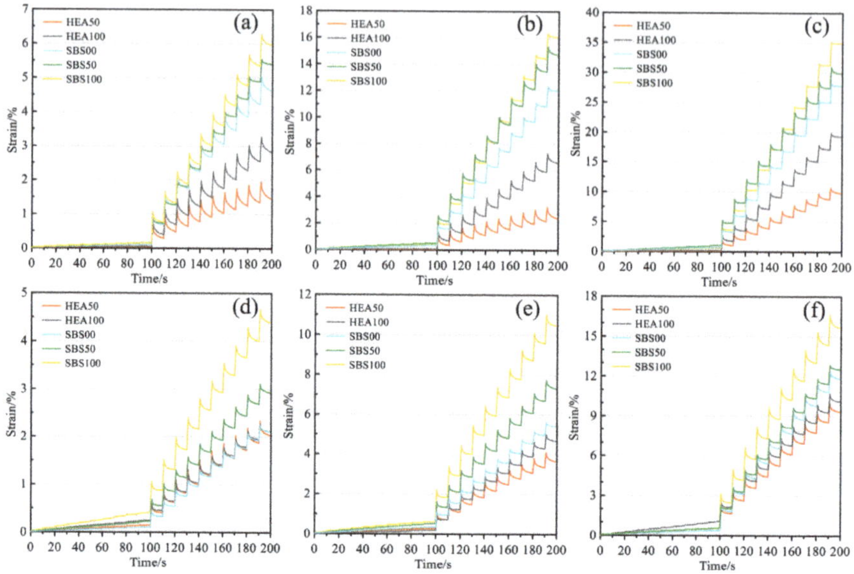

Figure 9. Strain curve: (**a**) 58 °C before TFOT; (**b**) 64 °C before TFOT; (**c**) 70 °C before TFOT; (**d**) 58 °C after TFOT; (**e**) 64 °C after TFOT; (**f**) 70 °C after TFOT.

3.4.2. Creep Compliance and Creep Recovery

Creep compliance reflects the high-temperature deformation capacity. The higher the value, the easier it is for the asphalt mastic to creep at high temperatures. The creep compliance of different salt storage asphalt mastics before and after aging is shown in Figure 10. In Figure 9a, the creep compliance increases with temperature because high temperature reduces the viscosity of the mastic and enhances the mobility of asphalt molecules [36]. The creep compliance of the HEA group is lower than that of the SBS group, indicating that HEA is more resistant to high-temperature creep. The creep compliance increases with the content of SMS, so SMS is not beneficial to the creep resistance of asphalt mastic. This may be due to the reduction in structural asphalt in asphalt mastic by SMSs [20,39]. Although limestone filler can give salt storage asphalt mastic better high-temperature performance, with the help of the high elastic agent, the creep compliance of HEA100 without limestone filler is still stronger than SBS00 without SMS. The influence of asphalt type on the high-temperature performance of the mastic is higher than that of SMS. The trend of creep compliance in Figure 9b–d is similar to that in Figure 9a. Comparing the four subplots in Figure 9, it can be seen that the aging effect can reduce the creep compliance, which is due to the reduction in light components in the aged asphalt mastic and the hardening of the asphalt mastic [36,56]. In addition, increasing the stress level will increase the creep compliance of the asphalt mastic, which is related to the increase in the cumulative deformation of the asphalt mastic caused by increasing the stress in Figure 9. Both increasing stress and increasing temperature will increase creep compliance, indicating that the effect of high load and high temperature has the same effect on the creep of asphalt mastic [57]. This equivalent effect is also confirmed in Figure 11.

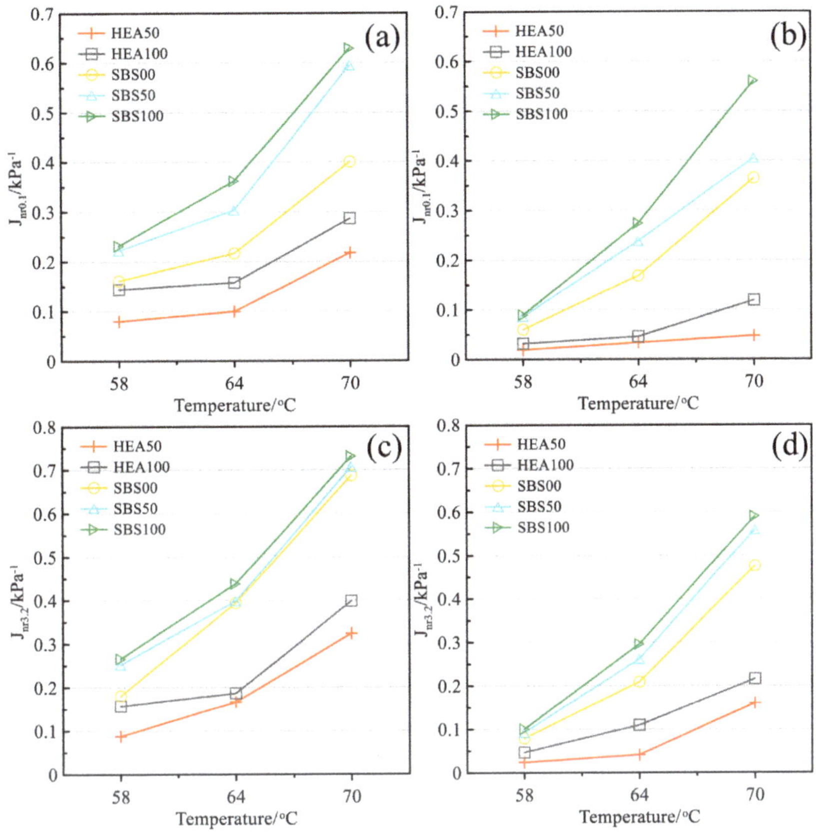

Figure 10. Creep compliance: (**a**) 0.1 kPa before TFOT; (**b**) 0.1 kPa after TFOT; (**c**) 3.2 kPa before TFOT; (**d**) 3.2 kPa after TFOT.

Creep recovery is the opposite of creep compliance and reflects the ability to resist high-temperature deformation. The higher the value, the better the deformation resistance of asphalt mastic [58]. The creep recovery of different types of salt storage asphalt mastics before and after aging is shown in Figure 11. In Figure 11a, the increase in temperature will lead to a decrease in the recovery of asphalt mastics, indicating that the increasing temperature will increase the viscous component of asphalt and reduce the elastic component, so the deformation recovery ability will decrease accordingly. The recovery rate of HEA50 and HEA100 salt storage asphalt mastics is much higher than other types of mastics. The recovery rate of the two groups of asphalt mastics decreased with the content of SMS, which was opposite to the creep compliance, but the reason was the same. The recovery rate of HEA100 was higher than that of SBS00. After aging, the $R_{3.2}$ index of HEA50 and HEA100 were 30.7% and 29.4% higher than that of SBS00 at 70 °C, respectively. Comparing Figure 11a–d, it is found that when the stress level increases from 0.1 to 3.2 kPa, the recovery rate decreases, indicating that the asphalt material can only exert a good elastic recovery ability within a certain external load [58]. So, this may be one of the reasons for limiting overloading. After aging, the recovery rate of all types of asphalt mastic increased significantly, because aging resulted in the transformation of light components into resins or asphaltenes. The heavy components have strong intermolecular forces, strong resistance to deformation, and good recovery performance [36].

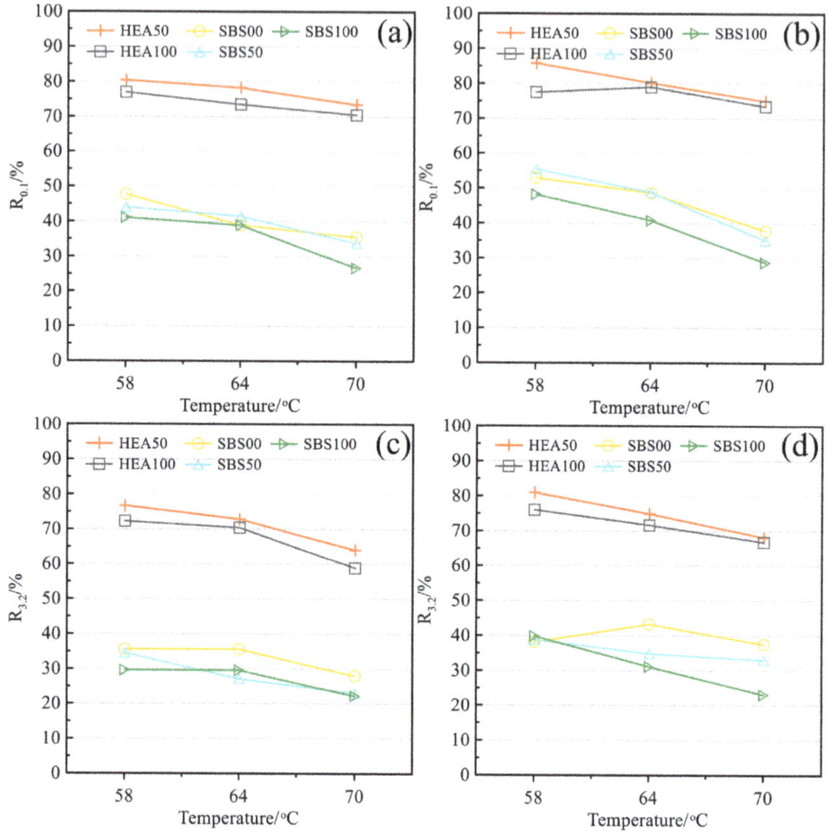

Figure 11. Recovery rate: (**a**) 0.1 kPa before TFOT; (**b**) 0.1 kPa after TFOT; (**c**) 3.2 kPa before TFOT; (**d**) 3.2 kPa after TFOT.

3.5. Analysis of BBR Test Results

3.5.1. Creep Stiffness and Creep Rate

Creep stiffness can reflect the creep resistance of asphalt mastic at low temperature. The creep stiffness of different salt storage asphalt mastics is shown in Figure 11a. The creep stiffness increases with the decrease in temperature, which is due to the hardening and brittleness of the asphalt mastic [57]. For example, the creep stiffness of HEA100 increased by 181.4% after the temperature was lowered from −12 to −18 °C. After the temperature was decreased from −18 to −24 °C, the creep stiffness increased by 99.2%. Comparing the creep stiffness of the HEA group and the SBS group, the creep stiffness of the HEA group is smaller than that of the SBS group. This shows that TPS not only increases the high-temperature performance of asphalt mastic but also improves the low-temperature performance [14,17,50]. The improvement in low-temperature performance may be due to the reinforcement and toughening effect of TPS in asphalt, which is consistent with the change in ductility. In addition, the creep stiffness was also affected by the SMS and the creep stiffness increased with the SMS content. As mentioned earlier, SMS increases the free asphalt content in the bituminous mastic. Generally speaking, as the free asphalt content increases, the asphalt mastic has a stronger low-temperature deformation ability and lower creep stiffness [59]. In this case, however, the creep stiffness increases with the SMS content. This may be because the SMS is distributed in the asphalt in a point-like manner, and with the increase in SMS, the interfacial phase increases, which causes the asphalt to become hard and brittle [40,60]. HEA decreased creep stiffness and SMS increased creep stiffness. Overall, the effect of HEA is more significant, which may be the reason why the creep compliance of HEA100 is lower than that of SBS00.

The creep rate reflects the ability of asphalt mastic to generate creep. The larger the value, the better the low-temperature deformation ability of asphalt mastic, and the less likely brittle fracture occurs [44,49,61]. The creep rates of different asphalt mastics are shown in Figure 12b. The creep rate of asphalt mastic increases with temperature, which is opposite to the trend of creep stiffness [57]. The m value of the HEA group was higher than that of the SBS group, indicating that the low-temperature performance of the HEA group was excellent. When the temperature is −12 °C, the effect of SMS content on creep rate is opposite to the *S* value. However, when the temperature is −18 °C and −24 °C, the effect of SMS on m is not obvious.

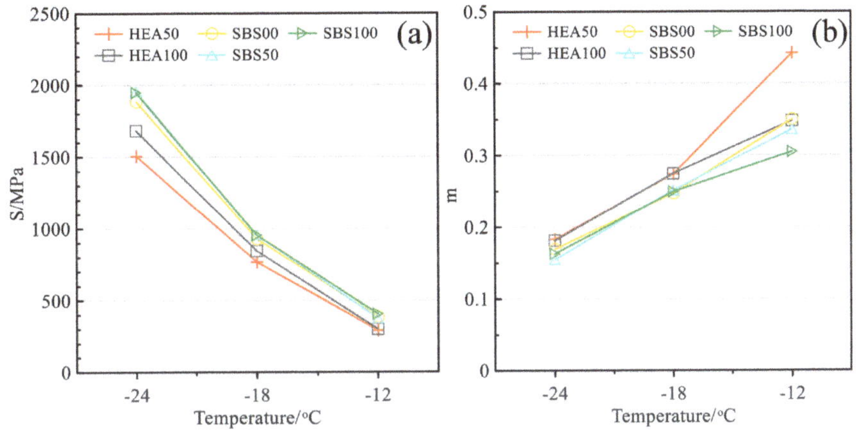

Figure 12. Low-temperature creep of asphalt mastic: (**a**) Creep stiffness. (**b**) Creep rate.

3.5.2. m/S

The S value is sensitive to the SMS, while the m value changes significantly with the SMS content only at −12 °C. Other studies have pointed out that there is a contradiction between the S value and the m value [36,47,62]. Using m/S can more accurately characterize the performance of salt storage asphalt mastic at low temperature [36,63]. The larger the m/S value, the better the flexibility of the asphalt mastic, which is beneficial to the low-temperature performance. The m/S of different salt storage asphalt mastics is shown in Figure 13. With the increase in the SMS content, the m/S value decreased, indicating that SMS reduced the low-temperature performance of asphalt mastic. This is consistent with the reduction in the ductility of asphalt mastic by SMS. As the temperature decreases, the m/S value decreases significantly, and the brittleness of asphalt mastic gradually manifests. The m/S value of HEA100 is also higher than that of SBS00, so although SMS has a negative effect on low-temperature performance, this negative effect can be completely covered by HEA.

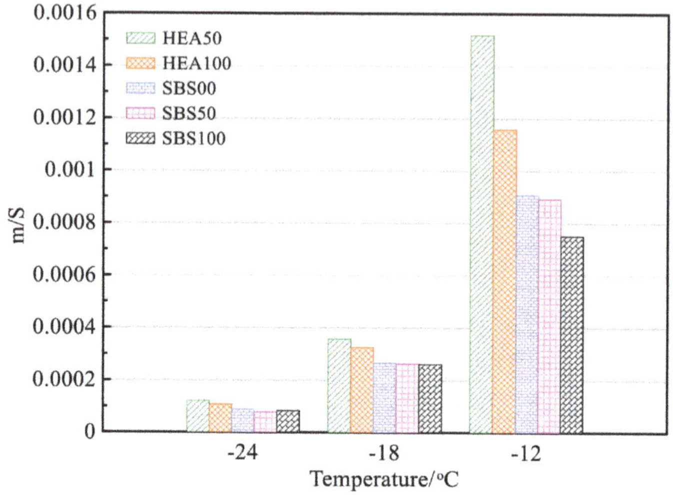

Figure 13. m/S of asphalt mastic.

4. Conclusions

The addition of slow-release SMS affects the performance of the salt storage asphalt mixture. However, the slow-release SMS mainly exists in the form of asphalt mastic in the asphalt mixture. Studying the effect of SMS on the performance of asphalt mastic has potential significance for improving the mechanical properties of a salt-storing asphalt mixture. For this reason, different types of salt storage asphalt mastic were prepared using SBS-modified asphalt and HEA as the asphalt matrix. The effects of asphalt matrix and SMS on the properties of asphalt mastic were analyzed through basic physical property tests and rheological property tests. The main conclusions are as follows:

(1) The routine test of asphalt mastic points out that the performance of HEA asphalt mastic is generally better than that of SBS. HEA asphalt mastic has lower penetration, higher softening point, ductility, and viscosity. With the increase in SMS content, the penetration of asphalt mastic increases, and the softening point and ductility decrease. The higher the SMS content, the lower the viscosity of the asphalt mastic.

(2) The changes in the G', G^*, rutting factor, and Z show that the content of SMS in asphalt will affect the stiffness and deformation properties of asphalt mastic. The higher the replacement rate of SMS, the less conducive to the elastic recovery performance of asphalt mastic. The phase angle is affected by SMS, but not significantly. The high-temperature performance of the asphalt mastic of the HEA group is better than that of the SBS group.

(3) MSCR results show that the addition of SMS increases the deformation of asphalt mastic under load. Under each stress condition, the order of the various asphalt mastic strains is SBS100 > SBS50 > SBS00 > HEA100 > HEA50. The high elasticity of HEA can well reduce the creep of salt storage asphalt mastic. The gain effect of TPS on creep recovery can compensate for the negative effect of SMS.

(4) The J_{nr} of the salt storage asphalt mastic increased with the content of SMS, while the R was the opposite. Although SMS hinders the high-temperature performance of asphalt mastic, HEA100 has better high-temperature performance than SBS00 with the help of high elastic agent TPS. The addition of TPS can make up for the adverse effect of SMS on the high-temperature performance of asphalt mastic.

(5) The creep stiffness increases with the content of SMS, and SMS reduces the creep rate and m/S of asphalt mastic. The larger the content of SMS, the worse the low-temperature crack resistance of the mastic. TPS can reduce the weakening effect of SMS on the low-temperature toughness of asphalt mastic.

Author Contributions: Conceptualization, Y.C.; methodology, Y.C.; validation, B.S.; formal analysis, Y.C.; investigation, Y.C., J.L. and X.L.; resources, B.S.; data curation, B.S.; writing—original draft preparation, Y.C.; writing—review and editing, Y.C., J.L. and F.Z.; visualization, F.Z., J.L. and X.L.; supervision, Z.L.; project administration, Z.L.; funding acquisition, Z.L. All authors have read and agreed to the published version of the manuscript.

Funding: This research was funded by the Fundamental Research Funds for the Central University, CHD (300102212701), and the Scientific Innovation Practice Project of Postgraduates of Chang'an University (300103722016).

Data Availability Statement: All data used during the study appear in the published article.

Conflicts of Interest: The authors declare no conflict of interest.

References

1. Sha, A.; Jiang, W.; Shan, J.; Wu, W.; Li, Y.; Zhang, S. Pavement structure and materials design for sea-crossing bridges and tunnel: Case study of the Hong Kong-Zhuhai-Macau Bridge. *J. Road Eng.* **2022**, *2*, 99–113. [CrossRef]
2. Pan, P.; Wu, S.; Xiao, Y.; Liu, G. A review on hydronic asphalt pavement for energy harvesting and snow melting. *Renew. Sust. Energ. Rev.* **2015**, *48*, 624–634. [CrossRef]
3. Luo, S.; Yang, X. Performance evaluation of high-elastic asphalt mixture containing deicing agent Mafilon. *Constr. Build. Mater.* **2015**, *94*, 494–501. [CrossRef]
4. Arabzadeh, A.; Ceylan, H.; Kim, S.; Gopalakrishnan, K.; Sassani, A. Superhydrophobic coatings on asphalt concrete surfaces: Toward smart solutions for winter pavement maintenance. *Transport. Res. Rec.* **2016**, *2551*, 10–17. [CrossRef]
5. Zhang, F.; Cao, Y.; Sha, A.; Lou, B.; Song, R.; Hu, X. Characterization of asphalt mixture using X-ray computed tomography scan technique after freeze-thaw cycle and microwave heating. *Constr. Build. Mater.* **2022**, *346*, 128435. [CrossRef]
6. Chen, Q.; Wang, C.; Yu, S.; Song, Z.; Fu, H.; An, T. Low-temperature mechanical properties of polyurethane-modified waterborne epoxy resin for pavement coating. *Int. J. Pavement Eng.* **2022**, 1–13. [CrossRef]
7. Muthumani, A.; Fay, L.; Akin, M. Correlating lab and field tests for evaluation of deicing and anti-icing chemicals: A review of potential approaches. *Cold Reg. Sci. Technol.* **2014**, *97*, 21–32. [CrossRef]
8. Tan, Y.; Zhang, C.; Xu, H.; Tian, D. Snow melting and deicing characteristics and pavement performance of active deicing and snow melting pavement. *China J. Highw. Transp.* **2019**, *32*, 1–17.
9. Zhu, X.; Zhang, Q.; Du, Z.; Wu, H.; Sun, Y. Snow-melting pavement design strategy with electric cable heating system balancing snow melting, energy conservation, and mechanical performance. *Resour. Conserv. Recycl.* **2022**, *177*, 105970. [CrossRef]
10. Joseph, W.; Daniels, E.; Mark, K. Hydronic heated pavement system performance using a solar water heating system with heat pipe evacuated tube solar collectors. *Sol. Energy* **2019**, *179*, 343–351.
11. Farcas, C.; Galao, O.; Navarro, R.; Zornoza, E.; Baeza, F.; Del, M.; Pla, R.; Garcés, P. Heating and de-icing function in conductive concrete and cement paste with the hybrid addition of carbon nanotubes and graphite products. *Smart Mater. Struct.* **2021**, *30*, 45010. [CrossRef]
12. Zhang, F.; Cao, Y.; Sha, A.; Wang, W.; Song, R.; Lou, B. Mechanism, rheology and self-healing properties of carbon nanotube modified asphalt. *Constr. Build. Mater.* **2022**, *346*, 128431. [CrossRef]
13. Gao, J.; Guo, H.; Wang, X.; Wang, P.; Wei, Y.; Wang, Z.; Huang, Y.; Yang, B. Microwave deicing for asphalt mixture containing steel wool fibers. *J. Clean. Prod.* **2019**, *206*, 1110–1122. [CrossRef]
14. Zhong, K.; Sun, M.; Chang, R. Performance evaluation of high-elastic/salt-storage asphalt mixture modified with Mafilon and rubber particles. *Constr. Build. Mater.* **2018**, *193*, 153–161. [CrossRef]
15. Varanasi, K.; Deng, T.; Smith, J. Frost formation and ice adhesion on superhydrophobic surfaces. *Appl. Phys. Lett.* **2010**, *97*, 92–102. [CrossRef]
16. Ryuji, A.; Jun, T.; Naruhiko, H. A study of winter road surface measures at airports in cold, snowy regions. *Int. J. Pavement Eng.* **2008**, *13*, 1–8.
17. Zhang, Y.; Liu, Z.; Shi, X. Development and use of salt-storage additives in asphalt pavement for anti-icing: Literature review. *J. Transp. Eng. B Pavements* **2021**, *147*, 3121002. [CrossRef]
18. Cao, Y.; Li, J.; Sha, A.; Liu, Z.; Zhang, F.; Li, X. A power-intensive piezoelectric energy harvester with efficient load utilization for road energy collection: Design, testing, and application. *J. Clean. Prod.* **2022**, *369*, 133287. [CrossRef]
19. Zhang, Y.; Shi, X. Laboratory evaluation of a sustainable additive for anti-icing asphalt. *Cold Reg. Sci. Technol.* **2021**, *189*, 103338. [CrossRef]
20. Liu, Z.; Sha, A.; Jiang, W. Advances in asphalt pavements containing salts: Additives, mixtures, performances, and evaluation. *China J. Highw. Transp.* **2019**, *32*, 18–31+72.
21. Tan, Y.; Sun, R.; Guo, M.; Zhong, Y.; Zhou, S. Research on deicing performance of asphalt mixture containing salt. *China J. Highw. Transp.* **2013**, *26*, 23–29.
22. Tan, Y.; Hou, M.; Shan, L.; Sun, R. Development of sustained release complex salt filler for asphalt pavement included salt. *J. Build. Mater.* **2014**, *17*, 256–260.

23. Yu, W.; Zhang, X.; Zhong, K. De-icing performance of asphalt mixture with snowmelt agent and high-elastic modified asphalt binder. *J. China Univ. Min. Technol.* **2015**, *44*, 912–916.
24. Guo, P.; Feng, Y.; Meng, X.; Meng, J.; Pan, W.; Gao, Y.; Liu, Y. Microscopic analysis of salt-storaged snow de-icing agent and its effect on water stability of mixture. *Mater. Rep.* **2020**, *34*, 6062–6065.
25. Wu, X.; Zhang, Z.; Zhu, J.; Li, Z. Selection and optimal mix ratio of carrier in anti-freeze materials. *J. Build. Mater.* **2022**, *25*, 278–284+293.
26. Dai, J.; Ma, F.; Fu, Z.; Li, C.; Jia, M.; Shi, K.; Wen, Y.; Wang, W. Applicability assessment of stearic acid/palmitic acid binary eutectic phase change material in cooling pavement. *Renew. Energ.* **2021**, *175*, 748–759. [CrossRef]
27. Liao, M.; Airey, G.; Chen, J. Mechanical properties of filler-asphalt mastics. *Int. J. Pavement Res. Technol.* **2013**, *6*, 576–581.
28. Moon, K.; Falchetto, A.; Park, J. Development of high performance asphalt mastic using fine taconite filler. *KSCE J. Civ. Eng.* **2014**, *18*, 1679–1687. [CrossRef]
29. Peng, C.; Yu, J.; Zhao, Z. Effects of a sodium chloride deicing additive on the rheological properties of asphalt mastic. *Road Mater. Pavement* **2015**, *17*, 382–395. [CrossRef]
30. Dou, H. Design and Pavement Performance of Ultra-Thin Snowmelt Salt Asphalt Mixture Pavement Overlay. Master's Thesis, Chang'an University, Xi'an, China, 2015.
31. *JTG E20-2011*; Standard Test Methods of Bitumen and Bituminous Mixtures for Highway Engineering. China Communications Press Co., Ltd.: Beijing, China, 2011.
32. Xing, H. Performance Research on Asphalt and Mixturemodified with SEBS. Master's Thesis, Chang'an University, Xi'an, China, 2018.
33. Cao, Y.; Sha, A.; Liu, Z.; Li, J.; Jiang, W. Energy output of piezoelectric transducers and pavements under simulated traffic load. *J. Clean. Prod.* **2021**, *279*, 123508. [CrossRef]
34. Lyu, D.; Liu, F.; Tian, H.; Wang, H.; Deng, J.; Li, M. Evaluating rheological properties of high-viscosity modified asphalt binder. *J. Mater. Sci. Eng.* **2021**, *39*, 820–825.
35. Tan, H.; Hu, S.; Liu, B.; Qin, R.; Tong, S.; Ren, S. Study on viscoelastic properties of rubber modified asphalt based on rheology. *China Civ. Eng. J.* **2017**, *50*, 115–122.
36. Yuan, D.; Jiang, W.; Xiao, J.; Lu, H.; Wu, W. Thermal oxygen aging effects on viscoelastic properties of high viscosity modified asphalt. *J. Chang'an Univ. Nat. Sci. Ed.* **2020**, *40*, 1–11.
37. Jiang, W.; Yuan, D.; Shan, J.; Ye, W.; Lu, H.; Sha, A. Experimental study of the performance of porous ultra-thin asphalt overlay. *Int. J. Pavement Eng.* **2022**, *23*, 2049–2061. [CrossRef]
38. Zhang, Z.; Luo, Y.; Zhao, F. Review of research on the effect of salt storage deicing material on the preformance of asphalt mixture. *Chem. Ind. Eng. Pro.* **2018**, *37*, 2282–2294.
39. Liu, Z.; Chen, S.; He, R.; Xing, M.; Bai, Y.; Dou, H. Investigation on the properties of asphalt mixtures containing antifreeze fillers. *J. Mater. Civil. Eng.* **2015**, *27*, 4014180. [CrossRef]
40. Tang, L.; Shi, J. Research on road performance and anti-icing effect of ice-resistant asphalt concrete. *J. China For. Highw.* **2013**, *33*, 307–309.
41. Meng, X.; Meng, J.; Pan, W.; Li, Z.; Gao, Y. Study on road performance and ice melting characteristics of salt preserving asphalt mixture. *N. Chem. Mater.* **2020**, *48*, 266–268+272.
42. Jiang, W.; Xiao, J.; Yuan, D.; Lu, H.; Xu, S.; Huang, Y. Design and experiment of thermoelectric asphalt pavements with power-generation and temperature-reduction functions. *Energ. Buildings.* **2018**, *169*, 39–47. [CrossRef]
43. Xing, M.; Wang, G.; Xia, H.; Li, Z.; Cui, Y.; Chen, H.; Wen, Y. Study on performance of environmental friendly compound snow-melting salted asphalt mortar. *Appl. Chem. Ind.* **2020**, *49*, 615–619+623.
44. Shi, K.; Fu, Z.; Song, R.; Liu, F.; Ma, F.; Dai, J. Waste chicken fat oil as a biomass regenerator to restore the performance of aged asphalt: Rheological properties and regeneration mechanism. *Road Mater. Pavement* **2021**, 1–25. [CrossRef]
45. Feng, D.; Yi, J.; Wang, D.; Chen, L. Impact of salt and freeze–thaw cycles on performance of asphalt mixtures in coastal frozen region of China. *Cold Reg. Sci. Technol.* **2010**, *62*, 34–41. [CrossRef]
46. Xu, O.; Xiang, S.; Yang, X.; Liu, Y. Estimation of the surface free energy and moisture susceptibility of asphalt mastic and aggregate system containing salt storage additive. *Constr. Build. Mater.* **2022**, *318*, 125814. [CrossRef]
47. Song, R.; Sha, A.; Shi, K.; Li, J.; Li, X.; Zhang, F. Polyphosphoric acid and plasticizer modified asphalt: Rheological properties and modification mechanism. *Constr. Build. Mater.* **2021**, *309*, 125158. [CrossRef]
48. Zhang, R.; You, Z.; Wang, H.; Ye, M.; Yap, Y.; Si, C. The impact of bio-oil as rejuvenator for aged asphalt binder. *Constr. Build. Mater.* **2019**, *196*, 134–143. [CrossRef]
49. Yuan, D.; Jiang, W.; Xiao, J.; Tong, Z.; Jia, M.; Shan, J.; Ogbon, A. Assessment of the aging process of finished product–modified asphalt binder and its aging mechanism. *J. Mater. Civil Eng.* **2022**, *34*, 4022174. [CrossRef]
50. Zhang, X.; Chen, H.; Barbieri, D.; Inge, H. Laboratory evaluation of mechanical properties of asphalt mixtures exposed to sodium chloride. *Transport. Res. Rec.* **2022**, *2676*, 90–98. [CrossRef]
51. Naderi, K.; Asgharzadeh, S.; Tabatabaee, N.; Partl, M. Evaluating aging properties of crumb rubber and styrene–butadiene–styrene modified binders: Using double logistic master curve model. *Transport. Res. Rec.* **2014**, *2444*, 110–119. [CrossRef]
52. Tang, N.; Lyu, Q.; Huang, W.; Lin, P.; Yan, C. Chemical and rheological evaluation of aging characteristics of terminal blend rubberized asphalt binder. *Constr. Build. Mater.* **2019**, *205*, 87–96. [CrossRef]

53. Wang, L.; Chen, G.; Xing, Y.; Hu, J.; Leng, B. Effect of aging on rheological properties of crumb rubber and sbs modified asphalt. *J. Build. Mater.* **2015**, *18*, 499–504.
54. Jia, M.; Sha, A.; Zhang, Z.; Li, J.; Yuan, D.; Jiang, W. Effect of organic reagents on high temperature rheological characteristics of organic rectorite modified asphalt. *Constr. Build. Mater.* **2019**, *227*, 116624. [CrossRef]
55. Wang, L.; Cui, S.; Chang, C. High temperature performance of warm mix crumb rubber modified asphalt based on rheological and viscoelastic theory. *Mater. Rep.* **2019**, *33*, 2386–2391.
56. Zhang, X.; Chen, H.; Inge, H. The mutual effect and reaction mechanism of bitumen and de-icing salt solution. *Constr. Build. Mater.* **2021**, *302*, 124213. [CrossRef]
57. Yuan, D.; Jiang, W.; Xiao, J.; Zhou, B.; Jia, M.; Wang, W. Comparison of rheological properties between SBS, rubber and high-viscosity modified asphalt binders. *J. Chang'an Univ. Nat. Sci. Ed.* **2020**, *40*, 135–142.
58. Yuan, D.; Xing, C.; Jiang, W.; Xiao, J.; Wu, W.; Li, P.; Li, Y. Viscoelastic behavior and phase structure of high-content sbs-modified asphalt. *Polymers* **2022**, *14*, 2476. [CrossRef]
59. Rebecca, S.; Hussain, U. Field evaluation of asphalt additives to control rutting and cracking. *Transport. Res. Rec.* **2003**, *1829*, 47–54.
60. Xia, H.; Zhao, X.; Wu, Y.; Yuan, T.; Song, L.; Yan, M.; Wang, F.; Chen, H. Preparation and performance of antifreeze adhesive materials for asphalt pavement. *Constr. Build. Mater.* **2020**, *258*, 119554. [CrossRef]
61. Jia, M.; Zhang, Z.; Liu, H.; Peng, B.; Zhang, H.; Lyu, W.; Zhang, Q.; Mao, Z. The synergistic effect of organic montmorillonite and thermoplastic polyurethane on properties of asphalt binder. *Constr. Build. Mater.* **2019**, *229*, 116867. [CrossRef]
62. Yu, J.; Zhao, Q.; Ye, F.; Song, Q. Low Temperature Rheological Performance Analysis of Rubber Modified Asphalt under Heat Aging Process. *J. Southwest Jiaotong Univ.* **2021**, *56*, 108–115.
63. Liu, S.; Cao, W.; Shang, S.; Qi, H.; Fang, J. Analysis and application of relationships between low-temperature rheological performance parameters of asphalt binders. *Constr. Build. Mater.* **2009**, *24*, 471–478. [CrossRef]

polymers

MDPI

Article

Evaluation of the Rheological Properties of Virgin and Aged Asphalt Blends

Tao Liu [1], Weidang Duan [1], Jialin Zhang [1], Qiuping Li [1], Jian Xu [2], Jie Wang [2,*], Yongchun Qin [2] and Rong Chang [2,*]

[1] Jiangxi Provincial Communications Investment Group Co., Ltd., Nanchang 330025, China
[2] Research Institute of Highway, Ministry of Transport, Beijing 100088, China
[*] Correspondence: j.wang@rioh.cn (J.W.); r.chang@rioh.cn (R.C.)

Abstract: To evaluate the effects of the source and admixture of aged asphalt on the rheological properties of reclaimed asphalt binders, the relative viscosity ($\Delta\eta$), relative rutting factor ($\Delta G^*/\sin\delta$), and relative fatigue factor ($\Delta G^*\sin\delta$) were selected as evaluation indicators based on the Strategic Highway Research Program (SHRP) tests to characterize the rheological properties of a reclaimed asphalt binder under medium- and high-temperature conditions. The results of the study showed that the viscosity, rutting factor, and fatigue factor of the reclaimed asphalt binder increased with the addition of aged asphalt; however, the effect of the source and admixture of aged asphalt could not be assessed. The relative viscosity, relative rutting factor, and relative fatigue factor are sensitive to the source, admixture, temperature, and aging conditions, which shows the superiority of these indicators. Moreover, the relative viscosity and relative rutting factor decreased linearly with increasing temperature under high-temperature conditions, while the relative fatigue factor increased linearly with increasing temperature under medium-temperature conditions. In addition, the linear trends of the three indicators were independent of the source and admixture of aged asphalt. These results indicate that the evaluation method used in this study can be used to assess the effects of virgin asphalt and aged asphalt on the rheological properties of reclaimed asphalt binders, and has the potential for application. The viscosity of recycled asphalt increases, and the rutting factor and fatigue factor both increase. The high-temperature stability of reclaimed asphalt is improved, and the fatigue crack resistance is weakened.

Keywords: road engineering; reclaimed asphalt binder; rheological properties; SHRP test; variance analysis

Citation: Liu, T.; Duan, W.; Zhang, J.; Li, Q.; Xu, J.; Wang, J.; Qin, Y.; Chang, R. Evaluation of the Rheological Properties of Virgin and Aged Asphalt Blends. *Polymers* **2022**, *14*, 3623. https://doi.org/10.3390/polym14173623

Academic Editors: Wei Jiang, Quantao Liu, Jose Norambuena-Contreras and Yue Huang

Received: 3 August 2022
Accepted: 31 August 2022
Published: 1 September 2022

1. Introduction

Reclaimed asphalt pavement (RAP) contains large amounts of aggregate and asphalt, which are potentially usable resources [1–4]. The recycling of RAP contributes to reductions in rock mining and aggregate production, with significant economic and environmental benefits [5,6]. Studies [2,7] have shown that the application of reclaimed asphalt mixtures in pavement subgrade construction can reduce greenhouse gas emissions by 20%, energy consumption by 16%, hazardous waste by 11% (RAP may leach toxic substances such as polycyclic aromatic hydrocarbons (PAHs) in the presence of rainwater in long-term stockpiles), and whole-life costs by 21%.

Asphalt is a viscoelastic material with excellent rheological properties [8,9]. After the blending of virgin and aged asphalt, the reclaimed asphalt binder becomes viscous and produces a large variation in rheology compared to the virgin asphalt [10–12]. RAP comes from a wide range of sources and has a very complex composition. Differences in the asphalt, aggregates, oil-to-rock ratio, gradation, and even admixtures may lead to significant differences in the rheology of the reclaimed asphalt binder or in the road performance of the recycled mix [11,13]. Numerous studies and engineering experiences have shown that pavements paved with poorly rheological asphalt are prone to high-temperature rutting, fatigue cracking, and other diseases [14,15]. In addition, the differences in the rheology of reclaimed asphalt binder will also lead to differences in the construction temperature of the mixture, which in turn will cause differences in the fuel required in the plant mix and greenhouse gas emissions [13,16].

Therefore, to optimize the design of reclaimed asphalt mixtures, reduce the reclaimed pavement issues, and prolong the service life, the rheological properties of reclaimed asphalt binders need to be evaluated effectively. At present, there are few studies on this subject, and the effects of different sources and admixtures of aged asphalt on the rheology of reclaimed asphalt binders under different temperatures and aging conditions have not been fully considered [10,17,18]. This study selects the Strategic Highway Research Program (SHRP) asphalt test method to quantitatively evaluate the effects of the source and admixture of aged asphalt on the high-temperature rheology of reclaimed asphalt binders using the relative viscosity, relative rutting factor, and relative fatigue factor.

2. Materials and Method

2.1. Materials

This study uses the Abson method (ASTM D 1856) to reclaim asphalt from asphalt mixtures. The asphalt mixture was extracted with a distillation device, and then the solvent in the extraction liquid was removed. The recovered asphalt samples were denoted by A, B, and C. The performance indexes are shown in Table 1. Asphalt with PG 90 was used for the virgin asphalt and denoted by N. The performance indicators of the virgin asphalt are shown in Table 2.

Table 1. The technical indicators of aged asphalt.

Category	Penetration (25 °C/0.1 mm)	Softening Point (°C)	Ductility (15 °C/cm)	Viscosity (135 °C/Pa·s)
Test method	T0604	T0606	T0605	T0625
A	27	73	3.2	2.99
B	31	71	4.6	2.67
C	35	67	6.5	2.23

Table 2. The technical indicators of virgin asphalt.

Indicators		Test Results	Specification
Penetration (25 °C/0.1 mm)		87	80~100
Softening point (°C)		46.5	≥44
Ductility (15 °C/cm)		>100	≥100
Viscosity (135 °C/Pa·s)		0.36	-
Residue after TFOT	Mass loss/%	−0.48	≤±0.8
	penetration ratio (25 °C/%)	63.5	≥57
	Ductility (10 °C/cm)	12.9	≥8

The aged and virgin asphalts were mixed uniformly at 135 °C, and the admixtures of the aged asphalts were 15% and 30%, respectively. It is worth mentioning that the admixture refers to the ratio of the mass of aged asphalt to the total mass of reclaimed asphalt binder. The thin-film oven test (TFOT) and pressure aging vessel (PAV) were used to simulate short-term aging and long-term aging, respectively. For the sake of simplicity, the following abbreviations are used in this study: W for unaged, D for short-term aging, and C for long-term aging. For example, A15W represents virgin asphalt mixed with 15% of aged asphalt A, which was not aged; B30D represents virgin asphalt mixed with 30% of aged asphalt B, which underwent short-term aging.

2.2. Asphalt Viscosity Evaluation Test

The Brookfield rotational viscosity test (T0625) was used to investigate the effects of the source and admixture of aged asphalt on the high-temperature rheology of reclaimed asphalt binders. The viscosities of virgin asphalt and reclaimed asphalt binders were tested under unaged and short-term aging conditions, respectively. Considering the temperature range of 130–160 °C for asphalt production and application, the test temperatures were set at 115 °C, 125 °C, 135 °C, 145 °C, 155 °C, and 165 °C.

2.3. Asphalt Viscoelasticity Evaluation Test

The rheological properties of virgin and reclaimed asphalt binders at high and medium temperatures were investigated using dynamic shear rheometer (DSR) tests (T0628). The test plate was a circular metal plate with a diameter of 25 mm. The test temperatures for the unaged and short-term aged asphalts were 52 °C, 58 °C, 64 °C, 70 °C, 76 °C, and 82 °C, and the test temperatures for long-term aged asphalt were 16 °C, 19 °C, 22 °C, 25 °C, 28 °C, and 31 °C. It should be noted that the test temperatures refer to the American Association of State Highway and Transportation Officials (AASHTO T315) regulations.

3. Results and Discussion

3.1. Asphalt Viscosity Evaluation Test Results and Analysis

The viscosities of the asphalts at each test temperature are listed in Table 3. Each asphalt sample was tested 3 times in parallel. The test results satisfy the allowable error of the repeatability test being 3.5% of the average value.

Table 3. Viscosity results of virgin and reclaimed asphalt binders.

Asphalt Types	Test Temperature (°C)					
	115	125	135	145	155	165
N_W, N_D	0.93, 1.46	0.66, 1.03	0.38, 0.68	0.29, 0.44	0.23, 0.32	0.19, 0.27
$A15_W$, $A30_W$	1.33, 1.87	0.94, 1.29	0.52, 0.72	0.39, 0.53	0.31, 0.42	0.25, 0.33
$B15_W$, $B30_W$	1.29, 1.72	0.89, 1.19	0.56, 0.68	0.38, 0.52	0.30, 0.39	0.24, 0.31
$C15_W$, $C30_W$	1.23, 1.61	0.85, 1.12	0.48, 0.65	0.36, 0.45	0.28, 0.34	0.23, 0.28
$A15_D$, $A30_D$	2.07, 2.83	1.42, 1.99	0.92, 1.23	0.59, 0.76	0.43, 0.55	0.35, 0.46
$B15_D$, $B30_D$	1.97, 2.61	1.38, 1.83	0.89, 1.14	0.56, 0.71	0.40, 0.51	0.33, 0.41
$C15_D$, $C30_D$	1.91, 2.49	1.31, 1.71	0.83, 1.08	0.54, 0.67	0.39, 0.46	0.32, 0.38

The viscosities of the virgin and reclaimed asphalt binders decreased gradually with increasing temperature, independent of whether they were aged or not. The viscosity of the reclaimed asphalt was significantly greater than that of the virgin asphalt, while the viscosity of the short-term aged asphalt was significantly greater than that of the virgin reclaimed asphalt.

To evaluate the influence of the source and admixture of the aged asphalt on the viscosity of the reclaimed asphalt, the relative viscosity Δ_η was used for the evaluation, which mainly characterizes the influence of the relative viscosity of the reclaimed asphalt on its viscosity for every 1% increase in the content of aged asphalt, calculated as Equation (1):

$$\Delta_\eta = \frac{\eta_{mix.r} - \eta_{new.r}}{x} \tag{1}$$

where Δ_η is the dimensionless viscosity of the reclaimed asphalt, $\eta_{mix.r}$ is the relative viscosity of the reclaimed asphalt, $\eta_{mix.r} = \eta_{mix}/\eta_{new}$, η_{mix} is the viscosity of the reclaimed asphalt (Pa·s), η_{new} is the viscosity of the virgin asphalt (Pa·s), $\eta_{new.r}$ is the relative viscosity of the virgin asphalt, $\eta_{new.r} = 1$, and x is the amount of aged asphalt blending.

The variance results for the Δ_η and viscosity η values of the reclaimed asphalt samples at different test temperatures and under different aging conditions are listed in Table 4. Usually, the significance level $\alpha = 0.05$.

Table 4. Analysis of variance results for reclaimed asphalt with Δ_η and η.

Category	Sources of Aged Asphalt	Admixture of Aged Asphalt	Test Temperature	Aging Conditions
Statistical probability *p*-value (Δ_η)	3.9×10^{-5}	0.004	3.3×10^{-7}	0.039
Statistical probability *p*-value (η)	0.851	0.100	2.9×10^{-4}	0.013

As can be seen from Table 4, the statistical probability *p*-value of η is less than 0.05 only for the test temperature and aging conditions, indicating that there is no significant difference in the effects of virgin asphalt and aged asphalt on the viscosity of reclaimed asphalt based on the η index. For Δ_η, the *p*-values for all four influencing factors are less than 0.05, indicating that assessing the viscoelasticity of reclaimed asphalt with Δ_η can identify the differences in these four factors. Therefore, the high-temperature rheology of the reclaimed asphalt is better assessed using Δ_η than the viscosity index.

The variation in relative viscosity Δ_η of the reclaimed asphalt versus temperature is shown in Figure 1. As can be seen from Figure 1, the Δ_η of the reclaimed asphalt gradually decreases with the increase in temperature, and after short-term aging the Δ_η also gradually decreases. When the temperature is 135 °C, the Δ_η values of $A15_W$, $B15_W$, and $C15_W$ are 2.46, 2.11, and 1.75, respectively, indicating that the relative viscosity increases by 2.46, 2.11, and 1.75 for each 1% increase in the content of aged asphalt under this test condition. However, under the same temperature conditions, the Δ_η of $B30_W$ is 2.63, which is not the same as that of $B15_W$, indicating that the viscosity of the reclaimed asphalt produces inconsistent changes under different aged asphalt admixtures, even if the aged asphalt is the same. Moreover, the different reclaimed asphalts at different test temperatures and different aging conditions produced similar test results as described above.

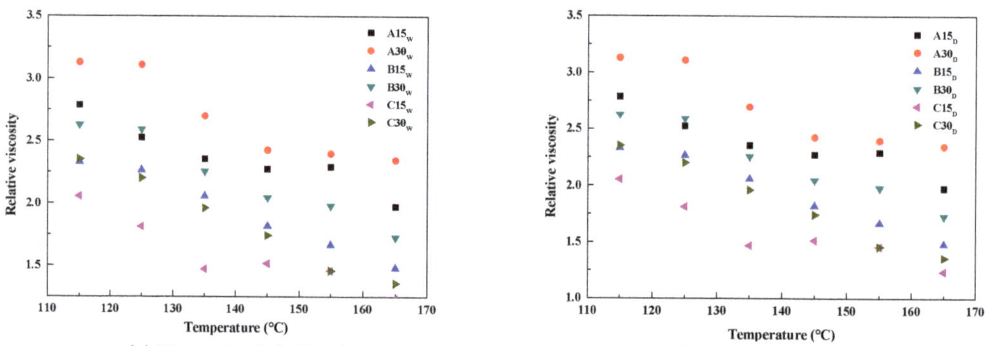

(**a**) Virgin Asphalt Blends (**b**) Aged Asphalt Blends

Figure 1. The results for the relative viscosity Δ_η versus temperature.

The regression analysis results show that the reclaimed asphalt Δ_η has a good linear relationship with the test temperature (T). The regression equations are all Δ_η = aT + b (a and b are the fitting parameters), as shown in Equations (2) and (3).

$$\Delta_\eta = \begin{cases} -0.0157T + 4.6776, R^2 = 0.92, A15_W \\ -0.0174T + 5.3464, R^2 = 0.97, A30_W \\ -0.0144T + 4.1637, R^2 = 0.92, B15_W \\ -0.0134T + 4.4118, R^2 = 0.87, B30_W \\ -0.0151T + 3.8304, R^2 = 0.97, C15_W \\ -0.02T + 4.8268, R^2 = 0.88, C30_W \end{cases} \quad (2)$$

$$\Delta_\eta = \begin{cases} -0.0138T + 4.2984, R^2 = 0.91, A15_D \\ -0.018T + 5.209, R^2 = 0.89, A30_D \\ -0.0179T + 4.4458, R^2 = 0.98, B15_D \\ -0.0186T + 4.8137, R^2 = 0.96, B30_D \\ -0.0146T + 3.6383, R^2 = 0.87, C15_D \\ -0.0212T + 4.8106, R^2 = 0.98, C30_D \end{cases} \quad (3)$$

From Equations (2) and (3), it can be seen that the effects of virgin asphalt and aged asphalt on the high-temperature rheology of the reclaimed asphalt can be characterized by the slope and intercept in the linear relationship equation. For example, the slope of $B30_W$ is −0.013, which is 23% and 33% smaller than for $A30_W$ and $C30_W$, respectively. In addition, the Δ_η of the reclaimed asphalt is always linearly related to the temperature, independent of the source, the admixture of the aged asphalt, or whether it undergoes short-term aging.

3.2. Asphalt Rutting Factor Test Results and Analysis

The DSR test results of the asphalts under different test conditions are shown in Figures 2–4.

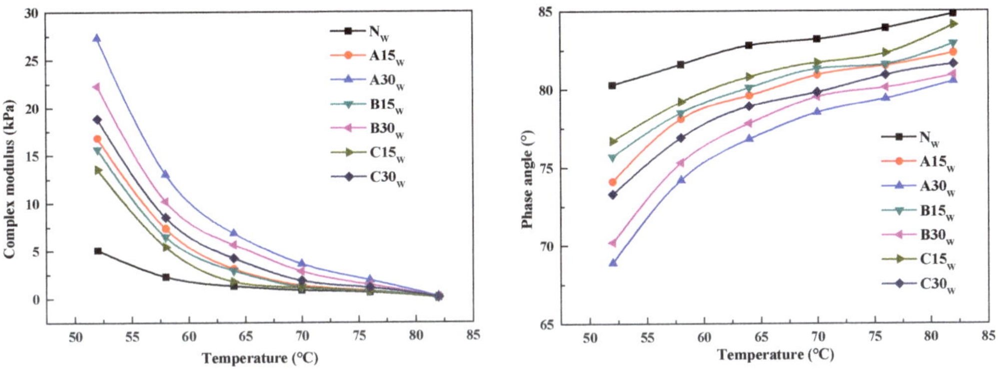

(**a**) G* of Virgin Asphalt (**b**) Phase angle of Virgin Asphalt

Figure 2. The results of the complex modulus and phase angle versus temperature for the unaged asphalt.

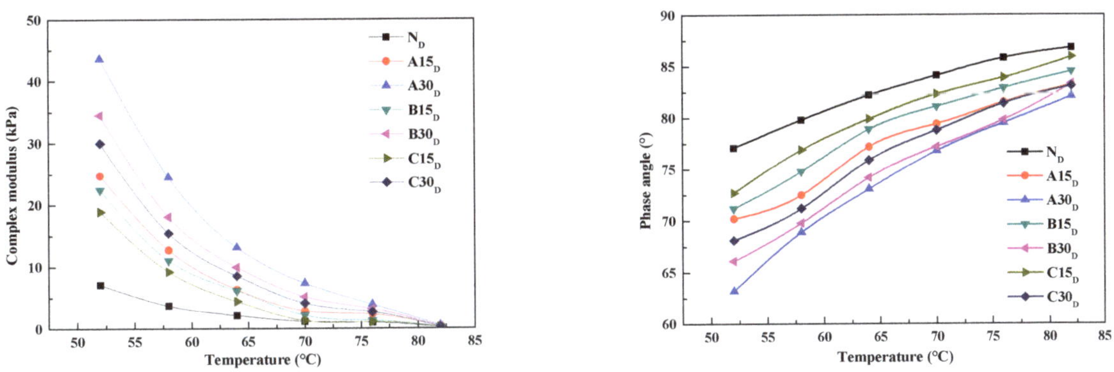

(**a**) G* of Asphalt (TFOT) (**b**) Phase angle of Asphalt(TFOT)

Figure 3. The results of the complex modulus and phase angle versus temperature for the short-term aged asphalt.

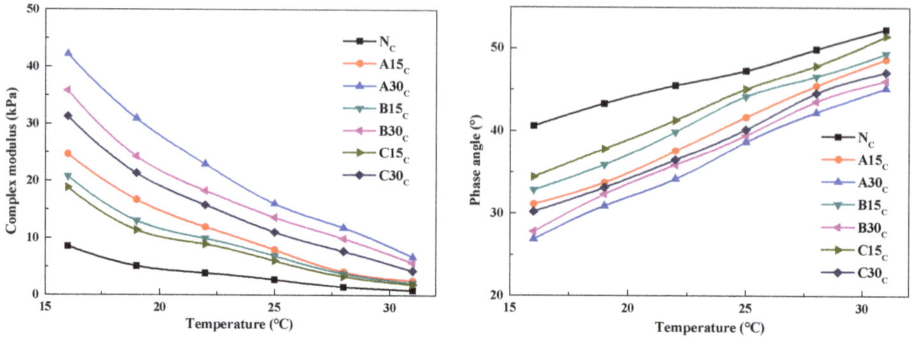

(**a**) G* of Asphalt (TFOT+PAV) (**b**) Phase angle of Asphalt(TFOT+PAV)

Figure 4. The results of the complex modulus and phase angle versus temperature for the long-term aged asphalt.

The complex modulus G* can describe the ability of the asphalt to resist deformation, and the δ can reflect the proportional relationship between the elastic and viscous parts of the asphalt. Generally speaking, the larger δ is, the more viscous the asphalt is. From Figures 2–4, it can be seen that as the temperature increases, G* decreases and δ increases, and the regularity is not related to the source of the aged asphalt, the admixture of the aged asphalt, or whether it has been aged. The G* of the reclaimed asphalt is greater than that of the virgin asphalt under each test temperature condition, and the δ of the reclaimed asphalt is less than that of the virgin asphalt, whereby the lower the temperature the more significant the result. In addition, the aged asphalt with the higher admixture content has a larger G* and smaller δ.

Asphalt under the long-term coupling effects of heat, oxygen, light, water, and load will experience serious aging, which will be manifested in the components as a decrease in aromatic content, an increase in asphalt content, and a macroscopic increase in hardness. From a viscoelastic point of view, the viscosity of asphalt decreases, the elasticity increases, and the asphalt changes from the sol–gel state to the gel state, which leads to a higher G* and lower δ.

Adding a different proportion of aged asphalt to the virgin asphalt can improve the high-temperature performance of recycled asphalt; that is, the ability of the asphalt to resist high-temperature deformation. This is manifested as an increase in G* and a decrease in δ. According to the changes of G* and δ, it is considered that adding virgin asphalt to the aged asphalt can restore the rheological properties of the aged asphalt mixture.

3.3. *Asphalt Rutting Factor Evaluation Test Results and Analysis*

The road rutting is the irrecoverable deformation of asphalt pavement under the coupling effect of load and high temperature, which can be evaluated by using the rutting factor (G*/sinδ). The variation curves of the G*/sinδ with temperature for the virgin and reclaimed asphalts are shown in Figure 5.

As shown in Figure 5, the G*/sinδ values of the virgin and reclaimed asphalts gradually decreased with the increase in temperature. The G*/sinδ of the reclaimed asphalt was larger than that of the virgin asphalt. The nonlinear regression analysis showed that the G*/sinδ had a good exponential relationship with the temperature, and the correlation coefficients were all above 0.90.

In order to evaluate the effects of the source and admixture of the aged asphalt on the rutting factor of the reclaimed asphalt, the dimensionless rutting factor $\Delta_{G^*/\sin\delta}$ was evaluated, and the calculation can be found in Equation (4):

$$\Delta_{G*/\sin\delta} = \frac{G*/\sin\delta_{mix.r} - G*/\sin\delta_{new.r}}{x} \quad (4)$$

where $\Delta_{G^*/\sin\delta}$ is the dimensionless rutting factor of the reclaimed asphalt, $G^*/\sin\delta_{mix.r}$ is the relative rutting factor of the reclaimed asphalt, $G^*/\sin\delta_{mix.r} = (G^*/\sin\delta_{mix})/(G^*/\sin\delta_{new})$, $G^*/\sin\delta_{mix}$ is the rutting factor of the reclaimed asphalt, $G^*/\sin\delta_{new}$ is the rutting factor of the virgin asphalt, $G^*/\sin\delta_{new.r}$ is the relative rutting factor of the virgin asphalt, $G^*/\sin\delta_{new.r} = 1$, and x is the amount of aged asphalt mixing.

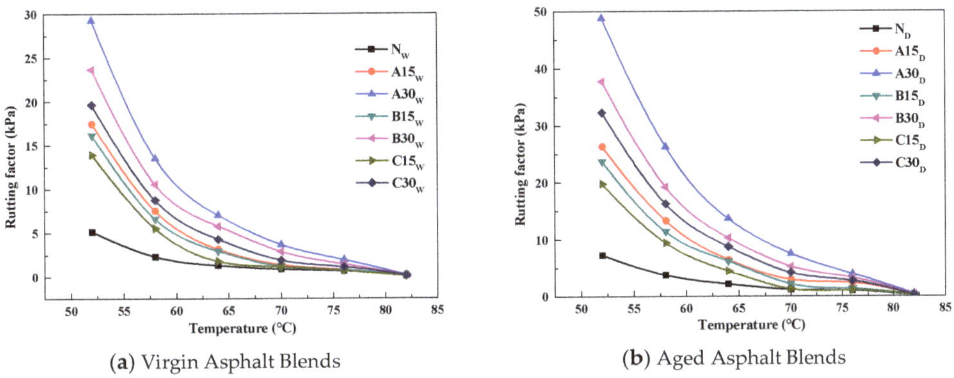

Figure 5. The results of the rutting factor versus temperature for unaged (**a**) and short-term aged (**b**) asphalts.

The variance results for $\Delta_{G^*/\sin\delta}$ and $G^*/\sin\delta$ for reclaimed asphalt at different test temperatures and under different aging conditions are shown in Table 5 with the significance level of $\alpha = 0.05$.

Table 5. Analysis of variance results for reclaimed asphalt $\Delta_{G^*/\sin\delta}$ and $G^*/\sin\delta$.

Category	Sources of Aged Asphalt	Admixture of Aged Asphalt	Test Temperature	Aging Conditions
Statistical probability p-value ($\Delta_{G^*/\sin\delta}$)	0.001	0.029	5.10×10^{-7}	0.031
Statistical probability p-value ($G^*/\sin\delta$)	0.533	0.056	1.35×10^{-5}	0.068

As can be seen from Table 5, the statistical probability p-value of $G^*/\sin\delta$ is only less than 0.05 under one test temperature, indicating that using $G^*/\sin\delta$ as an indicator to evaluate the high-temperature stability of the reclaimed asphalt under different aging conditions is unable to distinguish the difference between the virgin asphalt and aged asphalt. The four p-values of $\Delta_{G^*/\sin\delta}$ are less than 0.05, indicating that using $\Delta_{G^*/\sin\delta}$ as an indicator to evaluate the high-temperature performance of the reclaimed asphalt under different temperature and aging conditions can distinguish the differences between virgin asphalt and aged asphalt. Therefore, it is more reasonable to use $\Delta_{G^*/\sin\delta}$ as an indicator.

The variation in $\Delta_{G^*/\sin\delta}$ versus temperature for the reclaimed asphalt is shown in Figure 6. It can be found that the $\Delta_{G^*/\sin\delta}$ of the reclaimed asphalt decreases gradually with the increase in temperature. The $\Delta_{G^*/\sin\delta}$ values for $A30_W$, $B30_W$, and $C30_W$ at the test temperature of 58 °C were 16.0, 11.8, and 9.2, respectively, indicating that the relative rutting factors of the aged asphalt increased by 16.0, 11.8, and 9.2 for each 1% increase in the admixture of aged asphalt. Under the same temperature conditions, the relative rutting factor of $A15_W$ was 14.9, which was not the same as that of $A30_W$, indicating that every 1% increase in the admixture of aged asphalt produced inconsistent changes in the high-temperature performance of the reclaimed asphalt, even if the aged asphalt was from the same source. Different reclaimed asphalts at different test temperatures and different aging conditions will produce similar test results as above, which are similar to the relative viscosity test results.

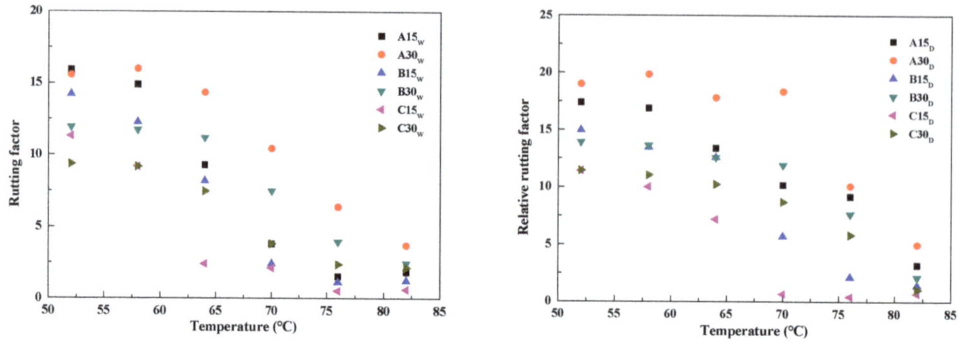

(**a**) Virgin Asphalt Blends (**b**) Aged Asphalt Blends

Figure 6. Variations of $\Delta_{G^*/\sin\delta}$ for reclaimed asphalt versus temperature.

The results of the regression analysis show that $\Delta_{G^*/\sin\delta}$ has a good linear relationship with the test temperature (T). The regression equations are all $\Delta_{G^*/\sin\delta}$ = aT + b (a and b are fitting parameters), as shown in Equations (5) and (6):

$$\Delta_{G*/\sin\delta} = \begin{cases} -0.5518T + 44.855, R^2 = 0.92, A15_W \\ -0.4383T + 40.44, R^2 = 0.92, A30_W \\ -0.4957T + 39.797, R^2 = 0.92, B15_W \\ -0.3548T + 31.908, R^2 = 0.91, B30_W \\ -0.3799T + 29.819, R^2 = 0.83, C15_W \\ -0.2861T + 24.891, R^2 = 0.92, C30_W \end{cases} \quad (5)$$

$$\Delta_{G*/\sin\delta} = \begin{cases} -0.4626T + 42.694, R^2 = 0.95, A15_D \\ -0.4702T + 46.508, R^2 = 0.76, A30_D \\ -0.5171T + 43.036, R^2 = 0.93, B15_D \\ -0.3696T + 35.064, R^2 = 0.81, B30_D \\ -0.4229T + 33.405, R^2 = 0.87, C15_D \\ -0.3281T + 30.029, R^2 = 0.86, C30_D \end{cases} \quad (6)$$

From Equations (5) and (6), it can be seen that the $\Delta_{G^*/\sin\delta}$ of the reclaimed asphalt is always linearly related to the test temperature, independent of the source and admixture of aged asphalt, and will not change after short-term aging.

3.4. *Asphalt Fatigue Factor Evaluation Test Results and Analysis*

An increase in rutting factor enhances the ability of the asphalt to resist permanent deformation under high-temperature conditions; however, a high rutting factor can lead to the asphalt being susceptible to cracking under low- and medium-temperature conditions. Therefore, the fatigue factor G*sinδ was introduced to characterize the ability of the asphalt to resist fatigue cracking under medium-temperature conditions after long-term aging. The variation curves of G*sinδ values with temperature for virgin and reclaimed asphalts after long-term aging are shown in Figure 7.

As can be seen from Figure 7, the G*sinδ gradually decreases as the temperature increases, and the G*sinδ of the reclaimed asphalt is larger than that of the virgin asphalt. The higher the amount of aged asphalt admixture, the larger the G*sinδ. The results of the nonlinear regression analysis showed that the G*sinδ was exponentially related to the temperature, and the correlation coefficients were all above 0.95.

Here, $\Delta_{G^*\sin\delta}$ was selected as the dimensionless fatigue factor indicator for the reclaimed asphalt, and the effect of the relative fatigue factor of the reclaimed asphalt on its resistance to fatigue cracking under medium-temperature conditions was evaluated.

The variance results for the reclaimed asphalt $\Delta_{G^*\sin\delta}$ and G*sinδ are shown in Table 6.

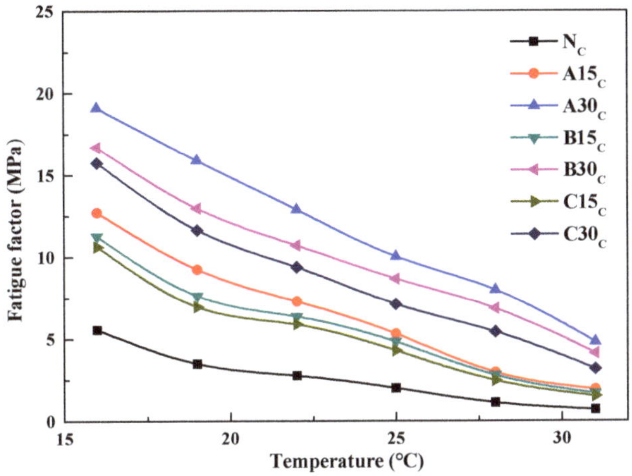

Figure 7. Variation of $\Delta_{G^{*}\sin\delta}$ for asphalt versus temperature.

Table 6. Analysis of variance results for reclaimed asphalt $\Delta_{G^{*}\sin\delta}$ and $G^{*}\sin\delta$.

Category	Sources of Aged Asphalt	Admixture of Aged Asphalt	Test Temperature
Statistical probability *p*-value ($\Delta_{G^{*}\sin\delta}$)	0.009	0.016	0.009
Statistical probability *p*-value ($G^{*}\sin\delta$)	0.524	0.003	2.87×10^{-3}

As can be seen from Table 6, the effect of the aged asphalt admixture on the fatigue resistance of reclaimed asphalt cannot be evaluated using $G^{*}\sin\delta$ as an indicator. The three *p*-values of $\Delta_{G^{*}\sin\delta}$ are less than 0.05, indicating that the effects of the source and admixture on the rheological properties of the reclaimed asphalt can be assessed using $\Delta_{G^{*}\sin\delta}$ as an indicator. Therefore, $\Delta_{G^{*}\sin\delta}$ is suitable for characterizing the mid-temperature rheology of aged asphalt. The variation in $\Delta_{G^{*}\sin\delta}$ of the reclaimed asphalt versus temperature is shown in Figure 8.

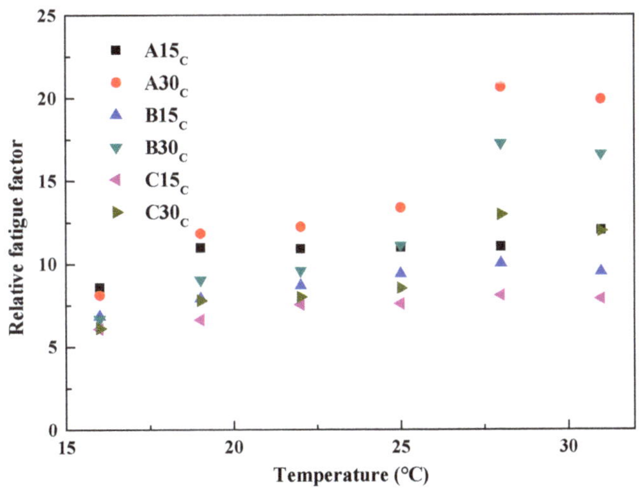

Figure 8. Variation of $\Delta_{G^{*}\sin\delta}$ for reclaimed asphalt versus temperature.

It can be seen that the $\Delta_{G^*\sin\delta}$ of the reclaimed asphalt increases gradually with the increase in temperature. Even if the source of the aged asphalt is the same, the fatigue resistance of the reclaimed asphalt will vary with every 1% increase in admixture. The different reclaimed asphalts at different temperatures produced similar test results, as described above, which were similar to the relative viscosity and relative rutting factor test results.

The regression analysis showed that the linear relationship between the $\Delta_{G^*\sin\delta}$ and temperature T is correlated well, and their regression equations are both $\Delta_{G^*\sin\delta} = \mathrm{a}T + \mathrm{b}$ (a and b are fitting parameters), the calculation equation for which is shown in (7):

$$\Delta_{G*\sin\delta} = \begin{cases} 0.1657T + 6.8668, R^2 = 0.67, A15_C \\ 0.823T - 4.9943, R^2 = 0.88, A30_C \\ 0.1941T + 4.1691, R^2 = 0.85, B15_C \\ 0.7204T - 5.2205, R^2 = 0.89, B30_C \\ 0.1282T + 4.2875, R^2 = 0.85, C15_C \\ 0.4313T - 0.9126, R^2 = 0.83, C30_C \end{cases} \quad (7)$$

From Equation (7), it can be seen that the effects of the source and admixture of the aged asphalt on the fatigue resistance of the reclaimed asphalt can be similarly expressed by the slope and intercept in the linear relationship equation.

4. Conclusions

The addition of the aged asphalt increases the viscosity, rutting factor, and fatigue factor of the reclaimed asphalt, indicating that the high-temperature stability of the reclaimed asphalt is enhanced but the fatigue cracking resistance is attenuated.

The effect of the aged asphalt on the viscosity of the reclaimed asphalt can be evaluated using Δ_η. The variance results showed that the value of Δ_η depends on the source and the admixture of aged asphalt. At high temperatures, Δ_η decreases linearly with increasing temperature, and its linear trend is independent of the source, the admixture, and whether it has been aged or not.

The effect of the aged asphalt on the viscoelasticity of reclaimed asphalt can be evaluated using $\Delta_{G^*/\sin\delta}$ and $\Delta_{G^*\sin\delta}$. The variance results showed that $\Delta_{G^*/\sin\delta}$ and $\Delta_{G^*\sin\delta}$ depend on the source and admixture of aged asphalt. At high temperatures, $\Delta_{G^*/\sin\delta}$ decreases linearly with increasing temperature, and at medium temperatures, $\Delta_{G^*\sin\delta}$ increases linearly with increasing temperature; both linear trends are independent of the source and admixture of aged asphalt.

Author Contributions: Conceptualization, T.L.; methodology, W.D.; software, Y.Q.; validation, J.Z.; formal analysis, Q.L.; resources, J.W.; data curation, J.X.; writing—original draft preparation, R.C. All authors have read and agreed to the published version of the manuscript.

Funding: This research was funded by [the basic scientific research of central institute: 2021-9045a].

Institutional Review Board Statement: Not applicable for studies not involving humans or animals.

Informed Consent Statement: Not applicable.

Conflicts of Interest: The authors declare no conflict of interest.

References

1. Ržek, L.; Turk, M.R.; Tušar, M. Increasing the rate of reclaimed asphalt in asphalt mixture by using alternative rejuvenator produced by tire pyrolysis. *Constr. Build. Mater.* **2020**, *232*, 177177. [CrossRef]
2. Giani, M.I.; Dotelli, G.; Brandini, N.; Zampori, L. Comparative life cycle assessment of asphalt pavements using reclaimed asphalt, warm mix technology and cold in-place recycling. *Resour. Conserv. Recycl.* **2015**, *104*, 224–238. [CrossRef]
3. Xing, C.; Jiang, W.; Li, M.; Wang, M.; Xiao, J.; Xu, Z. Application of atomic force microscopy in bitumen materials at the nanoscale: A review. *Constr. Build. Mater.* **2022**, *342*, 128059. [CrossRef]
4. Guo, M.; Tan, Y.; Hou, Y.; Wang, L.; Wang, Y. Improvement of evaluation indicator of interfacial interaction between asphalt binder and mineral fillers. *Constr. Build. Mater.* **2017**, *151*, 236–245. [CrossRef]

5. Yu, C.; Hu, K.; Yang, Q.; Wang, D.; Zhang, W.; Chen, G.; Kapyelata, C. Analysis of the Storage Stability Property of Carbon Nanotube/Recycled Polyethylene-Modified Asphalt Using Molecular Dynamics Simulations. *Polymers* **2021**, *13*, 1658. [CrossRef] [PubMed]
6. Hu, K.; Yu, C.; Yang, Q.; Chen, Y.; Chen, G.; Ma, R. Multi–scale enhancement mechanisms of graphene oxide on styrene–butadiene–styrene modified asphalt: An exploration from molecular dynamics simulations. *Mater. Des.* **2021**, *208*, 109901. [CrossRef]
7. Karlsson, R.; Isacsson, U. Material-Related Aspects of Asphalt Recycling—State-of-the-Art. *J. Mater. Civ. Eng.* **2006**, *18*, 81–92. [CrossRef]
8. Yu, C.H.; Hu, K.; Chen, G.X.; Chang, R.; Wang, Y. Molecular dynamics simulation and microscopic observation of compatibility and interphase of composited polymer modified asphalt with carbon nanotubes. *J. Zhejiang Univ. Sci. A* **2021**, *22*, 528–546. [CrossRef]
9. Yong, F.; Caihua, Y.; Kui, H.; Yujing, C.; Yu, L.; Taoli, Z. A study of the microscopic interaction mechanism of styrene–butadiene-styrene modified asphalt based on density functional theory. *Mol. Simul.* **2021**, *26*, 1–12. [CrossRef]
10. Jamshidi, A.; Hamzah, M.O.; Shahadan, Z. Selection of reclaimed asphalt pavement sources and contents for asphalt mix production based on asphalt binder rheological properties, fuel requirements and greenhouse gas emissions. *J. Clean. Prod.* **2012**, *23*, 20–27. [CrossRef]
11. Dony, A.; Colin, J.; Bruneau, D.; Drouadaine, I.; Navaro, J. Reclaimed asphalt concretes with high recycling rates: Changes in reclaimed binder properties according to rejuvenating agent. *Constr. Build. Mater.* **2013**, *41*, 175–181. [CrossRef]
12. Yu, X.; Leng, Z.; Wang, Y.; Lin, S. Characterization of the effect of foaming water content on the performance of foamed crumb rubber modified asphalt. *Constr. Build. Mater.* **2014**, *67*, 279–284. [CrossRef]
13. Suo, Z.; Yan, Q.; Ji, J.; Liu, X.; Chen, H.; Zhang, A. The aging behavior of reclaimed asphalt mixture with vegetable oil rejuvenators. *Constr. Build. Mater.* **2021**, *299*, 123811. [CrossRef]
14. Gao, J.; Yang, J.; Yu, D.; Jiang, Y.; Ruan, K.; Tao, W.; Sun, C.; Luo, L. Reducing the variability of multi-source reclaimed asphalt pavement materials: A practice in China. *Constr. Build. Mater.* **2021**, *278*, 122389. [CrossRef]
15. Rafiq, W.; Napiah, M.; Habib, N.Z.; Sutanto, M.H.; Alaloul, W.S.; Khan, M.I.; Musarat, M.A.; Memon, A.M. Modeling and design optimization of reclaimed asphalt pavement containing crude palm oil using response surface methodology. *Constr. Build. Mater.* **2021**, *291*, 123288. [CrossRef]
16. Shirodkar, P.; Mehta, Y.; Nolan, A.; Sonpal, K.; Norton, A.; Tomlinson, C.; Dubois, E.; Sullivan, P.; Sauber, R. A study to determine the degree of partial blending of reclaimed asphalt pavement (RAP) binder for high RAP hot mix asphalt. *Constr. Build. Mater.* **2011**, *25*, 150–155. [CrossRef]
17. MacEachern, M.; Sanchez, X.; Oh, W.T. Mechanical Properties of Aggregates for Roadbase Partially Replaced with Reclaimed Asphalt Shingles. *Int. J. Civ. Eng.* **2020**, *19*, 233–243. [CrossRef]
18. Alrajfi, E.; Ashteyat, A.M.; Murad, Y.Z. Shear behaviour of RC beams made with natural, recycled aggregate concrete and reclaimed asphalt aggregates under normal and elevated temperature. *J. Build. Eng.* **2021**, *40*, 102681. [CrossRef]

polymers

MDPI

Article

Evaluating the Effect of Polymer Modification on the Low-Temperature Rheological Properties of Asphalt Binder

Bagdat Teltayev [1,*], Erik Amirbayev [1] and Boris Radovskiy [2]

1 Kazakhstan Highway Research Institute, Almaty 050061, Kazakhstan; erik_neo@mail.ru
2 Radnat Consulting, Irvine, CA 90292, USA; b.radovskiy@att.net
* Correspondence: bagdatbt@yahoo.com; Tel.: +7-701-760-6701

Abstract: This paper investigates the viscoelastic properties of oxidized neat bitumen and three polymer-modified binders at low temperatures. The earlier proposed interrelated expressions for the relaxation modulus and for the creep compliance of bitumen binders are further developed. The results of creep testing of the binders on a bending beam rheometer at the six temperatures from −18 °C to −36 °C are presented. The results were analyzed using the equations developed for the relaxation modulus and the relaxation time spectrum. Viscosities at the low temperatures of tested binders were estimated. Approximate interrelations between the loss modulus and the relaxation spectrum were presented. The method for the determination of the glass transition temperature of a binder in terms of the relaxation time spectrum is proposed. The glass transition temperatures of tested binders were determined by the proposed method and compared with ones determined by the standard loss modulus-peak method.

Keywords: neat and polymer-modified bitumens; low temperatures; creep compliance; relaxation modulus; relaxation time spectrum; glass transition temperature

Citation: Teltayev, B.; Amirbayev, E.; Radovskiy, B. Evaluating the Effect of Polymer Modification on the Low-Temperature Rheological Properties of Asphalt Binder. *Polymers* **2022**, *14*, 2548. https://doi.org/10.3390/polym14132548

Academic Editors: Wei Jiang, Quantao Liu, Jose Norambuena-Contreras and Yue Huang

Received: 16 May 2022
Accepted: 15 June 2022
Published: 22 June 2022

1. Introduction

Bitumens are widely used in road paving because of their good adhesion to mineral aggregates and their viscoelastic properties. In paving applications, the bitumen should be resistant to climate and traffic loads, for which reason its rheological properties play a key role. It has to be stiff enough at high temperatures to resist rutting at local pavement temperature around 60 °C while it must remain soft and viscoelastic enough at low temperatures (from −20 °C to −45 °C) to resist thermal cracking. Those requirements are almost opposite, and most of the available neat bitumens would not provide all the needed characteristics together because bitumen is brittle in cold winters and softens readily in hot summers. Moreover, asphalt pavement at intermediate temperatures should be resistant to fatigue cracking from tensile and shear stresses under the action of repeated loading caused by traffic.

In order to enhance neat bitumen properties and widen the service temperature, bitumens are often modified by the addition of polymers. Polymer modification improves mechanical properties, decreases thermal susceptibility and permanent deformation (rutting), and increases resistance to low-temperature cracking. The most commonly used additives are copolymers, such as styrene butadiene styrene (SBS), ethylene vinyl acetate (EVA), styrene- ethylene-butylene-styrene (SEBS). The wide use of this type of polymer for modification is due to its thermoplastic nature at higher temperatures and its ability to form networks upon cooling. Particularly, when SBS is blended with bitumen, the elastomeric phase of the SBS copolymer absorbs the maltenes (oil fractions) from the bitumen and swells up to nine times its initial volume [1]. At SBS concentrations 5–7% by mass, a continuous polymer network (phase) is formed throughout the modified binder, significantly improving the bitumen properties at high and intermediate temperatures.

Until now, the researchers have not developed a convincing opinion on the positive and significant effect of modifying bitumens with polymers at low temperatures. Lu et al. [2] tested three bitumens blended with 6% SBS, EVA, SEBS, or EBA. They concluded that the effect of modification on the low-temperature properties of bituminous binders was small. Authors of [3,4] concluded that modified bitumens have better resistance to low temperature cracking compared to the unmodified ones while Peng [5] found that at −12 °C and −18 °C, the low-temperature stability of the modified bitumen is significantly increased, although at −24 °C it is slightly reduced. Lu et al. in work [6] reported that the glass transition temperature of a bitumen defined as the temperature of the peak loss modulus is reduced by polymer modification; but the results of creep tests performed using a BBR at temperatures of−15 °C, −20 °C, 25 °C, and −30 °C showed that polymer modification does not give beneficial effect; in some cases, especially for the limiting temperature at 0.3 m value, even adverse effect is found for polymer modification. In work [7], dense asphalt concrete samples prepared with bitumens modified with various amounts of SBS polymer were subjected to restrained cooling tests and at a standard cooling rate of 10 °C/h, no significant effect of modification was noted.

Although considerable research was undertaken in this area, the polymer-modified bitumen has still to be comprehensively characterized, due to the complex nature and interaction of the bitumen and polymer system. The present work focuses on the low-temperature rheological properties of the polymer-modified bitumen binders for road pavements.

The low-temperature transverse cracking of asphalt concrete pavements is major pavement distress commonly observed in regions affected by cold weather. Thermal cracking is induced by a rapid drop in temperature that tends to cause contraction and results in tensile stresses that may eventually reach the tensile strength of the material causing its fracture. The field performance data from test roads in Alberta, Manitoba, and Ontario (Canada), and Pennsylvania (USA) indicated that the binder is mostly responsible for the cold-weather cracking of asphalt pavements [8–12]. It became clear that only a fundamental understanding based on sound theory such as binder rheology might provide confidence for moving forward into a low-temperature cracking problem [13–20].

In a pioneering work, Monismith et al. [21] developed a calculation method for the thermally induced stress in the longitudinal direction of asphalt pavement as in an infinite viscoelastic beam. Boltzmann's superposition principle and the constitutive equation for linear viscoelastic material were applied to relate time-dependent stresses and strains. This approach was then widely used to estimate the development of thermal stresses [12,22–28]. It turned out that the problem of tensile strength determination for the binder and asphalt concrete was much more complicated, particularly including a ductile-to-brittle failure domain related to the peak to the tensile strength vs. temperature curve [16,24,29–32].

For many years, researchers have attempted to develop tests that can be incorporated into the binder pass/fail low-temperature specifications. Mandatory binder stiffness (S at 60 s < 300 MPa) and its slope (m value at 60 s, $d\log S/d\log t$ = m > 0.30) at the designed low pavement temperature were included in the standard specifications AASHTO M 320 and ASTM D 6816. Later proposed rheological low-temperature parameters include the difference ΔT between $T_{S=300}$ and $T_{m=0.30}$ [33]; stiffness at m-value = 0.3; Glover parameter that is a combination of storage modulus and a real part of complex viscosity and serves as a surrogate parameter for ductility [34]; Glover–Rowe parameter (G–R) which is the same parameter expressed in other terms [35].

Bitumen contains up to several thousand individual chemical components ranging from non-polar saturated alkanes to polar hetero-hydrocarbons [36]. Phenomenologically, it, therefore, seems natural to view a bitumen as a continuum of molecules with a gradual transition in molecular weight and polarity and with the corresponding continuous spectrum of relaxation times. The change in bitumen binder properties associated with the degree of packing does not occur instantaneously with change in temperature, but requires the passage of time. As the temperature is reduced, then the time scale for these rearrangements increases. In some temperature ranges, this time is of the order of the time scale of the experimental measurement, from minutes to hours. If the temperature is reduced still further, this time scale is increased so much that no further change in order can be observed over practical time scales. The temperature at which this transition occurs is the glass transition temperature.

It was shown that the glass transition temperature of bitumen binders is a reasonable predictor of the temperature close to which the asphalt pavement will thermally fracture [12,37,38]. Since low-temperature cracking is widespread pavement distress, the ability to determine bitumen binders is an important tool for asphalt researchers [18,29,39–45]. The objective of this paper is to increase the understanding of the low-temperature behavior of bitumen binders, particularly to estimate the glass transition temperature in terms of the relaxation time spectrum.

2. Theoretical Background

2.1. Linear Viscoelastic Rheological Characterization

To describe the rheological properties of binders we used the earlier proposed model [24,46,47] that includes expressions for tensile relaxation modulus $E(t)$ and for the tensile creep compliance $D(t)$:

$$E(t) = E_g \left[1 + \left(\frac{E_g t}{3\eta}\right)^b\right]^{-(1+1/b)}, \tag{1}$$

$$D(t) = \frac{1}{E_g}\left[1 + \left(\frac{E_g t}{3\eta}\right)^\beta\right]^{1/\beta}, \tag{2}$$

$$\beta = \left[\frac{1}{b} - \frac{\ln(\pi)}{\ln(2)} + 2\right]^{-1}, \tag{3}$$

where t is a time, s; E_g is tensile instantaneous modulus (tensile glassy modulus), Pa; η is the steady-state shear viscosity, Pa·s; b is a constant ($0 < b < 1$) governing the shape and width of the relaxation time spectrum.

The model [24,46,47] includes expressions (1) and (2) for tensile relaxation modulus $E(t)$ and for the tensile creep compliance $D(t)$. The constant β in Equation (2) was shown to be dependent on the thermal susceptibility of the bitumen binder based on SHELL testing data for 46 bitumens [24]. Equations (2) and (3) were derived by means of linear viscoelasticity in our monograph [46]. Expressions (1) and (2) for tensile relaxation modulus and for the tensile creep compliance match each other (Figure 1).

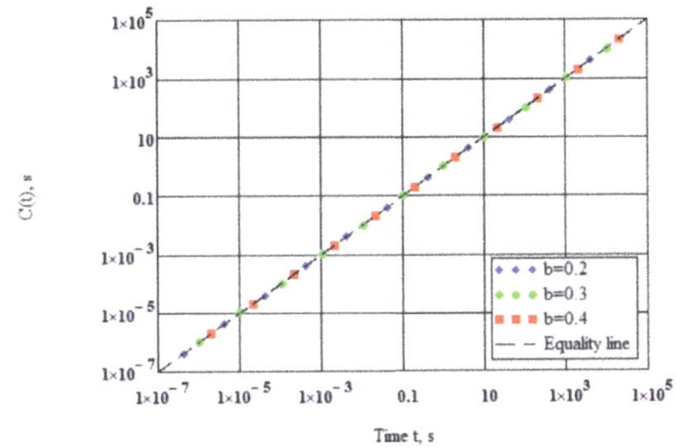

Figure 1. The convolution product of relaxation modulus and creep compliance.

As it is seen, Equations (1) and (3) include three parameters and all of them have physical meanings.

The relaxation and creep functions $E(t)$ and $D(t)$ are connected by the relation in form of the exact convolution integral [48]:

$$C(t) = \int_0^t E(\xi)\, D(t-\xi)\, d\xi \; = \; t. \tag{4}$$

The convolution product $C(t)$ of Equation (4) was calculated using Equations (1) and (2) at $3\eta/E_g = 0.01$ s for b = 0.2, 0.3, and 0.4. The results plotted in Figure 1 numerically confirm that the expressions (1) and (2) for relaxation modulus and creep compliance match each other.

The mean relative deviation of the convolution product $C(t)$ from the equality line in Figure 1 is around 1.4% for the time scale of twelve logarithmic decades. Supposing isotropy and incompressibility, Equations (1) and (2) for shear relaxation modulus $G(t)$ and for the shear creep compliance $J(t)$ can be written in the forms:

$$G(t) = G_g \left[1 + \left(\frac{G_g t}{\eta} \right)^b \right]^{-(1+1/b)}, \tag{5}$$

$$\mathrm{J}(t) = \frac{1}{G_g} \left[1 + \left(\frac{G_g t}{\eta} \right)^\beta \right]^{1/\beta}, \tag{6}$$

where G_g is the shear glassy modulus, MPa.

Complex modulus in shear can be determined from creep compliance Equation (6) using the Schwarzl and Struik method [49]:

$$|G^*|(\omega) = \frac{G_g}{\Gamma(1+m(\omega))} \left[1 + \left(\frac{G_g}{\eta\omega} \right)^\beta \right]^{-\frac{1}{\beta}}, \delta(\omega) = \frac{\pi}{2} m(\omega), \tag{7}$$

where $|G^*(\omega)|$ is the norm of complex modulus; $\delta(\omega)$ is phase angle; ω is the angular frequency, rad/s; $\Gamma(x)$ is the gamma function, and

$$m(\omega) = \frac{(G_g/\eta\omega)^\beta}{1 + (G_g/\eta\omega)^\beta}. \tag{8}$$

Here the Van der Poel-Koppelmann [13,50] conversion from time to frequency domain $t \to 1/\omega$ was applied to Equation (6) to derive Equation (7) using [49].

2.2. *Spectrum of Relaxation Times*

According to Bernstein's theorem [51], every monotonic function can be written as a sum of exponential decay functions $\exp(-t/\tau)$. The relaxation modulus of a viscoelastic liquid $E(t)$ is a continuous, decreasing function and thus it can be expressed in form of the integral transform [52]:

$$E(t) = \int_0^\infty H(\tau) \exp(-t/\tau) \frac{d\tau}{\tau} = \int_{-\infty}^\infty H(\tau) \exp(-t/\tau) d \ln \tau, \tag{9}$$

where $H(\tau)$ is the distribution function of relaxation times τ, shortly the relaxation–time spectrum.

If relaxation modulus $E(t)$ is given, the spectrum $H(\tau)$ can be found from the integral Equation (9) by inverting the Laplace transform. Applying the Widder's inverse Laplace transformation [53] leads to the following asymptotic formula for relaxation–time spectrum:

$$H_n(\tau) = \frac{(-1)^n (n\tau)^n}{(n-1)!} E^{(n)}(n\tau), \quad E^{(n)}(n\tau) = \left[\frac{d^n E(t)}{dt^n}\right]_{t=n\tau}, \tag{10}$$

where n is the degree of approximation (n = 1, 2, 3...).

As n becomes infinite [53], the right side of Equation (10) tends to the exact relaxation–time spectrum $H(\tau)$. The convergence rate depends on the relaxation modulus $E(t)$.

Substituting Equation (1) to Equation (10) leads to the following expressions for the first and second approximations of the spectrum:

$$H_1(\tau) = \frac{E_g(1+b)\left(\frac{E_g\tau}{3\eta}\right)^b}{\left[1+\left(\frac{E_g\tau}{3\eta}\right)^b\right]^{2+1/b}}, \tag{11}$$

$$H_2(\tau) = \frac{E_g(1+b)\left[(2+b)\left(\frac{2E_g\tau}{3\eta}\right)^b + 1 - b\right]\left(\frac{2E_g\tau}{3\eta}\right)^b}{\left[1+\left(\frac{2E_g\tau}{3\eta}\right)^b\right]^{3+1/b}}. \tag{12}$$

Figure 2 presents an example of calculated spectra for E_g = 2.460 × 10^9 Pa, b = 0.1914, η = 4.247 × 10^6 Pa·s in the first, second and third approximations. The maximum spectrum density in the first approximation is only 1.6% smaller than in the second and 1.8% smaller than in the third approximation.

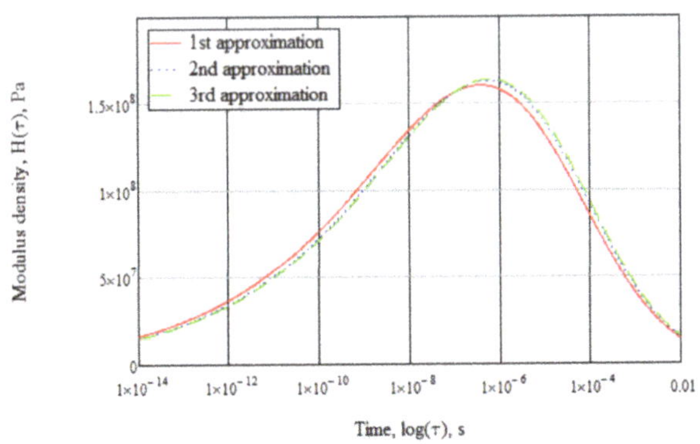

Figure 2. Comparison of calculated spectra.

Thus, the precision of simple first approximation Equation (11) is acceptable for our purposes. Moreover, Equation (11) has three intriguing features associated with it. First, analytically taking the integral one can obtain exactly

$$\int_0^\infty \frac{H_1(\tau)}{\tau} d\tau = E_g, \tag{13}$$

as it follows from Equation (9) at $t = 0$. The area between the curve $H_1(\tau)$ and axis $\ln\tau$ equals the instantaneous modulus, as it should.

Secondly, analytically taking the integral one can obtain exactly

$$\int_0^\infty H_1(\tau)d\tau = 3\eta, \tag{14}$$

as it should be [48]. The area between the curve $H_1(\tau)$ and axis τ equals the elongational viscosity $\eta_e = 3\eta$ [54].

Thirdly, differentiation leads to

$$d\ \log\ H_1(\tau)/d\ \log\ \tau = \frac{1+2b}{1+\left(\frac{E_g\tau}{3\eta}\right)^b} - 1 - b \tag{15}$$

It follows from Equation (15) that at $\tau = 0$ the slope $d\ \log\ H_1(\tau)/d\ \log\ \tau = b$ while when $\tau \to \infty$ the slope $d\ \log\ H_1(\tau)/d\ \log\ \tau = -(1+b)$. Obviously, the shape parameter b is related to the slopes of the relaxation spectrum. The slope $d\ \log\ H_1(\tau)/d\ \log\ \tau = b$ describes the low-relaxation time wing of the spectrum while the slope $d\ \log\ H_1(\tau)/d\ \log\ \tau = -(1+b)$ corresponds to the high-relaxation time wing (Figure 3).

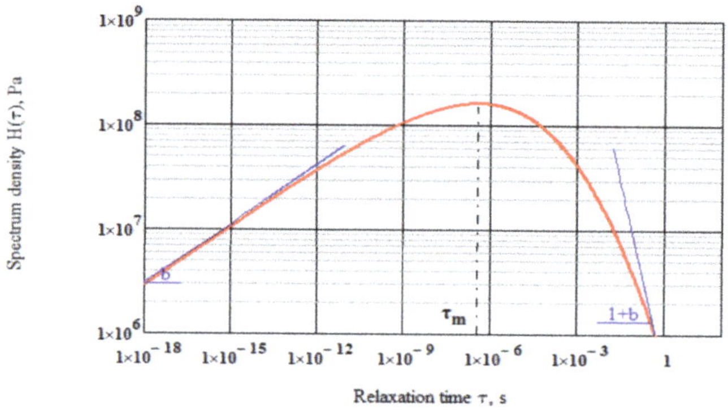

Figure 3. Spectrum and geometrical meaning of parameter b.

Spectrum density $H_1(\tau)$ peaks at the modal relaxation time:

$$\tau_m = \frac{3\eta}{E_g}\left(\frac{b}{1+b}\right)^{1/b}, \tag{16}$$

3. Materials and Methods

3.1. Binders

Four asphalt binders were tested. A neat bitumen of penetration grade BND 100/130 produced by direct oxidation from Siberian crude oil by Pavlodar petrochemical plant is commonly used in Kazakhstan paving industry. The second binder was the base bitumen BND 100/130 modified by the reactive ethylene terpolymer Elvaloy 4170 (Du Pont, NY, USA) in the amount of 1.4% by weight. The third binder was the base bitumen modified by the cationic bitumen emulsion of Butanol NS 198 (BASF, Ludwigshafen, Germany) in the amount of 3% by weight. The fourth binder was the base bitumen compounded with a flux (vacuum residue) from the same plant (flow time 82 s at 80 °C) in the amount of 20% by weight and modified by the polymer SBS L 30-01 (Sibur Co., Moscow, Russia)in the amount of 5% by weight.

Elvaloy 4170 is a chemically active copolymer of ethylene (71%) with butyl acrylate (20%) and glycidyl methacrylate (9%). Butanol NS 198 is a cationic, high molecular weight styrene butadiene dispersion designed for use in asphalt modification and waterproofing.

The content of solid polymers in Butanol is 64%. SBS L 30-01 A represents a linear block for copolymer of styrene (30%) and butadiene (70%). The molecular weights of the polymers Elvaloy and Butanol are in the range of 60,000–80,000 and 80,000–90,000, respectively. The structural formulas of SBS and Elvaloy polymers are shown in Figure 4.

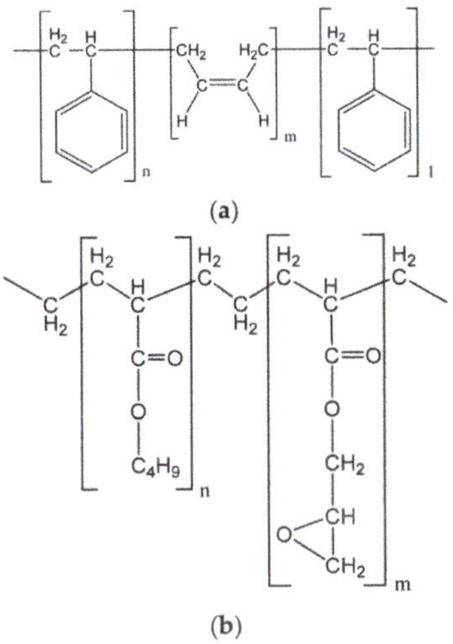

Figure 4. The structural formulas of (**a**) SBS and (**b**) Elvaloy polymers.

3.2. Preparation of Compounded and Modified Binders

Modification of the bitumen with polymers Elvaloy and Butanol was carried out in accordance with the normative documents of Kazakhstan [55] and [56], respectively. The polymers Elvaloy and Butanol were gradually added to the heated neat bitumen at 170 °C using a laboratory mixing device. Continuous mixing process of polymer–bitumen binders lasted for two hours, the next twelve hours the Elvaloy modified binder was conditioned at constant temperature of 170 °C.

Compounding of the base bitumen with the flux was performed by means of mixing with the rate of 450–500 rotations per minute at the constant temperature of 120 °C for 30 min [57]. Then, the compounded bitumen was gradually heated up to 180 °C and polymer SBS was gradually added. During the first two hours and next four hours, a mixing rate was equal to 1200 and 1800 rotations per minute, respectively.

3.3. Conventional Properties of Binders

Conventional properties of the binders were defined in the Research Laboratory of Kazakhstan Highway Research Institute according to the test specification ST PK 1373−2013 and they are presented in Table 1.

Table 1. Conventional properties of binders.

Property	Standard	Bitumen	Bitumen Modified by		
			Elvaloy	Butanol	Flux + SBS
Penetration, 25 °C (0.1 mm)	ASTM D5	103	86	88	73
Penetration index PI	EN 12591	−0.70	-	-	-
Ductility (cm), 25 °C (CM)	ASTM D113	>150	32	63	70
Softening point (°C)	ASTM D36	45.0	62.5	62.0	76.5
Fraass breaking point (°C)	EN 12593	−26.4	−25.8	−25.4	−23.0
Flash point (°C)	ASTM D92	265	250	265	-
Dynamic viscosity, 60 °C (Pa·s)	ASTM D2171	167	-	-	-
Kinematic viscosity, 135 °C (mm^2/s)	ASTM D445	394	-	-	-
Elastic recovery, 25 °C (%)	ASTM D6084	-	87	81	98

3.4. Rheological Testing

The binders were tested at low temperatures (−18 °C, −24 °C, −27 °C, −30 °C, −33 °C, and −36 °C) on the ATS Bending Beam Rheometer (BBR) according to the standard ASTM D 6648.

Prior to the rheological testing, the binders were aged using a Rolling Thin-Film Oven (RTFO) according to the test specification ASTM D2872 to simulate the short-term aging during asphalt mix manufacture. Then the binders were further aged in the Pressurized Aging Vessel (PAV) to simulate the long-term aging (ASTM D6521).

The binder samples for the testing had a shape of a beam with dimensions 6.25 × 12.5 × 125 mm. The duration of specimen conditioning prior to the testing was set to one hour. In the BBR creep test a constant load $P = 0.98$ N was applied at the midpoint of the simply supported binder beam for 240 s. The mid-span deflection $d(t)$ was constantly recorded. The creep stiffness $S(t)$ was automatically calculated from the equation:

$$S(t) = \frac{PL^3}{4wh^3\,d(t)}, \tag{17}$$

where L is the span of the beam (102 mm); w is the width of the beam (12.5 mm); h is the height of the beam (6.25 mm); $d(t)$ is maximum deflection of the beam at time t.

Only the observations at 8, 15, 30, 60, 120, and 240 s were employed in the present study.

4. Results and Discussion

4.1. Stiffness and Viscosity

As an example, the results of testing are shown in Figure 5 for the Elvaloy modified bitumen binder.

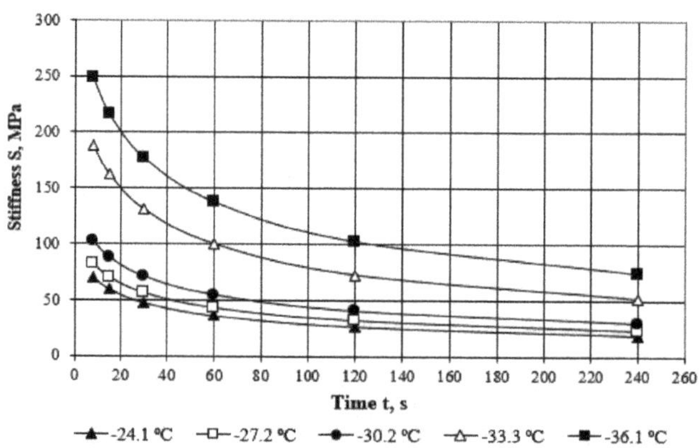

Figure 5. Time-dependent stiffness at different temperatures for Elvaloy modified binder.

To produce a master curve at a selected reference temperature T_r, Equation (2) combined with the Arrhenius time-temperature superposition function was used:

$$S(t) = \frac{1}{D(t)} = E_g\left[1 + \left(\frac{E_g t}{3\eta}\right)^{\beta}\right]^{-1/\beta}, \tag{18}$$

$$\eta = \eta_r \exp\left[\frac{\Delta H_a}{R}\left(\frac{1}{273 + T} - \frac{1}{273 + T_r}\right)\right], \tag{19}$$

where η_r is a viscosity at a reference temperature, Pa·s; ΔH_a is the flow activation energy, J/mol; R is the universal gas constant equal to 8.314 J/(mol·K).

The parameter β is related to b as before by Equation (3).

Based on our previous study [24,46], the instantaneous tensile modulus was assumed E_g = 2460 MPa for all tested binders. The reference temperature was selected close to the midrange of testing temperatures $T_r = -30$ °C. Using the Mathcad software, a nonlinear minimization algorithm was implemented to determine simultaneously the parameters η_r, ΔH_a, and b by minimizing the sum of squared deviations of data points from the master curve $S(t)$ Figure 6.

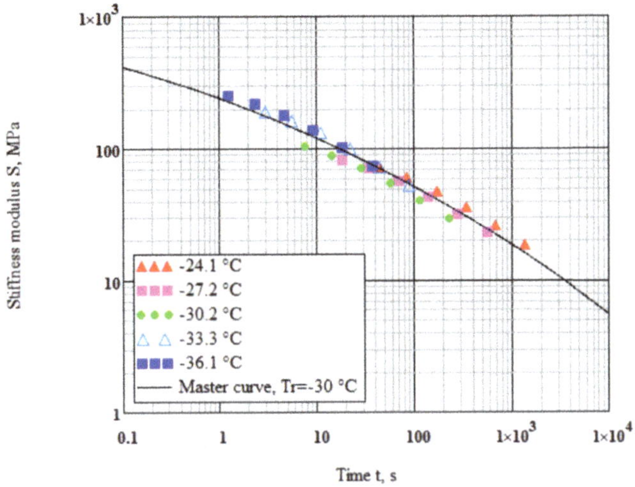

Figure 6. Master curve of stiffness as a function of time for the Elvaloy modified bitumen binder at $T_r = -30$ °C.

The obtained values of the parameters are given in Table 2. Figure 7 shows the viscosity as a function of temperature calculated using Equation (19).

Table 2. Values of the parameters for binders.

Binder	η at $T_r = -30$ °C, MPa·s	ΔH_a, J/mol	b
Bitumen	3.520×10^6	1.847×10^5	0.1418
Bitumen + Elvaloy	1.374×10^6	1.468×10^5	0.1412
Bitumen + Butanol	6.411×10^6	1.437×10^5	0.1452
Bitumen + Flux + SBS	8.143×10^6	1.833×10^5	0.1346

The slopes of the viscosity-temperature relationships to the temperature axis for the binders modified by the polymers Elvaloy and Butanol are almost equal and they are smaller than the slopes for the neat bitumen and the bitumen compounded by the flux and modified by the polymer SBS. This indicates the lower temperature susceptibility of the Elvaloy and Butanol-modified binders. In the range of the testing temperatures, the Elvaloy-modified binder has the smallest viscosity while the Butanol-modified binder (at temperatures higher than −27 °C) and the flux compounded and SBS-modified −27 °C) have the greatest one. The ability to estimate the viscosity of a binder at subzero temperatures indirectly from conventional BBR testing is a useful feature of the paper.

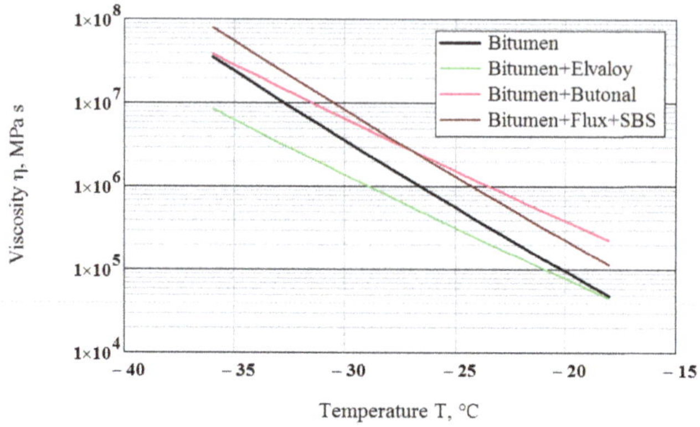

Figure 7. Viscosity–temperature relationships for the binders.

4.2. Glass Transition Temperature in Terms of Loss Modulus

Several researchers have shown that the glass transition temperature T_g of a bitumen binder is associated with the low-temperature cracking of a pavement [12,37,38]. The transition to a glassy state increases the brittleness of the binder, reducing its ability for stress relaxation and resulting in higher cracking susceptibility of an asphalt pavement. Researchers measured the glass transition temperature of bitumen binders by using three different techniques: dilatometry, calorimetry and rheological method-peak in the loss modulus versus temperature.

The classic method for the determination of the glass transition temperature is dilatometry. The temperature dependence upon cooling of the specific volume is determined by a suitable technique, and the temperature at the change in slope is taken as T_g at a given cooling rate [8,39–41]. Because of the need for precise measurements of small changes in volume with decreasing temperature, the dilatometric method is a difficult procedure to perform. Calorimetry was extensively employed, a peak in heat capacity being observed at the temperature T_g, depending on the heating rate [29,42,43].

Last year, the rheological dynamic measurements were conducted to estimate the glass transition temperature of bitumen binders. The data are collected over the temperature range at constant frequency and the loss modulus G″ (or E″) peak temperature is taken as T_g. The standard ASTM 1640 [58] recommends the testing frequency 1 Hz (ω = 6.28 rad/s). This standard admits other frequencies but they should be reported. Anderson and Marasteanu [41] used frequency 0.1 rad/s, Reinke and Engber [44] used 0.1–1.0 rad/s, Planche et al. [43] used 5 rad/s, Sun et al. used 10 rad/s [45]. Changing the time scale by a factor of 10 will generally result in a shift of about 8 °C for a typical amorphous material [58]. Anderson and Marasteanu [59] compared dilatometry, calorimetry, and the peak in the G″ and concluded that all methods give estimates of the glass transition temperature that are in relatively good agreement, given the different nature of the measurements and the time scale of the measurements.

In the present study, T_g is defined based on BBR testing as the temperature where the tensile loss modulus E'' peaks at frequency 1 rad/s. The tensile loss modulus E'' can be determined from tensile creep compliance $D(t)$ using the Schwarzl and Struik method [49] from the equation

$$E''(\omega) = \frac{E_g \sin\left(\frac{\pi}{2} m_E(\omega)\right)}{\Gamma(1+m_E(\omega))}\left[1+\left(\frac{E_g}{3\eta\omega}\right)^{\beta}\right]^{-\frac{1}{\beta}}, \tag{20}$$

where

$$m_E(\omega) = \frac{\left(E_g/3\eta\omega\right)^{\beta}}{1+\left(E_g/3\eta\omega\right)^{\beta}}.$$

Loss modulus $E''(\omega)$ was calculated from Equation (20) using Equation (19) and the parameters are shown in Table 2 at the frequency ω = 1 rad/s. When the loss modulus E'' is plotted versus temperature, the resulting curve exhibits a peak value (Figure 8). The temperature at this peak value can be interpreted as a glass transition temperature T_g. The temperatures at which the calculated E'' reached a maximum value were T_g = −45.4 °C for the neat bitumen, T_g = −52 °C for the Elvaloy-modified binder, T_g = −46.3 °C for the Butanol-modified binder and T_g = −46.2 °C for the flux-compounded and SBS-modified binder. Changing the "testing" frequency by a factor of 10 results in a T_g shift of 5.5–7 °C.

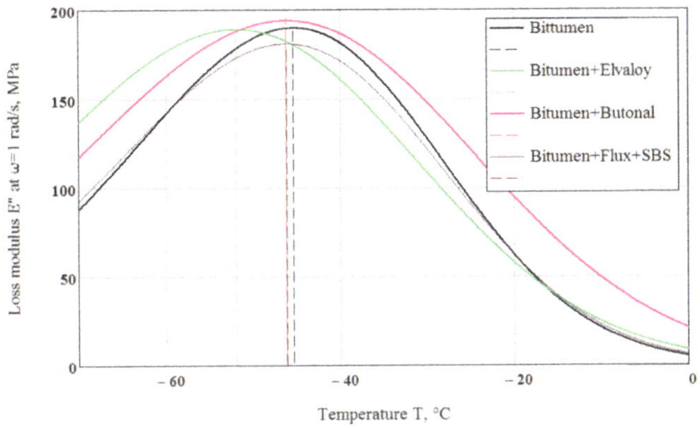

Figure 8. Calculation of the glass transition temperatures for the tested binders using the loss modulus peak method at frequency ω = 1 rad/s.

At low polymer concentration (not more than about 5%), the properties of the base bitumen and the compatibility of polymer with bitumen are very important. Modification with 3% Butanol primarily aims to improve the high-temperature properties of the bitumen and does not show a positive effect at subzero temperatures compared with the base neat bitumen in this study. Considerable improvement of high-temperature properties (softening point from 45 °C up to 76.5 °C) of the base bitumen, keeping its low-temperature properties, was achieved by means of compounding by flux and modifying by polymer SBS. Modification with 1.4% Elvaloy, which is intended to enhance the high-temperature properties as well, also improved the low-temperature properties. Modification with Elvaloy lowered the temperature susceptibility, lowered the low-temperature viscosity and reduced the glass transition temperature. Particularly, these results are important for countries with a sharp-continental climate, including Kazakhstan, which is the ninth-largest in the world where half of the territory requires the bitumen binder of Superpave grade PG 58−40.

4.3. Glass Transition Temperature in Terms of Relaxation Time Spectrum

4.3.1. Interrelations between Loss Modulus and Relaxation Spectrum

To increase the understanding of the glass transition temperature of the binder it would be of interest to express T_g in terms of the relaxation time spectrum. In the previous section, the widely accepted definition of the glass transition temperature was used, which consisted of taking the loss modulus (G'') maximum as a function of temperature at a given frequency ω. Loss modulus is related to the relaxation–time spectrum by an exact equation [48].

$$G''(\omega) = \int_0^\infty \frac{H(\tau)}{\tau} \cdot \frac{\omega\tau}{1+\omega^2\tau^2} d\tau. \tag{21}$$

The kernel of the loss modulus function

$$f(\tau) = \frac{\omega\tau}{1 + \omega^2\tau^2}, \tag{22}$$

is a crude approximation to the Dirac delta function $\delta_D(\tau)$. The function $f(\tau)$ reaches a maximum at $\tau = 1/\omega$. Since the integral

$$\int_0^\infty \frac{\omega\tau}{1 + \omega^2\tau^2} d\tau = \frac{\pi}{2}$$

is not unity as it is for the delta function $\delta_D(\tau)$, the function $f(\tau)$ must be multiplied by $2/\pi$ before being replaced by the delta function. Then

$$\frac{2}{\pi} \cdot \frac{\omega\tau}{1 + \omega^2\tau^2} \approx \delta(1 - 1/\omega). \tag{23}$$

If Equation (23) is substituted into Equation (21), and using the sifting property of the Dirac delta function yields the relation:

$$G''(\omega) \approx \frac{\pi}{2} H\left(\frac{1}{\omega}\right). \tag{24}$$

Applying the Van der Poel-Koppelmann conversion from the frequency domain to the time domain $\omega \to 1/\tau$ leads to the equation:

$$H(\tau) \approx \frac{2}{\pi} G''(\omega)|_{\omega \to 1/\tau}. \tag{25}$$

It follows from Equation (25) that a relaxation time spectrum $H(\tau)$ is approximately proportional to a loss modulus function $G''(\omega)$. Their shapes are similar and their maxima represent the concentration of the relaxation time spectrum or correspond to the relaxation frequency spectrum in a certain region of the logarithmic τ or ω scales.

4.3.2. Glass Transition Temperature

Calculated spectrum densities at the modal relaxation time $\tau_m = 1/\omega_m = 1$ s using Equation (11) and the parameters given in Table 2 are presented in Figure 9. The curves in Figures 8 and 9 look very similar with allowance for approximations in the derivation of Equations (11) and (25).

For a given modal frequency, e.g., $\omega_m = 1$ rad/s, loss modulus $G''(\omega)$ at a certain temperature has a maximum, e.g., $T_g = -45.4$ °C for bitumen (Figure 8). According to time–temperature superposition principle, it equivalently means that for the constant temperature −45.4 °C the loss modulus $G''(\omega)$ has a maximum at the modal frequency $\omega_m = 1$ rad/s. Similarly, in the relaxation time domain, the relaxation time spectrum $H(\tau)$ has a maximum at the modal relaxation time $\tau_m = 1/\omega_m$ [Equation (16)].

Substituting Equation (19) into Equation (16) leads to the relation:

$$\tau_m = \frac{3\eta_r}{E_g}\left(\frac{b}{1+b}\right)^{1/b} \exp\left[\frac{\Delta H_a}{R}\left(\frac{1}{273+T} - \frac{1}{273+T_r}\right)\right], \tag{26}$$

where the modal relaxation time τ_m corresponds to the arbitrary temperature T.

The ratio of the modal relaxation time τ_{mg} at a temperature T_g to the modal relaxation time τ_{mr} at a reference temperature T_r is

$$\frac{\tau_{mg}}{\tau_{mr}} = \exp\left[\frac{\Delta H_a}{R}\left(\frac{1}{273+T_g} - \frac{1}{273+T_r}\right)\right], \tag{27}$$

It follows from (27) that the fixed τ_{mg} glass transition temperature T_g can be calculated by equation:

$$T_g = \left[\frac{1}{273+T_r} + \frac{R}{\Delta H_a}\ln\left(\frac{\tau_{mg}}{\tau_{mr}}\right)\right]^{-1} - 273, \quad (28)$$

where τ_{mr} is the modal relaxation time at a reference temperature T_r:

$$\tau_{mr} = \frac{3\eta_r}{E_g}\left(\frac{b}{1+b}\right)^{1/b}, \quad (29)$$

For the fixed τ_{mg} = 1 s, using the parameters shown in Table 2, the calculated from Equation (28) glass transition temperature equals T_g = −45.8 °C for bitumen, T_g = −52.4 °C for the Elvaloy-modified binder and T_g = −46.7 °C for the Butanol-modified binder, which almost coincides with T_g determined from the peak of loss modulus at frequency ω = 1 rad/s (Figure 8). The calculated dependence of glass transition temperature on modal relaxation time is presented for the tested binders in Figure 10.

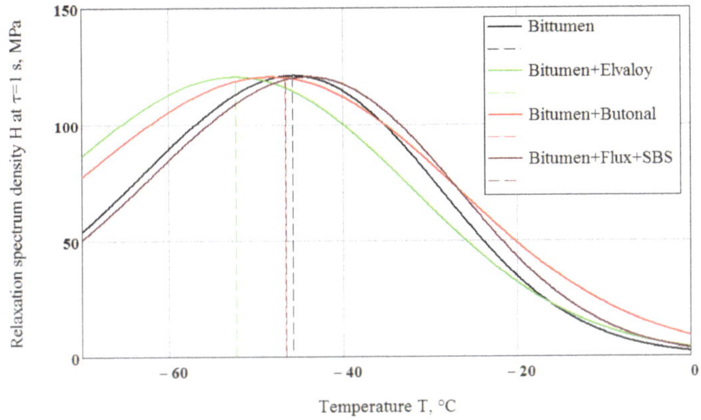

Figure 9. Definition of the glass transition temperatures for the binders using relaxation spectrum density at the modal relaxation time τ_m = 1 s.

The ability to estimate the glass transition temperature in terms of relaxation time spectrum is important for understanding the behavior of binders at low temperatures. The molecular mobility of liquids including bitumen binders expresses itself in a relaxation–time spectrum. The broader the molecular weight distribution is the broader the relaxation spectrum becomes [60,61]. When a liquid-cooled down, its volume reduces due to the translational molecular readjustments rather than due to their oscillating motions. When a temperature reaches the glass transition region, the speed of molecular adjustment becomes slower and no further change in order can be observed over a given time scale. Physically, the glass transition occurs at the temperature T_g when the root-mean-squared displacement of the particle of average molecular weight becomes smaller than the average size of that particle [62], on a given relaxation timescale (of the order of τ = 1 s, for example). Relaxation at a temperature lower than T_g occurs mostly due to the oscillating motion of molecules. Thus, the glass transition is the transformation of a disordered state with molecular mobility to an immobilized state of a similar structure by means of decreasing temperature. The transformation of liquid (e.g., a bitumen binder) is caused by a continuous increase in the modal relaxation time up to the given scaling time. It can be the time scale of practical observation or an experiment. Relating the bitumen binder properties or performance to the particular scaling time value requires more research.

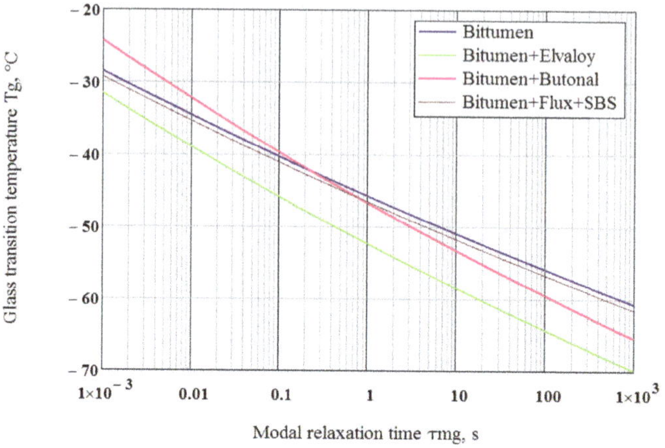

Figure 10. Dependence of glass transition temperature on modal relaxation time.

It is known that the addition of a polymer to a bitumen increases its viscosity, and one would expect a significant decrease in its low-temperature characteristics. However, the results of this work showed that when the bitumen is modified with the given amounts of selected polymers, a positive effect can be obtained. This can be explained by the lower glass transition temperatures of the elastic parts of the polymers (butyl acrylate in Elvaloy, butadiene in SBS and Butanol).

5. Conclusions

The main conclusions and findings based on the analysis presented in this paper are as follows:

- An approximate interrelation between a loss modulus and a relaxation–time spectrum was presented.
- The glass transition temperature of binders was calculated by the presented rheological method in terms of the relaxation–time spectrum. Calculations by this method showed that: (1) modification of bitumen with 3% Butanol NS 198 that primarily aims to improve the high-temperature properties does not show a positive effect at subzero temperatures compared with the base pure bitumen; modification of bitumen with 1.4% Elvaloy 4170, which is primarily intended to enhance the high-temperature properties, also improved the low-temperature properties; modification with Elvaloy lowered the temperature susceptibility of binder, lowered the low-temperature viscosity and lowered the glass transition temperature by about six degrees (from −45.4 °C to −52.0 °C).
- The proposed model for stiffness modulus (Equation (1)) enables the estimation of the viscosity of the binder at low temperature indirectly from conventional BBR testing. Viscosity values at temperature −30 °C were 3.520×10^6 MPa·s for the oxidized bitumen of penetration grade 100/130, 1.374×10^6 MPa·s for the bitumen modified by the polymer Elvaloy, 6.411×10^6 MPa·s for the bitumen modified by the polymer Butanol and 8.143×10^6 MPa·s for the bitumen compounded by flux and modified by the polymer SBS. The temperature susceptibility of the Elvaloy- and Butanol-modified binders was lower than for the neat bitumen and the flux compounded and SBS modified bitumen.
- In order to determine the optimal content of polymers in bitumens, it is recommended to continue the study of their rheological and other characteristics in the future by varying the technological conditions of modification.

Author Contributions: Conceptualization and writing—original draft preparation, investigation, B.R.; conceptualization and writing—original draft preparation, supervision, investigation, B.T.; investigation, E.A. All authors have read and agreed to the published version of the manuscript.

Funding: The work was performed under grant IRN AP 08857446 from the Committee of Science of the Ministry of Education and Science of the Republic of Kazakhstan. Agreement No. 230 dated 12 November 2020.

Institutional Review Board Statement: Not applicable.

Informed Consent Statement: Not applicable.

Data Availability Statement: Not applicable.

Conflicts of Interest: The authors declare no conflict of interest.

References

1. Walkering, C.P.; Vonk, W.C.; Whiteoak, C.D. Improved Asphalt Properties Using SBS modified Bitumens. *Shell Bitum. Rev.* **1992**, *66*, 9–11.
2. Lu, X.; Isacsson, U.; Ecblad, J. Influence of polymer modification on low temperature behavior of bituminous binders and mixtures. In Proceedings of the 6th RILEM Symposium PTEBM'03, Zurich, Switzerland, 14–16 April 2003; pp. 435–441.
3. Abdelaziz, A.; Ho, C.-H.; Snyder, M. Evaluating the influence of polymer-modified asphalt binders on low temperature properties. *MATEC Web Conf.* **2018**, *163*, 05012. [CrossRef]
4. Zhang, C.; Wang, H.; You, Z.; Gao, J.; Irfan, M. Performance test on Styrene-Butadiene-Styrene (SBS) modified asphalt based on the different evaluation methods. *Appl. Sci.* **2019**, *9*, 467. [CrossRef]
5. Peng, G. Study on high and low temperature properties of SBS modified asphalt with fasir. In Proceedings of the 2016 4th International Conference on Mechanical Materials and Manufacturing Engineering; Atlantis Press: Paris, France, 2016; pp. 490–492.
6. Lu, X.; Isacsson, U.; Ekblad, J. Influence of polymer modification on low temperature behavior of bituminous binder sand mixtures. *Mater. Struc.* **2003**, *36*, 652–656. [CrossRef]
7. Hesp, S.A.M.; Terlouw, T.; Vonk, W.C. Low temperature performance of SBS-modified asphalt mixes. *J. Assoc. Asph. Paving Technol.* **2000**, *69*, 540–574.
8. Fromm, H.J.; Phang, W.A. A study of transverse cracking of bituminous pavements. *J. Assoc. Asph. Paving Technol.* **1972**, *41*, 383–423.
9. McLeod, N.W. A 4-year survey of low temperature transverse pavement cracking on three Ontario test roads. *J. Assoc. Asph. Paving Technol.* **1972**, *41*, 424–493.
10. Kandhal, P.S. *Low-Temperature Ductility in Relation to Pavement Performance*; ASTM International: West Conshohocken, PA, USA, 1977; Volume 628, pp. 95–106.
11. Deme, I.J.; Young, F.D. Ste. Anne Test Road revisited twenty years later. *Proceed. Canad. Techn. Asph. Assoc.* **1987**, *32*, 254–283.
12. Bouldin, M.G.; Dongre, R.N.; Rowe, G.M.; Sharrock, M.J.; Anderson, D.A. Predicting thermal cracking of pavements from binder properties: Theoretical basis and field validation. *J. Assoc. Asph. Paving Technol.* **2000**, *69*, 455–496.
13. Vander Poel, C. A general system describing the viscoelastic properties of bitumen and its relation to routine test data. *J. Appl. Chem.* **1954**, *4*, 221–236. [CrossRef]
14. Di Benedetto, H.; Olard, F.; Sauzeat, C.; Delaporte, B. Linear viscoelastic behavior of bituminous materials: From binders to mixes. *Road Mat. Pavemement Des.* **2004**, *5*, 163–202. [CrossRef]
15. Secor, K.E.; Monismith, C.L. Analysis and interrelation of stress-strain-time data for asphalt concrete. *Transact. Soc. Rheol.* **1964**, *8*, 19–32. [CrossRef]
16. Heukelom, W. Observations on the rheology and fracture of bitumens and asphalt mixes. *J. Assoc. Asph. Paving Technol.* **1966**, *35*, 358–399.
17. Vinogradov, G.V.; Isayev, A.I.; Zolotarev, V.A.; Verebskaya, E.A. Rheological properties of road bitumens. *Rheol. Acta* **1977**, *16*, 266–281. [CrossRef]
18. Jongepier, R.; Kuilman, B. Characteristics of the rheology of bitumens. *J. Assoc. Asph. Paving Technol.* **1969**, *38*, 98–122.
19. Christensen, D.W.; Anderson, D.A. Interpretation of dynamic mechanical test data for paving grade asphalt cements. *J. Assoc. Asph. Paving Technol.* **1992**, *61*, 67–116.
20. Airey, G.D. Rheological Characteristics of Polymer Modified and Aged Bitumens. Ph.D. Thesis, University of Nottingham, Nottingham, UK, 1997.
21. Monismith, C.L.; Secor, G.A.; Secor, K.E. Temperature induced stresses and deformations in asphalt concrete. *J. Assoc. Asph. Paving Technol.* **1965**, *34*, 248–285.
22. Christison, J.T.; Murray, D.W.; Anderson, K.O. Stress prediction and low temperature fracture susceptibility of asphaltic concrete pavements. *J. Assoc. Asph. Paving Technol.* **1972**, *41*, 494–523.

23. Teltayev, B. Evaluation of low temperature cracking indicators of hot mix asphalt pavement. *Int. J. Pavem. Resear. Technol.* **2014**, *5*, 343–351.
24. Teltayev, B.; Radovskiy, B. Predicting thermal cracking of asphalt pavements from bitumen and mix properties. *Road Mater. Pav. Des.* **2018**, *19*, 1832–1847. [CrossRef]
25. Marasteanu, M.; Buttlar, W.; Bahia, H.; Williams, C.; Moon, K.H.; Teshale, E.Z.; Falchetto, A.C.; Turos, M.; Dave, E.; Paulino, G.; et al. *Investigation of Low Temperature Cracking in Asphalt Pavements, National Pooled Fund Study–Phase II 2012*; University of Wisconsin-Madison: Madison, WI, USA, 2012; 323p.
26. Farar, M.J.; Hajj, E.Y.; Planche, J.P.; Alavi, M.Z. A method to estimate the thermal stress build-up in an asphalt mixture from a single-cooling event, Road Mater. *Pav. Des.* **2013**, *14*, 201–211.
27. Wang, T.; Xiao, F.; Amirkhanian, S.; Zheng, M. A review on low temperature performances of rubberized asphalt materials. *Constr. Build. Mater.* **2017**, *145*, 483–505. [CrossRef]
28. Shields, D.H.; Zeng, M.; Kwok, R. Nonlinear Viscoelastic Behavior of Asphalt Concrete in Stress Relaxation. *J. Assoc. Asph. Paving Technol.* **1998**, *67*, 358–385.
29. Cheung, C.Y.; Cebon, D. Experimental study of pure bitumens in tension, compression, and shear. *J. Rheol.* **1997**, *41*, 45–73. [CrossRef]
30. Stock, A.F.; Arand, W. Low Temperature Cracking in Polymer Modified Binder. *J. Assoc. Asph. Paving Technol.* **1993**, *62*, 23–53.
31. Olard, F.; Di Benedetto, H.; Eckmann, B.; Vaniscote, J.-C. Low-temperature failure behavior of bituminous binder sand mixes. In Proceedings of the Soft, The 3rd Euro Mitumen and Euro Asphalt Congress, Vienne, Austria, 12–14 May 2004; pp. 1476–1490.
32. Molenaar, A.A.A.; Li, N. Prediction of compressive and tensile strength of asphalt concrete. *Inter. J. Pav. Res. Technol.* **2014**, *7*, 324–331.
33. Anderson, M.; King, G.; Hanson, D.; Blankenship, P. Evaluation of the relationship between asphalt binder properties and non-load related cracking. *J. Assoc. Asph. Paving Technol.* **2011**, *80*, 615–649.
34. Glover, C.J.; Davison, R.R.; Domke, C.H.; Ruan, Y.; Juristyarini, P.; Knorr, D.B.; Jung, S.H. *Development of a New Method for Assessing Asphalt Binder Durability with Field Validation*; Report No. FHWA/TX-05/1872-2; Texas Transportation Institute: College Station, TX, USA, 2005.
35. Rowe, G.M. Prepared Discussion presented in response to M. Anderson et al. *J. Assoc. Asph. Paving Technol.* **2011**, *80*, 649–663.
36. Speight, J.G. *The Chemistry and Technology of Petroleum*, 4th ed.; Taylor & Francis Group: New York, NY, USA, 2007.
37. Pink, H.S.; Merz, R.E.; Bosniack, D.S. Asphalt rheology: Experimental determination of dynamic moduli at low temperature. *J. Assoc. Asph. Paving Technol.* **1980**, *49*, 64–94.
38. Goodrich, J.L. Asphalt and polymer modified asphalt properties related to the performance of asphalt concrete mixes. *J. Assoc. Asph. Paving Technol.* **1988**, *57*, 116–176.
39. Schmidt, R.J.; Santucci, L.E. A practical method for determining the glass transition temperature of asphalt sand calculation of their low temperature viscosities. *J. Assoc. Asph. Paving Technol.* **1966**, *35*, 61–90.
40. Bahia, H.U.; Anderson, D.A. Glass transition behavior and physical hardening of asphalt binders. *J. Assoc. Asph. Paving Technol.* **1993**, *62*, 93–129.
41. Anderson, D.A.; Marasteanu, M.O.; Liu, Y. Dilatometric measurements of glass transition temperatures. In Proceedings of the Eurobitumen Workshop on Performance Related Properties for Bitumen Binder, Luxemburg, 3–6 May 1999.
42. Kriz, P.; Stastna, J.; Zanzotto, L. Glass transition and phase stability in asphalt binders. *Road Mater. Pav. Des.* **2008**, *9*, 37–65. [CrossRef]
43. Planche, J.-P.; Martin, D.; Claudy, P.; Letoffe, J.M.; Lesueur, D.; King, G.N. Evaluation of the low temperature properties of bituminous binders using calorimetry and rheology. In Proceedings of the Rheology of Bituminous Binders, European Workshop, Brussels, Belgium, 5–7 April 1995.
44. Reinke, G.H.; Engber, S.L. Impact of factors affecting determination of glass transition temperature using a dynamic shear rheometer. *J. Assoc. Asph. Paving Technol.* **2001**, *70*, 483–493.
45. Sun, Y.; Huang, B.; Edwin, G.; Chen, J.; Jia, X.; Ding, Y. Characterizing rheological behavior of asphalt binder over a complete range of pavement service loading frequency and temperature. *Constr. Build. Mater.* **2016**, *123*, 661–672. [CrossRef]
46. Radovskiy, B.; Teltayev, B. *Viscoelastic Properties of Asphalt Based on Penetration and Softening Point*; Springer: Cham, Switzerland, 2017.
47. Teltayev, B.B.; Amirbayev, E.D.; Radovskiy, B.S. Viscoelastic characteristics of blown bitumen at low temperatures. *Constr. Build. Mater.* **2018**, *189*, 54–61. [CrossRef]
48. Ferry, J.D. *Viscoelastic Properties of Polymers*; John Wiley & Sons Inc.: New York, NY, USA, 1980.
49. Schwarzl, F.R.L.; Struik, C.E. Analysis of relaxation measurements. *Adv. Molec. Relax. Proc.* **1968**, *1*, 201–255. [CrossRef]
50. Koppelmann, J. Über die Bestimmung des dynamischen Elastizitätsmoduls und des dynamischen Schubmoduls im Frequenzbereich von 10^{-5} bis 10^{-1} Hz. *Rheol. Acta.* **1958**, *1*, 20–28. [CrossRef]
51. Bernstein, S. Sur les fonctions absolument monotones. *Acta Math.* **1929**, *52*, 1–66. [CrossRef]
52. Gross, B. *Mathematical Structure of the Theories of Viscoelasticity*; Herman net Cie: Paris, France, 1953.
53. Widder, D.V. The Inversion of the Laplace Integral and the Related Moment Problem. *Transact. Amer. Math. Soc.* **1934**, *36*, 107–200. [CrossRef]

54. Riande, E.; Diaz-Calleja, R.; Prolongo, M.; Masegosa, R.; Salom, C. *Polymer Viscoelasticity: Stress and Strain in Practice*; Marcel Dekker: New York, NY, USA, 2000.
55. *RRK218-80-2010*; Recommendations for the Use of Elvaloy Polymer for the Modification of Bitumen and Asphalt Concrete. JSC Kazdor NII: Almaty, Kazakhstan, 2010.
56. *RRK218-93-2011*; Recommendations for the Use of Butonal NS Modifier in Road Construction. JSC Kazdor NII: Almaty, Kazakhstan, 2011.
57. Teltayev, B.; Seilkhanov, T.; Rossi, C.O.; Amirbayev, Y.; Begaliyeva, S. Low temperature resistance increase for bitumen by compounding with tar. *Appl Sci.* **2021**, *11*, 8579. [CrossRef]
58. *ASTM E1640-09*; Standard Test Method for Assignment of the Glass Transition Temperature by Dynamic Mechanical Analysis. West Conshohocken: Montgomery, PA, USA, 2019.
59. Anderson, D.A.; Marasteanu, M.O. Physical hardening of asphalt binders relative to their glass transition temperatures. *Transp. Res. Rec.* **1999**, *1661*, 27–34. [CrossRef]
60. De Gennes, P.G. *Scaling Concepts in Polymer Physics*; Cornell University Press Ithaca: New York, NY, USA, 1979.
61. Doi, M.; Edwards, S.F. *The Theory of Polymer Dynamics*; Clarendon Press: Oxford, UK, 1986.
62. Speedy, R.J. The hard sphere glass transition. *Mol. Phys.* **1998**, *95*, 169–178. [CrossRef]

MDPI
St. Alban-Anlage 66
4052 Basel
Switzerland
www.mdpi.com

Polymers Editorial Office
E-mail: polymers@mdpi.com
www.mdpi.com/journal/polymers

www.ingramcontent.com/pod-product-compliance
Lightning Source LLC
Chambersburg PA
CBHW061559220926
51850CB00005B/378